Earth's Climate

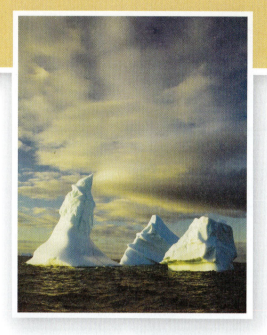

Earth's Climate

PAST AND FUTURE

Third Edition

WILLIAM F. RUDDIMAN

W. H. Freeman and Company · New York

Publisher: *Jessica Fiorillo*
Senior Acquisitions Editor: *Bill Minick*
Development Editor: *Brittany Murphy*
Assistant Editor: *Courtney Lyons*
Associate Director of Marketing: *Debbie Clare*
Marketing Assistant: *Samantha Zimbler*
Senior Media and Supplements Editor: *Amy Thorne*
Project Editor: *Kerry O'Shaughnessy*
Production Manager: *Julia DeRosa*
Text Designer: *Marsha Cohen*
Cover Designer: *Vicki Tomaselli*
Illustration Coordinator: *Janice Donnola*
Illustrations: *Robert L. Smith*
Photo Editors: *Hilary Newman, Ted Szczepanski*
Composition: *CodeMantra*
Printing and Binding: *RR Donnelley*

The "About the Author" photo is courtesy of Ginger Ruddiman.
Credit is also given to the following sources for the part and chapter opener photos
and the cover: **Cover:** P. A. Lawrence, LLC/Alamy; **Chapter openers:** John Sylvester/
Alamy; **Part 1:** Pal Hermansen/Steve Bloom Images/Alamy; **Part 2:** Jeffrey Kargel,
USGS/NASA JPL/AGU; **Part 3:** iStockphoto/Thinkstock; **Part 4:** Courtesy of Konrad
Steffen, CIRES; **Part 5:** Hemera/Thinkstock.

Library of Congress Control Number: 2013937645

ISBN-13: 978-1-4292-5525-7
ISBN-10: 1-4292-5525-0

W. H. Freeman and Company
41 Madison Avenue
New York, NY 10010
Houndmills, Basingstoke RG21 6XS, England
www.whfreeman.com

*To my many colleagues who have investigated
the science of climate change because they find it an endlessly
fascinating subject, and who in recent years have in some
cases done so under duress from those who refuse to
accept what the science is telling us*

About the Author

WILLIAM F. RUDDIMAN holds a 1964 undergraduate degree in geology from Williams College and a 1969 Ph.D. in marine geology from Columbia University. He worked at the U.S. Naval Oceanographic Office from 1969 until 1976 as a senior scientist/oceanographer and then returned to Lamont-Doherty Observatory as a senior research associate. He was associate director of the Oceans and Climate Division from 1982 to 1986 and an adjunct professor in the Department of Geology from 1982 to 1991. He moved to the University of Virginia in 1991 as a professor in the Department of Environmental Sciences and served as department chair from 1993 to 1996. At Virginia, he taught courses in climate change, physical geology, and marine geology. He has participated in fifteen oceanographic cruises and was co-chief on leg 94 of the Deep-Sea Drilling Project and leg 108 of the Ocean Drilling Project.

Professor Ruddiman's research interests have ranged across aspects of climate change on several time scales. His early research with Andrew McIntyre focused on orbital-scale climate change in and around the north and equatorial Atlantic Ocean. He was a member of the CLIMAP project from 1978 to 1984 and Project Director in 1980–1981. He was a member of the COHMAP project from 1980 until 1989 and served on the steering committee throughout that time. Along with other COHMAP scientists, he co-edited the 1993 volume *Global Climate Since the Last Glacial Maximum.*

In the late 1980s and the 1990s, his research focused mainly on the longer-term (tectonic-scale) physical and geochemical effects of uplift of the Tibetan Plateau and other high topography on regional and global climate. In 1997, he edited *Tectonic Uplift and Climate Change,* published by Plenum Press. His work on plateau uplift with colleagues Maureen Raymo, John Kutzbach, and Warren Prell has been featured in BBC and NOVA television documentaries.

Since 2001, his research has focused on the effects of early agriculture on greenhouse-gas concentrations. His early anthropogenic hypothesis proposes that releases of carbon dioxide and methane caused by farming drove an anomalous rise in greenhouse-gas concentrations during the last several thousand years. In 1995, his Princeton University Press book *Plows, Plagues, and Petroleum* won the Phi Beta Kappa award for best science book of the year.

Professor Ruddiman is a fellow of the Geological Society of America and the American Geophysical Union. In 2010, he received the Charles Lyell Medal from the Geological Society of London. In 2012, he received the Distinguished Career Award from the American Quaternary Association.

He lives on a hillside in the Shenandoah Valley with his wife, Ginger, and an ever-changing bunch of dogs. His hobbies include rock-wall building, gardening, and walking the Appalachian Mountains, remnants of the continental collisions that produced the super-continent Pangaea and the rocks now featured in his walls.

Brief Contents

Contents

Part V
Historical and Future Climate Change / 314

Chapter 16
Humans and Preindustrial Climate   317

Chapter 17
Climate Changes During the Last 1,000 Years   335

Chapter 18
Climatic Changes Since 1850   357

Preface

Several years after the second edition of *Earth's Climate* was published in 2007, the field of climate science has advanced far enough to warrant a new edition. During that time, an even stronger consensus has developed that rising greenhouse-gas concentrations in the atmosphere caused by human activities are the primary factor behind the warming of our planet. In several cases, changes during just these last few years have occurred faster than previously expected. Sea ice has retreated and thinned far more rapidly since 2006, the Greenland ice sheet has melted at an accelerating rate, and global sea level is rising faster than it did during the 1900s. This edition addresses this new evidence for anthropogenically driven global warming. It also addresses important new advances in understanding past climates.

New to This Edition

STRUCTURING AND UPDATING With one exception, the structure remains the same as in the second edition. Part 1 again covers introductory material, but Chapter 2 from the first edition ("Earth's Climate System Today") has been restored to this edition. This chapter provides a valuable summary of the basic operation of the climate system for use by instructors and students.

The formal treatments of oxygen and carbon isotopes are again placed in Appendices 1 and 2 so as not to break the flow of the text with these more technical treatments. Within the text, more functional definitions are provided (for example, more positive oxygen isotope ratios represent some combination of larger ice sheets and colder ocean temperatures).

Several chapters address significant advances in scientific understanding of complex problems that have occurred in recent years. Prominent advances include the cause of cyclic orbital-scale changes in ice sheet size (Chapter 11) and of millennial-scale changes in climate (Chapter 14). Responses of components of the climate system to global warming have also been updated to recent years (Chapter 17).

STREAMLINED TEXT All of the chapters move from introductory material to final conclusions along a clear logical path. At or near the end of each chapter or major section, "In Summary" statements in the text are marked by tan shading to make them easier to find.

Building on the Second Edition

The third edition retains several successful approaches from the first two editions.

MULTIDISCIPLINARY SCOPE The story of Earth's climate draws on many disciplines—geology, ecology, paleobotany, glaciology, oceanography, meteorology, biogeochemistry, climate modeling, atmospheric chemistry, and hydrology, among others. This range of disciplines is a large challenge, both to students for whom all of the science in this field is new, and also to instructors who may specialize in one or two of the many research fields and time scales of climate change. This text eases this challenge through logical, step-by-step explanations of critical material, accompanied by attractive color graphics that summarize key points and by lists of follow-up resources for background material.

FOLLOWING EARTH'S TIMELINE Like the previous editions, this edition explores the climatic responses of Earth's major systems (ice, water, air, vegetation, and land) as they developed through Earth's history. The structure again follows time's arrow, moving from the earliest known climates all the way to historical, modern, and future changes. The main reason for this structure is that this is the way Earth's climate actually developed, and so it is the most natural way to tell the story. In addition, shorter-term climate changes tend to ride on the back of longer-term changes, and the longer-term changes need to be understood first in order to provide context to those that are of more recent origin.

MYSTERY-SOLVING APPROACH Another advantage of organizing the book by time scale is that students see a coherent, integrated view of the evidence of climate change and the competing hypotheses posed to explain the evidence within each time scale. Because the central dynamics of science are the interactions among data, theory, and theory testing, this integrated approach makes science come alive. The evidence of past climatic changes summarized at the start of each chapter leads to the obvious question: What caused the observed changes? The chapters then describe and evaluate the hypotheses proposed to explain the observations. Students are invited to be detectives in the problem-solving process by weighing the proposed hypotheses against a range of data and other methods, including experiments with climate models. From my own teaching experience, the best students are generally intrigued to find out that so much work still remains to be done in this young field of science.

The major themes of the book remain the same as in the first edition:

- The causes (forcing) of climate change
- The natural response times of the many components of Earth's climate system
- Interactions and feedbacks among these numerous components
- The role of carbon as it moves within the climate system at each time scale

STRUCTURE The structure of this edition of *Earth's Climate* is similar to its predecessor. Part I surveys the field of climate science and the approaches used to unravel Earth's climatic history. Parts II through V describe how Earth's climate has changed at progressively shorter time scales: tectonic-scale and earlier changes in Part II; orbital-scale changes in Part III; deglacial and millennial changes in Part IV; and historical, recent, and future changes in Part V. Progressively more recent intervals receive increasingly detailed treatment because the climate changes can be resolved in finer detail. This approach also helps overcome the complexity of the climate system by focusing on the range of issues critical to each time scale.

ALTERNATIVE VERSIONS W. H. Freeman partners with CourseSmart to provide a low-cost subscription to the text in e-Book format. CourseSmart e-Books can be read online or offline, as well as via free apps for iPad, iPhone, iPod Touch, Android devices, and Kindle Fire. Students are able to take notes, highlight important text, and search for key words. For more information, or to purchase access to the text, visit www.coursesmart.com.

There is also a loose-leaf version of the text available at a discounted price. To order *Earth's Climate*, Third Edition, in loose-leaf, please use ISBN 1-4641-5282-9.

A Growing Audience for Earth's Climate

Adoptions of *Earth's Climate* have grown during the 13 years since the first edition was published. Initially, the book was a popular choice for upper undergraduate courses in earth science departments, often replacing or supplementing classical subjects like historical geology or sedimentology and stratigraphy. Adoptions also occurred in many related disciplines, such as environmental sciences, geography, ecology, botany, and oceanic and atmospheric sciences.

More recently, many instructors have realized that *Earth's Climate* can also be used in introductory-level courses for students who do not plan to major in science but want to satisfy a science requirement. Because climate change is in the news every week, student interest in such courses is running at a high level and these courses are growing. Introductory climate courses at several universities that have used the first two editions of *Earth's Climate* have attracted hundreds of students per term.

Using a book like this at the introductory level requires lecturing and testing at an appropriately generalized (conceptual) level and de-emphasizing some of the quantitative material. I think of students in introductory courses as members of a (it is hoped) attentive jury. They do not need to have prior knowledge of the subject, but they can reasonably be expected to follow the lines of argument carefully, understand the issues at hand, and draw basic conclusions from what they learn.

Instructors who are considering teaching such an introductory course should find the third edition helpful. As before, instructors can bypass the more advanced material set off in boxes titled Looking Deeper into Climate Science. The three other kinds of boxes (Climate Interactions and Feedbacks, Climate Debate, and Tools of Climate Science) are part of the basic text and should not be a major problem for students in introductory courses.

Given my experience teaching this course at the introductory level, I have written a short guide for instructors that can be found at www.whfreeman.com/earthsclimate3e.

Acknowledgments

I acknowledge the efforts of Bill Minick, senior acquisitions editor for the geosciences at W. H. Freeman. I also thank these people: development editor Brittany Murphy, assistant editor Courtney Lyons, media and supplements editor Amy Thorne, project editor Kerry O'Shaughnessy, copy editor Deborah Heimann, text designer Vicki Tomaselli, and illustration coordinator Janice Donnola. Once again, I especially thank Bob

Smith, who did the illustrations for the first two editions and again for this one. Scores of people who have adopted the book have thanked me for the availability of these figures for free downloading, and many of the figures appear in slides at talks during national meetings. I also thank everyone who reviewed chapters or parts of the book:

Mitchell W. Colgan
College of Charleston

David A. Fike
Washington University

Camille Holmgren
Buffalo State College

Suzanne O'Connell
Wesleyan University

Jason A. Rech
Miami University

Peter C. Ryan
Middlebury College

Stacey Verardo
George Mason University

James A. Wanket
Sacramento State University

Derek D. Wright
Lake Superior State University

My thanks also to the reviewers of the first and second editions of this book:

Paul Baker
Duke University

Subir Banerjee
University of Minnesota

Jay Benner
University of Texas, Austin

William B.N. Berry
University of California, Berkeley

Julia Cole
University of Arizona

Tom Crowley
Texas A&M University

P. Thompson Davis
Bentley College

Matthew Evans
The College of William and Mary

Bart Geerts
University of Wyoming

John Gosse
University of Kansas

Andrew Ingersoll
California Institute of Technology

Emily Ito
University of Minnesota

George Jacobsen
University of Maine

David Kemp
Lakehead University

Zhuangjie Li
California State University, Fullerton

Scott A. Mandia
Suffolk County Community College

Patricia Manley
Middlebury College

Isabel P. Montañez
University of California, Davis

R. Timothy Patterson
Carleton University

Maureen Raymo
Massachusetts Institute of Technology

Mike Retelle
Bates College

John Ridge
Tufts University

David Rind
Goddard Institute of Space Sciences

Lisa Sloan
University of California, Santa Cruz

Howard Spero
University of California, Davis

Alycia L. Stigall
Ohio University

Aondover Tarhule
University of Oklahoma

Lonnie Thompson
Byrd Polar Research Center

Stacey Verardo
George Mason University

Thompson Webb
Brown University

Al Verner
Mount Holyoke College

Herb Wright
University of Minnesota

Framework of Climate Science

Large changes in climate naturally pique our curiosity. In the recent past, the hot, dry interval that produced the Dust Bowl of the 1930s forced thousands of farmers from the Great Plains westward to California. A century earlier, air temperatures were cooler than now, and valley glaciers in Europe occupied positions well down the sides of mountains compared with their higher modern elevations. Considerably further back in time, 21,000 years ago, climate was so cold that enormous ice sheets covered Canada and northern Europe, and sea level was some 120 meters (~400 feet) lower than it is today because of the vast amount of ocean water stored on land as ice. Much further back, 100 million years ago, warmer conditions kept ice from forming on the face of Earth, even at the South Pole.

Changes in the distant past occurred for natural reasons, some of them well understood and others still in the process of being unraveled. Climate science is a young science, and many exciting discoveries lie ahead.

Today, Earth's climate is rapidly warming, and it is clear that humans are a major cause of this change. As scientists work to find out how large the warming will be, part of the answer will come from understanding changes that occurred in the past. The chapters in Part I provide a general framework for understanding climate change by addressing several questions:

part I

- What are the major components of Earth's climate system?

- How does climate change differ from day-to-day weather?

- What factors drive changes in Earth's climate?

- How do the many parts of Earth's climate system react to these driving forces and interact with each other?

- How do scientists study past climates as a way to project the changes that lie in our future?

Mountain glacier advances during the Little Ice Age. A view of the village of Argentiere in the valley of Chamonix during mid-1800's shows the Argentiere glacier extending far down the mountainside. (MARY EVANS PICTURE LIBRARY/THE IMAGE WORKS.)

Overview of Climate Science

Life exists nearly everywhere on Earth because the climate is favorable. We live in, on, and surrounded by the climate system: the air, land surfaces, oceans, ice, and vegetation. Climate change is an important thread in the tapestry of Earth's history, along with the evolution of life and the physical transformations of this planet.

But the study of climate also matters for a practical reason: it is relevant to the climatic changes we face in the near future. We have left an era when natural changes governed Earth's climate and have now entered a time when changes caused by human activity predominate. This emerging era is widely referred to as "the Anthropocene."

This chapter surveys the natural factors that cause Earth's climate to change. It also reviews how the field of climate science came into being, how scientists study climate, and how an understanding of the history of climate change can inform us about changes looming in our near future.

▶ Climate and Climate Change

Even from distant space, it is obvious that Earth is the only habitable planet in our solar system (Figure 1-1). More than 70% of its surface is a welcoming blue, the area covered by life-sustaining oceans. The remaining

FIGURE 1-1
The habitable planet

Even seen from distant space, most of Earth's surface looks inviting to life, especially its blue oceans and green forests, but also its brown deserts and white ice. All these areas are prominent parts of Earth's climate system. (NASA.)

30%, the land, is partly blanketed in green, darker in forested regions and lighter in regions where grass or shrubs predominate. Even the pale brown deserts and much of the white ice contain life.

Earth's favorable climate enabled our planet to evolve and sustain life. **Climate** is a broad composite of the average condition of a region, measured by its temperature, amount of rainfall or snowfall, snow and ice cover, wind direction and strength, and other factors. Climate change specifically applies to longer-term variations (years and longer), in contrast to the shorter fluctuations in **weather** that last hours, days, weeks, or a few months.

Earth's climate is highly favorable to life both in an overall, planetwide sense and at more regional scales. Earth's surface temperature averages a comfortable 15.5°C (60°F) and much of its surface ranges between 0° and 30°C (32° and 86°F) and can support life (Box 1-1).

Although we take Earth's habitability for granted, climate can change over time, and with it the degree to which life is possible, especially in vulnerable regions. During the several hundred years in which humans have been making scientific observations of climate, actual changes have been relatively small. Even so, climatic changes significant to human life have occurred. One striking example is the advance of valley glaciers that overran mountain farms and even some small villages in the European Alps and the mountains of Norway a few centuries ago because climate was slightly cooler than now. Those glaciers have since retreated to higher positions, as shown in the introduction to this part of the book.

Scientific studies reveal that historical changes in climate such as the advance and retreat of this glacier are tiny in comparison with the much larger changes that happened earlier in Earth's history. For example, at times in the distant past, ice covered much of the region that is now the Sahara Desert, and trees flourished in what are now Antarctica and Greenland.

1-1 Geologic Time

Understanding climatic changes that occurred in the past requires coming to terms with the enormous span of time over which Earth's climatic history has developed. Human life spans are generally measured in decades. The phases of our lives, such as childhood and adolescence, come and go in a few years, and our daily lives tend to focus mainly on needs and goals that we hope to satisfy within days or weeks.

Almost all of Earth's long history lies immensely far beyond this human perspective. Earth formed 4.55 billion years (Byr) ago (4,550,000,000 years!). Most of the earliest part of Earth's history is known only in a sketchy way. One reason for this gap in our knowledge is the climate system itself: the relentless action of air

Tools of Climate Science

Box 1-1

Temperature Scales

Three temperature scales are in common use in the world today. For day-to-day nonscientific purposes, most people in the United States use the Fahrenheit scale,

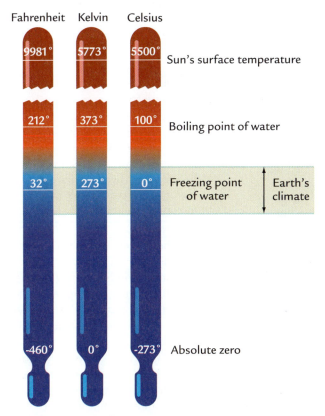

Fahrenheit	Kelvin	Celsius	
9981°	5773°	5500°	Sun's surface temperature
212°	373°	100°	Boiling point of water
32°	273°	0°	Freezing point of water
-460°	0°	-273°	Absolute zero

Earth's climate

Temperature scales Scientists use the Celsius and the Kelvin temperature scales to measure climate changes. Temperatures at Earth's surface vary mainly within a small range of −50°C to +30°C, just below and above the freezing point of water. (ADAPTED FROM W. F. KAUFMAN III AND N. F. COMINS, *DISCOVERING THE UNIVERSE*, 4TH ED., © 1996 BY W. H. FREEMAN AND COMPANY.)

developed by the German physicist Gabriel Fahrenheit. It measures temperature in degrees **Fahrenheit** (°F), with the freezing point of water at sea level set at 32°F and the boiling point at 212°F.

Most other countries in the world, and most scientists as well, routinely use the Celsius (or centigrade) scale developed by the Swedish astronomer Anders Celsius. It measures temperature in degrees **Celsius** (°C), with the scale set so that the freezing point of water is 0°C and the boiling point of water is 100°C.

These equations convert temperature values between the two scales:

$$T_C = 0.55 \, (T_F - 32) \qquad\qquad T_F = 1.8 T_C + 32$$

where T_F is the temperature in degrees Fahrenheit and T_C is the temperature in degrees Celsius.

Scientific calculations generally make use of a third temperature scale developed by the British physicist Lord Kelvin (William Thomson) and known as the **Kelvin** scale. This scale is divided into units of Kelvins (not degrees Kelvin). The lowest point on the Kelvin scale (absolute zero, or 0K) is the coldest temperature possible, the temperature at which motions of atomic particles effectively cease. The Kelvin scale does not have negative temperatures because no temperatures colder than 0K exist.

Temperatures above absolute zero on the Kelvin scale increase at the same rate as those on the Celsius scale, but with a constant offset. Absolute zero (0K) is equivalent to −273°C, and each 1K increase on the Kelvin scale above absolute zero is equivalent to a 1°C increase on the Celsius scale. As a result, 0°C is equivalent to 273K.

and water on Earth's surface has eroded away many of the early deposits that could have helped us reconstruct and understand more of this history.

This book focuses mainly on the last several hundred million years of Earth's history, equivalent to less than 10% of its total age. Our focus is limited in this way because many aspects of Earth's history are only vaguely known far back in the past. But more information becomes available from the younger part of the climatic record, and our chances of measuring and understanding climate change improve.

Even the last 10% of Earth's history covers time spans beyond imagining. The climate scientists who study records spanning hundreds of thousands to

hundreds of millions of years understand time only in a technical way—basically as a means of cataloging and filing information. Geologists often refer to these unimaginably old and long intervals as "deep time," hinting at their remoteness from any real understanding. Like the scientists who study climate change, you will learn in this book to catalog deep time in your own mental file, even if you cannot comprehend its vastness in a tangible way.

The plot of time on the left in Figure 1-2 shows that much of the focus of this book (Parts III through V) fits into a fraction of Earth's history too small even to show up on a simple linear scale. One way to overcome this problem is to start again with a plot

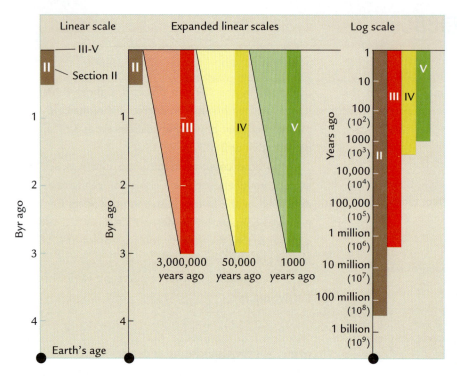

FIGURE 1-2
Earth history

Earth's age is 4.55 billion years. Most of the focus of this book fits into a very small fraction of this immense interval and can be represented only by a series of magnifications or by plotting time on a log scale that increases by factors of 10.

of Earth's full age, but progressively expand out and magnify (blow up) successively shorter intervals to show how they fit into the whole (Figure 1-2 center). Another method is to plot time on a logarithmic scale that increases by successive jumps of a factor of 10 (Figure 1-2 right). This kind of plot compresses the longer parts of the time scale and expands the shorter ones so that they all fit onto one plot.

1-2 How This Book Is Organized

Within its focus on the most recent 10% of Earth's age, this book is organized by time scale. Part II mainly covers climatic changes during the last several hundred million years, an interval during which dinosaurs appeared and later abruptly disappeared, while mammals evolved from primitive to diverse forms. Part III looks at the last 3 million years, the time span when our primitive ancestors evolved into us. Part IV explores changes over the last 50,000 years, an interval that spans large oscillations during the last major glaciation, the maximum development of that glaciation and the subsequent deglaciation that led to our present interglacial climate. Part V starts with the story of how our fully human ancestors initially lived a primitive hunting-and-gathering life and then developed agriculture, prior to the first human civilizations. Part V then focuses in from the last 1,000 years to the current industrial era and projects forward into the future.

This progression from longer to shorter time scales is a natural way to look at the climate system because faster changes at shorter time scales are embedded in and superimposed upon slower changes

at longer time scales. At the longest time scale, a slow warming between 300 and 100 million years (Myr) ago was followed by a gradual cooling during the last 100 Myr (Figure 1-3A). This gradual cooling led to the appearance of Antarctic ice and later to the massive northern hemisphere ice sheets that advanced and retreated many times during the last 3 million years at cycles of several tens of thousands of years (Figure 1-3B). Superimposed on these climatic cycles were shorter oscillations that lasted a few thousand years and were largest during times when climate was colder (Figure 1-3C). The last 1,000 years has been a time of relatively warm and stable climate, with much smaller oscillations (Figure 1-3D).

The way the climatic changes at these various time scales are linked is analogous to the way that cycles of daily heating and nighttime cooling are superimposed on the longer seasonal cycle of summer warmth and winter cold. To understand the extreme heat reached during a specific afternoon in July in the Northern Hemisphere, it first makes sense to consider that such an afternoon occurs in the larger context of the hottest season of the year, and then to factor in the additional contribution from daytime heating. For a similar reason, it makes sense to follow time's arrow and trace climate changes from older to younger eras, and from the larger cycles to the smaller ones superimposed upon them.

As the book progresses from older to younger time scales, you will notice a change in the level of information about past climate changes. In part, this development reflects a change in the amount of detail that can be retrieved from climatic records, called the **resolution**. Because older records tend to have lower

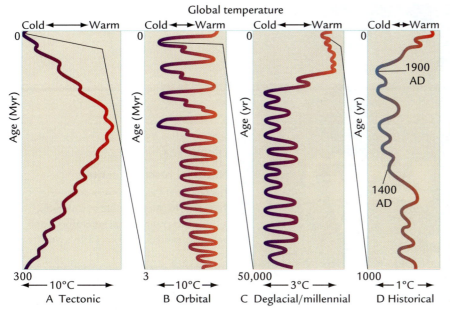

Global temperature

Cold ◄──► Warm   Cold ◄►Warm   Cold ◄──► Warm   Cold ◄►Warm

0   0   0   0

Age (Myr)   Age (Myr)   Age (yr)   Age (yr)

                                                   1900
                                                   AD

                                                   1400
                                                   AD

300   3   50,000   1000
◄── 10°C ──►   ◄─ 10°C ─►   ◄── 3°C ──►   ◄─ 1°C ─►
A Tectonic   B Orbital   C Deglacial/millennial   D Historical

FIGURE 1-3

Time scales of climate change

Changes in Earth's climate span several time scales, arrayed from longer to shorter: (A) the last 300 million years, (B) the last 3 million years, (C) the last 50,000 years, and (D) the last 1,000 years. Here progressively smaller changes in climate at successively shorter time scales are magnified out from the larger changes at longer time scales.

resolution, much of the focus of Part II of this book is on the longer-term average climatic states over millions of years, and on the way they differ from our climate today. By comparison, younger records tend to have progressively higher resolution, and Parts III through V look at successively shorter-term changes in climate that occur within intervals of thousands, hundreds, and finally even tens of years. We will examine the resolution issue more closely in Chapter 3.

▶ Development of Climate Science

As scientists began to discover examples of major climatic changes earlier in Earth's history, they were naturally curious about why these fluctuations happened. The few amateur scientists and university professors who studied climate in relative isolation during the nineteenth and early twentieth centuries have by now given way to thousands of researchers with backgrounds in geology, physics, chemistry, and biology working at universities, national laboratories, and research centers throughout the world (Figure 1-4). Today climate scientists use aircraft, ships, satellites, sophisticated new biological and chemical lab techniques, and high-powered computers, among other methods, to carry out their studies.

Studies of climate are incredibly wide-ranging. They vary according to the part of the climate system being studied, such as changes in air, water, vegetation, land surfaces, and ice. They also vary in the techniques used, including physical and chemical measurements of the properties of air, water, and ice and of life-forms fossilized in rocks; biological or botanical measurements of numerous kinds of life-forms; and computer simulations to model the behavior of air, water, and vegetation.

This huge diversity of studies covers a broad array of scientific disciplines. Some studies are directed at improving our understanding of the modern climate system: *meteorologists* study the circulation of the atmosphere; *oceanographers* explore the circulation of the ocean; *chemists* investigate the composition of the ocean, atmosphere, and land; *glaciologists* measure the behavior of ice; and *ecologists* analyze life-forms on land or in the water. Chapter 2 provides an overview of what we have learned about the operation of the climate system.

Other studies focus on changes in climate or climate-related phenomena in Earth's recent or more distant past: *geologists* explore the broader aspects of Earth's history; *geophysicists* investigate past changes in Earth's physical configuration (continents, oceans, and mountains); *geochemists* analyze past chemical changes in the ocean, air, or rocks; *paleoecologists* study past changes in vegetation and their role in the climate system; *climate modelers* evaluate possible causes of climate change; and *climate historians* explore written archives for information that will enable them to reconstruct past climates.

In recent decades, many studies of Earth's climatic history have crossed the traditional disciplinary boundaries and merged into an interdisciplinary approach referred to as "Earth system science" or "Earth system history." Such efforts recognize that the many parts of Earth's climate system are interconnected, and that investigators of climate must look at all of the parts in order to understand the whole. This book is an example of the **Earth system** approach.

In that regard, this book makes no special distinction between studies of Earth's past history and investigations of the current (or very recent) climatic record. Earth's climatic history is a continuum from

FIGURE 1-4
National research centers
The National Center for Atmospheric Research (NCAR) in Boulder, Colorado, is one of several national laboratories and university centers at which Earth's climate is studied. (SANDRA BAKER/ALAMY.)

the distant past to the present. The book is organized by time scale because that is the way Earth's climatic history has developed and will continue to develop in the future. Lessons learned about how the climate system has operated in the past can be applied directly to our understanding of the present and future, but the opposite is true as well. The broad term **climate science** refers to this vast *multidisciplinary* and *interdisciplinary* field of research, and to its linkage of the past, the present, and the future.

1-3 How Scientists Study Climate Change

Climate science moves forward by an interactive mix of observation and theory. Climate scientists gather and analyze data from the kinds of climatic archives reviewed in Chapter 3, and the results of this research are written up and published. Progress in science depends on the free exchange of ideas, and climate researchers publish in order to tell the scientific community what they have discovered.

Scientists who interpret their research results occasionally come up with a new **hypothesis** proposed to explain their observations. Other scientists evaluate these hypotheses, often discarding the ones that are less worthy because the predictions they make are contradicted by subsequent observations.

Any hypothesis that succeeds in explaining a wide array of observations over a period of time becomes a **theory**. Scientists continue to test theories by making additional observations, developing new techniques to analyze data, and devising models to simulate the operation of the climate system. Only a few theories survive years of repeated testing. These are sometimes called "unifying theories" and are generally regarded as close approximations to the "truth," but the testing continues.

Taken together, the many expanding efforts to understand climate change have led to a large-scale scientific revolution that has accelerated through the late 1900s and early 2000s. The mysteries of the climate system have yielded their secrets slowly, and many important questions still remain to be answered, but the revolution of knowledge to date has been immense, as this book will show.

This revolution has now expanded to a point where it is claiming a place alongside two great earlier revolutions of knowledge in Earth's history. The first was the development by Charles Darwin and others in the nineteenth century of the theory of **evolution**, which led to an understanding of the origin of the long sequence of life-forms that have appeared and disappeared during the history of this planet. The second was the synthesis during the 1960s and 1970s of the theory of **plate tectonics**, which has given us an understanding of the slow motions of the continents across Earth's surface through time, as well as associated phenomena such as volcanoes, earthquakes, and mountain ranges.

▶ Overview of the Climate System

Earth's **climate system** consists of air, water, ice, land, and vegetation. At the most basic level, changes in these components through time are analyzed in terms of *cause* and *effect*, or, in the words used by climate scientists, **forcing** and **response**. The term "forcing" refers to those factors that drive or cause changes; the responses are the resulting climatic shifts.

1-4 Components of the Climate System

Figure 1-5 provides an initial (and simplified) impression of the vast array of factors involved in studying Earth's climate. It shows the air, water, ice, land, and

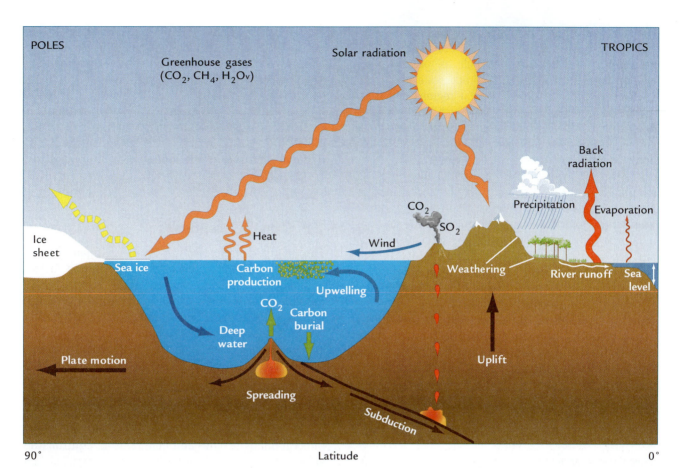

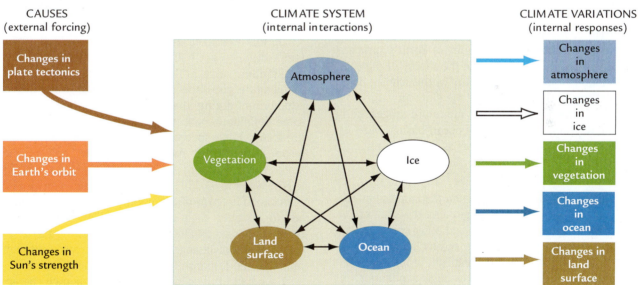

FIGURE 1-5

Earth's climate system and interactions of its components

Studies of Earth's climate cover a wide range of processes, indicated at the top. Climate scientists organize and simplify this complexity, as shown at the bottom. A small number of factors drive, or force, climate change. These factors cause interactions among the internal components of the climate system (air, water, ice, land surfaces, and vegetation). The results are the measurable variations known as climate responses.

vegetation that form the major components of the climate system, as well as some of the processes at work within the climate system, such as precipitation, evaporation, and winds. These processes extend from the warm tropics to the cold polar regions and from the Sun in outer space down into Earth's atmosphere, deep into its oceans, and even beneath its bedrock surface. All of these processes will be explored in this book.

The complexity of the top part of Figure 1-5 is simplified in the bottom part to provide an idea of how the climate system works. The relatively small number of external factors shown on the bottom left force (or drive) changes in the climate system, and the internal components of the climate system respond by changing and interacting in many ways (bottom center). The end result of all these interactions is a number of observed variations in climate (bottom right). This complexity can be thought of as the operation of a machine: the factors that drive climate change are the input, the climate system is the machine, and the variations in climate are the output.

1-5 Climate Forcing

Three fundamental kinds of climate forcing exist in the natural world:

- *Tectonic processes* generated by Earth's internal heat alter the basic geography of Earth's surface. These processes are part of the theory of plate tectonics, the unifying theory of the science of geology. Examples include the slow movement of continents, the uplift of mountain ranges, and the opening and closing of ocean basins. Most of these processes change very slowly over millions of years. The processes of plate tectonics are summarized in Part II of this book.

- *Changes in Earth's orbit around the Sun* also affect climate. These orbital changes alter the amount of solar **radiation** (sunlight and other energy) received on Earth by season and by latitude (from the nearly overhead sunlight in the warm low-latitude tropics to the low-angle sunlight or seasonal darkness at the cold high-latitude poles). Orbital changes occurring over tens to hundreds of thousands of years are the focus of Parts III and IV.

- *Changes in the strength of the Sun* also affect the amount of solar radiation arriving on Earth. One example appears in Chapter 5: the strength of the Sun has slowly increased throughout the 4.55 Byr of Earth's existence. In addition, shorter-term variations that occur over decades or longer are part of the focus of Part V.

A fourth factor capable of influencing climate, but not in a strict sense part of the natural climate system, is the effect of humans on climate, referred to as **anthropogenic forcing**. This forcing is an unintended by-product of agricultural, industrial, and other human activities, and it occurs through alterations of Earth's land surfaces and through additions of carbon dioxide (CO_2) and other **greenhouse gases**, sulfate particles, and soot to the atmosphere. Anthropogenic effects will be covered in Part V.

1-6 Climate Responses

The many components of Earth's climate system shown in Figure 1-5 respond to the driving factors with a characteristic **response time**, a measure of the time it takes to react fully to the imposed change. Consider the example shown in Figure 1-6A: a beaker of water placed above a typical laboratory Bunsen burner. The Bunsen burner represents an external climate forcing (such as the Sun's radiation), and the water temperature is the climatic response (such as the average temperature of Earth's surface). When the burner is turned on, it begins to heat the water. The water in the beaker gradually warms and after a long interval finally reaches and maintains an **equilibrium** value. The rate of warming is naturally rapid at first but slows as time passes and it nears its equilibrium state (Figure 1-6B).

In this example, the time it takes for the water temperature to get half of the remaining way to this equilibrium value is its response time. The water temperature rises the first 50% of the way toward equilibrium during the first response time, but the same definition continues to apply later in the warming trend. Each successive step takes one additional response time and moves the system half of the *remaining* way toward equilibrium: 50% $\left(\frac{1}{2}\right)$ to 75% $\left(\frac{3}{4}\right)$ to 87.5% $\left(\frac{7}{8}\right)$ to 93.75% $\left(\frac{15}{16}\right)$. This progression can also be understood in terms of the amount of the total response left to go after each step: $\frac{1}{2}$, $\frac{1}{4}$, $\frac{1}{8}$, $\frac{1}{16}$. This heating response has an exponential form. Note that the *absolute amount* of change in heating decreases through time, but the response time remains exactly the same.

Each part of the climate system has its own characteristic response time (Table 1-1), ranging from hours or days up to thousands of years. The atmosphere has a very fast response time, and significant changes can occur in just hours (as in daily cycles of heating and cooling). The land surface reacts more slowly, but it still shows large heating and cooling changes on time scales of hours to days to weeks. Beach sand can become too hot to walk on during a single summer afternoon, but it takes much longer to chill the upper layer of soil in winter to the point where it freezes.

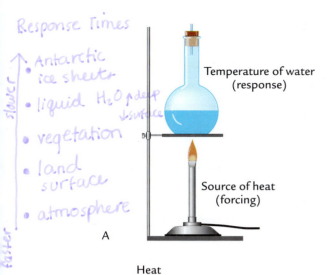

Response Times

↑ slower
• Antarctic ice sheets
• liquid H₂O deep ↓surface
• vegetation
• land surface
• atmosphere
↓ faster

Temperature of water (response)

Source of heat (forcing)

A

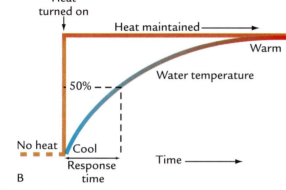

Heat turned on

Heat maintained ————————→

Warm

Water temperature

50% – – –

No heat | Cool

Response time

Time ————→

B

FIGURE 1-6
Response time

Earth's climate system has a response time, suggested conceptually by the reaction of a beaker of water to heating by a Bunsen burner. The response time is the rate at which water in the beaker warms toward an equilibrium temperature. (ADAPTED FROM J. IMBRIE, "A THEORETICAL FRAMEWORK FOR THE ICE AGES," *JOURNAL OF THE GEOLOGICAL SOCIETY* (LONDON) 142 [1985]: 417–32.)

Liquid water has a slower response time than air or land because it holds much more heat. The temperature response of shallow lakes or of the wind-stirred uppermost part of the ocean is measured in weeks to months. This is evident in the way lakes cool off seasonally, but never as fast as the land does. For ocean layers that lie more remote from interactions with the atmosphere, response times can range from decades for the shallow subsurface ocean layers up to many centuries for the deepest ocean.

Although a meter-thick layer of sea ice on polar oceans grows and melts in time periods ranging from months to years, thicker mountain glaciers react over longer time spans of decades or more. Massive (kilometers-thick) ice sheets like the one now covering the continent of Antarctica have the slowest response times in the climate system—many thousands of years, aptly captured in the familiar use of the word "glacial."

The response-time concept also applies to vegetation, an organic part of the climate system. Unseasonable frosts can kill leaves and grasses overnight, and abnormally hard freezes can do the same to the woody tissue of trees, with responses measured in just hours. On the other hand, seasonal spring greening of the landscape and normal autumn loss of leafy green material can take weeks or months to complete. Pioneering vegetation that occupies newly exposed ground (for example, bare ground left behind by melting glaciers) may take tens to hundreds of years or more to come to full development because of the slow dispersal of seeds and the time needed for them to germinate.

1-7 Time Scales of Forcing Versus Response

The parts of this book differ considerably in how the relationship between the forces that drive climate change and the responses of the climate system is treated. Several hypothetical cases shown in Figure 1-7 give a sense of these differences:

- *The forcing is very slow in comparison with the response of the climate system.* This case is equivalent to increasing the flame of the Bunsen burner in Figure 1-6A so slowly that the water temperature has no problem keeping pace with the gradual application of more heat. If the changes in climate forcing are very slow in comparison with the response time of the climate system, the system will simply passively track along with the forcing with a small but imperceptible lag (Figure 1-7A).

   This case is typical of many climate changes that occur over the long tectonic time scales discussed in Part II. For example, continents can be slowly carried by plate tectonic processes toward higher or lower latitudes at rates averaging about 1 degree of latitude (100 km or 60 miles) per million years. As the landmasses move toward lower latitudes, where incoming solar radiation is stronger, or toward higher latitudes, where it is weaker, temperatures over the continents will react to these slow changes in solar heating with imperceptibly tiny year-by-year responses. Because the response time of air over land is short (hours to weeks; see Table 1-1), the average temperature over the continents can easily keep pace with the slow changes in average overhead solar radiation over millions of years. Shorter-term changes also occur over tectonic time scales, but they are usually harder to resolve in older records.

Table 1-1 Response Times of Various Climate System Components		
Component	**Response Time (range)**	**Example**
FAST RESPONSES		
Atmosphere	Hours to weeks	Daily heating and cooling Gradual buildup of heat wave
Land surface	Hours to months	Daily heating of upper ground surface Midwinter freezing and thawing
Ocean surface	Days to months	Afternoon heating of upper few feet Warmest beach temperatures late in summer
Vegetation	Hours to decades/centuries	Sudden leaf kill by frost Slow growth of trees to maturity
Sea ice	Weeks to years	Late-winter maximum extent Historical changes near Iceland
SLOW RESPONSES		
Mountain glaciers	10–100 years	Widespread glacier retreat in 20th century
Deep ocean	100–1,500 years	Time to replace ocean deep water
Ice sheets	100–10,000 years	Advances/ retreats of ice sheet margins Growth/decay of entire ice sheet

- *The forcing is fast in comparison with the climate system's response.* At the other extreme, the response time of the climate system may be far slower than the duration of the forcing (Figure 1-7B). In this case, little or no response to the climate forcing occurs. This case is equivalent to turning the Bunsen burner on and off so quickly that the temperature of the water in the beaker has very little time to react.

One example of this extreme case is a total solar eclipse, which blocks Earth's only source of external heating for less than an hour. Air temperatures cool slightly during that brief interval, but then rise again. Volcanic eruptions are another example, such as the 1991 summer explosion of Mount Pinatubo in the Philippine Islands. Fine volcanic particles produced by that eruption blocked part of the Sun's radiation for several months and caused Earth's average temperature to fall by 0.5°C, but the cooling effect disappeared within a few years because fine volcanic particles only stay in the upper layers of the atmosphere for that long (Table 1-1).

- *The time scales of forcing and climate response are similar.* Between these two extremes is the more interesting case in which the time scales of the climate forcing and the climate system's responses fall within a similar range. This situation produces a more dynamic response of the climate system.

Consider a different experiment with the Bunsen burner and the beaker of water. This time, the Bunsen burner (again the source of climate forcing) is abruptly turned on, left on awhile, turned off, left off awhile, turned on again, and so on (Figure 1-7C). These changes cause the water to heat up, cool off, heat up again, and so on. The water temperature responds by cycling back and forth between two values, one toward the cold extreme with the flame off and one toward the warm extreme with the flame on. But the intervals of heating and cooling do not last long enough to allow the water time to reach either equilibrium temperature, as it did in the example shown in Figure 1-6B.

The difference between the two cases illustrated in Figures 1-7C and D shows that the frequency with which the flame is turned on and off determines the size of the water temperature response. Both examples assume the same equilibrium values (cold and warm) for the water temperature and the same position of the Bunsen burner relative to the beaker of water. The only difference in the forcing is the length of time the flame is left on or off. If the flame is switched on and off rapidly, the water temperature has less time to reach the equilibrium temperatures (hot or cold) and the size of the response is small (Figure 1-7D). But if the flame stays on or off for longer

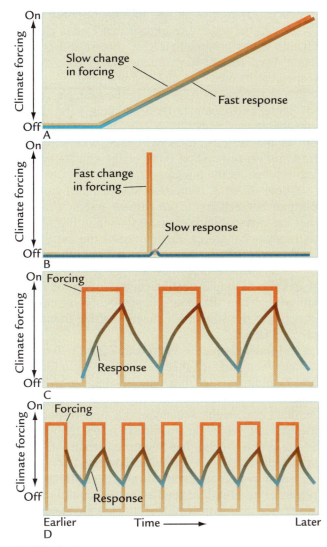

FIGURE 1-7

Rates of forcing vs. response

Climate responses depend on the relative rate of changes in climate forcing versus the response time of the climate system. (A) Very slow forcing may allow the climatic responses to track along with little delay. (B) Very fast changes in forcing may allow little reaction in slow climatic responses. (C, D) Roughly equal time scales of forcing and response allow varying amplitudes of response of the climate system.

intervals, the temperature of the water has time to reach values closer to the full equilibrium states (Figure 1-7C).

In the real world, climate forcing rarely acts in the on-or-off way implied by the examples in Figure 1-7. Instead, changes commonly occur in smooth continuous cycles. If we again use the Bunsen burner concept, this situation is analogous to keeping the burner flame (the climate forcing) on at all times, but slowly and cyclically varying its intensity (Figure 1-8). The

resulting cycles of warming and cooling of the water lag behind the shifts in the amount of heat applied, just as they did in Figure 1-7C and D.

Daily and seasonal changes provide familiar examples of this kind of forcing and response. In the Northern Hemisphere, the summer Sun is highest in the sky and therefore strongest at summer solstice on June 21, but the hottest air temperatures are not reached until July over the land and late August or early September over the ocean. Similarly, the coldest winter days occur over land in January or February, well after the weakest winter solstice Sun on December 21. Even during a single day, the strongest solar heating occurs near noon, but the warmest temperatures are not reached until the afternoon, hours later.

Even though the smooth cycles of forcing and response in Figure 1-8 look different from the cases examined in Figure 1-7, the underlying physical response of the beaker of water (or, by extension, of the climate system) remains exactly the same. The temperature of the water in the beaker continues to react at all times with the same characteristic response time defined earlier, and the rate of response of the climate system is once again fastest when the climate system is farthest from its equilibrium value.

The main difference now is that the climate forcing (the intensity of the flame from the Bunsen burner) is constantly changing, rather than holding at a single constant equilibrium value (as in Figure 1-6)

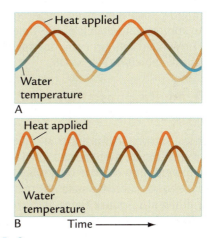

FIGURE 1-8

Cycles of forcing and response

Many kinds of climate forcing vary in a cyclical way and produce cyclic climate responses. The amplitude of the climate responses is related to the time they have to move toward equilibrium. (A) Climate changes are larger when the climate system has ample time to respond. (B) The same amplitude of forcing produces smaller climate changes if the climate system has less time to respond. (A AND B: ADAPTED FROM J. IMBRIE, "A THEORETICAL FRAMEWORK FOR THE ICE AGES," *JOURNAL OF THE GEOLOGICAL SOCIETY* (LONDON) 142 [1985]: 417–32.)

or switching between two equilibrium values in an alternating sequence (as in Figures 1-7C and D). These continuous changes in heating act as a "moving target." The climate system response (the water temperature) keeps chasing this moving target but can never catch up because the water temperature cannot respond quickly enough.

As was the case for the on-off changes shown in Figures 1-7C and D, the frequency with which these smooth cycles of forcing occur has a direct effect on the amplitude of the responses, an effect that is apparent in the differences between the cases shown in Figures 1-8A and B. If the forcing occurs in slower (longer) cycles, it produces a larger response (larger maxima and minima) because the climate system has more time to react before the forcing turns back in the opposite direction (see Figure 1-8A). In contrast, forcing that occurs in faster (shorter) cycles produces a smaller response because the climate system has less time to react before the forcing reverses direction (see Figure 1-8B). The two responses differ in size, even though the forcing moves back and forth between the same maximum and minimum values in both cases.

The relationships between forcing and response shown in Figure 1-8 are particularly useful for understanding the orbital-scale climatic changes explored in Parts III and IV of this book. Changes in incoming solar radiation due to changes in Earth's orbit occur over tens of thousands of years, and this also happens to be the response-time characteristic of the large ice sheets that grow and melt over these orbital time scales. This approximate match of the time scales of forcing and response sets up cyclic interactions very much like those shown in Figure 1-8.

1-8 Differing Response Rates and Climate System Interactions

The examples shown so far have summarized the response of the climate system by a single curve, as if the system were only capable of a single response. But Table 1-1 shows that the system has many components with different response times, each responding to climatic forcing at its own tempo.

One way to grasp the impact of these differences in response is to imagine that some change is abruptly imposed on the climate system from the outside (for example, a sudden strengthening of the Sun's radiation). Each part of the climate system will respond to this sudden increase in external heating in a way analogous to the beaker of water sitting over the Bunsen burner (Figure 1-6), but in this case at a tempo dictated by its own response time (Figure 1-9). The faster-responding parts of the climate system will warm up more quickly, and the slower-responding parts will do so more slowly.

We can apply this idea of differing response times to the case in which the forcing that causes climate change varies in smooth cycles (Figure 1-10). Here again, each part of the climate system will tend to respond at its own rate, and this will produce several different patterns of response. In the example shown in Figure 1-10, some faster-responding parts of the climate system respond so quickly to the climate forcing that they can track right along with it. In contrast, slower-responding parts of the climate system lag well behind the forcing.

These differing response rates can lead to complicated interactions in the climate system. Assume that the curve of initial climate forcing in Figure 1-10 represents changes in the amount of the Sun's heat reaching a particular region over cycles of thousands of years. Also assume that the fast-response curve represents the rapid heating of landmasses at lower latitudes, while the slow-response curve represents changes in the size of ice sheets that lag thousands of years later.

In this scenario, the asterisk in Figure 1-10 marks a time when large ice sheets have built up in Canada and Scandinavia (as has actually happened many times in the past, most recently 21,000 years ago). At this point in the sequence, the slow-responding ice has not yet begun to retreat, even though the heating from the Sun has begun to increase and the land well south of the ice sheets has begun to warm.

Given this situation, how do you think the air temperatures just south of the ice limits would respond? Would the air warm with the initial strengthening of

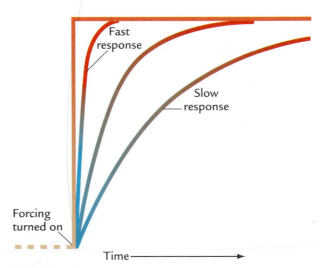

FIGURE 1-9
Variations in response times

An abrupt change in climate forcing will produce climate responses ranging from slow to fast within different components of the climate system, depending on their inherent response times.

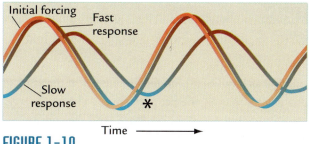

FIGURE 1-10
Variations in cycles of response
If the climate forcing occurs in cycles, it will produce differing cyclic responses in the climate system, with the fast responses tracking right along with the forcing cycles while the slower responses lag well behind.

the overhead Sun and heating of the land? If so, its response would track right behind the initial forcing curve in Figure 1-10.

Or would the air temperatures still be under the chilling influence of the large mass of ice lying just to the north and not begin to rise until the ice starts to retreat? In this case, the ice would in a sense be acting as a semi-independent player in the climate system by exerting an influence of its own on local climate. Although the ice initially acts as a slow climate response driven by slow changes in the Sun, it then exerts its own separate effect on climate.

Both of these explanations probably sound plausible, and they both are. The air temperatures just south of the ice sheets will be influenced *both* by the overhead Sun and the nearby ice. The actual timing of the air temperature response in such regions will fall somewhere in the middle, faster than the response of the ice but lagging behind the forcing from the Sun. As this example suggests, Earth's climate system is dynamic, with numerous interactions.

The response-time concept is directly relevant to projections of climate change in the near future. Part V of this book addresses the effects of humans on climate through the buildup of greenhouse gases, primarily CO_2 produced from burning fossil fuels (coal, oil, and natural gas). The changes in the next few centuries will be unusual because the large climate forcing caused by humans and the warming it will bring will arrive at a speed much faster than the large changes known from Earth's history. Within a few centuries, the fossil fuels that generate excess CO_2 in the atmosphere will be largely used up, CO_2 emissions will fall, and Earth's climate will begin to return toward its previous cooler state. But before that happens, Earth will face centuries of very substantial warmth, along with many other changes.

Scientists, and the public in general, want to know how large the disruption caused by these oncoming centuries of very high CO_2 concentrations will be, and

the answer to this question requires understanding the different response times of the major components in the climate system. Most parts of the system will begin to respond relatively quickly to the greenhouse-gas forcing, but others (those most closely tied to the ice sheets and deep ocean) will respond more sluggishly. A large part of the challenge facing climate scientists is to sort out these different responses and all their interactions.

1-9 Feedbacks in the Climate System

Another important kind of interaction in the climate system is the operation of **feedbacks**, processes that alter climate changes that are already underway, either by amplifying them (**positive feedbacks**) or by suppressing them (**negative feedbacks**). Figure 1-11 shows how feedbacks operate.

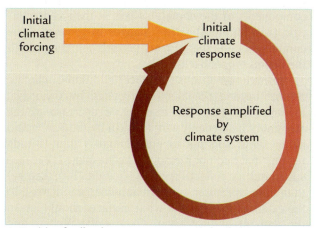

A Positive feedback

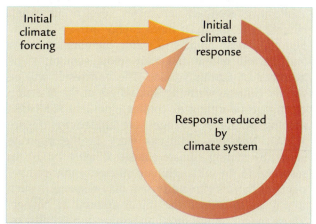

B Negative feedback

FIGURE 1-11
Climate feedbacks
(A) Positive feedbacks within the climate system amplify climate changes initially caused by external factors. (B) Negative feedbacks mute or suppress the initial changes.

Climate Interactions and Feedbacks

Box 1-2

Positive and Negative Feedbacks

The strength of a feedback on temperature, called the **feedback factor**, or f, is defined as:

$$f = \frac{\text{temperature change with feedback}}{\text{temperature change without feedback}}$$

where "temperature change" refers to the full equilibrium response

If f has a value of 1, no feedback exists. If the value of f is greater than 1, the net temperature change is larger than it would be without feedback, and the climate system has a positive feedback. If the value of f is less than 1, the temperature change is smaller than it would be in the absence of feedback, and the climate system has a negative feedback.

Assume that some external factor (again, perhaps a change in the strength of radiation from the Sun) causes Earth's climate to change. Those changes will consist of many responses among the various internal components of the climate system at their characteristic (and different) rates. Changes in some of these components will then further perturb climate through the action of feedbacks.

Positive feedbacks produce additional climate change beyond the amount triggered by the initial forcing (Box 1-2). For example, a decrease in the amount of heat energy sent to Earth by the Sun would allow snow and ice to spread across high-latitude regions that had not previously been covered. Because snow and ice reflect far more sunlight (heat energy) than bare ground or open ocean water, an increase in the extent of these bright white surfaces should lessen the amount of heat taken up in polar regions and amplify the climatic cooling in those regions.

Positive feedback processes also occur when climate is warming. If more energy arrives from the Sun and causes climate to warm, high-latitude snow and ice will retreat and allow more sunlight to be absorbed at Earth's surface. The end result will be further climatic warming. Regardless of the direction of change (warming or cooling), positive feedbacks amplify whatever changes are underway (Figure 1-11A). Negative feedbacks work in the opposite sense, by muting climate changes (Figure 1-11B). In response to an initial climate change, some components of Earth's climate system may respond in such a way as to reduce the initial amount of change. These two examples focus on temperature feedbacks. But other feedbacks can amplify or suppress other climatic responses, such as precipitation.

In Summary, the climate system is highly complex and can seem (to someone new to this subject) daunting. But an age-old wisdom suggests that seemingly insurmountable problems can be understood if they are broken down into smaller ("bite-sized") pieces. This book tries to simplify that complexity by first organizing climatic changes by time scale, and then exploring the most important climatic drivers (forcings) and responses, including of the interactions and feedbacks among the responses.

Key Terms

climate (p. 4)
weather (p. 4)
Fahrenheit (p. 5)
Celsius (p. 5)
Kelvin (p. 5)
resolution (p. 6)
Earth system (p. 7)
climate science (p. 8)
hypothesis (p. 8)
theory (p. 8)
evolution (p. 8)
plate tectonics (p. 8)
climate system (p. 8)

forcing (p. 8)
response (p. 8)
radiation (p. 10)
anthropogenic forcing (p. 10)
greenhouse gases (p. 10)
response time (p. 10)
equilibrium (p. 10)
feedbacks (p. 15)
positive feedback (p. 15)
negative feedback (p. 15)
feedback factor (p. 16)

Review Questions

1. How does climate differ from weather?
2. In what ways does climate science differ from traditional sciences such as chemistry and biology?
3. How does climate forcing differ from climate response?

4. In the example in which the Bunsen burner is lit and the beaker of water at first warms quickly and then more slowly, how does the response time of the water vary through time?

5. The climate system consists of many components with different response times. What is the total range of time scales over which these responses vary?

6. Do positive feedbacks always make the climate warmer?

Additional Resources

Basic Reading

1997. *Climate Change: State of Knowledge.* Washington, DC: Office of Science and Technology Policy.

Advanced Reading

Imbrie, J. 1985. "A Theoretical Framework for the Ice Ages." *Journal of the Geological Society* 142:417–32.

Chapter 2

Earth's Climate System Today

n this chapter, we examine Earth's climate system, which basically functions as a heat engine driven by the Sun. Earth receives and absorbs more heat in the tropics than at the poles, and the climate system works to compensate for this imbalance by transferring energy from low to high latitudes. The combination of solar heating and movement of heat energy determines the basic distributions of temperature, precipitation, ice, and vegetation on Earth.

We first look at Earth from the viewpoint of outer space as a sphere warmed by the Sun and characterized by global average properties. Then we explore how solar radiation is absorbed by the ocean and land at lower latitudes, transformed into several kinds of heat energy, and transported to higher altitudes and latitudes by wind and ocean currents. Next we examine the kinds of ice that accumulate at the high latitudes and high altitudes where little of this redistributed heat reaches. The chapter ends by reviewing characteristics of life that are relevant to climate.

Heating Earth

Earth's climate system is driven primarily by heat energy arriving from the Sun. Energy travels through space in the form of waves called **electromagnetic radiation**. These waves span many orders of magnitude in size, or wavelength, and this entire range of wave sizes is known as the **electromagnetic spectrum** (Figure 2-1).

The energy that drives Earth's climate system occupies only a narrow part of this spectrum. Much of the incoming radiation energy from the Sun consists of visible light at wavelengths between 0.4 and 0.7 µm (1 µm, or micrometer = 1 millionth of a meter), sometimes referred to as **shortwave radiation**. Some ultraviolet radiation from the Sun also enters Earth's atmosphere, but radiation at still shorter wavelengths (X rays and gamma rays, measured in nanometers, or billionths of a meter) does not affect climate.

2-1 Incoming Solar Radiation

The most basic way to look at Earth's climate system is to consider its average properties as a sphere. In this way, we reduce its three-dimensional complexities to a single global average value typical of the entire planet, as if we were space travelers looking at Earth from a great distance.

Radiation from the Sun arrives at the top of Earth's atmosphere with an average energy of 1,362 watts per square meter (W/m^2). These watts are the same units of energy used to measure the brightness (or more accurately the power) of a household light bulb. If Earth were a flat, one-sided disk directly facing the Sun, and if it had no atmosphere, 1,362 W/m^2 of solar radiation would fall evenly across its entire surface (Figure 2-2 top).

But Earth is a three-dimensional sphere, not a flat disk. A sphere has a surface area of $4\pi r^2$ (r being its radius) that is exactly four times larger than the surface area of a flat one-sided disk (πr^2). Because the same amount of incoming radiation must be distributed across this larger surface area, the average radiation received per unit of surface area on a sphere is only one-quarter as strong (1,362/4 = ~340 W/m^2). Another way of looking at this is to see that half of Earth's rotating surface is dark at any time because it faces away from the Sun at night, while the daytime side, warmed by the Sun, receives radiation at indirect angles, except at the one latitude in the tropics where the Sun is directly overhead in any given season (Figure 2-2 bottom).

The 340 W/m^2 of solar energy arrives at the top of the atmosphere, mainly in the form of visible radiation. About 70% of this shortwave radiation passes through Earth's atmosphere and enters the climate system (Figure 2-3). The other 30% is sent directly

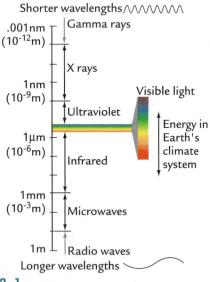

FIGURE 2-1
The electromagnetic spectrum
Energy moves through space in a wide range of wave forms that vary by wavelength. Energy from the Sun that heats Earth arrives mainly in the visible part of the spectrum. Energy radiated back from Earth's surface moves in the longer-wavelength infrared part of the spectrum. (MODIFIED FROM W. J. KAUFMAN III AND N. F. COMINS, *DISCOVERING THE UNIVERSE*, 4TH ED., © 1996 BY W. H. FREEMAN AND COMPANY.)

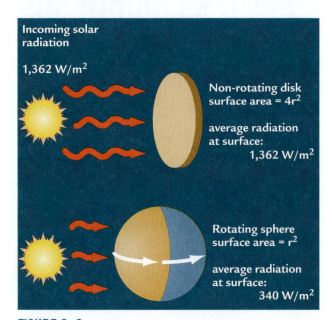

FIGURE 2-2
Average solar radiation on a disk and a sphere
The surface of a flat nonrotating disk that faces the Sun (top) receives exactly four times as much solar radiation per unit of area as the surface of a rotating sphere, such as Earth (bottom).

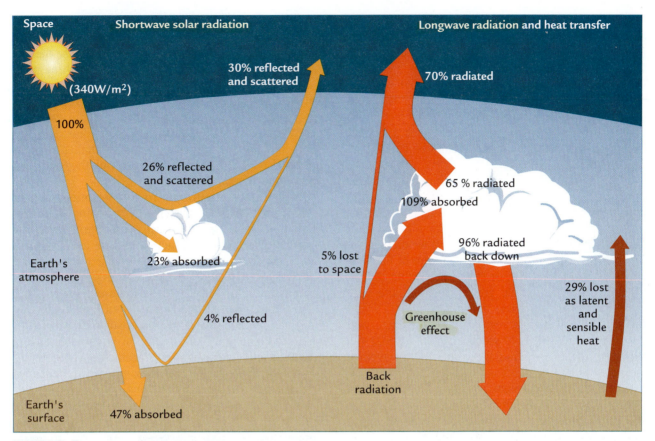

FIGURE 2-3

Earth's radiation budget

Solar radiation arriving at the top of Earth's atmosphere averages 340 W/m², indicated here as 100% (upper left). About 30% of the incoming radiation is reflected and scattered back to space, and the other 238 W/m² (70%) enters the climate system. Some of this entering radiation warms Earth's surface and causes it to radiate heat upward (right). The greenhouse effect (lower right) retains 95% of the heat radiated back from Earth's heated surface and warms Earth by 31°C.

(ADAPTED FROM T. E. GRAEDEL AND P. J. CRUTZEN, *ATMOSPHERE, CLIMATE, AND CHANGE,* SCIENTIFIC AMERICAN LIBRARY, © 1997 BY LUCENT TECHNOLOGIES, AFTER S. H. SCHNEIDER AND R. LONDER, *CO-EVOLUTION OF CLIMATE AND LIFE* [SAN FRANCISCO: SIERRA CLUB BOOKS, 1984], AND NATIONAL RESEARCH COUNCIL, *UNDERSTANDING CLIMATE CHANGE: A PROGRAM FOR ACTION* [WASHINGTON, D.C.: NATIONAL ACADEMY OF SCIENCES, 1975].)

back out into space after reflection (or scattering) by clouds, dust, and the more reflective regions at Earth's surface. As a result, the average amount of solar energy retained by Earth is 238 W/m² (0.7 × 340 W/m²).

Of the 70% of solar radiation that is retained within the climate system, about two-thirds is absorbed at Earth's surface and about one-third by clouds and water vapor in the atmosphere (see Figure 2-3). This absorbed radiation heats Earth and its lower atmosphere and provides energy that drives the climate system.

2-2 Receipt and Storage of Solar Heat

Because Earth is continually receiving heat from the Sun but is also maintaining a constant (or very nearly constant) temperature through time, it must be losing an equal amount of heat (238 W/m²) back to space. This heat loss, called **back radiation**, occurs at wavelengths lying in the infrared part of the electromagnetic spectrum (see Figure 2-1). Because it occurs at longer wavelengths (5–20 μm) than the incoming shortwave solar radiation, back radiation is also called **longwave radiation**.

Any object with a temperature above absolute zero (−273°C, or 0K) contains some amount of heat that is constantly being radiated away toward cooler regions. Radiated heat can come from such objects as red-hot burners on stovetops, but it is also emitted from objects not warm enough to glow, such as the asphalt pavement that emits shimmering ripples of heat on a summer day.

Although it may seem counterintuitive, even objects with temperatures *far* below the freezing point of water are radiators; that is, they emit some heat. The amount of heat radiated by an object increases with its temperature. The radiation emitted is proportional to T^4, where T is the absolute temperature of the object in Kelvins. Objects with temperatures of $-272°C$ (1K) emit at least a tiny bit of heat energy and so can technically be considered radiators!

What is the average radiating temperature of our own planet? One way to try to answer this question is to average the countless measurements of Earth's surface temperature made over many decades to derive its mean surface temperature. If we do this, we find that the average surface temperature of Earth as a whole is $+15°C$ (288K, or 59°F). This value sounds like a reasonable middle ground between the very large area of hot tropics, averaging 25°–30°C, and the much smaller polar regions, which average well below freezing (0°C). It seems reasonable that the *surface* of planet Earth radiates heat with this average temperature.

But there is a problem with this simple way of looking at radiator Earth. If we now take all available measurements from orbiting satellites and space stations high above Earth, they tell us our planet is sending heat out to space as if it had an average temperature of $-16°C$ (257K or 3°F). This value is 31°C colder than the $+15°C$ average temperature we are certain is correct for Earth's surface. If both sets of measurements are accurate, why are these values so different?

The reason for this discrepancy is the **greenhouse effect** (see Figure 2-3). Earth's atmosphere contains greenhouse gases that absorb 95% of the longwave back radiation emitted from the surface, thus making it impossible for most heat to escape into space. The trapped radiation is retained within the climate system and reradiated down to Earth's surface. This extra heat retained by the greenhouse effect makes Earth's surface temperature 31°C warmer than it would otherwise be.

In effect, measurements made by satellites and space stations in outer space cannot detect the radiation emitted directly from the warmer surface of Earth because of the muffling effect of the blanket of greenhouse gases and clouds. Instead, most of the heat actually radiated back to space is emitted from an average elevation of 5 kilometers, equivalent to the tops of many clouds—still well within the lowest layer of Earth's atmosphere (Box 2-1). These cold ($-16°C$) cloud tops emit radiation at an average value of about 238 W/m², exactly the level needed to offset the amount of solar radiation retained within Earth's climate system and keep it in balance.

The two main gases in Earth's atmosphere are nitrogen (N_2) at 78% of the total and oxygen (O_2) at 21%, but neither is a greenhouse gas because neither traps outgoing radiation. In contrast, the three most important greenhouse gases form very small fractions of the atmosphere. Water vapor (H_2O_v) averages less than 1% of a dry atmosphere, but it can range to above 3% in the moist tropics. Carbon dioxide (CO_2) and methane (CH_4) occur in much smaller concentrations of 0.04% and 0.00018%, but they are also important greenhouse gases.

Clouds also contribute to the retention of heat within the climate system by trapping outgoing radiation from Earth's surface. This role in warming Earth's climate works exactly opposite to the impact of clouds in reflecting incoming solar radiation and cooling our climate. The relative strength of these two competing roles varies with region and season.

Many important characteristics of Earth's climate, such as the amount of incoming sunlight, vary with latitude. Incoming solar radiation is stronger at low latitudes, where sunlight is concentrated more nearly overhead, than at high latitudes, where the Sun's rays strike Earth at a more indirect angle and cover a wider area (Figure 2-4). As a result, larger amounts of solar

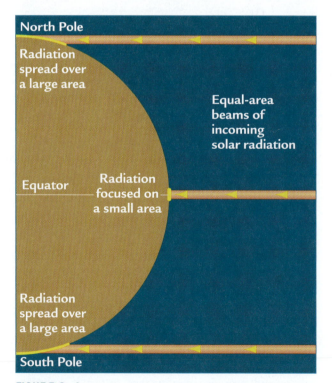

FIGURE 2-4

Unequal radiation on a sphere

More solar radiation falls on a unit area of Earth's surface near the equator than at the poles because of the more direct angle of incoming radiation. (ADAPTED FROM L. J. BATTAN, *FUNDAMENTALS OF METEOROLOGY* [ENGLEWOOD CLIFFS, N.J.: PRENTICE-HALL, 1979].)

Looking Deeper into Climate Science

Box 2-1

The Structure of Earth's Atmosphere

Earth's atmosphere is divided into four layers that extend far above its surface: the **troposphere,** which extends from the surface to between 8 and 18 kilometers; the **stratosphere,** extending up to 50 kilometers; and the overlying mesosphere (50–80 km) and thermosphere (above 80 km). Only the lowest two are critical in climatic changes.

The troposphere is both the layer in which we live and the layer in which most of Earth's weather happens. Storm systems that produce clouds and rainfall or snowfall are almost entirely confined to this layer. Dust or soot particles that are lifted by strong winds from Earth's surface into the lower parts of this layer are quickly removed by precipitation every few days or weeks. The troposphere is also the main layer in which we measure Earth's climate and its changes, particularly those at Earth's surface. As we will see later, about 80% of the gases that form Earth's atmosphere are contained within the troposphere.

Above the troposphere lies the stratosphere, a much more stable layer almost completely separated from the turbulent storms and other processes so common in the troposphere. Only the largest storms penetrate the stratosphere, and only its lowermost layer. Large volcanic eruptions occasionally throw small particles up out of the troposphere and into the stratosphere. Because no rain or snow falls in most of this layer, it may take years for gravity to pull these particles back to Earth's surface. The stratosphere forms 19.9% of Earth's atmosphere; the troposphere and stratosphere together account for 99.9% of its mass.

The stratosphere is also important to Earth's climate because it contains small amounts of oxygen (O_2) and ozone (O_3), which block ultraviolet radiation arriving from the Sun. This shielding effect accounts for a small fraction

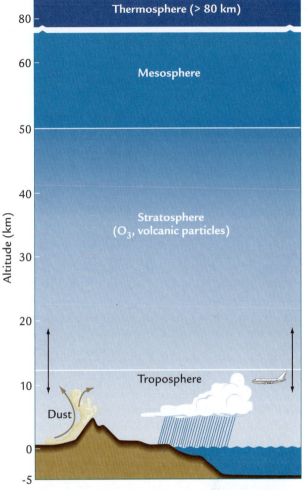

Structure of Earth's atmosphere Most day-to-day weather, as well as important changes in Earth's climate, occur in the lowermost layer of the atmosphere, called the troposphere, which varies in height from 8 to more than 18 km.

of the 30% reduction in incoming heat energy from the Sun. It also greatly reduces the exposure of life-forms on Earth to the harmful effects of ultraviolet radiation, which can cause skin cancers and genetic mutations.

radiation reach the same unit area of Earth's surface in the tropics than near the poles (Figure 2-5).

This unequal distribution of incoming solar radiation is aggravated by unequal absorption and reflection by Earth's surface at different latitudes. A smaller fraction of the incoming radiation is absorbed at higher latitudes than in the tropics mainly because (1) solar radiation arrives at a less direct angle (see

Figure 2-4) and (2) snow and ice surfaces at high latitudes reflect more radiation (see Figure 2-5).

The percentage of incoming radiation that is reflected rather than absorbed by a particular surface is referred to as its **albedo** (Table 2-1). Snow and ice surfaces at high latitudes have albedos ranging from 60% to 90%, with larger values typical of freshly formed snow and ice, and somewhat lower values

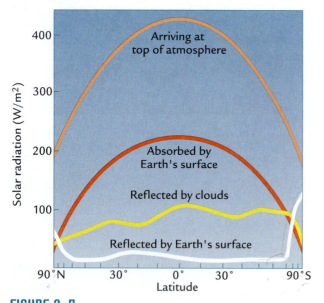

FIGURE 2-5

Pole-to-equator heating imbalances

Incoming radiation is higher in the tropics than at the poles. Less reflective surfaces at low latitudes absorb a larger fraction of incoming radiation, while highly reflective snow and ice surfaces at the poles reflect more, aggravating the pole-to-equator imbalance in absorbed radiation. (ADAPTED FROM R. G. BARRY AND R. J. CHORLEY, *ATMOSPHERE, WEATHER, AND CLIMATE*, 4TH ED. [NEW YORK: METHUEN, 1982].)

Table 2-1 Average Albedo Range of Surfaces	
Surfaces	**Albedo range (percent)**
Fresh snow or ice	60–90%
Old, melting snow	40–70
Clouds	40–90
Desert sand	30–50
Soil	5–30
Tundra	15–35
Grasslands	18–25
Forest	5–20
Water	5–10

Adapted from W. D. Sellers, *Physical Climatology* (Chicago: University of Chicago Press, 1965), and from R. G. Barry and R. J. Chorley, *Atmosphere, Weather, and Climate*, 4th ed. (New York: Methuen, 1982).

for snow or ice that contain dirt or are partly covered by pools of melted water. In contrast, snow-free land surfaces have much lower albedos (15–30%) and ice-free water reflects even less of the incoming radiation (below 5% when the Sun is overhead).

The albedo of any surface also varies with the angle at which incoming solar radiation arrives. For example, water reflects less than 5% of the radiation it receives from an overhead Sun, but a far higher fraction of the radiation from a Sun lying low in the sky (Figure 2-6). This same tendency holds true for the other surfaces. Because 70% of Earth's surface is low-albedo water, Earth's surface has an average albedo near 10%.

These factors combine to make Earth's surface more reflective near the poles than in the tropics. The Antarctic ice sheet and extensive areas of nearby sea ice in the Southern Hemisphere have very high albedos, as do the Arctic sea ice cover and the extensive winter snow cover in Eurasia and North America in

FIGURE 2-6

Sun angle controls heat absorption

All of Earth's surfaces (here water) absorb more solar radiation from an overhead Sun (A) than from a Sun lying low in the sky (B). (ADAPTED FROM L. J. BATTAN, *FUNDAMENTALS OF METEOROLOGY* [ENGLEWOOD CLIFFS, N.J.: PRENTICE-HALL, 1979].)

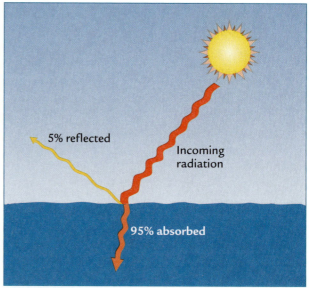

A Low latitude

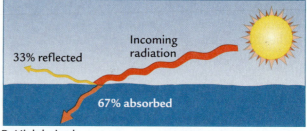

B High latitude

the Northern Hemisphere. Clouds counteract some of this imbalance in surface albedo by reflecting somewhat more solar radiation at low latitudes than at high latitudes.

The net result is that both polar regions absorb even less of the already low amounts of incoming solar radiation, while the tropics absorb even more of the already high radiation. The overall effect of these albedo differences is to increase the solar heating imbalance between the poles and the tropics (see Figure 2-5).

Radiation and albedo also vary seasonally. The 23.5° tilt of Earth's axis as it revolves around the Sun causes the Northern and Southern hemispheres to tilt alternately toward and away from the Sun, and this motion causes seasonal changes in solar radiation received in each hemisphere (Figure 2-7A). From our Earthbound perspective, we experience this orbital motion as a shift of the overhead Sun through the

tropics from a latitude of 23.5°N on June 21 to 23.5°S on December 21. This change in the Sun's angle results in large seasonal changes in the amounts of solar radiation (W/m^2) received on Earth (Figure 2-7B).

Accompanying these seasonal changes in solar heating are large seasonal changes in the albedo of Earth's surface with latitude (Figure 2-8). In the south polar region, the ice sheet over Antarctica remains intact through the year, but an extensive ring of sea ice surrounding Antarctica expands and contracts every year across an area of 16 million square kilometers. In contrast, the Arctic Ocean has a multiyear cover of sea ice that fluctuates much less in extent through the year, and the main seasonal albedo change in the Northern Hemisphere comes from the winter expansion and summer retreat of snow cover on Asia, Europe, and North America. Both of these changes in albedo play important roles in long-term climate change (Box 2-2).

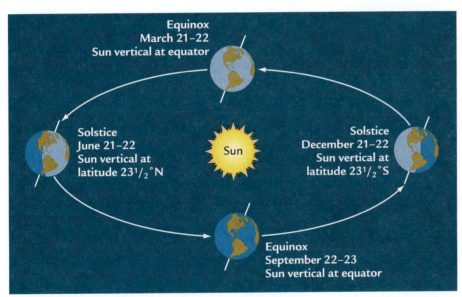

A Earth's orbit

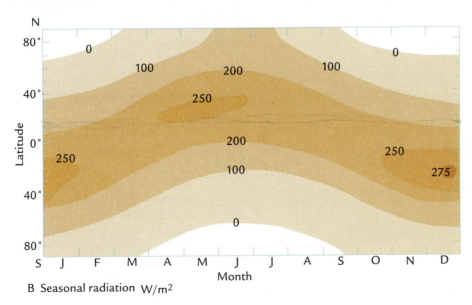

B Seasonal radiation W/m^2

FIGURE 2-7

Earth's tilt and seasonal radiation

The tilt of Earth's axis in its annual orbit around the Sun causes the Northern and Southern hemispheres to lean directly toward and then away from the Sun at different times of the year (A). This change in relative position causes seasonal shifts between the hemispheres in the amount of solar radiation received at Earth's surface (B). (A: ADAPTED FROM F. K. LUTGENS AND E. J. TARBUCK, *THE ATMOSPHERE* [ENGLEWOOD CLIFFS, N.J.: PRENTICE-HALL, 1992; B: ADAPTED FROM A. L. BERGER, "MILANKOVITCH THEORY AND CLIMATE," *REVIEWS OF GEOPHYSICS* 26 [1988]: 624–57.)

Climate Interactions and Feedbacks

Box 2-2

Albedo/Temperature

The large difference in albedo between regions of highly reflective snow or ice compared to land or water surfaces that absorb radiation is very important in climate change. Surface albedos can increase by 75% (from 15% to 90%) when snow-free land areas become covered with snow, and over oceans that become covered by sea ice. As a result, a surface that had previously absorbed most incoming radiation will now reflect it away, with significant implications for climate change.

Assume that climate abruptly cools, for any reason (for example, perhaps because of a decrease in the

output of the Sun). Part of the climate system's natural response to such a cooling is an increase in the land area covered by snow and in the ocean area covered by sea ice. The expansion of these bright, high-albedo surfaces will cause an increase in the percentage of incoming radiation reflected back out to space, and a decrease in the amount of heat absorbed at the surface.

The loss of absorbed heat in these regions will in turn cause the local climate to cool by an additional amount beyond the initial cooling. This is an example of the concept of positive feedback, introduced in Chapter 1. The same positive feedback process also works in the opposite direction: an initial warming will reduce the cover of snow and ice, increase the amount of heat absorbed by exposed land or water, and further warm the climate. Climate scientists estimate that any initial climate change will be amplified by about 40% by this positive feedback effect. An initial cooling of 1°C would be amplified to a total cooling of 1.4°C by this process, in the absence of other changes in the climate system.

Climate scientists call this positive feedback **albedo-temperature feedback.** In a larger sense, its net effect is to increase Earth's overall sensitivity to climate changes. The greater the area on Earth covered by snow and ice, the more sensitive the planet as a whole becomes to imposed changes in climate. Because most of the regional albedo contrast on Earth is localized at the equatorward limit of snow and sea ice, the albedo-temperature feedback most strongly affects climate at latitudes near and poleward of these limits.

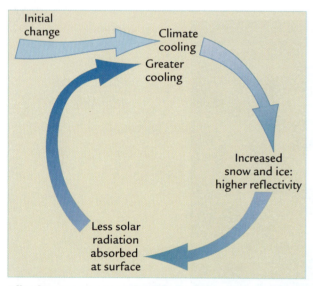

Albedo-temperature feedback When climate cools, the increased extent of reflective snow and ice increases the albedo of Earth's surface in high-latitude regions, causing further cooling by positive feedback. The same feedback process amplifies climate warming.

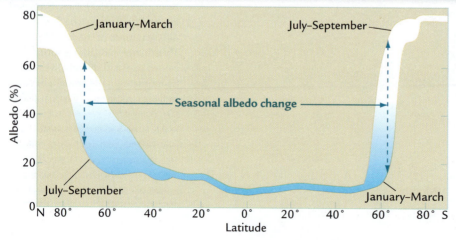

FIGURE 2-8

Albedo changes with season

Average albedo increases in the Northern Hemisphere in winter (January–March) mainly because of increased snow cover over land and also because of more extensive sea ice. Albedo increases in the Southern Hemisphere's winter (July–September) because of more extensive sea ice. (ADAPTED FROM G. KUKLA AND D. ROBINSON, "ANNUAL CYCLE OF SURFACE ALBEDO," *MONTHLY WEATHER REVIEW* 108 [1980]: 56–68.)

Clouds also affect the regional receipt of heat at Earth's surface. Areas such as subtropical deserts that are free of clouds receive far more solar radiation than do most parts of the oceans, especially subpolar oceans where frequent storms produce nearly continuous cloud cover. The cloud cover over tropical rain forests also reduces the receipt of radiation there. But areas without clouds also back-radiate more heat from Earth's surface than do areas with heavy cloud cover.

Water is the key to Earth's climate system (Box 2-3). Absorption and storage of solar heat are strongly affected by the presence of liquid water because of its high **heat capacity,** a measure of the ability of a material to absorb heat. Heat energy is measured in units of **calories** (one calorie is the amount of heat required to raise the temperature of 1 gram of water by 1°C). Heat capacity is the product of the density (in g/cm³) of a heat-absorbing material and its **specific heat,** the number of calories absorbed as the temperature of 1 gram of this material increases by 1°C:

$$\text{Heat capacity} = \text{Density} \times \text{Specific heat}$$
$$(\text{cal/cm}^3) \qquad (\text{g/cm}^3) \qquad (\text{cal/g})$$

The specific heat of water is 1, higher by far than any of Earth's other surfaces. The ratios of the heat capacities of water:ice:air:land are 60:5:2:1. Much of the heat capacity of air is linked to the water vapor it contains. Likewise, much of the heat capacity of land is due to the small amount of water held in the soil.

The low-latitude oceans are Earth's main storage tanks of solar heat. Sunlight penetrates into and directly heats the upper tens of meters of the ocean, especially in the tropics, where the radiation arrives from a Sun high in the sky. Equally important, winds blowing across the ocean's surface stir the upper layers and rapidly mix solar heat as deep as 100 meters (Figure 2-9). In contrast, even though tropical and subtropical landmasses generally become very hot under the strong Sun, they are not capable of storing much heat because heat is conducted down into soil or rock at very slow rates (see Figure 2-9 bottom).

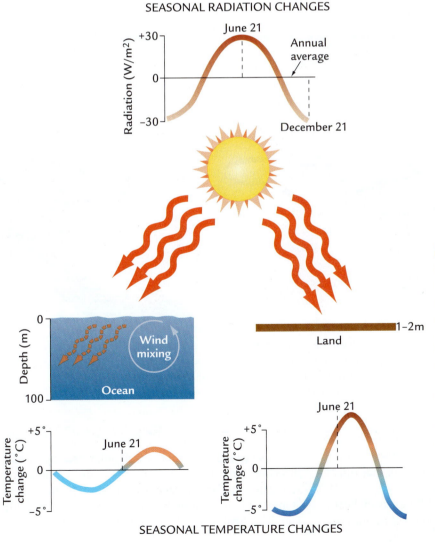

SEASONAL RADIATION CHANGES

SEASONAL TEMPERATURE CHANGES

FIGURE 2-9

Difference in heating of land and oceans

During the seasonal cycle of solar radiation (top), ocean surfaces heat and cool slowly and only by small amounts because temperature changes are mixed through a layer 100 m thick (lower left). In contrast, land surfaces heat and cool quickly and strongly because of their low capacity to conduct and store heat (lower right). (ADAPTED FROM J. E. KUTZBACH AND T. WEBB III, "LATE QUATERNARY CLIMATIC AND VEGETATIONAL CHANGE IN EASTERN NORTH AMERICA: CONCEPTS, MODELS, AND DATA," IN *QUATERNARY LANDSCAPES*, ED. L. C. K. SHANE AND E. J. CUSHING [MINNEAPOLIS: UNIVERSITY OF MINNESOTA PRESS, 1991].)

Climate Interactions and Feedbacks

Box 2-3

Water in the Climate System

Water is a critical part of the miracle of life on Earth. Of all the planets in the solar system, only here on Earth can water exist in three forms: as a liquid (in oceans and lakes), as a frozen solid (ice), and as a gas (water vapor). The essence of our good fortune comes from the fact that water not only accounts for the largest fraction of our bodies and those of most other organisms, but is also the medium we drink for survival, the substance we swim in and skate or ski on for amusement, and part of the air we breathe from day to day without ill effect.

Water is also a vital component of the climate cycle. The oceans contain over 97% of Earth's water, with just 2% held in glacial ice and less than 1% in all the rest of Earth's reservoirs. The **hydrologic cycle,** or continual recycling of water among all these reservoirs, including the much smaller amounts held in the atmosphere, in plants,

in lakes, in rivers, and in soil, is vital to the operation of the climate system, as we will see in this and future chapters.

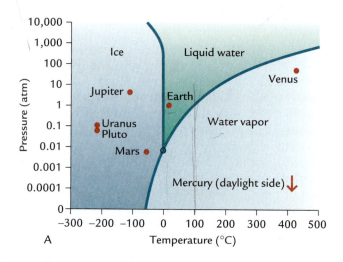

B

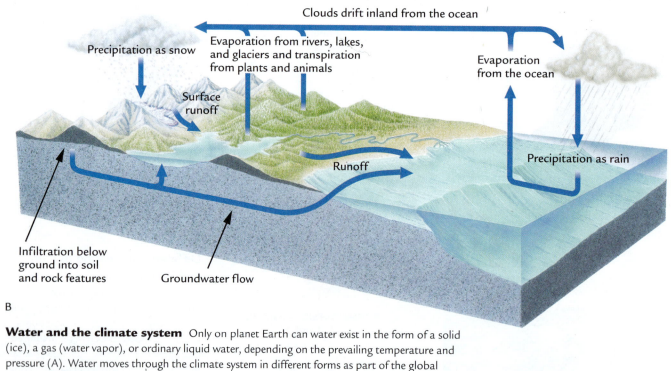

Water and the climate system   Only on planet Earth can water exist in the form of a solid (ice), a gas (water vapor), or ordinary liquid water, depending on the prevailing temperature and pressure (A). Water moves through the climate system in different forms as part of the global water cycle (B). (D. MERRITTS ET AL., *ENVIRONMENTAL GEOLOGY*, © 1997 BY W. H. FREEMAN AND COMPANY.)

As a result, the large amount of heat stored in the low-latitude ocean provides most of the fuel that runs Earth's climate system.

One consequence of the large difference in heat capacity between land and water is evident in Earth's

response to seasonal changes in solar heating: a map showing only the seasonal *range* of temperatures at Earth's surface succeeds in outlining the shapes of most of its continents and oceans (Figure 2-10). The low heat capacity of landmasses allows them to respond strongly to seasonal changes in heating, while the high heat capacity and strong wind mixing of the ocean's upper layers limit its seasonal response (see Figure 2-9 bottom). The largest temperature changes occur over the largest, most continental landmass, Asia, with lesser changes on smaller continents such as Australia (see Figure 2-10). Small seasonal changes also characterize the tropics, which receive consistently strong solar radiation throughout the year and remain warm.

Land and ocean surfaces also differ markedly in their *rates* of response to seasonal heating changes (see Figure 2-9). Interior regions of large continents heat up and cool off quickly because of their low heat capacity, reaching maximum (or minimum) seasonal temperatures about one month after the seasonal radiation maximum (or minimum). In contrast, the upper ocean responds much more slowly because of its high heat capacity, reaching its extreme temperature responses only two or more months after the June 21 and December 21 solar radiation peaks in each hemisphere.

These differences in amplitude and timing of response between land surfaces and the upper ocean layers are referred to as differences in **thermal inertia**. The fast-responding land has a low thermal inertia; the slower-responding upper layers of the ocean have a high thermal inertia.

2-3 Heat Transformation

The heat energy received and stored in the climate system is exchanged among water, land, and air through several processes. As we saw earlier, some of the absorbed heat is lost from Earth's warm surface by longwave back radiation, but most back radiation is trapped by greenhouse gases and radiated back down toward Earth's surface (see Figure 2-3).

Two other important kinds of heat transfer occur within the climate system. One process involves the transfer of **sensible heat** by moving air. Sensible heat is the product of the temperature of the air and its specific heat. It is also the heat that a person directly senses as it is carried along in moving air masses. Surfaces heated by the Sun warm the lowermost layer of the overlying atmosphere. The heated air expands in volume like any heated gas, becomes lighter (less dense), rises higher in the atmosphere, and carries sensible heat along with it in a large-scale process known as **convection**.

Convection of sensible heat by air is analogous to what occurs in a pot of water heated from below on a stovetop: the stovetop heating warms the lower layers of water and causes them to expand, rise, and transfer heat upward (Figure 2-11 top). The same process happens when air is heated and rises. Sensible

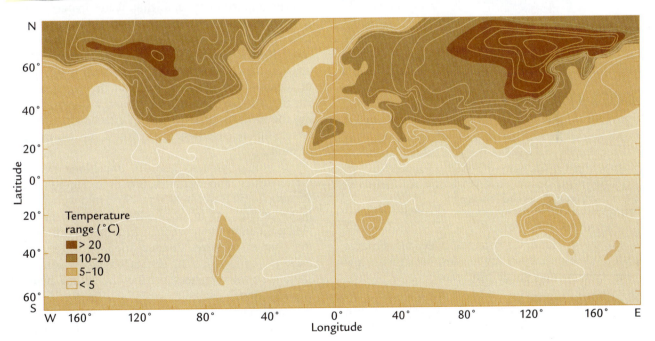

FIGURE 2-10

Sensitivity of solar heating to land vs. ocean

The total change in mean daily surface temperature between summer and winter is greatest over large landmasses and much smaller over oceans and small continents. The locations of most continents and oceans can be detected on this map from their temperature responses alone.

(ADAPTED FROM A. S. MONIN, "ROLE OF OCEANS IN CLIMATE MODELS," IN *PHYSICAL BASIS OF CLIMATE AND CLIMATE MODELING*, REPORT NO. 16, GARP PUBLICATION SERIES [GENEVA: WORLD METEOROLOGICAL ORGANIZATION, 1975].)

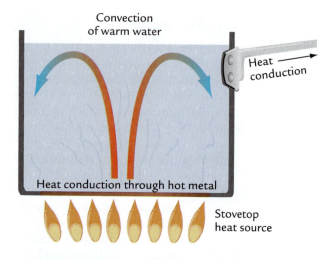

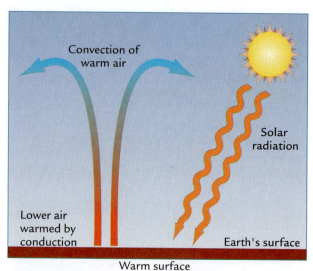

FIGURE 2-11
Convection of heat

A kettle heated on a stovetop conducts heat to the lower layers of water, which then rise and convect heat upward (top). Similarly, a land surface warmed by the Sun transfers heat by conduction to the lower layers of the atmosphere, which then rise and convect heat upward (bottom).

heating at low latitudes is largest over land surfaces in summer, especially in dry regions, because the low water content and small heat conductance of land surfaces allow them to warm to much higher temperatures than the oceans and they transfer this heat to the overlying air (Figure 2-11 bottom). As air and water move horizontally across Earth's surface, they transport sensible heat.

The second form of heat transfer within the climate system involves the movement of latent heat. This more powerful process of heat transfer also depends on the convective movement of air, but in this case, the heat carried by the air is temporarily hidden in latent form as water vapor. Transfer of this

latent heat occurs in two steps: (1) initial evaporation of water and storage of heat in water vapor and (2) later release of stored heat during condensation and precipitation, usually far from the site of initial evaporation.

To understand this process, we take a closer look at the behavior of water (Figure 2-12). We learned earlier that it takes 1 calorie of heat energy to raise the temperature of 1 gram of water by 1°C within the range of 0°–100°C. These stored calories are all available to be returned from the water to the air for each 1°C that the water subsequently cools.

But additional changes happen when water changes state, either to ice (a solid) or to water vapor (a gas). Large amounts of latent heat are stored during the warming process that transforms ice to water or water to water vapor, and this stored heat is available for later release during the cooling process that transforms water vapor back to water or water back to ice (see Figure 2-12).

When ice warms to a temperature of 0°C, any addition of heat (usually from the atmosphere) causes it to melt. Melting of solid ice to form liquid water requires a large input of energy (80 calories per gram of ice). During the melting process, the temperature of the ice holds at 0°C, because all the additional energy is being used for melting.

The opposite happens when water freezes: caloric energy (heat) is liberated. Water chilled to 0°C by loss of heat to the atmosphere begins to freeze if still more heat is extracted. The freezing process liberates the same 80 calories of heat energy per gram of water that had been stored in the water as the ice melted. Because the heat energy liberated during the freezing process was in effect hidden in the water, this released energy is called the **latent heat of melting**.

A similar process governs transitions between liquid water and water vapor (see Figure 2-12). Turning liquid water heated to 100°C into water vapor (gas) requires the input of 540 calories per gram of water, again with no change in temperature during the change of state. Condensation of water vapor back to liquid water in clouds or fog liberates the same amount of energy, called the **latent heat of vaporization**.

Despite what Figure 2-12 implies, water does not have to boil to be turned into vapor. Evaporation occurs across the entire range of temperatures at which liquid water exists, including all the temperatures of the surface ocean and the lower atmosphere. As with boiling, the change of state from liquid to vapor at lower temperatures requires the addition of a large amount of heat energy, and this energy is stored in the water vapor in latent form for later release.

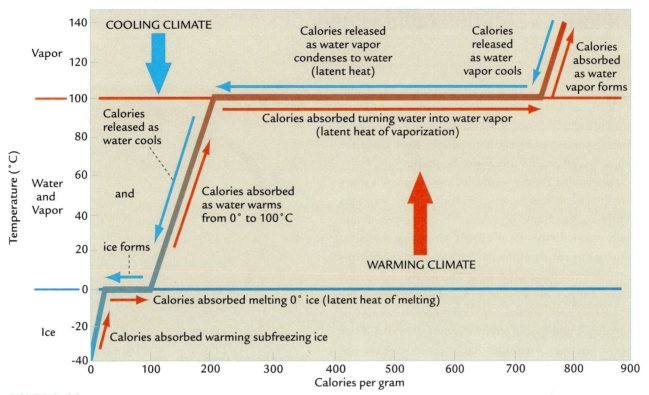

FIGURE 2-12

Heat transformations

Heat calories are absorbed from the atmosphere when ice, water, or water vapor is warmed. This heat is released back to the atmosphere during cooling. Heat is also absorbed and stored during changes of state from ice to liquid water or from liquid water to water vapor and this stored latent heat is available for release when water vapor condenses to water or when water freezes to form ice. (ADAPTED FROM H. V. THURMAN, *INTRODUCTION TO OCEANOGRAPHY*, 6TH ED. [NEW YORK: MACMILLAN, 1991].)

The amount of water vapor that can be held in air is limited, much like the amount of sugar that can be dissolved in coffee. Attempts to add more water vapor to fully saturated air will cause condensation (the formation of dew or raindrop nuclei at the so-called **dew point** temperature). This limit of full saturation, measured in grams of water per cubic meter of air and called the **saturation vapor density**, roughly doubles for each 10°C increase in air temperature (Figure 2-13). Warm tropical air at 30°C can hold almost ten times as much water vapor as cold polar air masses near or below 0°C. As a result of this relationship, water vapor is an important positive feedback in the climate system (Box 2-4).

Evaporation of water from Earth's surface in warmer regions stores excess heat energy in the warm atmosphere. This energy stored in water vapor is carried along with the moving air, both vertically and horizontally. When condensation and precipitation occur, the stored latent energy is released as heat, far from the site of evaporation. The average parcel of water vapor stays in the air for 11 days and travels over 1,000 kilometers, about the distance from Los Angeles to Denver.

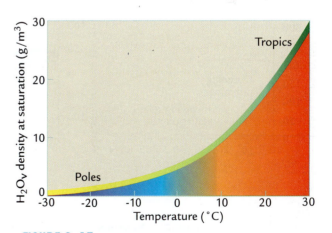

FIGURE 2-13

Water vapor content of air

Warm air is capable of holding almost ten times as much water vapor (H_2O_v) as cold air. (ADAPTED FROM L. J. BATTAN, *FUNDAMENTALS OF METEOROLOGY* [ENGLEWOOD CLIFFS, N.J.: PRENTICE-HALL, 1979].)

Climate Interactions and Feedbacks

Box 2-4

Water Vapor

Water vapor, Earth's major greenhouse gas, varies in concentration from 0.2% in very dry air to over 3% in humid tropical air. This strong dependence on temperature produces an important positive feedback in the climate system called **water vapor feedback.**

Assume that climate warms for any reason. A warmer atmosphere can hold more water vapor, and the increased greenhouse gas traps more heat. This large greenhouse effect in turn further warms Earth, amplifying the initial warming through a positive feedback loop. The same positive feedback process works when the climate cools: initial cooling reduces the amount of water vapor held in the atmosphere and produces additional cooling because of the reduced greenhouse effect. It is estimated that direct positive feedback from water vapor can double the size of an initial climate change under current climatic conditions. This estimate is based on the action of water vapor in a clear sky; it ignores the more complicated effects that occur when water vapor condenses and forms clouds.

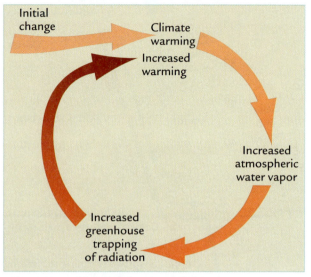

Water vapor feedback When climate warms, the atmosphere is able to hold more water vapor (the major greenhouse gas in the atmosphere), and the increase in water vapor leads to further warming by means of a positive feedback. This feedback works in reverse during cooling.

▶ Heat Transfer in Earth's Atmosphere

Tropical and subtropical latitudes below 35° have a net excess of incoming solar radiation over outgoing back radiation, while at latitudes higher than 35° the opposite is true (Figure 2-14). Most of this excess heat is stored within a thin upper layer of the tropical ocean, and it is this heating imbalance that drives the general circulation of the Earth's atmosphere and

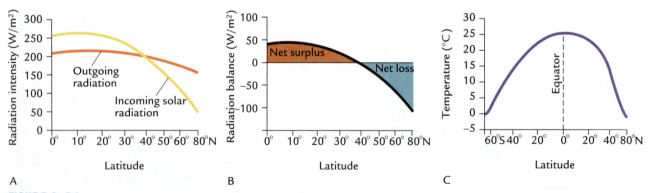

A   B   C

FIGURE 2-14
Unequal heating of tropics and poles
Incoming solar radiation is strongest near the equator, while radiation emitted back to space is more evenly distributed between the tropics and poles (A). The resulting radiation surplus in the tropics and deficit at the poles (B) creates temperature imbalances (C) that drive the circulation of the atmosphere and oceans. (D. MERRITTS ET AL., *ENVIRONMENTAL GEOLOGY*, © 1997 BY W. H. FREEMAN AND COMPANY.)

oceans. Roughly two-thirds of the net transport of heat from the Earth's equator toward its poles occurs in the lower atmosphere.

2-4 Overcoming Stable Layering in the Atmosphere

Both sensible and latent heating are associated with upward motion or convection in the atmosphere. But how high does this convected air go, and why? Air is highly compressible, and most of the mass of the atmosphere is held close to Earth's surface by gravity, with 50% of the air molecules below 7 kilometers and 75% below 10 kilometers (Figure 2-15). Under the pull of gravity, each layer of the atmosphere presses down on the layers below, compressing them and increasing their density. As a result, atmospheric pressure—pressure exerted by the weight of the overlying column of air—increases toward the lower elevations.

Opposing this tendency of dense air to be held close to Earth's surface by gravity is a natural tendency of air to flow from the high pressures near Earth's surface to the lower pressures at higher elevations. In effect, the compressed air at lower elevations has nowhere to go and so it pushes back against the overlying layers. These two opposing forces, the downward pull of gravity and the resistance directed upward, tend to remain in a stable but delicate balance that limits the amount of vertical air motion.

Over limited areas, however, parcels of air will rise if they become less dense than the surrounding air. The major way this process occurs is by the

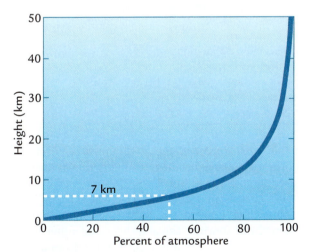

FIGURE 2-15

Distribution of air with elevation

Each layer of air in Earth's atmosphere presses down on the underlying layers, increasing the pressure on the lower layers and at the surface. Most of the mass of Earth's atmosphere lies at lower elevations. (MODIFIED FROM R. G. BARRY AND R. J. CHORLEY, *ATMOSPHERE, WEATHER, AND CLIMATE,* 4TH ED. [NEW YORK: METHUEN, 1982].)

warming of Earth's surface and its lower atmosphere. Heating causes the lower layers of air to expand, become lower in density, and rise buoyantly in the atmosphere (like a balloon full of hot air).

As these parcels of heated air rise to higher elevations, processes come into play that involve changes in temperature and in the density of the moving air due to changes in pressure. Climate scientists refer to these changes as **adiabatic** processes. In the case of rising parcels of air, the lower pressures encountered at higher elevations result in an additional expansion of the air beyond the amount originally caused by heating at Earth's surface. But expansion requires the expenditure, or loss, of heat energy, and no source of heat exists to replace the lost heat. As a result, the rising air parcels begin to cool and increase in density, especially in comparison with the surrounding air that is not rising. Eventually, the parcel stops rising at the level where its density matches that of the surrounding air. This upward loss of heat is a dry process that is independent of the amount of water vapor in the air.

Latent heating is the second process that can destabilize the atmosphere, and it occurs as a wet process driven by water vapor, which weighs roughly a third less than the mixture of gases that form Earth's atmosphere. Evaporation adds water vapor to the atmosphere at low elevations and causes a net decrease in the density of the air, which can then rise to higher elevations. As before, the moist air rises, expands, and at some point cools to the temperature at which it becomes fully saturated with water vapor. Then condensation begins. Condensation releases latent heat, which partially opposes the cooling of the rising air parcel due to its expansion. The release of this heat in the air parcels also makes them more buoyant (less dense) and allows them to rise still higher in the troposphere (sometimes to 10 to 15 km). Eventually, the cumulative loss of most of the water vapor in the parcels stops the release of latent heat that had kept them rising.

The rate at which Earth's atmosphere cools with elevation is called the **lapse rate**. This rate ranges from 5°C/km to as high as 9.8°C/km but typically averages 6.5°C/km both at middle latitudes and for the planet as a whole. Localized lapse rates are highest in very dry regions where the only factor at work on rising air parcels is adiabatic expansion and cooling. For the areas of Earth's surface where rising air carries water vapor, the release of latent heat at higher elevations warms the air at upper levels and reduces the lapse rate to values near the lower end of this range.

2-5 Tropical-Subtropical Atmospheric Circulation

The large-scale circulation in Earth's atmosphere that transports heat from low to high latitudes is

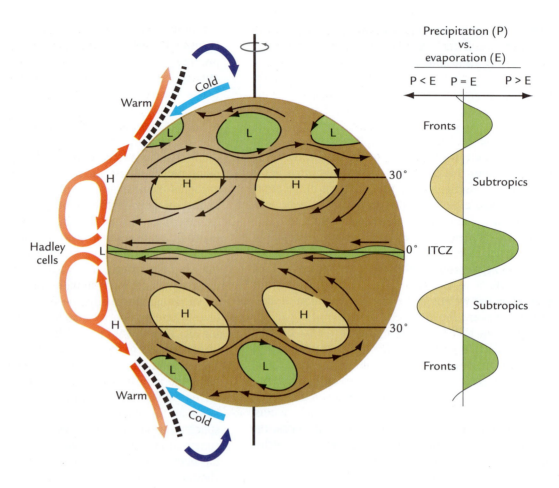

FIGURE 2-16

General circulation of the atmosphere

Heated air rises in the tropics at the intertropical convergence zone (ITCZ) and sinks in the subtropics as part of the large-scale Hadley cell flow, which transports heat away from the equator (left). Additional poleward heat transfer occurs along moving weather systems (called "fronts") at middle and higher latitudes, with warm air rising and moving poleward and cold air sinking and moving equatorward (right). Rising air in the tropics causes a net excess of precipitation over evaporation, while dry air sinking in the subtropics produces more evaporation than precipitation. Higher latitudes tend to have small excesses of precipitation over evaporation. (ADAPTED FROM S. H. SCHNEIDER AND R. LONDER, *CO-EVOLUTION OF CLIMATE AND LIFE* [SAN FRANCISCO: SIERRA CLUB BOOKS, 1984]; E. BRYANT, *CLIMATE PROCESS AND CHANGE* [CAMBRIDGE: CAMBRIDGE UNIVERSITY PRESS, 1998]; AND J. P. PEIXOTO AND M. A. KETTANI, "THE CONTROL OF THE WATER CYCLE," *SCIENTIFIC AMERICAN,* APRIL 1973.)

summarized in Figure 2-16. The red arrows on the left show heat being transported by warm air, with blue arrows indicating movement of cold air. The large-scale patterns of precipitation and evaporation resulting from these air motions are summarized on the right. This summary figure is important to the discussion that follows, so refer to it often as we go along.

Tropical heating drives a giant tropical circulation pattern called the **Hadley cell** (see Figure 2-16 left). Warm air rises in giant columns marked by towering, puffy (cumulonimbus) clouds created by evaporation of water vapor from tropical oceans and subsequent condensation at high altitudes. Condensation produces a narrow zone of high rainfall in the rising part of the Hadley cell near the equator. The rising motion

in the tropical part of the Hadley cell represents an enormous transfer of heat through the atmosphere from low to high altitudes.

Air parcels that rise and lose water vapor in the tropics move toward the subtropics in both hemispheres, transporting sensible heat and other energy from lower to higher latitudes. In the subtropics, this air sinks toward the surface near 30° latitude. The sinking air is then warmed by the increasing pressure of the atmosphere at lower elevations (another adiabatic process), and it gradually becomes even drier and able to hold still more water vapor. This Hadley cell flow prevents condensation from occurring in much of the subtropics and makes these latitudes a zone of low average precipitation and high evaporation, in regions such as the Sahara Desert.

The Hadley cell circulation is completed at Earth's surface, where trade winds from both hemispheres blow from the subtropics toward the tropics and replace the rising air. As warm dry air carried by the trade winds passes over the tropical ocean, it continually extracts water vapor from the sea surface. The region near the equator where the northern and southern trade winds meet is called the **intertropical convergence zone (ITCZ)**. Water vapor carried by the trade winds contributes to the rising air motion and abundant rainfall along the ITCZ.

Viewed on a daily or weekly basis, the actual circulation at low latitudes consists of small-scale cloud systems that develop explosively in limited regions. But when the circulation is averaged over seasons or years, it takes the form shown by the schematic Hadley cell in Figure 2-16. This flow is an important part of the poleward transfer of heat, with a net upward movement and release of latent heat in the tropics followed by a net horizontal transfer of sensible heat to the subtropics at high elevations.

These large-scale movements of masses of air alter the pressure (weight of air) at Earth's surface.

Air movement upward and away from the tropics reduces the weight of the column of overlying air and produces low surface pressures near the ITCZ. Air moving into the subtropics and then down toward Earth's surface increases the pressure there.

Because solar heating is the basic driving force behind the Hadley cell circulation, the seasonal shifts of the Sun between hemispheres also affect the location of the ITCZ. It moves northward during the Northern Hemisphere's summer (June to September) and southward during the Southern Hemisphere's summer (December to March). The slow thermal response of the land and oceans causes the seasonal shifts of the ITCZ to lag more than a month behind those of the Sun.

Important seasonal transfers of heat between the tropical ocean and land, called **monsoons**, arise from the fact that water responds more slowly than land to these seasonal changes in solar heating because of its larger heat capacity and high thermal inertia. The *summer monsoon* circulation is basically an in-and-up flow of moist air that produces precipitation. The strong, direct solar radiation in summer at low and middle latitudes heats Earth's surface (Figure 2-17A).

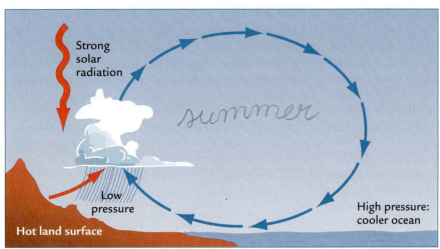

A Summer monsoon

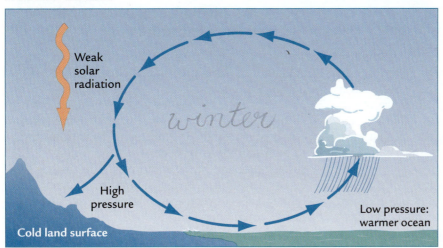

B Winter monsoon

FIGURE 2-17

Monsoonal circulations

In summer, more rapid heating of land surfaces than of the ocean produces rising motion over the continents and draws moist air in from the ocean, producing precipitation over land (A). In winter, more rapid cooling of the land surfaces than of the ocean produces sinking motion over the continents and sends cold dry air out over the warmer ocean, shifting most winter precipitation out to sea (B).

Because soil contains relatively little water, land surfaces have low thermal inertia and heat up quickly. The ocean, with its much higher thermal inertia, absorbs the heat, mixes it through a layer up to 100 meters thick, and warms up far more slowly and to a smaller extent than the land.

These different responses to solar heating set in motion a large-scale land-sea circulation, the monsoon. Initially, dry air heated rapidly over the continental interior rises, and the upward movement of this mass of air produces a region of low pressure over the land. Air is then drawn in toward the low-pressure region from the cooler oceans. The moist air coming in from the oceans is slowly heated and joins in the prevailing upward motion.

As the moist air rises, it cools and its water vapor condenses. Condensation produces heavy precipitation and releases substantial amounts of latent heat, which fuels an even more powerful upward motion (see Figure 2-17A). This net in-and-up circulation in summer monsoons is a two-stage process: initially a dry process due to the rising of sensible heat and later a wet process linked to ocean moisture and release of latent heat.

The strongest summer monsoon circulations on Earth today occur over India. Heating of the large high landmass of southern Asia focuses a strong wet summer monsoon against the Himalaya Mountains (see Chapter 7).

The *winter monsoon* circulation is the reverse of the summer monsoon. The basic flow is a down-and-out motion of cold, dry air from land to sea (Figure 2-17B). In winter, the Sun's radiation is weaker, and land surfaces cool by back radiation. Because of differences in thermal inertia, land surfaces cool faster and more intensely than the oceans. Air cooled over the land sinks toward the surface and creates a region of high pressure where the extra mass of air piles up. Air flows outward from this cell toward the oceans at lower levels. Because the sinking air holds little moisture, the outflow from the continents to the oceans is cold and dry.

These near-surface monsoon circulations are part of a much larger circulation that links the subtropical continents and oceans (Figure 2-18). In summer, the upward motion prevailing over land produces an excess of atmospheric mass (and high pressures) near 10 kilometers of altitude. This high pressure results in a high-level flow of air out toward the adjacent oceans, and this flow produces regions of high pressure in the lower atmosphere over the subtropical ocean in summer (Figure 2-18A).

In winter, the slow thermal response of the oceans keeps them relatively warm, and they provide heat to the cold air flowing out from the land. The heat gained by the atmosphere results in rising motion and

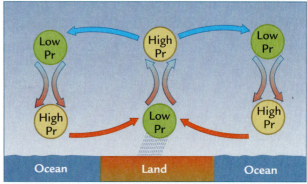

A Summer

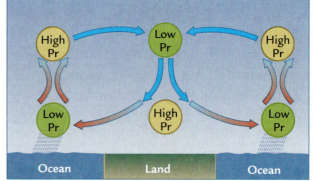

B Winter

FIGURE 2-18
Large-scale monsoon circulations

Air motion associated with monsoon circulations at larger scales is upward over the land and downward over the ocean in summer, but the exact reverse in winter. Precipitation is heaviest in regions of low pressure (Pr) and upward motion. (ADAPTED FROM J. E. KUTZBACH AND T. WEBB III, "LATE QUATERNARY CLIMATIC AND VEGETATIONAL CHANGE IN EASTERN NORTH AMERICA: CONCEPTS, MODELS, AND DATA," IN *QUATERNARY LANDSCAPES*, ED. L. C. K. SHANE AND E. J. CUSHING [MINNEAPOLIS: UNIVERSITY OF MINNESOTA PRESS, 1991].)

increased precipitation, and the upward movement of air produces strong low-pressure cells over the oceans at higher latitudes (Figure 2-18B).

2-6 Atmospheric Circulation at Middle and High Latitudes

The giant Hadley cells are a simple and convenient summary of basic atmospheric circulation across that half of Earth's surface area lying between 30°S and 30°N latitude (see Figure 2-16). The circulation at latitudes above 30° is more difficult to summarize.

To understand the transfer of heat from middle to high latitudes, we begin with the regions of high surface pressure in the subtropics. The air that sinks to Earth's surface in the subtropical branch of the Hadley cell converges or piles up there, creating a nearly continuous band of high surface pressures in the subtropics (Figure 2-19). Superimposed on this

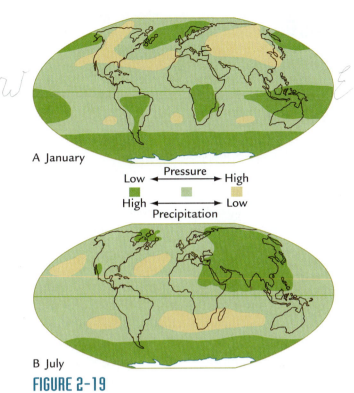

A January

Low ⟵ **Pressure** ⟶ High

High ⟵ ⟶ Low
Precipitation

B July

FIGURE 2-19

Seasonal pressure patterns

A band of high surface pressure occurs in the subtropics in both hemispheres in both seasons. During summer, this zone is interrupted over land (especially Asia) by areas of low pressure produced by summer monsoon circulations. (ADAPTED FROM E. BRYANT, *CLIMATE PROCESS AND CHANGE* [CAMBRIDGE: CAMBRIDGE UNIVERSITY PRESS, 1998].)

basic pattern is a tendency for the monsoonal flow of air from land to sea in summer to produce oval-shaped cells of high pressure over the subtropical oceans (see Figure 2-17A).

Because air naturally flows away from regions of higher pressures toward areas where pressure is lower, the subtropical high-pressure cells send air moving outward near Earth's surface in all directions. Ultimately, however, the path taken by this air is not directly from regions of high to low pressure. Earth's rotation deflects the path of this air in a *relative* sense (Box 2-5), and this deflection produces a clockwise spin of air around subtropical highs in the Northern Hemisphere and a counterclockwise spin in the Southern Hemisphere.

Returning to the summary map of Earth's circulation in Figure 2-16, notice that the trade winds that flow out from the subtropical high-pressure zones toward the equator in the Northern Hemisphere are deflected to the west by Earth's rotation and given a net easterly trajectory: from northeast to southwest. These are the trade winds that move toward the tropics.

The same Coriolis deflection turns air flowing poleward from the subtropical highs toward the east,

resulting in a net southwest-to-northeast ("westerly") flow. This surface flow of warm air out of the subtropics transports heat into high latitudes where the radiation balance is negative (see Figure 2-14B). Much of this heat transport in the atmosphere is ultimately tied to the large transfer of latent heat from the ocean to the atmosphere in the tropics.

Toward higher middle latitudes in both hemispheres, the circulation of the lower atmosphere is a complex zone of transition between the prevailing warm flow coming out of the subtropics and a much colder equatorward flow from higher latitudes (see Figure 2-16). The surface flow at these latitudes is dominated by an ever-changing procession of high- and low-pressure cells moving from west to east, separated by *frontal zones,* or regions near Earth's surface where large changes in temperature occur over short distances in association with fast-moving air. Both the poleward movement of warm air and the equatorward movement of cold air along these frontal systems have the effect of warming the higher latitudes in a net sense, adding a major contribution to heat redistribution on Earth.

Precipitation generally exceeds evaporation in the temperate middle latitudes for two reasons: the cooler air temperatures reduce the rates of evaporation, and the precipitation associated with moving low-pressure cells is large.

Because the low-pressure cells move rapidly from west to east across the middle latitudes, they encounter and interact strongly with the topography of the land. Air flowing in from the ocean tends to carry large amounts of water vapor. As this air encounters mountain ranges that block its flow, it is forced to rise to higher elevations, and it cools. Water vapor condenses from the cooling air and produces heavy precipitation on the sides of mountains that face upwind toward warm oceans, such as the Olympic Mountains of Washington State. This is referred to as **orographic precipitation** (Figure 2-20).

Air that has been stripped of much of its water vapor then sinks on the downwind side of the mountains. As it moves to lower elevations, it is compressed and warmed, and it gains in capacity to store even more water vapor without condensation occurring. As a result, the lee, or *rain shadow*, sides of mountain ranges are areas of lower precipitation. This process also reinforces the natural tendency of mid-continental regions far from the oceanic sources of moisture, such as the Great Plains of the United States, to be dry.

At higher elevations in the mid-latitude atmosphere, winds flow more steadily from west to east. Narrow ribbons of faster flow called **jet streams** occur at altitudes of 5 to 10 kilometers in two regions: a persistent but weaker jet near 30° latitude in the

Looking Deeper into Climate Science

The Coriolis Effect

The direction of movement of fluids (air and water) is complicated by Earth's rotation. One way to visualize this effect is to contrast the actual motion of a person during a single day at either pole in comparison with that of a person on the equator. A person standing exactly at one of the poles spins around in place once each day, without ever moving from that point. In contrast, a person standing on the equator and facing east (the direction of Earth's rotation) zooms through space at 500 meters per second around the 40,000 km of Earth's circumference once each day, all the while facing east and not spinning at all. Earth's rotation accounts for this shift from a spinning motion at the poles to a one-way trip through space at the equator.

To appreciate the effect Earth's rotation has on moving objects, we can track the movement of an airplane across the Northern Hemisphere. If the flight sets out from a point P on Earth's surface and flies in a straight line in relation to the stars, it will appear to an Earthbound observer located under the path of the airplane starting at point P_1 to be moving toward the northeast. Several hours later, with the plane still moving in exactly the same direction *in relation to the stars,* Earth's coordinate system (its reference directions of north, south, east, and west) will have rotated out from under it, and the airplane will now appear from an Earthbound perspective at point P_2 to be moving to the southeast rather than the northeast. In effect, the rotation of Earth's coordinates out from under the airplane makes it appear to have turned to the right, *although it has not really turned at all.*

Earth's rotation has the same effect on moving air and water. Air naturally tends to flow from regions of high to low pressure, but rotation causes its motion to appear to be deflected to the right (again, from an Earthbound perspective). This *apparent* deflection is called the **Coriolis effect,** after the French engineer who first described it.

In the Northern Hemisphere, air moving from a zone of high pressure to low pressure always appears to be deflected to the right. This deflection causes air moving outward from a high-pressure cell to acquire a net clockwise spin, and air moving in toward a low-pressure region acquires a counterclockwise spin for the same reason.

Both the direction of deflection and the spinning of air around the highs and lows are reversed in the Southern Hemisphere because the direction of Earth's rotation viewed from the perspective of the South Pole is exactly opposite that of the view from the North Pole. At the equator, where Earth's rotation has no net spinning motion, the Coriolis effect drops to zero.

---

subtropics, near the sinking branch of the Hadley cell; and a more mobile jet that wanders between latitudes 30° and 60° above the moving high- and low-pressure cells. The jet at middle latitudes is especially strong in winter. Almost hidden by these prevailing west-to-east motions and the meandering paths of these jet streams is a net transport of heat and water vapor from low to high latitudes. Poleward meanders in the

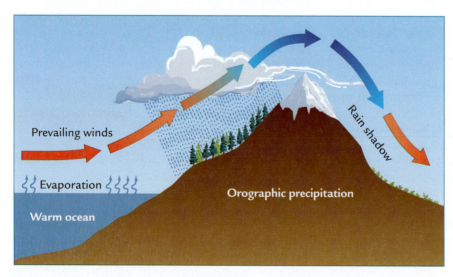

FIGURE 2-20
Orographic precipitation

As moist air masses driven up against mountains rise and cool, water vapor condenses and produces precipitation. The air masses subsiding on the downwind side of the high topography warm, retain water vapor, and suppress precipitation. (ADAPTED FROM F. PRESS AND R. SIEVER, *UNDERSTANDING EARTH,* 2ND ED., © 1998 BY W. H. FREEMAN AND COMPANY.)

Prevailing winds

Evaporation

Warm ocean

Rain shadow

Orographic precipitation

Box 2-5

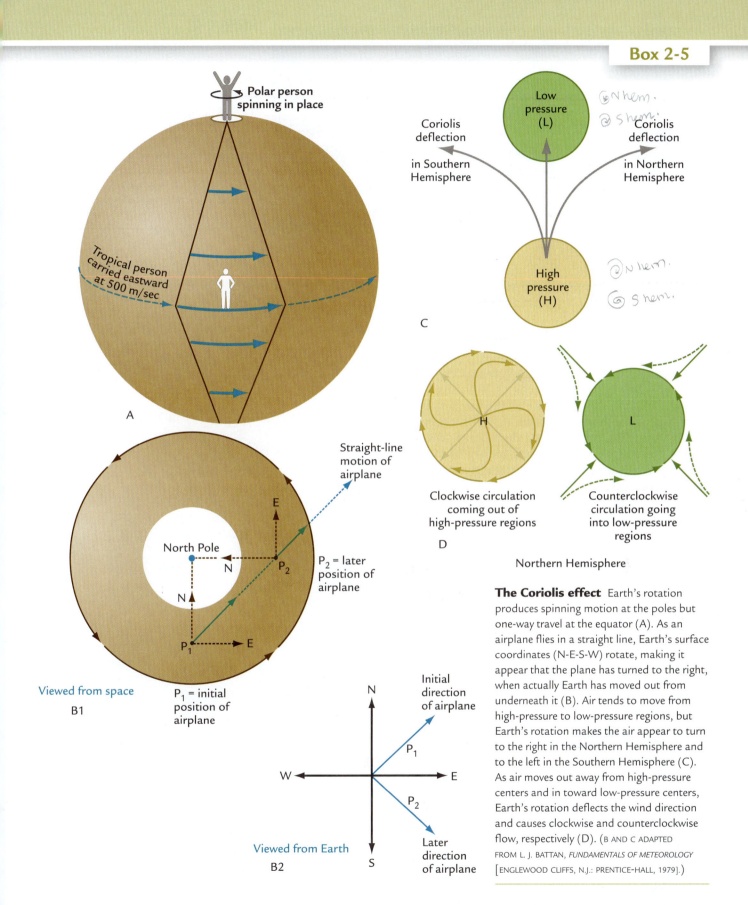

Polar person spinning in place

Tropical person carried eastward at 500 m/sec

A

North Pole

P₂ = later position of airplane

Straight-line motion of airplane

P₁ = initial position of airplane

N   E

Viewed from space

B1

Initial direction of airplane

N

W   E

S

Later direction of airplane

Viewed from Earth

B2

Coriolis deflection in Southern Hemisphere

Low pressure (L)

Coriolis deflection in Northern Hemisphere

High pressure (H)

C

Clockwise circulation coming out of high-pressure regions

H

Counterclockwise circulation going into low-pressure regions

L

D

Northern Hemisphere

The Coriolis effect Earth's rotation produces spinning motion at the poles but one-way travel at the equator (A). As an airplane flies in a straight line, Earth's surface coordinates (N-E-S-W) rotate, making it appear that the plane has turned to the right, when actually Earth has moved out from underneath it (B). Air tends to move from high-pressure to low-pressure regions, but Earth's rotation makes the air appear to turn to the right in the Northern Hemisphere and to the left in the Southern Hemisphere (C). As air moves out away from high-pressure centers and in toward low-pressure centers, Earth's rotation deflects the wind direction and causes clockwise and counterclockwise flow, respectively (D). (B AND C ADAPTED FROM L. J. BATTAN, *FUNDAMENTALS OF METEOROLOGY* [ENGLEWOOD CLIFFS, N.J.: PRENTICE-HALL, 1979].)

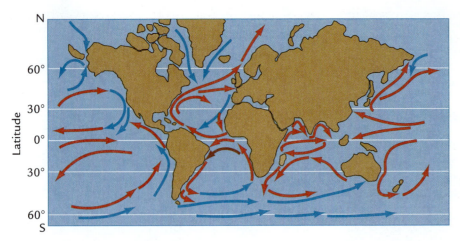

FIGURE 2-21
Surface ocean circulation

The surface flow of the oceans is organized into strong wind-driven currents. These currents encircle large spinning gyres in the subtropical oceans. Currents moving out of the tropics carry heat poleward, while currents moving away from the poles carry cold water equatorward. (MODIFIED FROM S. STANLEY, *EARTH SYSTEMS HISTORY*, © 1999 BY W. H. FREEMAN AND COMPANY.)

jet stream carry warm air to the north, and equatorward meanders carry cold air south.

Heat Transfer in Earth's Oceans

The uppermost layer of the ocean is heated by solar radiation. Like air, water expands as it warms and becomes less dense, but in this case the warmest layers are already at the top of the ocean, so they simply float on top of the colder, denser deep ocean. Winds mix the stored solar heat to maximum depths of 100 meters, a small fraction of the 4,000-meter average depth of the oceans. Some of this warm water is transported from the tropics toward the poles, and this poleward flow carries about half as much heat as is transported by the atmosphere.

2-7 The Surface Ocean

Most of the surface circulation of the oceans is driven by winds, and one of the most prominent results is huge **gyres** of water at subtropical latitudes (Figure 2-21). These spinning gyres are mainly the result of an initial push (or drag) of the winds on the ocean surface and of the Coriolis deflection of the moving water (see Box 2-5).

Blowing wind exerts a force on the upper layer of the ocean and sets it in motion in the same direction as the wind. The Coriolis effect turns this surface flow of water to the right in the Northern Hemisphere (and to the left in the Southern Hemisphere). The top layer of water in turn pushes (or drags) the underlying layers, which are deflected a little farther to the right than the surface layer and are also slowed by friction. This process continues down into the water column to a depth of about 100 meters, creating a downward spiral of water gradually deflected farther and farther to the right (Figure 2-22). The net transport of water in this entire 100-meter layer of ocean water is 90°

to the right of the wind in the Northern Hemisphere (and to the left south of the equator).

In the North Atlantic Ocean, the prevailing low-altitude winds are the tropical trade winds and mid-latitude westerlies. Trade winds blowing toward the southwest push shallow waters toward the northwest, and mid-latitude westerlies blowing toward the northeast push surface water to the southeast (Figure 2-23). Together these winds drive the uppermost layer

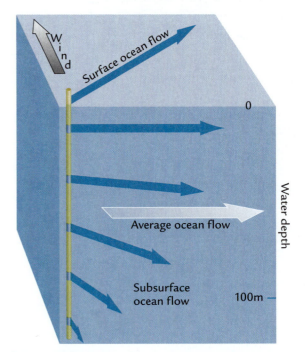

FIGURE 2-22
Effect of surface winds on the ocean

In the Northern Hemisphere, low-level winds drive surface waters to the right of the direction in which the wind is moving. Subsurface water is turned progressively farther to the right, and the net transport of the upper layer of water is 90° to the right of the direction of the wind. (MODIFIED FROM D. MERRITTS ET AL., *ENVIRONMENTAL GEOLOGY*, © 1997 BY W. H. FREEMAN AND COMPANY.)

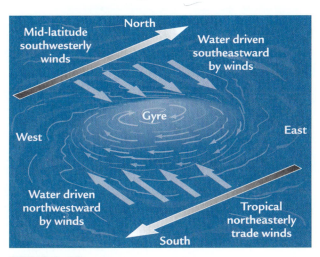

FIGURE 2-23
Subtropical gyres

In the Northern Hemisphere, mid-latitude southwesterly winds and tropical northeasterly trade winds drive warm water toward the centers of subtropical gyres, forming a thick lens of warm water that circulates in a clockwise gyre.

of water into the centers of the subtropical gyres and pile up a lens of warm water.

Sea level in the center of this lens sits 2 meters higher than the surrounding ocean. Water that flows away from this lens is turned to the right by the Coriolis deflection, and this creates a huge subtropical ocean gyre spinning in a clockwise direction (counterclockwise for gyres in the Southern Hemisphere). The edges of the continents also play a role in forming these gyres by acting as boundaries that contain the flow within individual ocean basins.

Subtropical gyres extend all the way to depths of 600 to 1,000 meters. In the North Atlantic, most of the water moving into the deeper parts of the gyre comes from its northern margin, where the prevailing westerly winds are strong enough to push large volumes of water toward the south just underneath the lens of warm surface water.

Viewed over long intervals of time, most of the flow in subtropical gyres consists of water moving around in giant spirals. The volume of water circulating is enormous, about 100 times the transport of all Earth's rivers flowing into the oceans. Almost hidden in this recirculation is a much smaller amount moving *through* the gyres and carrying heat toward high latitudes.

The prevailing flow toward the equator in the deeper parts of the gyres must be balanced by a return flow toward the poles, and this flow is concentrated in narrow regions along the western gyre margins. In the North Atlantic, the poleward transport occurs in the **Gulf Stream** and its continuation, the **North Atlantic Drift** (see Figure 2-21). As the Gulf

Stream emerges from the Gulf of Mexico, it forms a narrowly concentrated outflow of warm salty water headed north.

Another factor that affects poleward heat transport only in the North Atlantic Ocean is a giant vertical circulation cell linked to the deeper circulation of the ocean (Figure 2-24). A large volume of surface water sinks to depths below 2 kilometers in the higher latitudes of the North Atlantic, and this sinking water must be balanced by a compensating inflow of surface water from the south.

The effects of this circulation cell are felt even beyond the North Atlantic. The surface circulation typical of most oceans carries heat from the warm equator to the cold poles, as partial compensation for the imbalances set up by uneven solar heating and heat absorption at the surface. This normal equator-to-pole flow is reversed in the South Atlantic, where the net direction of heat transport is from south to north, *against* the planet's temperature gradient (see Figure 2-21). Temperate water flows into the South Atlantic from the middle latitudes of the Indian and Pacific oceans, moves northward across the equator, and sinks in the high-latitude North Atlantic.

The net northward transport of heat in the North Atlantic is often referred to as a "conveyor belt" (see Figure 2-24), but the overall flow has also been

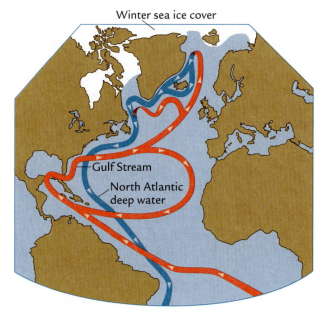

FIGURE 2-24
Sinking of surface water

Warm salty water flowing northward in the North Atlantic Ocean chills and sinks north of Iceland and in the Labrador Sea, between North America and Greenland. This cold deep water flows south out of the Atlantic at depths of 2 to 4 km. (MODIFIED FROM D. MERRITTS ET AL., *ENVIRONMENTAL GEOLOGY,* © 1997 BY W. H. FREEMAN AND COMPANY.)

compared to an airport baggage carousel. Most of the luggage (heat) spins around and around the carousel (warm water recirculating in the subtropical gyre), while only a small amount enters the carousel (comes across the equator from the South Atlantic) or is removed (heads farther north in the Gulf Stream and its continuation).

This warm, northward-moving water in the Atlantic transfers a huge amount of heat to the atmosphere. At latitudes above 50°N, the large temperature contrast between the warm North Atlantic waters and the cold overlying air produces a loss of sensible heat from the ocean to the atmosphere that is comparable to the amount of heat delivered locally by incoming solar radiation.

The fundamental circulation and heat transport of the oceans are less well understood than those of the atmosphere, mainly because of the difficulty of maintaining long-term monitoring stations at sea. The closer oceanographers look at the surface flow, the more complicated it turns out to be, with smaller gyres of water recirculating within larger gyres. At even smaller scales, spinning cells of water 100 kilometers wide move erratically across the ocean and transport large amounts of heat. These are analogous to moving low-pressure and high-pressure cells in the atmosphere.

2-8 Deep-Ocean Circulation

The poleward flow of warm water that counters some of Earth's heat imbalance occurs above the **thermocline**, a zone of rapid temperature change between warm upper layers and cold water filling the deeper ocean basins. Actually, two thermoclines exist: (1) a deeper permanent portion that is maintained throughout the year and (2) a shallower portion that changes as a result of seasonal heating by the Sun (Figure 2-25).

We've just seen how the warm poleward flow above the thermocline in the North Atlantic is balanced by sinking of cold water at high latitudes and movement of this cold deep water toward the equator. This overturning circulation is called the **thermohaline flow.** This term refers to the two main processes that control formation of deep water: temperature ("thermo-") and salinity ("-haline," from the same root as "halite," a synonym for rock salt).

Deep waters form and sink because they become more dense than the underlying water, as a result of several mechanisms. Seawater contains dissolved salt (on average near 35 parts per thousand, or ‰, by mass), and this salt content, or **salinity**, makes it 3.5% more dense than freshwater. The density of ocean water can be increased at lower latitudes when the atmosphere evaporates freshwater as water vapor, leaving the remaining water saltier (denser). The

density of seawater can also can be increased at high latitudes by formation of sea ice during **salt rejection**, a process that stores freshwater in sea ice and leaves the salt behind.

Another way to increase the density of ocean water is by cooling it. Saltwater is slightly compressible, which means it loses volume and gains density when it is cooled by the atmosphere. Cooling can occur either because warm ocean water is carried poleward into cooler regions or because colder air masses move to lower latitudes.

Salinity and temperature often work together to raise the density of water. Initially, evaporation or formation of sea ice increases the salinity and density of surface waters as a kind of preconditioning process. Cold air then cools the water, further increases its density, and causes it to sink.

The large-scale thermohaline flow in the deep ocean is closely linked to temperature and its effect on density. The very dense waters filling the deepest ocean basins today form at higher, colder latitudes.

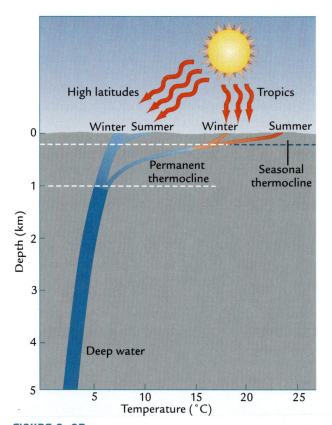

FIGURE 2-25
Thermoclines

The permanent thermocline (100–1,000 m) separates cold deep water from shallower layers affected by changes in Earth's surface temperature. Shallow seasonal thermoclines (0–100 m) vary in response to seasonal solar heating of the upper ocean layers.

The progressively less dense waters that fill successively shallower depths of the ocean form at less frigid latitudes farther from the poles.

Most of the deepest ocean of the world is filled by water delivered from just two regions, the high-latitude North Atlantic Ocean and the Southern Ocean, near Antarctica (Figure 2-26). No deep water forms in the high latitudes of the Pacific Ocean today because the salinity of the surface water is too low and the surface waters there are not sufficiently dense.

The surface waters that sink and form deep water in the North Atlantic initially acquire high salinity in the dry subtropics as a result of strong evaporation. Some of the water vapor taken from the tropical Atlantic Ocean is exported westward over the low mountains of Central America into the Pacific Ocean, leaving the Atlantic saltier than equivalent latitudes in the Pacific. Some of the salty water left in the Atlantic is carried northward by the Gulf Stream and North Atlantic Drift. Frigid air masses from the surrounding continents then extract sensible heat from the water in winter and further increase its density to the point where it sinks.

This water mass is called **North Atlantic deep water**. Sinking occurs in two regions in the North Atlantic, one north of Iceland, the other east of Labrador (see Figure 2-24). Together these two sources of deep water fill the Atlantic Ocean between depths of 2 and 4 kilometers. This flow moves southward with a total volume fifteen times the combined flow of all the world's rivers. Eventually, much of this flow rises toward the sea surface in the Southern Ocean and joins the waters circling eastward around Antarctica.

An even colder and denser water mass forms in the Antarctic region and flows northward in the Atlantic below 4 kilometers (see Figure 2-26). This water mass, called **Antarctic bottom water**, fills the deep Pacific and Indian oceans. Some forms near the Antarctic coast when seawater is chilled by very cold air masses and as a result of salt rejection as sea ice forms. Some also forms by intense cooling in gaps in the extensive sea-ice cover well away from Antarctica.

Two smaller water masses are prominent at intermediate depths of the North Atlantic. **Antarctic intermediate water** forms far north of Antarctica at latitudes 45°–50°S. This water is warmer and less dense than North Atlantic deep water and flows northward above it at depths above 1.75 kilometers. **Mediterranean overflow water** forms in the subtropical Mediterranean Sea as a result of winter chilling of surface waters with a very high salt content caused by strong evaporation. Water vapor extracted by subtropical evaporation leaves Mediterranean waters much saltier than normal ocean water.

Deep ocean water is one of the slowest-responding parts of the climate system. On average, it takes more than 1,000 years for a parcel of water that leaves the surface and sinks into the deep ocean to emerge back at the ocean's surface. The oldest and slowest-moving deep water is found in the Pacific Ocean, while the Atlantic has younger, faster-moving water. The long journey of water through the deep ocean keeps much of it out of touch with the climatic changes that affect the atmosphere and surface ocean.

With all this water sinking into the deep ocean, how does it get back to the surface? It turns out that

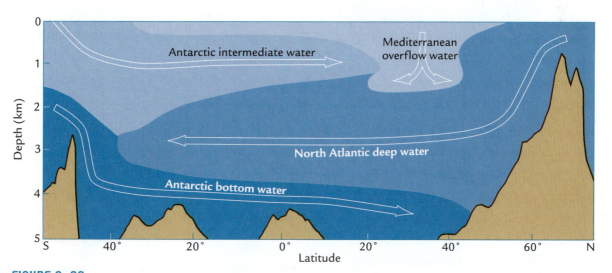

FIGURE 2-26
Deep Atlantic circulation
Water filling the North Atlantic basin comes from sources in the high-latitude North Atlantic, the Southern Ocean near Antarctica, and (at shallower depths) the Mediterranean Sea. (ADAPTED FROM E. BERNER AND R. BERNER, *GLOBAL ENVIRONMENT* [ENGLEWOOD CLIFFS, N.J.: PRENTICE-HALL, 1996].)

climate scientists don't really know the answer to this question very well. In the Atlantic, some of the southward-flowing North Atlantic deep water rises to the surface near Antarctica, and very strong winds mix it into the upper layers of the Southern Ocean. But what happens to the large volume of Antarctic water moving north in the deep Pacific and Indian Oceans, where no deep water forms?

For decades a widely accepted explanation has been that deep water injected into the ocean in specific regions gradually mixes into the central ocean basins and slowly moves upward across the thermocline and into the warm surface waters. But this highly diffuse return flow has been difficult to detect because it is spread across such a large area. Recent measurements show that this upward diffusion of water is too slow to account for much of the return flow.

In addition, a process called **upwelling**, rapid upward movement of subsurface water from intermediate depths, occurs in two other kinds of ocean regions. Both upwelling processes are initiated by surface winds and aided by the Coriolis effect:

- When surface winds in the Northern Hemisphere blow parallel to coastlines along the path shown in Figure 2-27A, they push water away from the land. To replace surface water pushed offshore, water rises from below. The upwelling water is cooler than the nearby surface water that has remained at the surface and has been warmed by the Sun.

- A second kind of upwelling occurs along the equator, especially in the eastern end of ocean basins (Figure 2-27B). Trade winds push surface

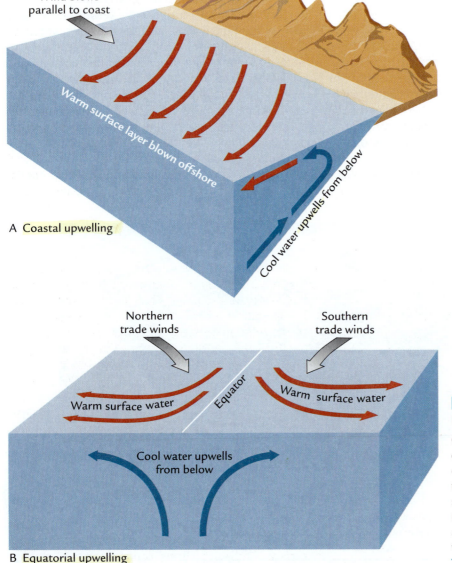

Wind blows parallel to coast

Warm surface layer blown offshore

Cool water upwells from below

A Coastal upwelling

Northern trade winds

Southern trade winds

Warm surface water   Equator   Warm surface water

Cool water upwells from below

B Equatorial upwelling

FIGURE 2-27
Upwelling

Cool subsurface water rises along coastal margins (A), where winds drive warm water offshore, and near the equator (B), where winds drive surface waters away from the equator. (MODIFIED FROM D. MERRITTS ET AL., *ENVIRONMENTAL GEOLOGY*, © 1997 BY W. H. FREEMAN AND COMPANY.)

waters away from the equator. Warm surface water is driven northward in the Northern Hemisphere and southward in the Southern Hemisphere by the opposing Coriolis deflections north and south of the equator. The movement of warm water away from the equator causes upwelling of cooler water from below.

▶ Ice on Earth

Despite the redistribution of heat by air and water, temperatures cold enough to keep water frozen solid occur at high altitudes and latitudes. Ice is one of the most important components of the climate system because its properties are so different from those of air, water, and land.

2-9 Sea Ice

Although freshwater freezes at 0°C, typical seawater resists freezing until it is cooled to −1.9°C. As sea ice forms, it rejects almost all the salt in the seawater. Because sea ice is less dense than seawater, it floats on top of the salty ocean.

When sea ice forms, it seals off the underlying ocean from interaction with the atmosphere. This change is vital to regional climates. Without an ice cover, high-latitude oceans transfer large amounts of heat to the atmosphere, especially in winter, when air temperatures are low (Figure 2-28A). This heat transfer keeps temperatures in the lower atmosphere close to those of the ocean surface (near 0°C).

But if an ice cover is present, this heat release stops, and the reflective ice surface absorbs little incoming solar radiation. Because of these changes, winter air temperatures can cool by 30°C or more in

regions that develop a sea-ice cover (Figure 2-28B). In effect, an ice-covered ocean behaves like a snow-covered continent. This change forms a prominent part of the albedo-temperature feedback process examined in Box 2-2.

Many ocean surfaces are only partially ice-covered. Gaps ("leads") produced in the ice by changing winds allow some heat exchange with the atmosphere and moderate the climate effects of a full sea-ice cover. Also, in summer, meltwater pools may form on the ice surface, and this water, along with a gradual darkening of the melting ice, may absorb more solar radiation.

The formation and melting of sea ice are driven mainly by seasonal changes in solar heating. In the Southern Ocean, most of the sea ice melts and forms again every year, over an area comparable in size to the entire Antarctic continent it surrounds. This annual ice cover averages 1 meter in thickness, except where strong winds cause the ice to buckle and pile up in ridges. In contrast, the landmasses surrounding the Arctic Ocean constrain the movement of sea ice and allow it to persist for 4 or 5 years. Older sea ice in the center of the Arctic may reach 4 meters in thickness, while annually formed ice around the margins is about 1 meter thick.

Recall that large inputs and extractions of heat calories from the atmosphere are required to form and melt sea ice, and the cycle of freezing and melting also involves exchanges of heat with the slow-responding ocean because of its high heat capacity. For these reasons, seasonal extremes in sea-ice cover lag well behind the seasonal extremes of heating by solar radiation. The maximum extent of sea ice is usually reached in the spring, the minimum extent in the autumn.

2-10 Glacial Ice

Glacier ice occurs mainly on land, in two forms. Mountain glaciers are found in mountain valleys at high elevations (Figure 2-29 top). Because glaciers can exist only where mean annual temperatures are below freezing, mountain glaciers near the equator are restricted to elevations above 5 kilometers (Figure 2-29 bottom). At higher and colder latitudes, mountain glaciers may reach down to sea level. Typical mountain glaciers are a few kilometers in length and tens to hundreds of meters in width and thickness. They typically flow down mountain valleys, constrained on both sides by rock walls.

Continental ice sheets are a much larger form of glacier ice, typically hundreds to thousands of kilometers in horizontal extent and 1 to 4 kilometers in thickness. The two existing ice sheets, which cover most of Antarctica and Greenland, represent roughly 3% of Earth's total surface area and 11% of its land surface. These ice sheets contain some 32 million

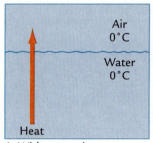
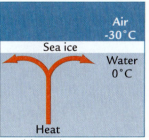

FIGURE 2-28

Effect of sea ice on climate

Whereas heat can escape from an unfrozen ocean surface (A), a cover of sea ice (B) stops the release of heat from the ocean to the atmosphere in winter and causes air temperatures to cool by as much as 30°C.

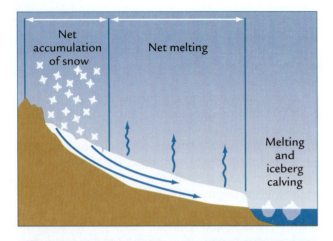

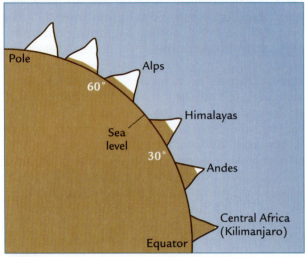

FIGURE 2-29
Mountain glaciers

Mountain glaciers accumulate snow at colder, higher elevations. The snow turns to ice, flows to lower elevations where temperatures are warmer, and melts (top). Mountain glaciers can exist near sea level at high latitudes but survive only at elevations above several kilometers in the warm tropics (bottom). (MODIFIED FROM F. PRESS AND R. SIEVER, *UNDERSTANDING EARTH*, 2ND ED., © 1998 BY W. H. FREEMAN AND COMPANY.)

cubic kilometers of ice, equivalent to about 70 meters of sea level change.

Continental ice sheets indeed have dimensions comparable to those of sheets—usually more than 1,000 times as wide as they are thick—but their surfaces do have structure (Figure 2-30 top). The highest regions on the ice sheets are rounded **ice domes**, with the elevations sloping gently away in all directions. Domes may be connected by high broad ridges with gentle sags called **ice saddles**. On the sides of ice sheets, ice flows in fast-moving **ice streams** from which **ice lobes** protrude beyond the general ice margins.

These great masses of ice depress the underlying bedrock below the elevation it would have had if no ice were present (Figure 2-30 bottom). About 30% of the total ice thickness sits below the original (undepressed) bedrock level; the other 70% protrudes into the atmosphere as a broad smooth ice plateau.

Snow that falls on the higher parts of mountain glaciers and ice sheets gradually recrystallizes into ice. The ice then moves toward lower elevations under the force of gravity. Ice in mountain glaciers is affected by gravity because of the steep slopes of mountain valleys. Continental ice sheets are high plateaus, and gravity moves ice from higher to lower elevations.

Ice deforms and moves in the upper 50 meters of a glacier in a brittle way: fracturing and forming crevasses, emitting loud cracking noises in the process. Below about 50 meters, ice deforms more gradually, by slow plastic flow. Parcels of ice may take hundreds to thousands of years to travel through mountain glaciers. For the central domes of ice sheets, where the flow is directed deep into the center of the ice mass, the trip may take tens of thousands of years.

As ice flows, its layers are stretched and thinned. If the ice is very cold (below −30°C), it behaves in a stiff manner, and its sloping edges can be relatively steep. Impurities such as dust also tend to stiffen the ice. If the ice is somewhat warmer (close to the freezing point), it is more plastic—softer and easier to deform—and it will tend to relax into gentler slopes.

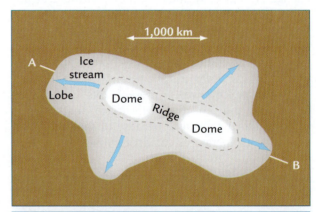

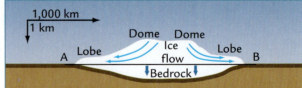

FIGURE 2-30
Continental ice sheets

The central portions of large continent-sized ice sheets have high central domes connected by ridges. Ice streams on the flanks carry ice to lobes protruding from the ice margins (top). In cross section, snow accumulates on the high part of an ice sheet, turns to ice, and flows to the lower margins (bottom).

Ice also moves by mechanisms favorable to sliding on its basal layers. Large amounts of water can accumulate at the base of mountain glaciers, causing them to **surge** down valleys at rates far in excess of their normal movement. Lobes of continental ice sheets also move by sliding along their bases. In regions where ice streams occur, the ice may move several meters per day, or 100 to 1,000 times faster than the rest of the ice mass.

Ice streams occur for two reasons. First, the pressure from the weight of the overlying ice may cause some of the ice at the bottom to melt and create a thin layer of water on which the ice can slide. Second, this water may percolate into and saturate soft unconsolidated sediments lying beneath the outer margins of the ice sheet, causing them to lose their mechanical strength or cohesiveness. These water-lubricated sediments provide a slippery *deformable bed* on which the overlying ice can easily slide.

Under certain conditions, **ice shelves** may form over shallow ocean embayments, and several shelves exist today on the margins of Antarctica. In these regions, gravity pulls ice out of the interior of the continent to the embayments, where it spreads out in shelves tens to hundreds of meters thick. Bedrock surrounding these embayments and at the shallow depths below it provides friction that keeps the ice from sliding away into the ocean. Immense **tabular icebergs** occasionally break off from these shelves and float away. Icebergs the size of Manhattan Island have broken off from both Greenland and Antarctica in recent years.

The bottom of the western part of the Antarctic ice sheet lies below sea level, and this portion is called the West Antarctic **marine ice sheet**. Because marine ice sheets have bases lying below sea level, they are highly vulnerable to sea level changes and respond to changes in climate much more quickly than ice sheets that sit higher on the land.

Mountain glaciers and continental ice sheets ultimately exist for the same reason: the overall rate of snow falling across the entire ice mass equals or exceeds the overall rate at which ice is lost by melting and other means. Climate scientists analyze the conditions over present glaciers and ice sheets in terms of their **mass balance**, the average rate at which ice either grows or shrinks every year. The concept of mass balance can also be applied to different portions of glaciers: mass balances are positive at upper elevations, where **accumulation** of snow and ice dominates, but negative at lower elevations, where rapid **ablation** (loss of ice) occurs.

Ice accumulation occurs in regions where temperatures are cold enough not just to cause frozen precipitation but also to allow new-fallen snow to persist through the warm summer season. For mountain glaciers, subfreezing temperatures occur on the highest parts of mountains, where the air is coldest.

For continent-sized ice sheets, parts of which often exist at sea level, the cold required to sustain ice is found at high polar latitudes and on the high parts of the ice sheets.

Ablation of glacial ice by melting occurs when temperatures exceed the freezing point. Melting can occur because of absorption of solar radiation or by uptake of sensible or latent heat delivered by warm air masses and by summer rain moving across the ice. Ablation can also occur by **calving**, the shedding of icebergs to the ocean or to lakes. Calving differs from the other processes of ablation in that icebergs leave the main ice mass and move elsewhere to melt, often in an environment much warmer than that near the ice sheet.

The boundary between the high-elevation region of positive ice mass balance and the lower area of net loss of ice mass occurs at a mean annual temperature near −10°C for ice sheets but closer to 0°C for mountain glaciers. The mass balance at high elevations on the Greenland and Antarctic ice sheets is positive because of the absence of melting to offset the slow accumulation of snow (Figure 2-31). At elevations above 1 to 2 kilometers, air is so cold that it contains little water vapor. Although the precipitation that falls on these higher parts of the ice sheets is all snow, accumulation rates are low. This is especially the case for the frigid Antarctic continent, centered on the South Pole and

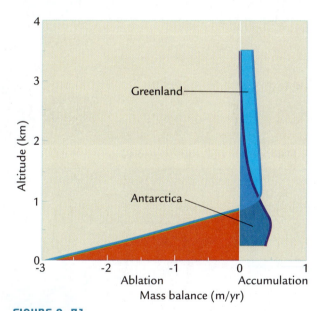

FIGURE 2-31
Ice mass balance

Snow accumulates on the upper parts of ice sheets where melting does not occur. At lower elevations, accumulation is overwhelmed by net loss of ice by ablation due to melting (in Greenland) and to calving of icebergs (in Antarctica). The units shown convert snow to equivalent thicknesses of ice (in meters). (ADAPTED FROM J. OERLEMANS, "THE ROLE OF ICE SHEETS IN THE PLEISTOCENE CLIMATE," *NORSK GEOLOGISK TIDSSKRIFT* 71 [1991]: 155–161.)

surrounded by an ice-covered ocean. The mass balance is more positive on the sides of these ice sheets, where air masses carry more moisture and cause more snowfall, yet ablation is not strong.

The mass balance on ice sheets is negative at lower elevations, usually because mean annual temperatures above 0°C accelerate the rate of melting. In Greenland, some of the low-elevation precipitation also falls as rain rather than snow, which further promotes ablation. In Antarctica, freezing conditions persist at sea level even in summer, and no melting occurs. The Antarctic ice sheet loses mass mainly by calving icebergs into the ocean.

Earth's Biosphere

To this point, we have examined only the physical side of the climate system, expressed mainly by variations in temperature, precipitation, winds, and pressure, but these physical parts of the climate system also interact with its organic parts (Earth's **biosphere**). Many of these interactions result from the movement of carbon (C) through the climate system and in turn affect the distribution of heat on Earth.

Carbon moves among and resides in several major reservoirs. The amount of carbon in each reservoir is typically quantified in gigatons (billions of tons, or 10^{15} grams) of carbon. Relatively small amounts of carbon reside in the atmosphere, the surface ocean, and vegetation; a slightly larger reservoir resides in soils, a much larger reservoir in the deep ocean, and a huge reservoir in rocks and sediments (Figure 2-32A).

Carbon takes different chemical forms in these different reservoirs. In the atmosphere, it is a gas (CO_2). Carbon in land vegetation is organic, as is most carbon in soils, while that in the ocean is mostly inorganic, occurring as dissolved ions (atoms carrying positive or negative charges). Despite these

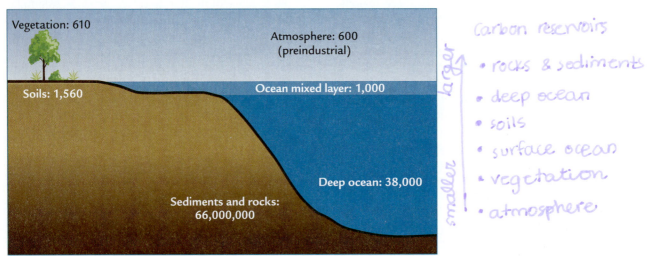

A Major carbon reservoirs (gigatons)

carbon reservoirs
- *rocks & sediments*
- *deep ocean*
- *soils*
- *surface ocean*
- *vegetation*
- *atmosphere*

larger ← → smaller

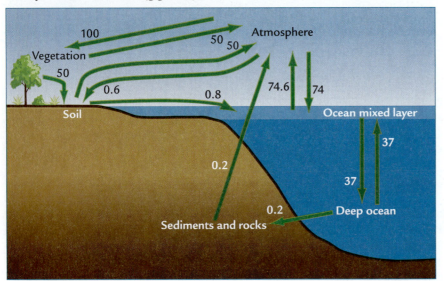

B Carbon exchange rates (gigatons/year)

FIGURE 2-32
The carbon cycle

The major carbon reservoirs on Earth vary widely in size (A) and exchange carbon at differing rates (B). Larger reservoirs (rocks, the deep ocean) exchange carbon much more slowly than smaller reservoirs (air, vegetation, the surface ocean). (ADAPTED FROM J. HOREL AND J. GEISLER, *GLOBAL ENVIRONMENTAL CHANGE* [NEW YORK: JOHN WILEY, 1997], AND FROM NATIONAL RESEARCH COUNCIL BOARD ON ATMOSPHERIC SCIENCES AND CLIMATE, *CHANGING CLIMATE*, REPORT OF THE CARBON DIOXIDE ASSESSMENT COMMITTEE [WASHINGTON, D.C.: NATIONAL ACADEMY PRESS, 1993].)

differences in form, carbon is exchanged freely among all the reservoirs, changing back and forth between organic and inorganic forms as it moves.

Rates of carbon exchange among reservoirs vary widely (Figure 2-32B). In general, the sizes of the reservoirs are inversely related to their rates of carbon exchange. The small surface reservoirs (the atmosphere, surface ocean, and vegetation) exchange all their carbon with one another in just a few years. The much larger deep-ocean reservoir is partly isolated from the surface reservoirs by the thermocline and exchanges carbon with the surface ocean and the atmosphere over hundreds of years. The carbon buried in sediments and rocks interacts very slowly, moving in and out of the surface reservoirs only over hundreds of thousands of years or longer.

Plants grow on land if the conditions necessary for **photosynthesis** (the production of plant matter) are met: sunlight is needed to provide energy, and nutrients (mainly phosphorus and nitrogen) provide food for plant growth (Figure 2-33). With these conditions satisfied, plants draw CO_2 from the air and water and from the soil to create new organic matter, while oxygen is liberated to the atmosphere:

$$6CO_2 + 6H_2O \xrightarrow[\text{Oxidation}]{\text{Photosynthesis}} \underset{\text{Plants (organic C)}}{C_6H_{12}O_6 + 6O_2}$$

During the time that plants grow by photosynthesis, and also during the time they are mature but no longer growing, plants take the water they need from the soil and give it back to the atmosphere (see Figure 2-33). This process, known as **transpiration** (and also as "respiration"), is a highly efficient way to return water vapor to the atmosphere, and it can occur at much faster rates than ordinary evaporation from vegetation-free ground.

After plants die (either by seasonal dieback or by reaching the end of their natural lifetime), oxygen is consumed in destroying their organic matter during **oxidation**. Oxidation can occur either through rapid burning (in fires) or by slow decomposition in the presence of oxygen, with the same ultimate result. Through either process, oxidation converts organic carbon back to inorganic form, as shown by the equation above.

2-11 Response of the Biosphere to the Physical Climate System

Trees, shrubs, and other plants accomplish most photosynthesis on land. CO_2 and sunlight are usually available over the continents, but the distributions of temperatures and rainfall critical to photosynthesis and plant life vary widely. To a large extent, rainfall (Figure 2-34 top) determines the total amount of

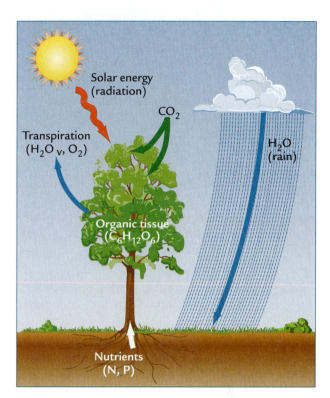

FIGURE 2-33
Photosynthesis on land

Plant life on land uses sunlight, CO_2 and H_2O from the atmosphere, and soil nutrients in the process of photosynthesis. Plants also return water vapor (H_2O_v) and oxygen (O_2) to the atmosphere by transpiration. (ADAPTED FROM F. T. MACKENZIE, *OUR CHANGING PLANET* [ENGLEWOOD CLIFFS, N.J.: PRENTICE-HALL, 1998].)

organic (live) matter present, called the **biomass**, and the predominant types of vegetation and associated organisms, called **biomes** (Figure 2-34 bottom).

The tendency for rainfall to be abundant along the ITCZ produces tropical rain forest biomes with dense biomasses. Toward the dry subtropics, rain forests grade into **savanna** (scattered trees in a grassland setting) and then to the sparse scrub vegetation typical of deserts. Total biomass decreases along with precipitation toward the subtropics.

Large-biomass **hardwood forest** (maple, oak, hickory, and other leaf-bearing trees) occurs in the wetter portions of the middle latitudes (eastern North America, Europe, Asia) and on the upwind side of mountain ranges facing the ocean, while low-biomass biomes such as grasslands and desert scrub are found in drier interior regions in the rain shadow of mountain ranges (see Figure 2-20). **Conifer forest** (spruce and other trees with needles) dominates toward the higher latitudes of the Northern Hemisphere, but the fringes of the Arctic Ocean are surrounded by a wide band of scrubby **tundra** vegetation with low biomass

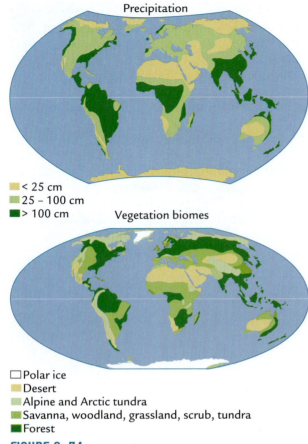

Precipitation

< 25 cm
25 – 100 cm
> 100 cm

Vegetation biomes

Polar ice
Desert
Alpine and Arctic tundra
Savanna, woodland, grassland, scrub, tundra
Forest

FIGURE 2-34

Precipitation and vegetation

Global precipitation (top) is highest in the tropics and along mountain slopes that receive moisture-bearing winds from the ocean and lowest in subtropical deserts and over polar ice. Vegetation biomes (bottom) largely reflect the patterns of precipitation, with high-biomass forests in regions of high precipitation and low evaporation. (TOP: ADAPTED FROM L. J. BATTAN, *FUNDAMENTALS OF METEOROLOGY* [ENGLEWOOD CLIFFS, N.J.: PRENTICE-HALL, 1979]; BOTTOM: ADAPTED FROM E. BRYANT, *CLIMATE PROCESS AND CHANGE* [CAMBRIDGE: CAMBRIDGE UNIVERSITY PRESS, 1998].)

above ground and large amounts of carbon stored below ground. Ice sheets are free of life forms except cold-tolerant bacteria.

Life in the oceans depends on a different combination of the same factors as on land. Obviously, water is abundantly available in the oceans, and CO_2 is plentiful in surface waters that exchange CO_2 with the atmosphere. In addition, light from the Sun is widely available in the upper layers of the ocean into which it penetrates (Figure 2-35).

With all these conditions favorable to photosynthesis, why isn't the surface ocean an enormous photosynthesis machine? The answer is a lack of the nutrients nitrogen (N) and phosphorus (P). Nutrient food sources are scarce in most parts of the surface ocean.

A floating form of microscopic plant life called **phytoplankton** lives in the surface layers of the ocean and uses sunlight for photosynthesis. These minute organisms extract nutrients and incorporate them in the soft organic tissues of their bodies. Phytoplankton have short life spans (days to weeks), and when they die, they sink to deeper waters, leaving the surface layer depleted of nutrients. Thus the rates of photosynthesis in these sunlit surface waters are limited.

Initially near the surface, and mainly later at depths well below the surface, the slow decay and oxidation of the soft tissues of these sinking organisms releases nitrogen and phosphorus back into ocean water. Because most of these nutrients are released into and stay in the deeper ocean, their scarcity in surface waters limits the amount of life that can exist across most ocean areas.

In the few parts of the surface ocean where upwelling occurs, nutrients are more plentiful, and they result in greater **productivity**, or rates of photosynthesis, by phytoplankton (Figure 2-36). Wind-driven upwelling along some coastal margins returns nutrients to the surface from below and supplements nutrients that were delivered to continental shelves by rivers and resuspended during storms. As a result, surface waters near continental margins tend to be relatively productive. Upwelling in the eastern equatorial Pacific and Atlantic oceans also increases rates of photosynthesis and productivity in those regions.

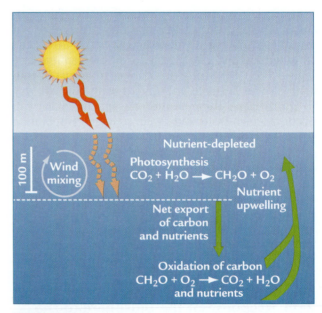

100 m

Wind mixing

Nutrient-depleted

Photosynthesis
$CO_2 + H_2O \rightarrow CH_2O + O_2$

Nutrient upwelling

Net export of carbon and nutrients

Oxidation of carbon
$CH_2O + O_2 \rightarrow CO_2 + H_2O$
and nutrients

FIGURE 2-35

Photosynthesis in the ocean

Sunlight penetrating the surface ocean causes photosynthesis by microscopic plants. As they die, their nutrient-bearing organic tissue descends to the seafloor. Oxidation of this tissue at depth returns nutrients and inorganic carbon to the surface ocean in regions of upwelling.

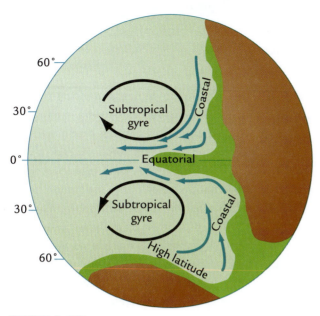

FIGURE 2-36
Ocean productivity

The greatest amount of photosynthesis in the surface ocean occurs along shallow continental margins and in coastal, equatorial, and high-latitude regions where nutrients upwell from below.

The Southern Ocean around Antarctica is another productive region (see Figure 2-36). Deep water from the Atlantic flows toward the surface in this area, bringing nutrients up from below. Strong winds mix these nutrients into the surface layers, producing the rich biomass of the Southern Ocean. Because sunlight is not plentiful, and because the long season of sea-ice cover limits the amount of time in which photosynthesis can occur, nutrients in the surface waters of the Southern Ocean are never depleted, even when productivity is highest.

2-12 Effects of the Biosphere on the Climate System

Life affects climate in many ways. One way is by providing positive feedback to physical processes that affect climate (Box 2-6). A second way is through changes in the amount of greenhouse gases in the atmosphere, especially carbon dioxide (CO_2) and methane (CH_4).

As we will see in later parts of this book, *all* the exchanges of carbon shown in Figure 2-32 have affected atmospheric CO_2 and climate, but at different time scales. The slow movement of carbon into and out of rock reservoirs affects atmospheric CO_2 and climate on the tectonic (million-year or longer) time scale, explored in Part II. Somewhat faster exchanges between the surface and deep ocean reservoirs affect climate on the orbital (ten-thousand-year) time scale,

examined in Parts III and IV. Rapid exchanges among the surface reservoirs (vegetation, the ocean, and the atmosphere) affect climate over the shorter time scales reviewed in Parts IV and V.

Atmospheric CO_2 trends measured over recent decades show two superimposed effects (Figure 2-37A). Each year, a small drop in CO_2 values occurs in April–May and a comparable rise the following September–October. This oscillation reflects cycling of vegetation in the Northern Hemisphere: CO_2 is taken from the air by plant photosynthesis every spring and released by oxidation every autumn. The signal follows the tempo of the Northern rather than the Southern Hemisphere because most of Earth's land (and land vegetation) lies north of the equator.

The second trend evident in the CO_2 curve is its gradual overall increase (see Figure 2-37A). This increase results mainly from burning of fossil-fuel

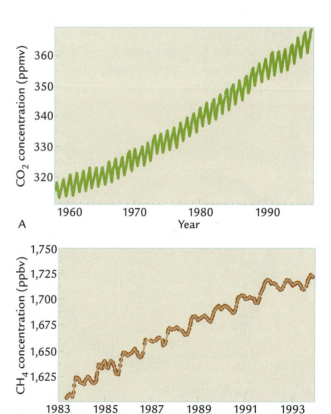

FIGURE 2-37
Recent increases in carbon dioxide and methane

Instrument measurements record rapid rises of the greenhouse gases CO_2 (A) and CH_4 (B) in recent years. The gases are measured in parts per million by volume (ppmv) and parts per billion by volume (ppbv). (A: ADAPTED FROM H. H. FRIEDLI ET AL., "ICE CORE RECORD OF THE $^{13}C/^{12}C$ RATIO OF ATMOSPHERIC CO_2 IN THE PAST TWO CENTURIES," *NATURE* 324 [1986]: 237–38; B: AFTER M. A. K. KHALIL AND R. A. RASMUSSEN, "ATMOSPHERIC METHANE: TRENDS OVER THE LAST 10,000 YEARS," *ATMOSPHERIC ENVIRONMENT* 21 [1987]: 2445–52.)

Climate Interactions and Feedbacks

Box 2-6

Vegetation-Climate Feedbacks

The type of vegetation covering a land region can affect its average albedo. The two major types of vegetation in the Arctic, spruce forest and tundra, interact in different ways with freshly fallen snow and produce surfaces with very different albedos. Snow that falls on tundra covers what little scrub vegetation exists and creates a high-albedo surface that reflects most incoming solar radiation. Snow that falls on spruce forests is blown from the trees and falls to the ground, allowing the dark-green surface of the exposed treetops to absorb most incoming solar radiation.

When climate cools, these contrasts in albedo produce an important positive feedback. Tundra gradually advances southward and replaces spruce forest, expanding Earth's high-albedo surface area. With more solar radiation reflected from this surface, the reduction in absorbed heat leads to further cooling. This process is called **vegetation-albedo feedback**. This same positive feedback works in the opposite direction during times of climate warming: as forest replaces tundra, more solar heat is absorbed, and the climate warms even more.

A second type of vegetation feedback depends on the way vegetation recycles water. Land vegetation draws water needed for photosynthesis from the ground. Some of the water is handed off to the atmosphere as water vapor during times when the plants are actively extracting CO_2 from the air. Trees transpire much larger volumes of water vapor than grass or desert scrub, and this contrast is responsible for the second kind of positive feedback. When climate becomes wetter, forests gradually replace grasslands in some regions. The trees transpire more water vapor back to the atmosphere, thus increasing the amount available for rainfall. This positive feedback (called **vegetation-precipitation feedback**) works in the reverse sense when climate dries.

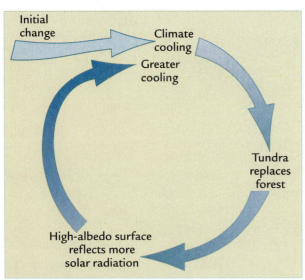

A   Vegetation-albedo feedback

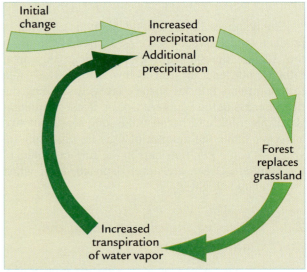

B   Vegetation-precipitation feedback

Vegetation-climate feedbacks   When high-latitude climate cools, replacement of spruce forest by tundra raises the reflectivity (albedo) of the land in winter and causes additional cooling as a positive feedback (A). When climate becomes wetter, replacement of grasslands by trees increases the release of water vapor back to the atmosphere and causes increases in local rainfall as a positive feedback (B).

carbon and secondarily from **deforestation** (clearing of vegetation from the land), which releases carbon into the atmosphere through burning and oxidation. The rapid increase in consumption of fossil fuels by humans over the last two centuries has tapped into huge reservoirs of coal, oil, and gas in rocks that naturally release their carbon at slow rates and has greatly accelerated these rates (see Part V).

Methane (CH_4) is a second important atmospheric greenhouse gas, although far less plentiful than CO_2. It has many sources, including swampy lowland bogs, rice paddies, the stomachs and bowels of cows digesting vegetation, and termites. Common to all these CH_4 sources is the decay of organic matter in an oxygen-free environment. By the early 21st century, methane concentrations in the atmosphere had risen by well over a factor of 2 above their natural (preindustrial) level to above 1,700 parts per billion (Figure 2-37B). This recent increase in methane is the result of human activities (see Part V).

As we noted earlier, both CO_2 and CH_4 trap part of Earth's back radiation, keep the heat in the atmosphere, and make Earth warmer than it would otherwise be. This warming in turn activates the positive feedback effect of water vapor (H_2O_v), the most important greenhouse gas. The combined effects of these three greenhouse gases in the recent past and near future are the focus of the last part of this book.

Key Terms

electromagnetic radiation (p. 20)

electromagnetic spectrum (p. 20)

shortwave radiation (p. 20)

back radiation (p. 21)

longwave radiation (p. 21)

greenhouse effect (p. 22)

troposphere (p. 23)

stratosphere (p. 23)

albedo (p. 23)

albedo-temperature feedback (p. 26)

heat capacity (p. 27)

calories (p. 27)

specific heat (p. 27)

hydrologic cycle (p. 28)

thermal inertia (p. 29)

sensible heat (p. 29)

convection (p. 29)

latent heat (p. 30)

latent heat of melting (p. 30)

latent heat of vaporization (p. 30)

dew point (p. 31)

saturation vapor density (p. 31)

water vapor feedback (p. 32)

adiabatic (p. 33)

lapse rate (p. 33)

Hadley cell (p. 34)

intertropical convergence zone (ITCZ) (p. 35)

monsoon (p. 35)

orographic precipitation (p. 37)

jet streams (p. 37)

Coriolis effect (p. 38)

gyres (p. 40)

Gulf Stream (p. 41)

North Atlantic Drift (p. 41)

thermocline (p. 42)

thermohaline flow (p. 42)

salinity (p. 42)

salt rejection (p. 42)

North Atlantic deep water (p. 43)

Antarctic bottom water (p. 43)

Antarctic intermediate water (p. 43)

Mediterranean overflow water (p. 43)

upwelling (p. 44)

mountain glaciers (p. 45)

continental ice sheets (p. 45)

ice domes (p. 46)

ice saddles (p. 46)

ice streams (p. 46)

ice lobes (p. 46)

surge (p. 47)

ice shelves (p. 47)

tabular icebergs (p. 47)

marine ice sheet (p. 47)

mass balance (p. 47)

accumulation (p. 47)

ablation (p. 47)

calving (p. 47)

biosphere (p. 48)

photosynthesis (p. 49)

transpiration (p. 49)

oxidation (p. 49)

biomass (p. 49)

biomes (p. 49)

savanna (p. 49)

hardwood forest (p. 49)

conifer forest (p. 49)

tundra (p. 49)

phytoplankton (p. 50)

productivity (p. 50)

vegetation-albedo feedback (p. 52)

vegetation-precipitation feedback (p. 52)

deforestation (p. 53)

Review Questions

1. How does solar radiation arriving on Earth differ from the back radiation emitted by Earth?

2. What kind of radiation is trapped by greenhouse gases? What is the effect on Earth's climate?

3. What different and opposing roles do clouds play in the climate system?

4. How does reflection of solar radiation from Earth's surface add to the effects of uneven solar heating in creating a pole-to-equator heat imbalance?

5. What processes cause air to rise from Earth's surface?

6. What causes the monsoon circulation to reverse from summer to winter?

7. Describe the main pathway by which heat in the atmosphere is transported toward the poles.

8. Why does rain fall on the sides of mountains in the path of winds from nearby oceans?

9. How do low-level winds create spinning gyres in the subtropical oceans?

10. Why does deep water form today at higher latitudes?

11. What effect does the formation of sea ice have on the overlying atmosphere?

12. What parts of ice sheets gain and lose mass? Why?

13. How closely does land vegetation (and total biomass) follow global precipitation trends?

14. What regions of the ocean are most productive? Why?

15. Describe two positive feedback processes discussed in this chapter.

Additional Resources

Barry, R. G., and Chorley, R. J. 2009. *Atmosphere, Weather, and Climate*. New York: Routledge.

Thurman, H. V. 1997. *Introductory Oceanography*. New Jersey: Prentice Hall.

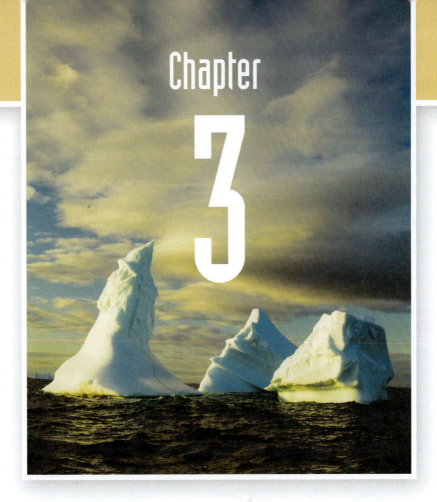

Chapter

3

Climate Archives, Data, and Models

Climate scientists use a wide range of techniques to extract and reconstruct Earth's climatic history. Much of this history is recorded in four archives: sediments, ice, corals, and trees. In this chapter, we first examine each of these major archives. Then we explore how their climate records are dated, the way climate history is recorded in each archive, and the resolution each archive yields.

The interpretation of climate data is complemented by the use of climate models. Hypotheses of climate change can be tested by comparing results from models against "ground-truth" climatic data. In this chapter, we describe physical models that simulate the circulation of Earth's atmosphere and ocean, and then we examine the concept behind models used to track mass movements of chemical tracers through the climate system.

Climate Archives, Dating, and Resolution

Like written chronicles of human history, climate archives hold stories of climate change for those who can read them. For the immense span of Earth's history prior to the invention of instruments in recent centuries, sediments, ice, corals, and trees are the major climatic archives.

3-1 Types of Archives

Although relatively recent climate changes can be studied in an array of archives, sedimentary debris deposited by water is the major climate archive on Earth for over 99% of geologic time.

SEDIMENTS Rainfall and the runoff it produces erode rocks exposed on the continents and transport eroded debris in streams and rivers in both physical (granular) and chemical (dissolved) forms. The sediments are eventually deposited in quieter waters where layers of sediment are laid down in undisturbed succession. Most sediment is carried to the ocean, either right after it is eroded or after initial deposition on land followed by long intervals of erosion and redeposition. Sediment delivered to the seafloor may persist there for tens of millions of years until destroyed by tectonic processes. The relentless action of these two processes, erosion and tectonic activity, decreases the likelihood of preservation of older sedimentary records as time passes.

For intervals prior to the last 170 million years, all surviving sedimentary records come from the continents. Under favorable conditions, sediments may be preserved in the regions shown in Figure 3-1: continental basins that contain lakes; shallow interior seas that at times flood low-lying land; lens-shaped piles of sediment along continental shelves (the barely submerged coasts of continents); and steeper continental slopes leading to the deep ocean.

Sediments are most useful as climate archives if deposition is uninterrupted. Major disturbances come from wave action reaching several meters below sea level and from occasional large storms that reach tens of meters deep in the water column and erode previously deposited layers. In addition, sediments deposited on steep continental slopes are vulnerable to being dislodged by disturbances such as earthquakes.

In the longer term, sea level change is a major factor that interrupts sediment deposition. Through time, the sea has moved up and down along the continental margins over a total vertical range of at least 200 meters. Sediments are deposited on the upper margins when sea level is high (such as now), but these deposits are often eroded by waves and storms and carried to the deep sea when sea level subsequently falls.

All these factors ultimately determine the types of climate records preserved in sediment archives (Figure 3-1). Sediments deposited on continental shelves when sea level is high form lens-shaped units separated by distinct surfaces where erosion has occurred. Deposition is relatively continuous within some of these coastal-margin sequences, especially in regions where rivers deliver the most sediment. Sediments deposited in interior seas within the continents during times when the ocean floods low-lying regions form thin continuous sequences covering wide areas.

Sediments deposited in lakes in tectonic depressions are less common because the tectonic activity that generates these structures is localized. The sediments in these basins generally conform to the structural framework of the bedrock depressions, and deposition tends to be most continuous in the deeper

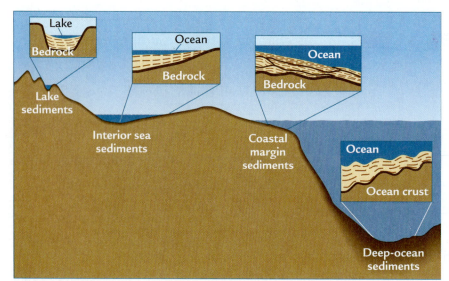

FIGURE 3-1
Sediment archives

Layered sediments are major climate archives on all time scales. The insets show typical sediment layering in sediment archives from land and sea.

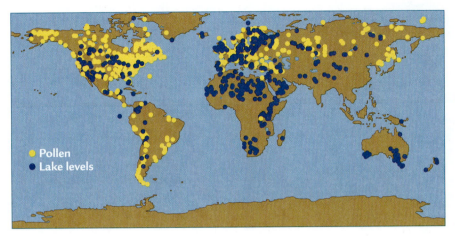

FIGURE 3-2
Lake cores

Hundreds of cores have been taken from small lakes and analyzed for records of changes in pollen (vegetation) and lake level over the last several thousand years. (NATIONAL PALEOCLIMATE DATA CENTER, NGDC, BOULDER, CO.)

parts of lakes. Lake sediments also fill widespread depressions left behind by melting glaciers during their retreat from regions like northern North America during the last 20,000 years (Figure 3-2).

Ice and wind are also powerful agents of sediment erosion and transport in many regions. After ice sheets reach maximum size and then begin to retreat, they leave behind long curving ridges called **moraines**. Moraines are made up of a jumbled mix of unsorted debris carried by ice, ranging from large boulders to very fine clay. When an ice sheet readvances over moraines deposited by earlier ice advances, it erodes the earlier debris and incorporates it into the newer deposits. Unraveling a climate history from this kind of record is like trying to decipher repeated episodes of writing that have been largely erased on a blackboard. By contrast, the coarse debris carried to the ocean and dropped by melting icebergs into the underlying sediments can survive in its original layer-by-layer form because of the more protected deep-ocean environment.

Strong winds weather rocks and form fine sediment particles in regions with dry climates. Winds create sand dunes that slowly migrate across desert areas, but continuous reworking of the sand-sized particles complicates efforts to use the dunes as climate archives. Winds also pick up smaller silt-sized grains and transport them beyond their original sources. In regions where the winds weaken, the silt is deposited in sequences called **loess**. Loess deposits can be excellent climate repositories of the last 3 million years, especially in China (Figure 3-3). Finer sediments carried off the continents by winds and deposited in ocean sediments are also useful indicators of climate.

OCEAN SEDIMENTS For the portion of geologic time younger than 150 million years, climate scientists have access to an additional climate archive: sediments preserved in ocean basins. Sediments from the older and more deeply buried part of this record have for decades

FIGURE 3-3
Windblown loess

Strong winds have deposited thick layers of silt-sized grains in central China during the last 3 million years. The total thickness of these loess deposits can reach several hundred meters. In some regions, people have carved out homes in the loess cliffs. (JULIA WATERLOW/EYE UBIQUITOUS/CORBIS.)

been retrieved by the *JOIDES Resolution,* capable of drilling into and recovering sequences several kilometers

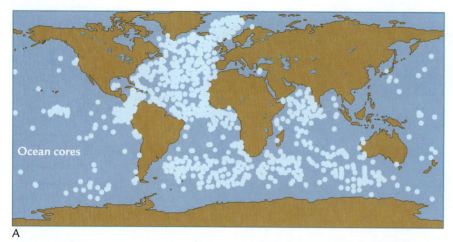

A

B   C

D

FIGURE 3-4
Ocean drilling
(A) Hundreds of ocean sediment cores are archives of past climatic changes. (B, C, D) The longest cores have been retrieved by drilling operations on the *JOIDES Resolution,* run by the international Ocean Drilling Program. (A: NATIONAL PALEOCLIMATE DATA CENTER, NGDC, BOULDER, CO; B, C, D: COURTESY INTEGRATED OCEAN DRILLING PROGRAM, UNITED STATES IMPLEMENTING ORGANIZATION [IODP-USIO].)

thick (Figure 3-4). Shorter sequences recovered by other coring techniques have been retrieved from much of the 70% of Earth's surface covered by oceans.

Because the deep ocean is generally a quiet place with relatively continuous deposition, it usually yields climate records of higher quality than records from land, where water, ice, and wind repeatedly erode deposits. Yet even some deep-sea sediments are subject to disturbances such as dislodgment from steep slopes, physical erosion and reworking by currents on the sea floor, and chemical dissolution by corrosive water in deeper basins. Despite these problems, many ocean basins have been sites of continuous sediment deposition for tens of millions of years. Deposition of sediments is usually much slower in the ocean than on land, but rates are higher in regions that receive influxes eroded from nearby continents, in areas beneath productive surface waters, and at sites situated high above the corrosive bottom waters that dissolve more vulnerable sediment components in the deeper ocean.

GLACIAL ICE At the very cold temperatures found at high latitudes and high altitudes, annual deposition of snow can pile up in continuous sequences of ice that range in thickness from small mountain glaciers tens to hundreds of meters thick to much larger continent-sized ice sheets several kilometers thick (Figure 3-5).

Ice-core archives contain many kinds of climatic information, although limited geographically to the few regions where ice survives (Figure 3-6).

Ice recovered from the Antarctic ice sheet now dates back to 800,000 years, and ice from the Greenland ice sheet dates to just beyond 125,000 years. Future ice drilling will likely extend the records in Antarctica even farther back in time. In contrast, most small glaciers that exist today in mountain valleys (even in the high-elevation tropics) record only the last 10,000 years or less of climate change. Deposition rates range from a few centimeters per year in the coldest and driest areas to meters per year in less frigid regions with greater snowfall.

OTHER CLIMATE ARCHIVES In areas of sufficient rainfall, groundwater percolating through soil and bedrock dissolves and redeposits limestone (calcite, or $CaCO_3$) in caves. These deposits contain records of climate over intervals that can extend back several hundred thousand years.

Trees are valuable climate archives for the interval of the last few tens, hundreds, or (in exceptional cases) thousands of years. The outer softwood layers of many kinds of trees are deposited in millimeter-thick layers that soon turn into hardwood. These annual layers are best developed in mid-latitude and

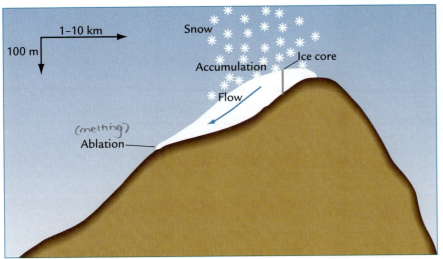

A Mountain glaciers

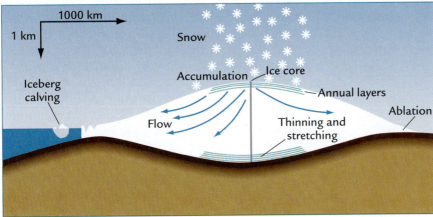

B Continental ice sheets

FIGURE 3-5
Ice archives

Ice is an important archive of many climate signals. Ice cores retrieve climate records extending back thousands of years in small mountain glaciers (A) to as much as hundreds of thousands of years in continent-sized ice sheets (B).

high-latitude regions that experience large seasonal climate changes, but some tropical regions have come under more intensive investigation (see Figure 3-6).

In clear sunlit waters at tropical and subtropical latitudes, *corals* form annual bands of calcium carbonate ($CaCO_3$) or magnesium carbonate ($MgCO_3$) that hold geochemical information about climate (see Figure 3-6). Individual corals may live for time spans of years to tens or even hundreds of years.

Within the last few thousand years, people have kept **historical archives** of climate-related phenomena. Examples include the time of blooming of cherry trees in Japan, the success or failure of grape and grain harvests in Europe, and the number of days with extensive sea ice in regions such as Iceland and Hudson Bay in Canada. These records precede (and in most cases overlap) the **instrumental records** of the last 100 to 200 years. The first thermometers for

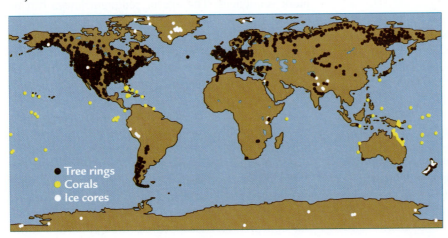

FIGURE 3-6
Ice cores, corals, and tree rings

Ice cores, corals, and tree rings are archives of climate change in more recent Earth history. (NATIONAL PALEOCLIMATE DATA CENTER, NGDC, BOULDER, CO.)

FIGURE 3-7

Instrument measurements

Instruments that have been used to measure climate range from the primitive thermometers of the seventeenth century to the multiple sensors flown aboard the TOPEX/Poseidon satellite. (NASA.)

measuring climate appeared in the eighteenth century, but human ingenuity has now devised instruments that measure climate remotely from space (Figure 3-7).

3-2 Dating Climate Records

Climate records in older sedimentary archives are dated by a two-step process. First, scientists use the technique of radiometric dating to measure the decay of radioactive isotopes in rocks. (Isotopes are forms of a chemical element that have the same atomic number but differ in mass.) Dates are obtained on hard crystalline igneous rocks that once were molten and then cooled to solid form. In the second step, dates obtained from the igneous rocks provide constraints on the ages of sedimentary rocks that occur in layers between the igneous rocks and form the main archives of Earth's early climate history.

RADIOMETRIC DATING AND CORRELATION Radiometric dating is based on the radioactive decay of a **parent isotope** to a **daughter isotope**. The parent is an unstable radioactive isotope of one element, and radioactive decay transforms it into the stable isotope of another element (the daughter). This decay occurs at a known rate, the *decay constant*, which is a measure of the likelihood of one parent-to-daughter decay per amount of parent present per unit of time. This rate of decay in effect forms a clock with which to measure age.

An event of some kind is required to start this clock ticking. The igneous rock most commonly used

for dating is basalt, which cools quickly from molten outpourings of lava. The event that starts the clock ticking is the cooling of this material to the point where neither the parent nor the daughter isotope can migrate in or out of the molten mass. At this point, the rock forms a **closed system**, one in which the only changes occurring are caused by internal radioactive decay.

In the simplest example of a closed system, the decay of a parent to a daughter produces the changes shown in Figure 3-8: the parent decays away exponentially, while the daughter shows an exactly opposite (and compensating) exponential increase in abundance. The **half-life** is a convenient measure of the rate at which this process occurs: one half-life is the time needed for half the parent that was present previously to decay to the daughter isotope. The first half-life reduces the parent to half its initial abundance, the second reduces it to half of that half (one-quarter), and so on. Radioactive isotopes remain useful for many half-lives after the clock is set, until at some point too little of the parent is left to be measured reliably.

Because radioactive parents have a very wide range of half-lives, each is most useful over a different part of Earth's history (Table 3-1). The very long and slow decay series from uranium (U) to lead (Pb) is useful for rocks that are nearly as old as Earth itself. The somewhat faster decay from potassium (K) to argon (Ar) is widely used for dating within much of Earth's history.

Several factors can complicate radiometric dating. Unlike the simple case shown in Figure 3-8, the initial abundance of the daughter isotope is rarely zero: usually some amount of daughter product was already present in the igneous rock at the time the decay clock was set, and the dating procedure needs to correct for that amount. Other problems arise when the system does not remain fully closed to the migration of parent or daughter isotopes.

If both igneous and sedimentary rocks are present in a specific region, the igneous rocks can be used to constrain the ages of the intervening sediment sequences. The age of sediment layers can be obtained from the nearby igneous rocks based on which of them is older or younger than the other. For example, a layer of igneous rock that spreads across the top of a layer of sediment must postdate the time the sediment was deposited and so it provides a minimum age for that layer.

In actual practice, it is rare to find enough igneous rock in any one location to date sediments in this way. Instead, sediment sequences are dated by a combination of radiometric dating followed by correlation using fossils or other features in the sediments. The fossil correlation method relies on the fact that a *unique and unrepeated* sequence of organisms has appeared and disappeared through Earth's entire

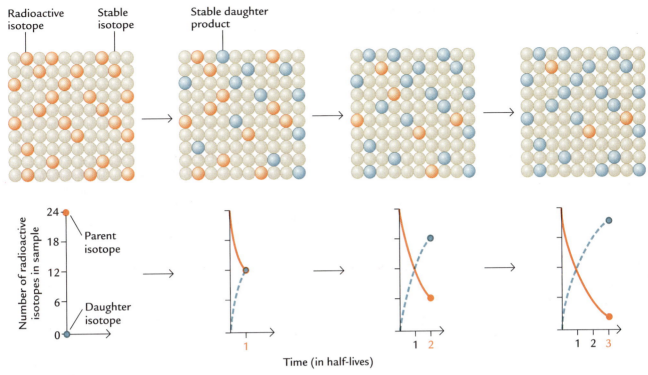

FIGURE 3-8
Radioactive decay

Time is determined by measuring the gradual decay of a radioactive parent isotope to a daughter isotope. The half-life is the time needed for half the parent to decay to the daughter. The relative abundances of parent and daughter isotopes follow the trends shown at the bottom. (D. MERRITTS ET AL., *ENVIRONMENTAL GEOLOGY*, © 1997 BY W. H. FREEMAN AND COMPANY.)

history and has left fossilized remains. The most useful fossils are those that are shortest-lived but geographically most widespread. If the brief existence of these species can be reliably dated in sediments in a few key areas based on nearby igneous rocks, the ages that are obtained can be transferred to sediments in other regions that contain the same short-lived fossils but lack igneous rocks. Other physical or chemical features within sedimentary rocks that show

distinctively varying patterns (such as the presence of distinctive layers of volcanic ash) can be used in a similar way.

RADIOCARBON In the younger geologic record, radio-carbon dating is widely used to date lake sediments and other kinds of carbon-bearing archives. Neutrons that constantly stream into Earth's atmosphere from space convert ^{14}N (nitrogen gas) to ^{14}C (an unstable

Table 3-1 Radioactive Decay Used to Date Climate Records

Parent isotope	Daughter isotope	Half-life	Useful for ages:	Useful for dating:
Rubidium-87 (^{87}Rb)	Strontium-87 (^{87}Sr)	47 Byr	100 Myr	Granites
Uranium-238 (^{238}U)	Lead-206 (^{206}Pb)	4.5 Byr	>100 Myr	Many rocks
Uranium-235 (^{235}U)	Lead-207 (^{207}Pb)	0.7 Byr	>100 Myr	Many rocks
Potassium-40 (^{40}K)	Argon-40 (^{40}Ar)	1.3 Byr	>100,000 years	Basalts
Thorium 230 (^{230}Th)	Radon-226* (^{226}Ra)	75,000 years	<400,000 years	Corals
Carbon-14 (^{14}C)	Nitrogen-14* (^{14}N)	5,780 years	<50,000 years	Anything that contains carbon

*Daughter in this case is a gas that has escaped and cannot be measured.

isotope of carbon). Vegetable and animal life forms on Earth extract this carbon from the atmosphere to build both their hard shells and soft tissue, and a small part of the carbon they extract is the radioactive ^{14}C isotope.

The death of the plant or animal then closes off carbon exchanges with the atmosphere and starts the decay clock ticking. The ^{14}C parent decays to the ^{14}N daughter, a gas that escapes to the atmosphere. The amount of ^{14}C that has been lost by the time a sample is analyzed can be determined by measuring a different isotope of carbon that is stable and has not been removed by radioactive decay. Because half of the original amount of ^{14}C (relative to the stable isotope of carbon) is lost by radioactive decay every 5,780 years, radiocarbon dating is most useful over five or six half-lives, back to about 30,000 years ago. With optimal materials and technology, it can be applied over the last 50,000 years or more (see Table 3-1).

Another technique relies on the same uranium (U) decay series used to date igneous rocks (see Table 3-1), but uses it in a different way to date corals. Ocean corals incorporate a small amount of ^{234}U and ^{238}U from seawater into their shells (substituting it for calcium), but no ^{230}Th. When the corals die, the parent (^{238}U) slowly decays and produces ^{230}Th in the coral skeleton. In this case, however, the daughter product (^{230}Th) is not stable but radioactively decays away with a half-life of 75,000 years. Gradually the amount of ^{230}Th present in the coral moves toward a level that reflects a balance between the slow decay of the parent U and the faster loss of the daughter ^{230}Th. The clock provided by the Th/U ratio is useful for dating over the last few hundred thousand years. This technique is also used for dating stalactite and stalagmite deposits in caves.

Counting Annual Layers Some climate repositories contain annual layers that form because of seasonal changes in the accumulation of distinctive materials. These annual layers can be used to date archives by simply counting back in time year-by-year from the present.

The most visible forms of annual layering in ice (mountain glaciers and ice sheets) are alternations between darker layers that contain dust blown in from continental source regions during the dry cold windy season, and lighter layers marking the warmer part of the year with little or no dust (Figure 3-9A). These dark/light couplets form annual layers that are easily visible in the upper parts of glacial ice, but gradually thin deeper in the ice to the point that they cannot easily be detected.

Sediments in some lakes contain annual couplets called **varves** (Figure 3-9B). These layers are particularly common in the deeper parts of lakes containing little or no life-sustaining oxygen. The lack of oxygen

suppresses or eliminates bottom-dwelling organisms that would otherwise obliterate the thin annual layers by their burrowing or other physical activity. Varve couplets usually result from seasonal alternations between deposition of light-hued mineral-rich debris and darker sediment rich in organic material.

In regions of well-marked seasonal variations of climate, trees produce annual layers called **tree rings** (Figure 3-9C). These rings are alternations between thick layers of lighter wood tissue (cellulose) formed by rapid growth in spring, and thin dark layers marking cessation of growth in autumn and winter. Because most individual trees live no more than a few hundred years, the time span over which this dating technique can be used is limited, but in some areas distinctive year-to-year variations in tree ring thickness can be used to splice records from younger trees with records from older trees whose fossil trunks are found on the landscape.

In tropical oceans, corals record seasonal changes in the texture of the calcite ($CaCO_3$) incorporated in their skeletons (Figure 3-9D). The lighter parts of these **coral bands** are laid down during intervals of fast growth, and the darker layers when growth slows. Individual corals dated in this way rarely live more than a few decades or at most a few hundred years, but older records may be spliced with younger ones (as in the case of tree rings).

Correlating Records with Orbital Cycles Another way to date climate records is to make use of the characteristic imprints of variations in Earth's solar orbit in a "tuning" exercise. Changes in Earth's orbit around the Sun alter the amount of solar radiation received by season and by latitude. The timing of these orbital variations is known very accurately from astronomical calculations (see Part III), and the physical processes that link these orbital changes to climatic responses on Earth have become increasingly well understood in recent decades. The two most prominent examples are changes in the strength of low-latitude monsoons and the cyclical growth and decay of high-latitude ice sheets. Because of these relationships, climate scientists can date many of Earth's climatic responses by linking them to the well-dated external driver provided by the orbital variations. This technique provides scientists with accurate absolute dating of Earth's climatic responses over many millions of years.

Internal Chronometers In some cases, counting annual layers and orbital tuning can serve a similar purpose much farther back in time. Even in the absence of radiometric dates of absolute age (in years before the present), some climate archives contain *internal chronometers* with which climate scientists can measure *elapsed time* (duration in years).

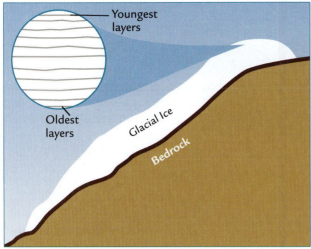

A Annual ice layers

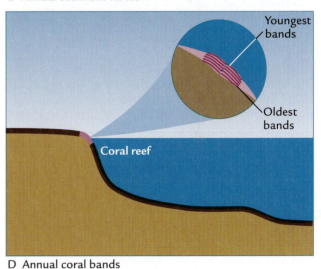

B Annual sediment varves

C Annual tree rings

D Annual coral bands

FIGURE 3-9

Annual layering

Four kinds of climate archives have annually deposited layers that can be used to date the climate records they contain: ice, varved lake sediments, trees, and corals.

For example, annual varves deposited in lake sediments millions of years ago still survive today in a few protected regions. Determining the actual age of these sequences by counting varves back in time from the present is impossible because varves have not been deposited since the time the lakes dried out. However, the varves supply an internal chronometer with which to count the years that elapsed during the interval when they were deposited.

3-3 Climatic Resolution

The extent to which information in climatic archives can be resolved depends mainly on the interplay between two factors: (1) the amount of disturbance of the sedimentary record by various processes soon after deposition, and (2) the rate at which the record

is buried beneath additional sediments and thereby protected from further disturbance.

SEDIMENT ARCHIVES Most sedimentary archives used for climate studies form in *low-energy* marine environments undisturbed by turbulent waves and storms. The primary disturbance after particles settle on the seafloor is physical stirring by deep-dwelling organisms (Figure 3-10). Organisms living on the sediment surface thoroughly mix the uppermost layers. A smaller number of animals burrow deep into the sediments, but do so only infrequently, and subsurface sediments are increasingly protected from most disturbances as they are gradually buried. Eventually the sediments pass beneath the region of active mixing and become part of the permanent sedimentary record.

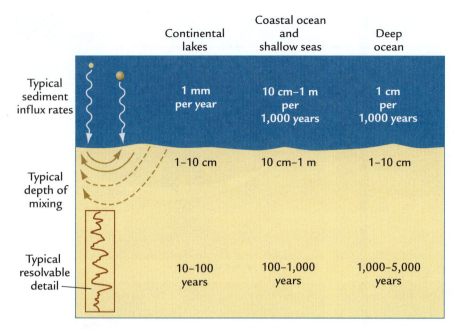

FIGURE 3-10
Sediment mixing

The degree of resolution of climate records in sediment archives is related to the rate of deposition (and burial) of sediment and to the amount of activity of organisms burrowing into the sediments.

Typical rates of sediment deposition range from as much as meters per year in coastal marine sequences to millimeters per year in lakes to millimeters per thousand years in some deep-sea sediments. Rates can vary locally around these average values by a factor of 10 because of factors such as the amount of sediment supplied locally by rivers or redistributed by currents.

The degree of disturbance by organisms that move across and burrow into the sediment surface varies with different environments. In highly productive coastal regions, large organisms burrow tens of centimeters or even meters down into the sediment. Relatively unproductive deep ocean basins have fewer and smaller bottom-dwelling organisms that typically burrow down no more than a few centimeters. Most lakes also have fewer and shallower burrowers. As a result, the resolution of sedimentary records also varies with environment. Lakes usually have the best resolution and deep-ocean sediments the poorest, although locally enhanced deposition can improve resolution.

After particles pass through the upper layers, no further mixing occurs unless erosion re-exposes the sequence to the sediment-water interface. Increased pressure and loss of water caused by deep burial of sediments gradually compact the sediment layers and turn them into soft rock and then harder rock, but do not dramatically reduce the resolution they can provide.

ICE CORES Annual layers of snow are visible at the surfaces of many mountain glaciers and rapidly deposited ice sheets (see Figure 3-9A). As the snow is buried and slowly recrystallized into ice, annual layers remain

resolvable to a depth that depends on their initial thickness at the time of deposition. Below this level, the layering is lost. In cores from the ice sheet on Greenland, where deposition of snow is rapid, the annual layering may remain detectible tens of thousands of years into the past. In the polar ice sheet covering eastern Antarctica, where only a small amount of snow/ice accumulates each year, annual layering may not be visible even at the ice surface.

TREE RINGS AND CORALS At middle and high latitudes where trees produce annual layers, tree rings become a permanent record of annual climate change, unless they are later disturbed by fire or by sporadic boring by insects or excavation by birds. Similarly, $CaCO_3$ bands in corals form a permanent record of seasonal to annual climate change.

The types of climate archives, the maximum time span of the records they contain, and the highest resolution achievable in each archive are summarized in Figure 3-11 in a log time scale that changes by powers of 10. Also shown at the top are the time spans covered by the major parts of this book.

◢ Climatic Data

Climate archives contain indicators of past climate referred to as **climate proxies.** Climate scientists use the term *proxy* (meaning "substitute") because the process of extracting climate signals from these indicators is not direct. Scientists must first determine the mechanism by which climate signals are recorded by the proxy indicators in order to use them to decipher climate changes.

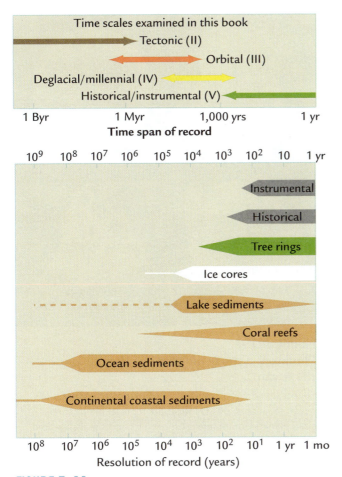

FIGURE 3-11
Resolution of climate records

Climate archives vary widely in the length of the records they contain and in the degree of resolution they yield. A log scale (changing by powers of 10) is needed to show all geologic time in a single plot. (ADAPTED FROM J. C. BERNABO, *PROXY DATA: NATURE'S RECORD OF PAST CLIMATES* [WASHINGTON, D.C.: NATIONAL OCEANIC AND ATMOSPHERIC ADMINISTRATION, 1978].)

FIGURE 3-12
Past vegetation

For older geologic intervals, climate on the continents can be inferred from distinctive vegetation. The remains of trees similar to modern palms are found in rocks from Wyoming dating to 45 million years ago. Today, winters in Wyoming are much too frigid to permit palm trees. (CHIP CLARK, SMITHSONIAN.)

The two climate proxies that are most commonly used are (1) **biotic proxies**, which are based on changes in composition of plant and animal groups, and (2) **geological-geochemical proxies**, which are measurements of mass movements of materials through the climate system, either as discrete (physical) particles or in dissolved (chemical) form.

3-4 Biotic Data

Because no seafloor older than 170 million years exists, basin-wide reconstructions of earlier oceanic environments are not possible. As a result, fossil remains from the continents are the main climate proxy for older tectonic-scale intervals. Most of the organisms that have ever existed on Earth are now extinct, and the further back in time we look, the less recognizable the fossils appear. Using biotic proxies

to reconstruct past climates over longer tectonic time scales often requires a reliance on the general resemblance of past forms to their modern counterparts, either in general appearance or in specific features that can be measured.

Because fossil remains of plants tend to be more numerous than those of animals in geologic records from continents, vegetation plays an important role in the reconstruction of ancient climates. Often the presence of a single critical temperature-sensitive form can serve as a useful climate indicator. For example, warmer climates tens of millions of years ago at high northern latitudes are inferred from the presence of palm-like trees (Figure 3-12).

For the younger continental record, climate scientists more commonly use the relative abundance of climate-sensitive vegetation indicated by pollen assemblages deposited in sediments (Figure 3-13).

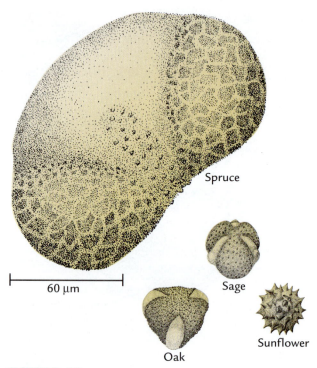

FIGURE 3-13
Pollen: a proxy indicator of climate on land

For younger intervals, climate on land can be reconstructed from changes in the relative abundance of distinctive types of pollen. For comparison, small grains of sand are 60 μm or larger in diameter. (COURTESY OF ALLEN M. SOLOMON, CHIEF SCIENTIST, CAPE ARAGO ECOLOGICAL CONSULTANTS.)

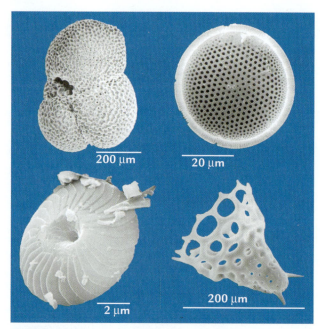

FIGURE 3-14
Plankton: a proxy indicator of climate in the ocean

Four types of shelled remains of plankton are common in ocean sediments: $CaCO_3$ shells are represented by sand-sized planktic foraminifera (upper left) and small clay-sized coccoliths (lower left); SiO_2 shells include silt-sized diatoms (upper right) and sand-sized radiolaria (lower right). For comparison, small grains of sand are 60 μm or larger in diameter. (MODIFIED FROM W. F. RUDDIMAN, "CLIMATE STUDIES IN OCEAN CORES," IN *PALEOCLIMATE ANALYSIS AND MODELING*, ED. A. D. HECHT [NEW YORK: JOHN WILEY, 1977].)

Minute pollen grains are produced in vast numbers by vegetation, distributed mostly by wind, and deposited in lakes, where they are preserved in oxygen-poor waters (see Figure 3-2). Pollen can be identified initially by major vegetation type (trees, grass, and shrubs) and then further subdivided (spruce trees indicate cold climates, oak trees indicate warmth). Larger vegetation remains that cannot have been carried far from their points of origin are also examined to make sure that the pollen in a lake sequence is representative of the nearby vegetation. These larger **macrofossils** include cones, seeds, and leaves.

In the oceans, four major groups of shell-forming animal and plant **plankton** are used for climate reconstructions (Figure 3-14). Two groups form shells made of calcite ($CaCO_3$). Globular sand-sized animals called **planktic foraminifera** (upper left) inhabit the upper layers of the ocean. Much smaller spherical algae called Coccolithophoridae secrete tiny plates called **coccoliths** (lower left) in sunlit waters. Two other groups of hard-shelled plankton secrete shells of opaline silica ($SiO_2 \cdot H_2O$) and tend to thrive in productive, nutrient-rich surface waters. **Diatoms** (upper right) are silt-sized plant plankton shaped like pillboxes or needles. **Radiolaria** (lower right) are

sand-sized animals with ornate shells often resembling pre-modern (Prussian) military helmets.

Sediments rich in $CaCO_3$ fossils occur in open-ocean waters at depths above 3,500–4,000 meters (Figure 3-15). Below that level, corrosive bottom waters dissolve calcite shells. SiO_2-shelled diatoms inhabit deltas and other coastal areas and extract silica from river water flowing off the land, but their abundance along the coasts is masked by the influx of mud eroded from the land. Radiolaria and diatoms are abundant in Antarctic and equatorial regions where highly productive waters upwell from below and influxes of continental debris are minimal.

Plankton and pollen share traits that make them especially useful as climate proxies. Both are widely distributed: plankton live in all oceans, and pollen are produced everywhere on continents except under ice sheets. Also, because fossil remains of these two groups are so abundant in sediments (usually thousands in a tablespoon-sized sample), their relative abundances can be determined with a much higher degree of accuracy than those fossil types that are

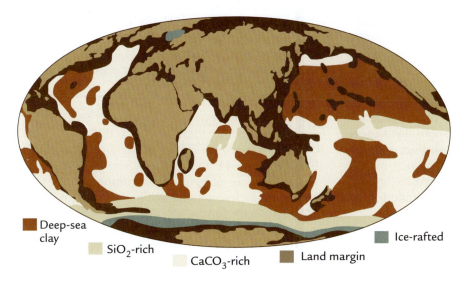

FIGURE 3-15

Distribution of ocean sediments

The predominant type of sediment on the seafloor of the world ocean today varies regionally, with ice-rafted sediment in polar areas, SiO$_2$-rich sediment in productive areas, CaCO$_3$-rich sediment on higher rises and ridges, and windblown deep-sea silt and clay in basins far from continents. Coastal regions contain mainly debris from the land. (MODIFIED FROM W. H. BERGER, "DEEP-SEA SEDIMENTATION," IN *THE GEOLOGY OF CONTINENTAL MARGINS*, ED. C. A. BURKE AND C. L. DRAKE [NEW YORK: SPRINGER-VERLAG, 1974].)

only present sporadically. Populations of plankton and pollen in different areas also tend to be dominated by a small number of species with well-defined climate preferences. The only other organisms with comparable ranges and abundance are insects, which less often leave fossil remains.

Most species of plankton and vegetation that live today have been present on Earth for hundreds of thousands to millions of years. The climatic preferences of these modern species can be determined by comparing their present distributions to measurements of current climate. These modern climate preferences can then be used to reconstruct past climates from fossil assemblages with great accuracy in sediment archives as old as a few million years or more.

3-5 Geological and Geochemical Data

Mass movements of materials through the climate system are tied to processes of erosion, transport, and deposition, mainly by water but also by ice and wind. Most climate studies of the older parts of Earth's history rely on physical debris deposited in sedimentary archives on the continents as the main proxy for inferring past climates. For example, sediment textures can tell us about erosion and subsequent deposition of unsorted debris by ancient ice sheets in cold environments, sand dunes moving across deserts under extremely arid conditions, and deposition by water in moist environments. Although these sediment types are useful for drawing broad inferences about climate, poor dating control and the prevalence of erosion make detailed study of many older continental records difficult, and alteration of these deposits increases with the passage of time.

In contrast, ocean sediment deposition in many areas spanning the last 170 million years was relatively continuous and is better dated. As a result, the distribution of sediment types that carry distinctive information about climate can be mapped, and changes in their patterns of deposition can be quantified as **burial fluxes** (measures of the mass of sediment deposited per unit area per unit time).

Sediment is eroded from the land and deposited in ocean basins in two forms. One type is debris eroded and transported as discrete particles or grains as a result of **physical weathering**, the process by which water, wind, and ice physically detach pieces of bedrock and reduce them to smaller fragments. An example shown in Figure 3-16 is coarse **ice-rafted debris** (sand and gravel) eroded by ice sheets and delivered by icebergs that melt in ocean waters. Other examples

FIGURE 3-16
Sediment particles

Deep-ocean sediments contain granular debris from land that reveals the climate of the source region. For example, sand-sized grains of quartz and other minerals rafted in from ice sheets by icebergs indicate cold climates. (COURTESY OF GERARD BOND, LAMONT-DOHERTY EARTH OBSERVATORY OF COLUMBIA UNIVERSITY.)

include finer **eolian sediments** (silts and clays) lifted from the continents and blown to the ocean by winds, as well as the wide range of grain sizes in **fluvial sediments** carried by rivers to the ocean.

Geological and geochemical techniques can further unravel the original sources of sediments formed by physical weathering. Microscope counts of sand-sized grains from marine sediments can distinguish different sources on the basis of distinct mineral types. In recent years, geochemical analyses of distinctive elements and isotopes have become widely used to trace mineral grains back to specific source regions on the continents.

The second major way of removing sediments from the land is by chemical weathering and subsequent transport of dissolved ions (charged ions or compounds) to the oceans in rivers (Figure 3-17). Chemical weathering occurs mainly in two ways: (1) by dissolution, in which carbonate rocks (such as limestone, made of $CaCO_3$) and evaporite rocks (such as rock salt, made of $NaCl$) are dissolved in water, and (2) by hydrolysis, in which the weathering process adds water to minerals derived from continental rocks made of silicates, such as basalts and granites.

Both processes depend on the fact that atmospheric CO_2 and rain (H_2O) combine in soils and rock crevices to form carbonic acid (H_2CO_3), a weak acid that attacks rocks chemically. After weathering, rivers carry off many dissolved materials, including ions (Ca^{+2}, Mg^{+2}, Na^{+1}, K^{+1}, Sr^{+2}, Cd^{+2}, Al^{+4}, and Cl^{-1}), and ion complexes (HCO_3^-, CO_3^{-2}, and $SiO(OH)_2$).

Some of these dissolved ions (Si^{+4}, Ca^{+2}, and CO_3^{-2}) are used by plankton to form their shells (see Figures 3-14 and 3-17). A small fraction ends up in the shells of **benthic foraminifera**, sand-sized animals that live on the seafloor and form calcite ($CaCO_3$) shells from Ca^{+2} and CO_3^{-2} ions in deep waters. Because the shells made of calcite and opal ($SiO_2 \cdot H_2O$) that are preserved in ocean sediments are the end products of chemical weathering on land and ion transport in rivers, they are useful for tracking large-scale fluxes of calcium, silicon, carbon, and oxygen over time.

A wide range of important climatic data is also stored in the isotopes of elements in the calcite shells of planktic organisms and benthic foraminifera. Cases that will be examined in detail in later chapters include isotopes of oxygen, which record changes in the global volume of ice and in local ocean

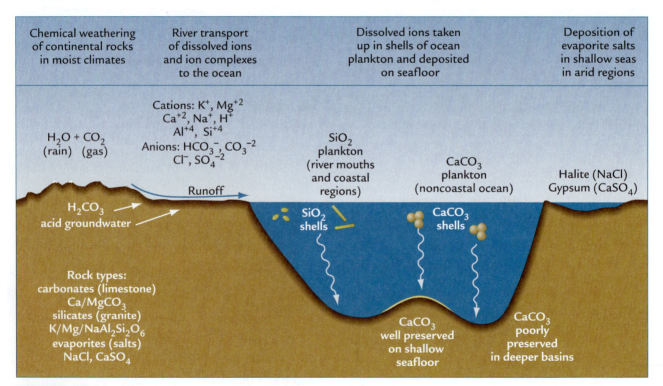

FIGURE 3-17
Chemical weathering, transport, and deposition
Chemical weathering slowly attacks rocks on land and sends dissolved ions into rivers for transport to the ocean. Ocean plankton incorporate some of the dissolved ions in their shells, which fall to the seafloor and form part of the geologic record. Some dissolved ions are also deposited in shallow salty evaporating pools on continental margins where the climate is dry.

temperatures (Chapters 7 and 10); and isotopes of carbon, which trace movements of organic material among reservoirs on the continents, in the air, and in the ocean (Chapter 11).

Additional geochemical proxies gradually become available within the younger part of Earth's long climatic history. For almost the last million years, ice cores contain samples of ancient air that reveal changes in concentrations of the greenhouse gases carbon dioxide (CO_2) and methane over orbital time scales (see Chapter 1). Other important proxies in ice cores include changes in the thickness of snow deposited (related to the temperature and moisture content of the air), in the amount of dust delivered by winds from various continents, and in isotopes of oxygen and hydrogen that measure air temperatures over the ice sheet.

Cave deposits contain records of groundwater derived from atmospheric precipitation. Changes in the chemical composition of this water reflect changes in the original sources of the water vapor, the atmospheric transport path to the site of precipitation, and the groundwater environment (Chapter 11). Sedimentary deposits in lakes record not only changes in pollen, but also climatically driven fluctuations in lake levels (Chapters 13 and 14) and other chemical tracers now under active investigation.

Trees record the amount of cellulose deposited in each annual layer (determined from the width and density of tree rings), providing an index of changes in precipitation during the rainy season in arid regions and changes in summer temperatures in cold regions (Chapter 17). Annual coral bands contain a wide range of chemical information, including ratios of isotopes of oxygen that record changes in temperature and precipitation (Chapter 17).

Climate Models

Scientists who extract records from Earth's climate archives frequently discover new trends that were previously unknown, and they often propose new explanations for these trends. Some of their proposed explanations become hypotheses that are then tested using climate models because models provide an objective way to try to quantify ideas.

In this section, we examine two kinds of numerical (computer) models used by climate scientists. **Physical climate models** emphasize the physical operation of the climate system, particularly the circulation of the atmosphere and ocean but also interactions with global vegetation (biology) and with atmospheric trace gases (chemistry). **Geochemical models** track the movement of distinctive chemical tracers through the climate system.

3-6 Physical Climate Models

Most physical models are constructed to simulate the operation of the climate system as it exists today (recall Chapter 2). The simulation of modern climate is called the **control case**. Models must be capable of simulating modern climate reasonably well in order to be trusted for exploring past climates.

Simulations of past climates occur in a three-step process (Figure 3-18). The first step is to choose the experiment to be run by specifying the input to the model. One or more aspects of the model's representation of the modern world is altered from its present form to reflect changes known to have occurred in the past. For example, the level of CO_2 in the model atmosphere might be increased or decreased, the height of

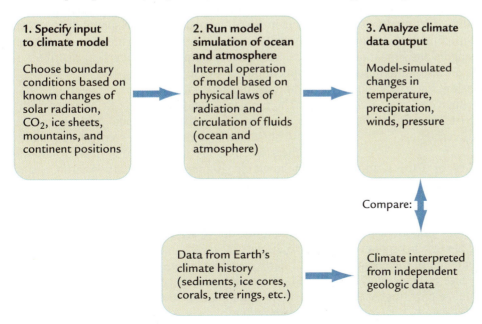

1. Specify input to climate model

Choose boundary conditions based on known changes of solar radiation, CO_2, ice sheets, mountains, and continent positions

2. Run model simulation of ocean and atmosphere

Internal operation of model based on physical laws of radiation and circulation of fluids (ocean and atmosphere)

3. Analyze climate data output

Model-simulated changes in temperature, precipitation, winds, pressure

Compare:

Data from Earth's climate history (sediments, ice cores, corals, tree rings, etc.)

Climate interpreted from independent geologic data

FIGURE 3-18
Data-model comparisons

Models of Earth's climate are constructed to simulate present circulation. Then changes based on Earth's history (different CO_2 levels, ice sheet sizes, or mountain elevations) are inserted into the model, and simulations of past climates are run. Finally, the climate output is compared with independent geologic data.

mountains raised or lowered, ice sheets removed or added, or the position of continents moved around. The features that are altered to test hypotheses of climate change are called **boundary conditions**.

The second step is the actual operation of the model. Physical laws that drive the flow of heat energy through Earth's climate system are embedded in the internal workings of the model. When an experiment is run, these laws come into play during the **climate simulation**.

The third step is to analyze the **climate data output** that emerges from the experiment. The climatic data produced in a simulation can then be used to evaluate the hypotheses being tested. For example, does a specific change in boundary condition cited in a hypothesis (changes in atmospheric CO_2 level, mountain elevation, or continental position) affect climate in the way the hypothesis proposed?

Often climate data output can be tested against independent geologic data that played no part in the experimental design (see Figure 3-18). For example, if a model run simulates stronger winds in a specific region for a particular interval of geologic time, scientists can sample sediment cores from that area to check whether or not larger particles of windblown dust were deposited in the locations indicated by the simulation.

Mismatches between geologic data and the climate data output from physical circulation models may imply any of several possible problems: perhaps key boundary conditions were specified incorrectly or were omitted from the experiment; maybe the model does not adequately simulate some part of the climate system; or else the geologic data used for comparison to the model output may have been misinterpreted. Despite this range of possible problems, the main cause of data-model mismatches is often obvious enough to lead to useful refinements in boundary conditions, in data interpretation, or in model construction. The science of reconstructing past climates moves ahead best when the strengths and limitations of both the data and the models are constantly tested against each other. The review below starts with models of atmospheric circulation, then looks at ocean models, and finally briefly reviews physical models that simulate changes in ice and vegetation.

ATMOSPHERIC MODELS Models of Earth's atmosphere vary widely in complexity. Simpler models are less expensive to run and can simulate the evolution of climate over long intervals of time (thousands of years), but they lack or oversimplify important parts of the climate system. Complex models incorporate a more complete physical representation of the climate system, but they do so at the cost of being slower, more expensive, and able to simulate only brief snapshots of climate spanning a few years.

One-dimensional "column" models are the simplest kind of physical model of the atmosphere. They simulate a single vertical column of air that represents the average structure of the atmosphere of the entire planet. This air column is divided into layers that are closely spaced near Earth's surface and more widely spaced at higher elevations. Each layer contains climatically important constituents, such as greenhouse gases. Earth's surface is represented by a global average value that has the combined properties of water, land and ice weighted according to their present fraction of Earth's surface. One-dimensional (1-D) models offer a way of gaining an initial understanding of climatic effects of changes in concentrations of greenhouse gases and of airborne particles called **aerosols**, such as volcanic ash and dust.

Two-dimensional (2-D) models are a step toward a more complete representation of the climate system. One type of 2-D model includes an atmosphere with many vertical layers but adds a second horizontal dimension that represents Earth's physical properties averaged by latitude. Adding a second dimension (even in a simplified average way) makes it possible to use these models to simulate processes that vary from pole to equator because of the presence of snow and ice at higher latitudes. Because 2-D models can simulate long intervals of time quickly and inexpensively, they are used to explore longer-term interactions among the surface ocean, sea ice, and land. They are also used in combination with models of slowly changing ice sheets (see Chapter 10).

Three-dimensional **atmospheric general circulation models (A-GCMs)** provide still more complete numerical representations and simulations of the climate system. These 3-D models incorporate many key features of the real world: the spatial distribution of land, water, and ice; the elevation of mountains and ice sheets; the amount and vertical distribution of greenhouse gases in the atmosphere; and seasonal variations in solar radiation.

The boundary conditions for A-GCM experiments are specified for hundreds of model **grid boxes**, like those shown in Figure 3-19. The vertical boundaries of the grid boxes are laid out along lines of latitude and longitude at (and above) Earth's surface, and the box size shrinks near the poles because lines of longitude converge there. The horizontal boundaries of the grid boxes divide the atmosphere along lines of equal altitude above sea level. Models generally have 10 to 20 vertical layers that are more closely spaced near Earth's surface because the interactions with the land, water, and ice surfaces in the lower atmosphere are more complex than the smoother flow higher in the atmosphere.

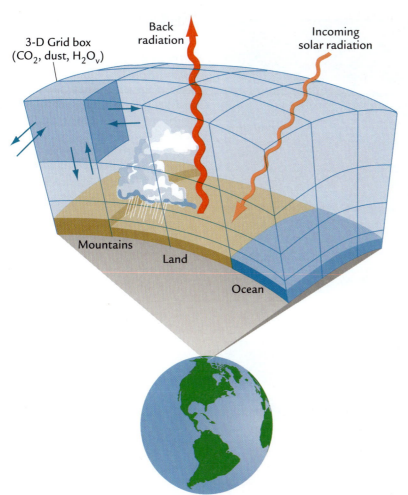

Back radiation

Incoming solar radiation

3-D Grid box (CO_2, dust, H_2O_v)

Mountains

Land

Ocean

FIGURE 3-19
3-D GCMs

General circulation models (GCMs) are full 3-D representations of Earth's surface and atmosphere, represented by individual grid boxes. Representations of Earth's surface within each grid box are entirely land, ocean, or ice, depending on which is most abundant in that area. (ADAPTED FROM W. F. RUDDIMAN AND J. E. KUTZBACH, "PLATEAU UPLIFT AND CLIMATE CHANGE," *SCIENTIFIC AMERICAN* 264 [1991]: 66–75.)

The operation of A-GCMs incorporates the physical laws and equations that govern the circulation of Earth's atmosphere: the fluid motion of air; the conservation of mass, energy, and other properties; and gas laws covering the expansion and contraction of air. The individual grid boxes in A-GCMs all interact with their immediate neighbors in three dimensions.

Model runs begin with the atmosphere in a state of rest. After solar heating causes air to begin to move, the model is run long enough for the atmosphere to reach a state of equilibrium (Figure 3-20). Equilibrium occurs when the long-term drift in the simulated climate data reaches minimal levels. The oscillations that remain are analogous to short-term changes in weather over days and weeks.

Running climate experiments on current-generation A-GCMs requires a simulation of at least 20 years of climate. The first 15 years of the simulation—the "spin-up" interval—allow the model to reach a state of equilibrium, or near-equilibrium. The last 5 years of the simulation generate the climate data used as the output of the model. For the control-case simulation of modern climate, the climate data output from the A-GCM is compared with regional instrumental measurements of temperature, precipitation, pressure, and winds in the present climate system averaged over the last several decades (for example, Figure 3-21). Areas of major disagreement between the model

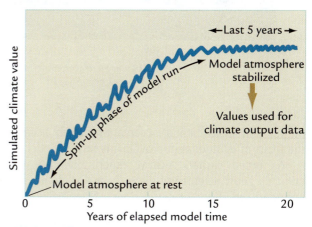

Simulated climate value

←Last 5 years→

Spin-up phase of model run

Model atmosphere stabilized

Values used for climate output data

Model atmosphere at rest

Years of elapsed model time

0   5   10   15   20

FIGURE 3-20
Model equilibrium

Atmospheric GCMs require about 15 years of simulated climate change before they arrive at an equilibrium state. The final 5 years or more of the simulation are then averaged for use as the climate data output.

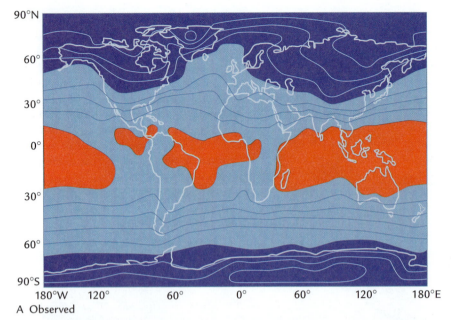

A Observed

FIGURE 3-21
Control-case simulations

GCMs are developed by testing how well they reproduce modern climate (temperature, precipitation, and winds) based on present boundary conditions (CO_2, mountains, and land-sea distribution). This case compares observed January surface temperatures (A) with model-simulated values (B). (ADAPTED FROM J. HANSEN ET AL., "EFFICIENT THREE-DIMENSIONAL GLOBAL MODELS FOR CLIMATE STUDIES: MODELS I AND II," *MONTHLY WEATHER REVIEW* 111 [1983]: 609–62.)

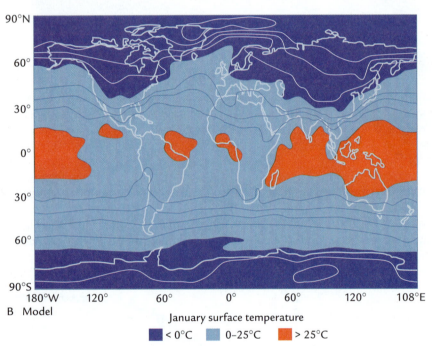

B Model

January surface temperature

< 0°C   0–25°C   > 25°C

output and instrumental observations often become the focus of additional improvements in the model.

As noted earlier, A-GCM experiments on past climates allow scientists to specify major changes in boundary conditions on the basis of geological evidence from Earth's history. In one approach, called a **sensitivity test**, just one boundary condition at a time is altered in relation to the present configuration. When the output of such an experiment is compared with the output from the modern control case, the differences in climate between the two runs isolate and reveal the unique impact caused by the change in that one boundary condition.

In contrast, a climate **reconstruction** requires changing all known boundary conditions at the same time in order to try to simulate the full state of the climate system at some time in the past. This approach is more demanding than a sensitivity test because all of the potentially critical boundary conditions are rarely known well enough to specify as input to the simulation. This method is used mainly to study glacial maximum and deglacial climates of the last 20,000 years, an interval for which numerous records dated by ^{14}C methods exist (see Chapters 13 and 14).

Every year or two, the power of the world's best computers increases by a factor of 10. Over time, this increase in computing power has gradually reduced the horizontal size of the grid boxes used in A-GCMs. Twenty to thirty years ago, typical grid boxes were 8° of latitude by 10° of longitude, or as much as 1,000 kilometers on a side. More recently, A-GCM grid boxes have been reduced to as little as 1° of latitude by longitude, or 100 kilometers on a side. As a result, the models now resolve in considerable detail the coastal outlines of continents (including narrow isthmuses), as well as small seas, some ocean islands, and even a few of the larger lakes.

The shrinking size of grid boxes has also improved the way elevation is represented in A-GCMs. Although low-resolution models captured the underlying bulges of broad high plateaus and ice sheets, they smoothed high but narrow mountain ranges such as the Andes into lower-elevation blobs. Higher-resolution models increasingly distinguish the extent and height of these narrower features.

Increasing computer power has also allowed modelers to include more aspects of the climate system in recent A-GCMs. Features of the climate system such as soil moisture levels or vegetation types that once had to be fixed at modern values (not allowed to interact with the model's atmosphere) are now commonly included as interactive components.

The modeling process is not always a steady one-way march toward success. Initial attempts to include new components in models can be somewhat crude and in some cases can make the resulting climate simulations less realistic than those obtained from previous models that had held those components fixed at modern values. But as the model representations of the new processes become more realistic, so do the simulations. In any case, it is preferable to have as much of the climate system as possible represented interactively in the models, rather than artificially held constant at modern values.

Ocean general circulation models (O-GCMs) are at a somewhat more primitive stage of development than atmospheric GCMs (A-GCMs). One reason is that climate researchers know much less about the modern circulation of the oceans, especially critical processes such as the brief but intense episodes when deep water forms at high latitudes. As a result, scientists do not have as well defined a modern target for ocean models to reproduce.

Three-dimensional O-GCMs are similar to A-GCMs (Figure 3-22). The lower boundary of the boxes is the seafloor, broken into flat stair steps between different regions. The upper boundary of the ocean model is the air-sea boundary. The horizontal grid boxes that subdivide the ocean typically cover 3° to 4° of latitude and longitude. The dozen or so vertical layers in the ocean are more closely spaced near the sea surface, where the flow is faster and complex interactions with the atmosphere occur, than at greater depth, where the ocean flow is slower and more isolated. Typical climate data output from O-GCM experiments includes ocean temperature, salinity, and sea-ice extent.

Like atmospheric models, most ocean GCMs have limitations imposed by the their grid box size. They cannot capture the shape of very small openings, such as the modern mouth of the Mediterranean Sea at the Strait of Gibraltar. These narrow openings can be important in the large-scale circulation of the ocean and important to the success of some ocean-model simulations. Most ocean models also cannot yet resolve details of flow in narrow swift currents such as the Gulf Stream.

Models that include the full structure of the ocean are not directly coupled to atmospheric models. The problem is that air and water respond to climate changes at different rates that impose different computational demands. Ocean models can ignore

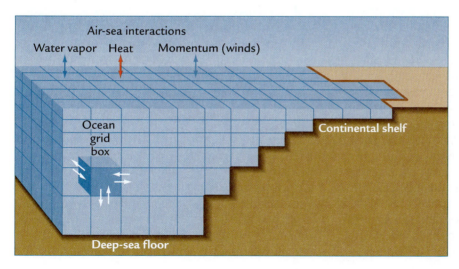

FIGURE 3-22
Ocean GCMs

Ocean models use 3-D grid boxes that represent the basic shapes of ocean basins. Exchanges of water, heat, and momentum between the ocean and the atmosphere occur at the sea surface.

interactions that occur on a daily cycle because these short-term changes have negligible effects on most ocean circulation. As a result, O-GCMs generally only need to calculate changes over time steps separated by a month or more. In contrast, daily changes are critical to models of the fast-responding atmosphere. Therefore, A-GCMs need to calculate changes in time steps separated by just a few hours, and the cost of simulating the same amount of "model time" is more expensive.

This basic incompatibility between the two kinds of models can be overcome by an approach known as *asynchronous coupling*. This procedure involves an ongoing series of alternating runs, first using the atmosphere to drive the ocean, then using the ocean to drive the atmosphere, and then back again. Because the ocean and atmosphere continuously exchange heat, water and water vapor, and wind-driven momentum, switching back and forth between the ocean and atmosphere models keeps the two systems that are being modeled from getting too far out of touch with each other. With the atmospheric model run only at selected intervals, the computer does not have to make short-term calculations of the atmospheric circulation through the entire simulation, and the overall simulation can progress much faster. Recent models have coupled the ocean more directly to a simplified version of the circulation of the atmosphere. Coupled models are called A/O-GCMs.

ICE SHEET MODELS Continent-sized ice sheets slowly grow and shrink over thousands to tens of thousands of years. A-GCMs can simulate the instantaneous effects that these high, broad, reflective masses of ice have on the rest of the climate system, including the circulation of the nearby atmosphere and ocean. The output from an A-GCM run spanning a few years of simulated time can also be examined to see if an ice sheet accumulated or lost mass during the brief simulation. The answer tells modelers whether the ice in various regions would have slowly melted or grown or remained at constant size under the climate conditions specified.

A-GCM's can only reproduce short-lived snapshots of the circulation of the atmosphere and cannot simulate the slow evolution of ice sheets over long intervals of time. To learn about this longer-term response, climate scientists create physical models of the ice sheets. One simplified type of ice-sheet model has two dimensions, one vertical and the other showing average variations with latitude but omitting any representation of longitude. These 2-D ice sheet models have been used to simulate the growth and decay of ice sheets in the Northern Hemisphere over tens of thousands of years in response to changes in solar radiation caused by changes in Earth's orbit.

The models simulate features such as changes in ice accumulation and melting with ice elevation, flow of layers within the ice, and depression of underlying bedrock by the weight of the ice. The 2-D ice sheet models can also be linked to 2-D atmospheric circulation models to simulate interactions among the ice sheets, atmosphere, and land surface. Some ice sheet models are three-dimensional, with the ice accumulating on a specified land surface (such as Antarctica) divided into grid boxes 50 to 100 kilometers on a side.

Hybrid models have also been developed to simulate longer-term climatic responses, particularly those of the ice sheets. These models are called EMICs (Earth system Models of Intermediate Complexity). They sacrifice certain computationally demanding aspects from A-GCMs (such as higher grid box resolution) in order to achieve much longer simulations spanning hundreds to thousands or tens of thousands of years. Most EMICs include submodels for the atmosphere, ocean, sea ice, land surfaces, and the biosphere (vegetation). EMICs can also attempt to simulate features like the growth or melting of ice sheets.

VEGETATION MODELS Vegetation is an active component within the climate system, and the representation of vegetation in climate models has progressed through several stages. Early A-GCMs either ignored vegetation entirely or held to a representation of modern vegetation that did not interact with the changes in climate simulated by the model.

Recent models incorporate vegetation in an interactive way. One approach works in two steps. First, climate data (mainly changes in temperature and precipitation) derived as output from an A-GCM or A/O-GCM experiment are used as input to a vegetation model that simulates the resulting changes in vegetation. Then the simulated changes in vegetation are used as input to another A-GCM or A/O-GCM experiment that simulates the additional climatic feedback effects caused by the *changes* in vegetation (primarily increases or decreases in recycling of water vapor and in reflectivity of Earth's surface). Another approach is to embed a vegetation submodel within the main model more directly.

3-7 Geochemical Models

Geochemical models are used to follow the movements of Earth's chemical materials (called **geochemical tracers**) through the climate system. Unlike physical circulation models, most geochemical models do not reproduce the physical processes that govern the flow of air and water. Instead, the models focus on the sources, rates of transfer, and ultimate depositional fate of two major sedimentary components: those particles that result from physical weathering (wind, water, and ice), and the dissolved ions produced by chemical weathering (dissolution or hydrolysis). Movements of these tracers can be evaluated as long as they are not created or destroyed by radioactive decay along the way. Geochemical models can also trace exchanges of biogeochemical materials such as carbon or oxygen isotopes that cycle back and forth among the atmosphere, ocean, ice, and vegetation.

ONE-WAY TRANSFER MODELS The most basic kind of geochemical model tracks transfers of material from their sources to their ultimate sites of deposition, such as debris eroded from the land and deposited in ocean sediments (Figure 3-23). If this deposited material has distinctive geochemical characteristics, it can be analyzed and its abundance can be quantified in terms of a flux rate, defined as its rate of burial in that sedimentary archive. For example, scientists can quantify the rate of influx of coarse ice-rafted debris from continental sources to high-latitude North Atlantic Ocean sediments by separating out all sediment that is sand-sized or larger and dissolving the shells of fossil plankton to concentrate the mineral grains. This analysis quantifies a process directly related to climate—changes in the delivery of continental minerals by icebergs.

This analysis can be carried one step further by using microscopes to count different types of ice-rafted grains such as volcanic debris, quartz, and limestone. The composition of some of these grains

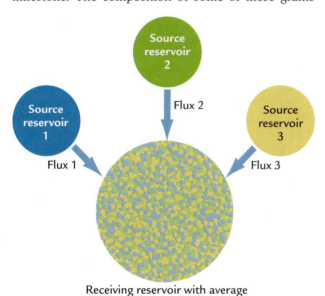

Receiving reservoir with average tagged value of three inputs

FIGURE 3-23

One-way transfers

Geologists and geochemists often need to distinguish the separate contributions of several sources (usually linked to weathering of continental rocks) to a single depositional archive (such as ocean sediments).

provides a general idea of their source regions: for example, in the North Atlantic, volcanic debris from Iceland and quartz and limestone grains from Europe or North America. Further subdivisions can be made by geochemical analyses of the individual grains for their isotopic composition or other distinctive characteristics. This level of analysis can help climate scientists determine which specific region within those continents was the source of particular grains.

A more complicated situation arises if the material examined is fine-grained and derived from multiple sources. For example, fine silt and clay deposited in the middle latitudes of the North Atlantic Ocean could have been ice-rafted from North America or Europe, blown in from North Africa by dust storms, or moved long distances within the ocean by deep currents. Although it is easy to measure the total accumulation rate of fine sediment per unit of time, it is not practical to try to separate out the origin of individual particles.

When physical subdivision of fine material is impossible, chemical analyses offer an alternative, especially if each source of fine sediment is marked with a distinctive chemical value. One widely used marker of this kind is the ratio of isotopes of a specific element, which may differ from source to source. These different inputs combine to determine the average value of the fine-grained sediment (see Figure 3-23). The goal of this kind of analysis is to understand how the individual fluxes combine to create the average value observed.

CHEMICAL RESERVOIRS A different modeling approach is used for geochemical tracers that are transported in dissolved form and move among different **reservoirs**, including the atmosphere, ocean, ice, vegetation, and sediments. The models that trace these movements among reservoirs are called **mass balance models**. The ocean is the most important reservoir because it receives almost all of the erosional products from the continents, it interacts with all of the other reservoirs, and it eventually deposits the tracers in well-preserved sedimentary archives.

The ocean reservoir is in a sense analogous to a bathtub (Figure 3-24). It gradually receives the inputs of geochemical tracers in the same way that water slowly drips from a faucet into a large tub, and it loses geochemical-tracer outputs like water slowly leaking through a partially open drain. The tracer stays in the ocean for a specific amount of time, the same way water does in a drippy, leaky tub.

If the flux rates of a tracer into and out of a particular reservoir (the ocean) are equal, the system is said to be at steady state, with no net gain or loss of the tracer in the reservoir. By analogy, if the drip from the faucet and the leak down the drain are perfectly

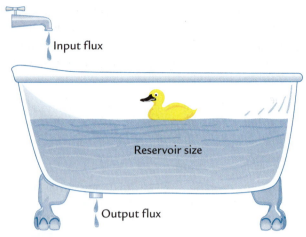

FIGURE 3-24
Geochemical reservoirs and fluxes
Geochemical reservoirs are like bathtubs with the faucet and drain both left partly open. The faucet delivers the input flux, the drain takes away the output flux, and the balance between the input and output determines the water level in the tub (reservoir). At steady state, input and output are in balance, and the water level in the tub remains constant.

balanced, the water level in the tub will stay the same, even though new water continually enters the tub and older water leaves.

The **residence time** is the average time it takes for a geochemical tracer to pass through a reservoir. In the tub analogy, the residence time is the time the average molecule of water takes to pass from the faucet to the drain. For a reservoir at steady state (a tub with an unchanging water level), the residence time is:

$$\text{Residence time} = \frac{\text{Reservoir size}}{\text{Flux rate in (or out)}}$$

RESERVOIR-EXCHANGE MODELS The methods discussed to this point have been based on one-way mass transfers in which geochemical tracers leave the interactive climate system by being buried in sea-floor sediments and isolated out of touch with other reservoirs for millions of years. Another important exchange is the movement of a geochemical tracer back and forth between two (or more) reservoirs (Figure 3-25). In this case, most of the tracer never comes permanently to rest in either reservoir. Instead, the process that interests scientists is the amount of movement between the reservoirs through time. As before, each tracer is naturally tagged with a distinctive value, but in this case it moves back and forth between a larger reservoir (usually the ocean) and a smaller one (often ice sheets or vegetation). The history of exchanges is usually detected in the sediment record recovered from the larger reservoir (the

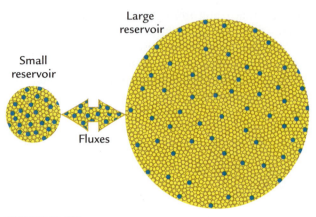

FIGURE 3-25
Reservoir exchange models
Some geochemical models are designed to track reversible exchanges of important components such as water and carbon as they cycle between smaller reservoirs such as ice sheets or vegetation and the larger ocean reservoir.

ocean), but the goal is to monitor changes in size of the smaller reservoirs (the volume of ice or the amount of vegetation).

One example is the transfer of water between the ocean and ice sheets on orbital time scales (discussed in Chapters 10, 12, and 13). Exchanges of water between the relatively small reservoir stored in ice sheets on land and the much larger reservoir left behind in the ocean can be tracked by using the fact that the isotopic composition of oxygen in the H_2O molecules that are deposited in ice sheets differs from the average composition of the molecules left in the ocean. Measurements of the oxygen isotope composition of the ocean in shells of plankton provide a way to estimate past changes in the volume of ice stored on land (see Appendix 1).

Another useful application is the movement of distinctive kinds of carbon among its many reservoirs. Because terrestrial carbon (vegetation) is enriched in one isotope of carbon compared to the average in the ocean, net transfers of terrestrial carbon from land to sea (or vice versa) can be detected by examining changes in the carbon isotope composition of the ocean recorded in the shells of calcite ($CaCO_3$) organisms buried in ocean sediments (see Chapter 11 and Appendix 2).

Key Terms

moraines (p. 57)

loess (p. 57)

historical archives (p. 59)

instrumental records (p. 59)

radiometric dating (p. 60)

parent isotope (p. 60)

daughter isotope (p. 60)

closed system (p. 60)

half-life (p. 60)

radiocarbon dating
(p. 61)

varves (p. 62)

tree rings (p. 62)

coral bands (p. 62)

climate proxies (p. 64)

biotic proxies (p. 65)

geological-geochemical
proxies (p. 65)

macrofossils (p. 66)

plankton (p. 66)

planktic foraminifera
(p. 66)

coccoliths (p. 66)

diatoms (p. 66)

radiolaria (p. 66)

burial fluxes (p. 67)

physical weathering
(p. 67)

ice-rafted debris (p. 67)

eolian sediments
(p. 68)

fluvial sediments (p. 68)

chemical weathering
(p. 68)

dissolution (p. 68)

hydrolysis (p. 68)

benthic foraminifera
(p. 68)

physical climate models
(p. 69)

geochemical models
(p. 69)

control case (p. 69)

boundary conditions
(p. 70)

climate simulation (p. 70)

climate data output
(p. 70)

aerosols (p. 70)

atmospheric general
circulation models
(A-GCMs) (p. 70)

grid boxes (p. 70)

sensitivity test (p. 72)

reconstruction (p. 72)

ocean general
circulation models
(O-GCMs) (p. 73)

geochemical tracers
(p. 74)

reservoirs (p. 75)

mass balance models
(p. 75)

residence time (p. 76)

Review Questions

1. Why does the importance of different climate archives change for different time scales?

2. Why are ocean sediments and ice cores especially important archives of climate?

3. How does the method of dating climate records vary with the type of archive?

4. How does the resolution from sedimentary archives vary with depositional environment?

5. Which two major groups of organisms are most important to climate reconstructions over the past several million years?

6. Describe how the products derived from physical and chemical weathering provide different kinds of information about the climate system.

7. Describe two ways the performance of climate models is evaluated.

8. Why aren't models of the atmosphere and ocean allowed to interact continuously?

9. Describe two features that make the ocean useful in geochemical mass balance models.

Additional Resources

Basic Reading

National Climatic Data Center Web site. "World Data Center for Paleoclimatology—Data Sets Listing." http://www.ngdc.noaa.gov/paleo/datalist.html. Last accessed March 17, 2013.
Contains climate data of all kinds, as well as the locations of sites containing each type of data. Maintained by the World Data Center for Paleoclimatology in Boulder, Colorado.

Advanced Reading

Bradley, R. S. 1998. *Paleoclimatology: Reconstructing Climates of the Quaternary.* International Geophysics Series, vol. 64. San Diego: Harcourt Academic Press.

Cronin, T. M. 2010. *Paleoclimates: Understanding Climate Change Past and Future.* New York: Columbia University Press.

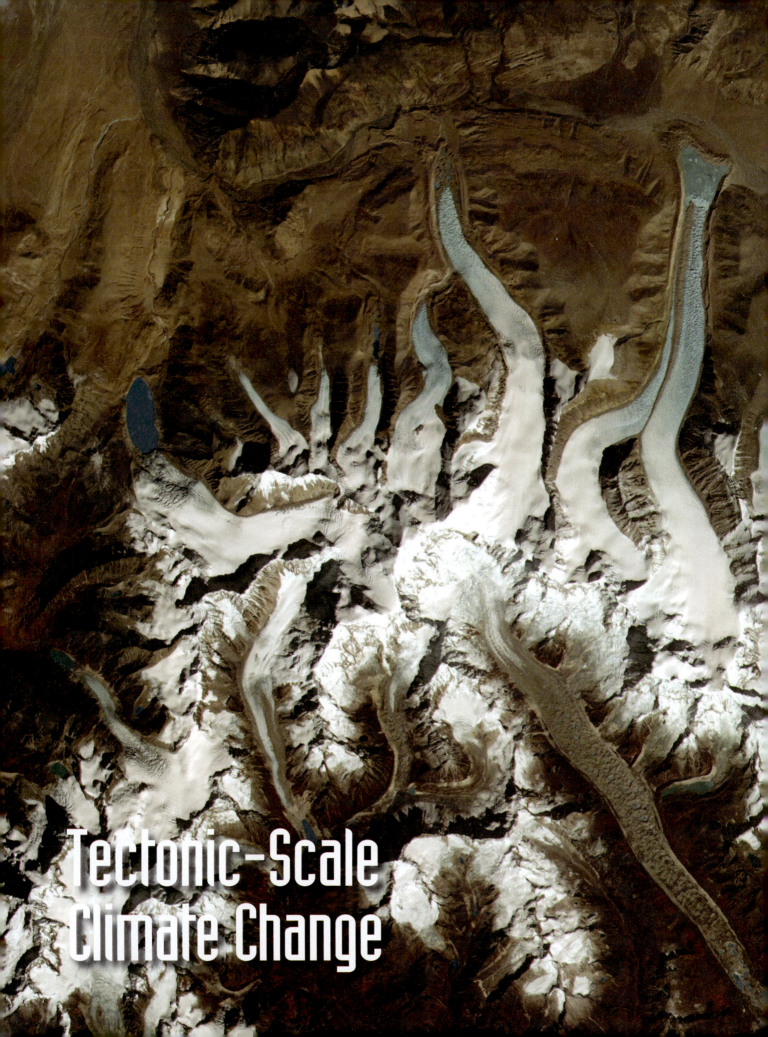

Tectonic-Scale
Climate Change

In this part, we examine the longest time scale of climate change on Earth, the *tectonic scale*. Tectonic processes driven by Earth's internal heat have altered Earth's geography and its climate for billions of years. Prior to the last several hundred million years, the record of climate change is relatively sparse, and the configurations of continents and oceans are generally not known with a high degree of accuracy. Yet we do know that Earth's climate for the most part remained relatively moderate, neither freezing completely (although coming close once) nor heating up enough to boil away its oceans. Earth appears to have had a built-in thermostat that allowed it to avoid those extremes.

With greater knowledge about the locations and vertical movements of the continents toward the present, greater insights into climatic cause-and-effect relationships emerge. We know that climate has varied between times when major ice sheets were present somewhere on Earth (such as the last several million years) and times when no ice sheets were present (about 100 million years ago). These long-term changes, and their causes, are the primary focus of Part II. Evidence of climate change over this long span of Earth's history comes mainly from sediments preserved on the continents and their margins.

We explore these basic questions about Earth's tectonic-scale climate history:

▷ Why has Earth remained habitable throughout its history?

▷ Did Earth ever freeze completely?

▷ What explains changes in Earth's climate over the last several hundred million years?

▷ Why was Earth ice-free even in polar regions 100 Myr ago?

▷ What are the causes and climatic effects of changes in sea level through time?

▷ How did the apocalyptic asteroid impact 65 Myr ago affect climate?

▷ What caused the extremely warm interval near 55 Myr ago?

▷ Why did Earth's climate cool during the last 50 Myr?

part II

Past glaciations and continental positions.

During Earth's 4.55-billion-year history, intervals when large continental ice sheets were present alternated with times when they were not (left). The earliest history of these changes is poorly defined because few ancient records are preserved. The movements of continents in relation to ocean basins are well known only for the last several hundred million years (right). (GLOBES ADAPTED FROM D. MERRITTS ET AL., ENVIRONMENTAL GEOLOGY, © 1997 BY W. H. FREEMAN AND COMPANY.)

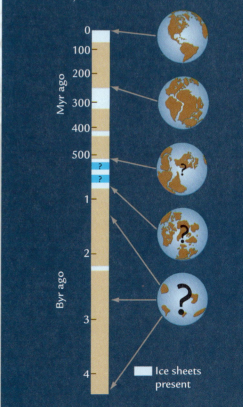

Ice sheets present

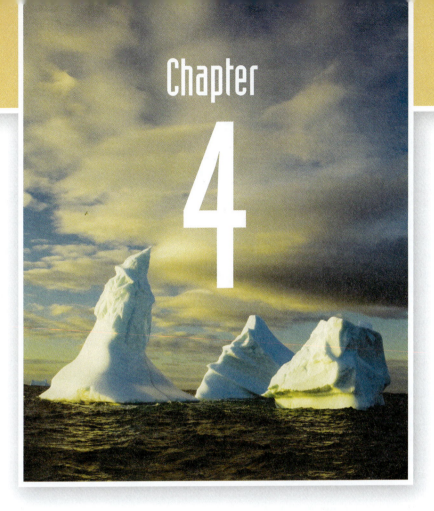

CO₂ and Long-Term Climate

Why is Earth habitable? The answer to this question at first seems obvious: Earth is just the right distance from the Sun for solar heating to keep it comfortable for life. With a mean temperature near 15°C (59–60°F) and relatively limited geographic variation around that average, life can flourish almost everywhere on this planet.

But this answer proves insufficient when we ask why Earth has remained habitable for most of the 4.55 billion years of its existence. Over that immense interval of time, our Sun has slowly increased in strength by 25% to 30%, a trend that should have produced a very large climatic warming. Yet somehow Earth's climate has varied only within relatively narrow limits. Our planet's continuing habitability seems to require some kind of natural thermostat that allows its climate to warm up but not overheat during **greenhouse eras** (times when no ice sheets are present) and to cool off but not freeze solid during **icehouse eras** (times when ice sheets are present). This chapter describes the search for Earth's thermostat.

Greenhouse Worlds

The first clue that a factor other than distance to the Sun is involved in Earth's habitability comes from comparing it to Venus, another "terrestrial" planet with a similar overall chemical composition (Figure 4-1). Venus is a very hot planet with a mean surface temperature of 460°C, and it is located just 72% as far from the Sun as Earth is.

The average amount of solar radiation sent to each planet varies *inversely* with the square of its distance from the Sun $(1/d^2)$. Based on this relationship, Venus receives almost twice (1.93 times) as much solar radiation as Earth does:

$$\frac{Earth}{Venus} \quad \frac{(1)^2}{(0.72)^2} = \frac{1}{0.518} = 1.93$$

At first, this calculation might seem to confirm that climate depends entirely on distance from the Sun: because Venus is closer to the Sun, its surface is hotter. In fact, however, this is not the real answer because most of the Sun's radiation never makes it down to Venus's surface. Its upper atmosphere is shrouded in thick sulfuric acid clouds that reflect 80% of the incoming radiation and allow only 20% to reach the surface. In contrast, clouds on Earth reflect just 26% of the incoming radiation, allowing the other 74% to reach its surface.

This large difference in average albedo (the percentage of incoming radiation reflected back to space) between the atmospheres of the two planets almost exactly reverses the relative amounts of solar energy that actually reach their surfaces. Even though Venus receives almost twice as much incoming solar energy at the top of its atmosphere, its higher albedo reduces the amount that reaches its surface to just over half that received on Earth:

$$1.93 \times \frac{0.20}{0.74} = 0.52$$

With less incoming solar radiation, how can Venus be so much hotter? The answer is that Venus has an atmosphere 90 times as dense as that of Earth, and 96% of its atmosphere is composed of carbon dioxide (CO_2), a greenhouse gas that is very effective in trapping radiation. Some sunlight does penetrate the thick atmosphere and heat the surface, which causes Venus to emit long wave radiation, just as Earth does. But most of the back radiation never leaves the atmosphere of Venus because the CO_2 gas traps it and retains it as internal heat.

In contrast, much less of the energy radiated back from Earth's surface is trapped by water vapor, CO_2, and other greenhouse gases (recall Chapter 2).

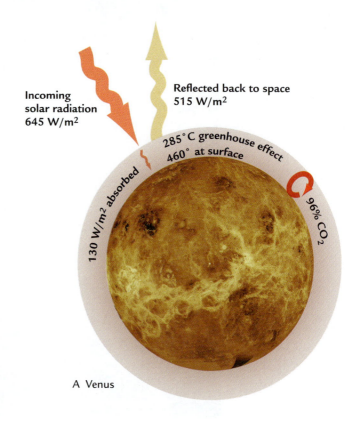

Incoming solar radiation 645 W/m²

Reflected back to space 515 W/m²

130 W/m² absorbed

285°C greenhouse effect 460° at surface

96% CO_2

A Venus

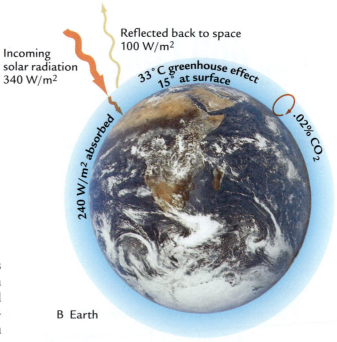

Incoming solar radiation 340 W/m²

Reflected back to space 100 W/m²

240 W/m² absorbed

33°C greenhouse effect 15° at surface

.02% CO_2

B Earth

FIGURE 4-1
Why is Venus hot?

Venus (A) receives almost twice as much solar radiation as Earth (B), but the dense cloud cover on that planet permits less radiation to penetrate to its surface. Yet Venus is much hotter than Earth because its CO_2-enriched atmosphere creates a much stronger greenhouse effect that traps much more heat. (NASA.)

In Summary, the main reason Venus is so hot compared to Earth is not its closer proximity to the Sun, but its far greater concentrations of heat-trapping greenhouse gases.

Because Venus and Earth both formed as rocky planets in the inner part of our solar system, they contain nearly equal amounts of carbon. Yet the two planets store their carbon in very different reservoirs. Most of Earth's carbon is tied up in its rocks (some of it as coal, oil, and natural gas), while relatively little resides in the atmosphere. Combined with water vapor and other greenhouse gases, the net greenhouse heating of Earth's atmosphere is relatively small—about 32°C (although that difference keeps Earth from freezing solid). In marked contrast, almost all the carbon on Venus resides in its atmosphere as CO$_2$ and produces an enormous net greenhouse warming (285°C), without any significant contribution from water vapor.

This comparison shows how vital greenhouse gases can be to the climate of planets. It also highlights the fact that Earth's comfortably small greenhouse effect is an important factor in its present habitability.

▶ The Faint Young Sun Paradox

By studying the evolution of stars in the universe, astronomers have recreated the history of our own Sun over the 4.55 billion year existence of our solar system. Throughout this interval, the Sun's interior has been the site of an ongoing nuclear reaction that fuses nuclei of hydrogen (H) together to form helium (He). Models developed by astronomers indicate that this process has caused our Sun to expand and gradually become brighter. These models indicate that the earliest Sun shone 25% to 30% more faintly than today, and that its luminosity, or brightness, then slowly increased to its current strength.

This insight from the field of astronomy creates an intriguing problem for climate scientists. A relatively small decrease in our Sun's present strength would cause all the water on Earth to freeze, despite the warming effect from greenhouse gases. If all our oceans and lakes were to freeze, their bright snow and ice surfaces would reflect more solar radiation, and they would be difficult to melt. One-dimensional numerical climate models that simulate the mean climate of the entire planet (recall Chapter 2) suggest that the combination of a weak Sun and greenhouse gas levels at their present values would have kept

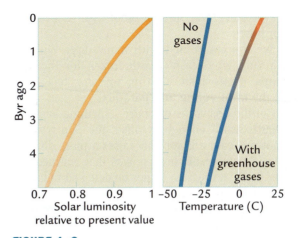

FIGURE 4-2
The faint young Sun paradox
Astrophysical models of the Sun's evolution indicate it was 25% to 30% weaker early in Earth's history (left). Climate model simulations show that the weaker Sun would have resulted in a completely frozen Earth for more than half of its early history if the atmosphere had the same composition as it does today (right). (ADAPTED FROM D. MERRITTS ET AL., *ENVIRONMENTAL GEOLOGY*, ©1997 BY W. H. FREEMAN AND COMPANY.)

Earth completely frozen for the first 3 billion years of its existence (Figure 4-2).

Yet evidence left in Earth's sedimentary deposits shows that Earth was not frozen for its first 3 billion years. Although the first half-billion years of Earth's existence have left no record, evidence of Earth's climatic history gradually becomes more complete after that time and toward the present. Most sedimentary rocks (recall Chapter 3) are made up of particles that were eroded from other rocks, reworked by running water, and transported to a site of deposition. The prevalence of water-deposited sedimentary rocks throughout Earth's early history is direct evidence that Earth was not frozen.

The first evidence of ice-deposited sediments occurs in rocks dated to about 2.3 billion years ago, but these deposits could have been the result of ice sheets in polar regions, similar to those that occur today, and not evidence of a completely frozen planet. As summarized at the end of this chapter, a debate is currently under way as to how close Earth's climate came to a nearly frozen condition during several intervals between 750 and 580 million years ago, but for most of Earth's history the sedimentary evidence leaves no doubt that most of the water on Earth has remained unfrozen.

This conclusion is supported by the continued presence of life on Earth. Primitive life-forms date back to at least 3.5 billion years ago, and their presence on Earth is incompatible with a completely frozen planet at that time. The succession of ever more complex life-forms that have continuously occupied

Earth ever since add further proof against extreme cold (or heat).

So we are confronted with a mystery: With so weak a Sun, why wasn't Earth frozen for the first two-thirds of its history? This mystery has been named the **faint young Sun paradox.**

Part of the answer is obvious: something kept the early Earth warm enough to offset the Sun's weakness, but this easy answer only raises a more difficult problem. Whatever the process was that warmed the younger Earth, it must no longer be doing so today, or at least not as actively as it once did. If this same warming process had continued working at full strength right through the entire 4.55 billion years of Earth's history, it would have combined with the steadily increasing warmth from the strengthening Sun (Figure 4-2) to overheat Earth and make it uninhabitable. Yet that has not happened: somehow Earth has stayed in a moderate temperature range throughout the entire interval when the Sun's brightness was increasing.

The most likely solution to the faint young Sun paradox requires a process that works in the same way a **thermostat** works in a house. When outside temperatures fall in winter, the thermostat detects the cooling and turns on a heat source that keeps the house warm. When temperatures become too hot outside in summer, the thermostat activates a cooling source that keeps the house cool. The thermostat moderates extreme swings in temperature. Such a thermostat must have been at work through Earth's history, warming its climate very early on when it would otherwise have frozen under a weak Sun, and later on cutting back on the extra heat as the Sun strengthened.

One possibility is that greenhouse gases have been part of the mechanism that acts as Earth's thermostat. Our present concentrations of greenhouse gases do not provide enough warming to have counteracted the effects of a weak early Sun, but if these gases were more abundant earlier in Earth's history and subsequently decreased in abundance, that would provide a thermostat-like control.

Our earlier comparison of Earth and Venus lends credibility to this explanation. Earth's carbon is mainly stored in its rocks, while carbon on Venus is mostly in its atmosphere. If carbon can reside in different reservoirs on different planets, why couldn't it move among reservoirs during the history of a single planet? More specifically, could the early Earth have held more carbon in its atmosphere (like Venus), and then transferred it to its rocks later in its history?

▶ Carbon Exchanges between Rocks and the Atmosphere

To understand how carbon may have shifted among Earth's reservoirs, we can examine the present carbon cycle (Figure 4-3A). Small amounts of carbon exist in the atmosphere, in the surface ocean, and in vegetation, along with a slightly larger reservoir in soils, a much larger reservoir in the deep ocean, and an immensely large reservoir in rocks and sediments. Carbon storage in these reservoirs is measured in billions of tons ("gigatons").

The rates of carbon exchange among these reservoirs vary widely (Figure 4-3B). In general, an inverse relationship exists between the size of a reservoir and the rate at which it exchanges carbon. The smaller reservoirs (the atmosphere, surface ocean, and vegetation) all exchange carbon relatively quickly, while the huge rock reservoir gains and loses carbon much more slowly. Because of the combined effects of small reservoir sizes and rapid exchange rates, carbon can cycle through the surface reservoirs in a few years or decades, but takes much longer to move through the larger and deeper reservoirs.

Because all of these reservoirs exchange carbon with the atmosphere, each has the potential to alter atmospheric CO_2 concentrations and affect Earth's climate. The relative importance of each carbon reservoir in Earth's climate history varies according to the time scale under consideration. In this chapter, we are concerned with very gradual climate changes over tens to hundreds of millions of years. Over these very long (tectonic) time scales, the slow carbon exchanges between the rocks and the surface reservoirs are the source of changes in the amount of CO_2 in the atmosphere.

4-1 Volcanic Input of Carbon from Rocks to the Atmosphere

Carbon cycles constantly but slowly between Earth's interior and its surface. It moves from the deep rock reservoir to the surface mainly as CO_2 gas produced during volcanic eruptions and in the activity of hot springs (Figure 4-4).

The present rate of natural carbon input to the atmosphere from the rock reservoir is estimated at approximately 0.15 gigatons of carbon per year (see Figure 4-3B). This value is uncertain by a factor of at least 2 because volcanic explosions are irregular in time and because the amount of CO_2 released varies with each eruption. As we will see later, this natural rate of carbon input is roughly balanced by a similar rate of natural removal. This balance between natural input and removal rates helped to keep the size of the "natural" (preindustrial) atmospheric carbon reservoir at ~600 gigatons.

But how likely is it that this balance could have persisted over immensely long intervals of geologic time? We can evaluate this question by a simple thought experiment. Using the reservoir concept

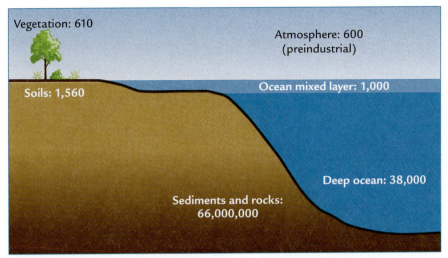

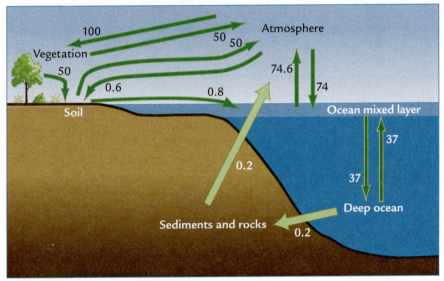

A Major carbon reservoirs (gigatons)

B Carbon exchange rates (gigatons/year)

FIGURE 4-3
Carbon exchanges with Earth's rocks

The largest reservoir of carbon on Earth lies in its rocks, not in its atmosphere, vegetation, or ocean (A). All of Earth's reservoirs exchange carbon (B). Over intervals of millions of years, slow exchanges among the rock and surface reservoirs can cause large changes in atmospheric CO_2 levels. (ADAPTED FROM J. HOREL AND J. GEISLER, *GLOBAL ENVIRONMENTAL CHANGE* [NEW YORK: JOHN WILEY, 1997], AND FROM NATIONAL RESEARCH COUNCIL BOARD ON ATMOSPHERIC SCIENCES AND CLIMATE, *CHANGING CLIMATE*, REPORT OF THE CARBON DIOXIDE ASSESSMENT COMMITTEE [WASHINGTON, D.C.: NATIONAL ACADEMY PRESS, 1993].)

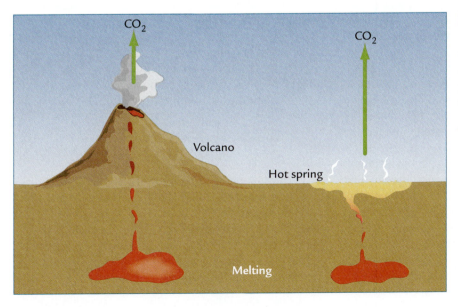

FIGURE 4-4
Input of CO_2 from volcanoes

CO_2 enters Earth's atmosphere from deep in its interior through release of gases in volcanoes and at hot springs such as those at Yellowstone National Park in Wyoming.

introduced in Chapter 3, we can calculate how long it would take for the atmospheric CO_2 level to fall to zero if all volcanic release of carbon from Earth's interior to the atmosphere abruptly ceased, but carbon continued to be removed from the atmosphere at the same rate as before.

The answer, 4,000 years, is derived by dividing the preindustrial atmospheric carbon reservoir of 600 gigatons by an annual rate of carbon removal of 0.15 gigaton. This number, although far beyond the length of a human lifetime, is remarkably brief in the context of the several billion years of Earth's existence. It tells us that changes in volcanic input persisting over that relatively "small" span of time could have had a drastic effect on the CO_2 content of our atmosphere.

In actuality, the atmosphere is not really this vulnerable because rapid exchanges of carbon occur continuously between the atmosphere and the other surface or near-surface carbon reservoirs. As a result, these reservoirs in effect act as a single unit over long intervals, and their rapid exchanges have the effect of slowing and reducing the impact of the loss of carbon from Earth's interior.

In this hypothetical example of a sudden cessation of volcanic CO_2 input, the actual scenario might initially develop more like this: As CO_2 levels in the atmosphere begin to fall, the other surface reservoirs (vegetation, surface ocean, soils) begin to surrender some of their carbon to the atmosphere, slowing its rate of loss. The combined size of the near-surface reservoirs (atmosphere, vegetation, soil, and surface ocean) is 3,700 gigatons, more than six times larger than the atmospheric reservoir alone. As a result, it would take roughly 24,700 years after volcanism ceased for these reservoirs to lose all their carbon (3,700 gigatons divided by 0.15 gigaton/yr).

Over longer time spans of centuries, the large deep-ocean carbon reservoir would also play a growing role. If all of the surface reservoirs were losing carbon, the deep ocean would deliver some of its ample supply to the surface ocean, from which it would be redistributed among the atmosphere, the vegetation, and the soil. If we take into account the deep ocean, the total size of these combined reservoirs amounts to 41,700 gigatons. In this case, it would take 278,000 years for a total shutdown of volcanic carbon input to deplete them completely (41,700 gigatons divided by 0.15 gigaton/yr).

At this point, it might seem as if we have shown that Earth's surface reservoirs, and particularly the atmosphere, are actually *not* particularly vulnerable to the slow changes in the amount of carbon coming out of (or going into) its large rock reservoirs. But this conclusion would be incorrect. Compared to Earth's unimaginably old age of 4.55 billion years (4,550,000,000 years!), even the long time span of

278,000 years is still just the blink of an eye. Because Earth is so incredibly old, plenty of time is still available for the slow carbon exchanges with Earth's deeper and larger rock reservoirs to alter the amount of carbon in the surface reservoirs and the atmosphere.

With Earth's great antiquity taken into account, it is still amazing that over this immense span of time its volcanoes have somehow managed to keep delivering just enough carbon from Earth's interior to keep the atmosphere from running out of CO_2 and freezing the planet, but not so much as to overheat and boil it. Even more amazing is the fact that this balancing act had to be maintained as the faint young Sun was slowly increasing in strength. A crude analogy for this long-term balancing act would be a tightrope walker who has to stay balanced on a narrow wire that slopes uphill over a very long distance.

We noted earlier that this balancing act appears to require some kind of natural thermostat to moderate Earth's temperature. Could the rate of volcanic input of CO_2 from Earth's interior have varied in such a way as to function as the thermostat? The answer is no. The basic operating principle of a thermostat is that it first *reacts* to external changes and then *acts* to moderate their effects: a thermostat detects the chill of a cold night and sends a signal that turns on the heat.

Volcanic processes do not operate in this way. The volcanic activity that has occurred on Earth throughout its history has been driven mainly by heat sources located deep in its interior and generally well removed from contact with (and reactions to) the climate system. Climatically driven changes in surface temperature penetrate only the outermost tens to hundreds of meters of the land or seafloor, and their amplitude is greatly reduced compared to those at the surface. As a result, climate changes confined to Earth's surface have no physical mechanism by which they can alter deep-seated processes acting across most of Earth's interior. Without such a link, volcanic processes cannot act as a thermostat controlling CO_2 delivery to the surface.

Earth's thermostat must lie elsewhere. It must be found in a process that responds directly to the climate conditions at Earth's surface.

4-2 Removal of CO_2 from the Atmosphere by Chemical Weathering

To avoid a long-term buildup of CO_2 levels, the ongoing CO_2 input to the atmosphere by volcanoes must be countered by CO_2 removal. The major long-term process of CO_2 removal is tied to chemical weathering of continental rocks (see Chapter 3). Two major types of chemical weathering occur on continents: *hydrolysis* and *dissolution*.

Hydrolysis is the main mechanism for removing CO_2 from the atmosphere. The three key ingredients in the process of hydrolysis are the minerals that make up typical continental rocks, water derived from rain, and CO_2 derived from the atmosphere (Figure 4-5).

Most of the continental crust consists of rocks such as granite, made of **silicate minerals** like quartz and feldspar. Silicate minerals typically are made up of positively charged cations (Na^{+1}, K^{+1}, Fe^{+2}, Mg^{+2}, Al^{+3}, and Ca^{+2}) that are chemically bonded to negatively charged SiO_4 (silicate) structures. These silicate minerals are slowly attacked by groundwater containing carbonic acid (H_2CO_3) formed by combining atmospheric CO_2 with rainwater.

Part of the weathered rock is chemically converted to clay minerals (compounds of Si, Al, O, and H) in soils. Chemical weathering also produces several types of dissolved ions and ion complexes, including HCO_3^{-1}, CO_3^{-2}, H_2SiO_4, and H^{+1}. These ions are carried by rivers to the ocean, where most are incorporated in the shells of planktic organisms (see Figure 4-5).

Dozens of chemical equations describe the process of chemical weathering—in fact, at least one equation for each of the many types of silicate minerals found on continents. The part of these processes that is most important to the carbon system can be represented by these reactions:

$$\underset{\text{Rain}}{H_2O} \quad + \quad \underset{\text{From atmosphere}}{CO_2}$$
$$\searrow \qquad \swarrow$$
$$\underset{\substack{\text{Silicate rock}\\(\text{continents})}}{CaSiO_3} + \underset{\substack{\text{Carbonic acid}\\(\text{soil})}}{H_2CO_3} \rightarrow \underset{\text{Shells of organisms}}{CaCO_3} + SiO_2 + H_2O$$

For simplicity, the many kinds of continental rocks and minerals are represented in this equation by one silicate mineral, $CaSiO_3$ (wollastonite). Carbon dioxide (CO_2) is removed from the atmosphere, incorporated in groundwater to form carbonic acid in soils, used in the chemical weathering of $CaSiO_3$, and deposited in the $CaCO_3$ shells of marine organisms. This reaction is a shorthand summary of the way chemical weathering removes CO_2 from the atmosphere and buries it in ocean sediments. This process acts slowly but persistently over long intervals of geologic time and accounts for 80% of the 0.15 gigatons of carbon buried each year in ocean sediments.

It is important to distinguish weathering of silicates by hydrolysis from the action of dissolution, the other kind of weathering. Dissolution is the process that eats away at limestone bedrock and in some areas forms caves. Again, rainwater and CO_2 combine in soils to form carbonic acid (H_2CO_3) and attack limestone bedrock ($CaCO_3$), and the dissolved ions created by dissolution again flow to the ocean in rivers. Dissolution can also be summarized by these simple reactions:

$$\underset{\text{Rain}}{H_2O} \quad + \quad \underset{\text{From atmosphere}}{CO_2}$$
$$\searrow \qquad \swarrow$$
$$\underset{\substack{\text{Limestone}\\\text{rock}}}{CaCO_3} + \underset{\text{In soils}}{H_2CO_3} \rightarrow \underset{\substack{\text{Shells of}\\\text{organisms}}}{CaCO_3} + H_2O + \underset{\substack{\text{Returned to}\\\text{atmosphere}}}{CO_2}$$

Dissolution of limestone proceeds at much faster rates than hydrolysis of silicates. Similar to hydrolysis, dissolution extracts CO_2 from the atmosphere to attack rock. But unlike the weathering of silicate rocks, limestone weathering results in no net removal of CO_2 from the atmosphere. In the relatively short interval of time it takes for the dissolved HCO_3^{-1} and CO_3^{-2} ions to reach the sea and become incorporated in the shells of organisms, all of the CO_2 is returned to the atmosphere.

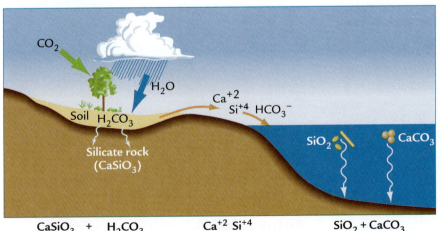

CaSiO₃ + H₂CO₃	Ca⁺² Si⁺⁴ HCO₃⁻	SiO₂ + CaCO₃
Silicate bedrock / Carbonic acid in soils	Ions dissolved in river water	Shells of ocean plankton
Weathering on land	Transport in rivers	Deposition in ocean

FIGURE 4-5
Chemical weathering removes atmospheric CO₂

Chemical weathering of silica-rich rocks on the continents removes CO_2 from the atmosphere, and part of the carbon is later stored in the shells of marine plankton and buried in ocean sediments.

In Summary, slow weathering of granite and other silicate rocks on the continents by hydrolysis is the main way that CO_2 is pulled out of the atmosphere over very long time scales. In the context of Earth's delicate long-term balancing act, the rate of removal of carbon by chemical weathering must have very nearly balanced the rate of carbon input from volcanoes. If these rates had not been very nearly equal, the system would have gotten out of balance and caused drastic changes in CO_2 levels and climate.

The existence of this delicate balance does not imply that either the (volcanic) CO_2 input rate or the (weathering) CO_2 removal rate remained absolutely constant through time. Yet the fact of Earth's long-term habitability requires that the rates of input and output must have always remained fairly closely balanced even though they varied.

How has this near-perfect balance been possible? As we noted earlier, a thermostat is needed to provide such a balance. In our search for Earth's thermostat within its carbon system, we have ruled out volcanic input of CO_2. The other possibility left is chemical weathering. If the rate of chemical weathering is sensitive to climate, it may be able to act as Earth's thermostat.

Climatic Factors that Control Chemical Weathering

Decades of laboratory experiments and many field studies have shown that rates of chemical weathering are influenced by three environmental factors: temperature, precipitation, and vegetation. These factors all act in a mutually reinforcing way to affect the intensity of chemical weathering.

Laboratory experiments have shown that higher temperatures cause faster weathering of individual silicate minerals. This trend is consistent with many temperature-dependent chemical reactions in water and other aqueous solutions. Weathering rates roughly double for each 10°C increase in temperature.

Unfortunately, it is difficult to transfer these laboratory results to studies of the real Earth in all its complexity. So far, laboratory experiments have examined only a few of the many silicate minerals that are common enough in Earth's crust to be important contributors to the overall rate of silicate weathering on a global scale. Natural chemical weathering rates are also difficult to determine in field studies because of the complicating effects from carbonate dissolution. Because dissolution occurs many times faster than hydrolysis, the ions flowing down rivers from actively eroding terrain are usually dominated by those derived from limestone dissolution, which does not control CO_2 levels in Earth's atmosphere, rather

than from hydrolysis of silicates, which does control long-term CO_2 concentrations. Another problem with field studies is that agriculture and industrial activities have disturbed the natural chemistry of most of Earth's rivers, even in many remote regions.

Still, we can apply the laboratory rule of thumb that says that silicate weathering rates double for each 10°C increase in temperature across the roughly 30°C range of mean annual temperatures found on Earth's surface (Figure 4-6A). Based on this relationship, rates of silicate weathering should increase by

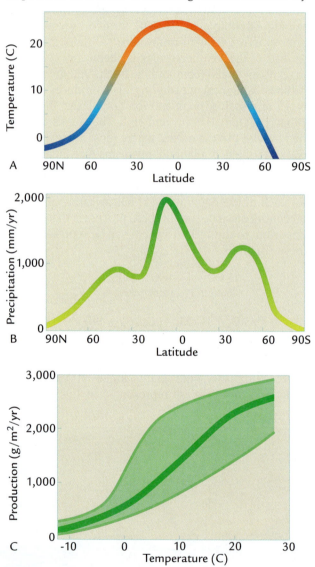

FIGURE 4-6

Climate controls on chemical weathering

Temperature (A) and precipitation (B) both show a general trend from high values in warm (low) latitudes to low values in cold (high) latitudes. The total amount of vegetation produced per year increases with temperature (C), as well as with precipitation. (A AND B: ADAPTED FROM R. G. BARRY AND R. J. CHORLEY, *ATMOSPHERE, WEATHER, AND CLIMATE,* 4TH ED. [NEW YORK: METHUEN, 1982]; C: ADAPTED FROM R. L. SMITH AND T. M. SMITH, *ELEMENTS OF ECOLOGY* [MENLO PARK: ADDISON WESLEY LONGMAN, 1998].)

a factor of at least 8 ($2 \times 2 \times 2$) from the cold polar regions to the hot equatorial latitudes.

The second major control on weathering is precipitation (Figure 4-6B). Increased rainfall boosts the level of groundwater held in soils, and the water combines with CO_2 to form carbonic acid and enhance the weathering process.

Temperature and precipitation are so closely linked in Earth's climate system that it is often difficult to measure their separate contributions to chemical weathering. The heaviest rainfall on Earth occurs in the tropics because warm tropical air holds much more moisture than cool high-latitude air. Polar regions have very little precipitation because the atmosphere holds so little water.

This close relationship breaks down in some regions. For example, lower precipitation in many subtropical regions greatly reduces chemical weathering, even though the relatively warm temperatures in those areas would otherwise favor it. Despite these complications, temperature and precipitation generally act together. A warmer Earth is likely to be a wetter Earth, and both factors tend to act together to intensify chemical weathering.

Vegetation also enhances chemical weathering. Plants extract CO_2 from the atmosphere through the process of photosynthesis, and deliver it to soils, where it combines with groundwater to form carbonic acid. Although H_2CO_3 is a weak acid, it enhances the rate of chemical breakdown of minerals. Scientists estimate that the presence of vegetation on land can increase the rate of chemical weathering by a factor of 2 to 10 compared to the rates typical of land that lacks vegetation.

Vegetation is closely linked to precipitation and temperature (see Chapter 2). Dense rain forests occur in regions with year-round rainfall, open forest or savannas in areas with a short dry season, grasslands and steppes in places with a long dry season, and deserts or semi-deserts in areas with little or no rainfall. Each step in the direction of greater rainfall is a step toward more vegetation and more total carbon biomass stored in vegetation and soils.

In addition, the rate of production of carbon by photosynthesis across the planet is broadly correlated with temperature (Figure 4-6C). Cold ice-covered regions obviously produce little plant matter, and seasonally or permanently frozen (but ice-free) polar regions produce only sparse tundra vegetation. In comparison, production of carbon in warmer mid-latitude and tropical regions is much greater.

▶ Is Chemical Weathering Earth's Thermostat?

Now we have in hand a physical mechanism that could act as Earth's thermostat and moderate long-term

climate: the **chemical weathering thermostat**. The global rate of chemical weathering is analogous to a thermostat because it reacts to (depends on) the average state of Earth's climate and then alters that state by regulating the rate at which CO_2 is removed from the atmosphere.

Consider what would happen if Earth's climate began to warm (Figure 4-7A). Any initial climate change (for any reason) toward a warmer, moister, more heavily vegetated Earth should enhance chemical weathering of silicate minerals, but the faster weathering in this greenhouse world should then speed up the rate of removal of CO_2 from the atmosphere. The result should be a slow negative feedback that reduces the CO_2 concentration in the atmosphere and moderates the size of the imposed warming.

The opposite sequence should happen if Earth's climate began to cool (Figure 4-7B). Icehouse climates are typically cold, dry, and more sparsely vegetated,

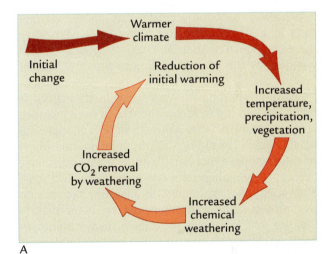

A

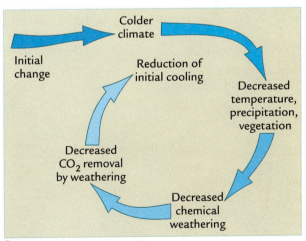

B

FIGURE 4-7

Negative feedback from chemical weathering

Chemical weathering acts as a negative climate feedback by reducing the intensity of an imposed climate warming (A) and cooling (B).

with more extensive snow and ice. An initial climate change toward a colder, drier, less vegetated Earth should reduce chemical weathering and slow the rate of removal of CO_2 from the atmosphere. By leaving more CO_2 in the atmosphere, slower CO_2 removal should reduce the effect of the initial push toward climate cooling.

The functioning of these long-term negative feedbacks does not mean that no climate change occurs at all. Any process that initially acts to warm Earth succeeds in doing so, but by an amount smaller than would have been the case without the negative feedback. Conversely, any process that initially acts to cool Earth succeeds in doing so, but also to a reduced degree. The existence of a climate-dependent negative feedback due to chemical weathering was proposed in 1981 by the geochemist James Walker and his colleagues Paul Hays and James Kastings.

How do we apply this concept to the mystery of the faint young Sun paradox? Recall that we need to account for a global thermostat that made it warmer early in Earth's history to counter the weakness of the early Sun, but that later throttled back on that warming as the strengthening Sun provided greater heat.

Earth's environment very early in its history is poorly known but is widely thought to have included active volcanism that caused large emissions of volatile gases (including CO_2) from its interior. Many scientists believe that Earth's surface may even have been entirely molten for a few hundred million years after 4.55 billion years ago. In addition, ancient craters preserved on our moon and on other planets indicate that Earth was once under heavy bombardment by asteroids, meteors, and comets, and these collisions may have triggered greater volcanism as well. Radioactive elements deep in Earth's interior

also released heat that could have increased the amount of volcanism. Increased volcanic activity would have delivered more CO_2 to the atmosphere and helped to make Earth hotter. As noted earlier, however, it is very unlikely that volcanism is the thermostat responsible for maintaining Earth's moderate climate through all 4.55 billion years of its existence.

Chemical weathering is a more promising explanation. The weakness of the young Sun would have tended to make the early Earth cooler than it is today, and the rate of CO_2 removal from the atmosphere by weathering would have been slower because of the lower temperatures. In addition, early continents are thought to have covered a smaller area than they do today. The smaller area of the continents would also have favored slower CO_2 removal from the atmosphere by weathering because less rock surface was available to weather. Slower rates of weathering would have left more CO_2 in the atmosphere over much of Earth's early history (Figure 4-8A). The warmth produced by this CO_2-enriched atmosphere could have countered most of the cooling caused by the smaller amount of incoming solar radiation.

Then, as Earth began to receive more radiation from the brightening Sun, its surface warmed and the rate of chemical weathering gradually increased. Faster chemical weathering began to draw more CO_2 out of the atmosphere, and the resulting drop in atmospheric CO_2 levels provided a cooling effect that counteracted the gradual increase in solar warming and kept Earth's temperatures moderate (Figure 4-8B). The centerpiece of this explanation is that the slow warming of Earth by the strengthening Sun would have caused changes in weathering rates that moderated changes in climate.

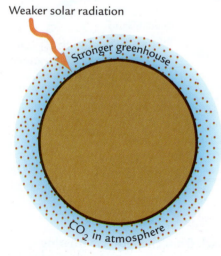

A Early Earth

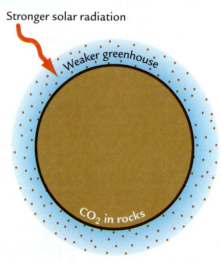

B Modern Earth

FIGURE 4-8
Earth's thermostat

A plausible explanation of the faint young Sun paradox is that the weakness of the early Sun was compensated for by a stronger CO_2 greenhouse effect in the atmosphere (A). Later, when the Sun strengthened, increased chemical weathering deposited the excess atmospheric greenhouse carbon in rocks, and the weakened greenhouse effect kept Earth's temperatures moderate (B). (ADAPTED FROM W. BROECKER AND T.-H. PENG, *GREENHOUSE PUZZLES* [NEW YORK: ELDIGIO PRESS, 1993].)

In Summary, chemical weathering is an excellent candidate for Earth's thermostat. Even though water vapor is a more important greenhouse gas than CO$_2$, it could not have been the source of this thermostat-like action because it amplifies rather than moderates climatic changes (see Chapter 2, Box 2-4).

If chemical weathering is Earth's thermostat, we face still another question: What happened to all the CO$_2$ that once resided in the atmosphere and kept Earth warm? The most likely answer is found by looking at the size of the carbon reservoirs in Figure 4-3: the carbon removed from today's atmosphere by weathering is buried in ocean sediments that eventually turn into rocks. The same process would also have been at work in the past, and over time it would have caused a slow but massive transfer of carbon from the atmosphere to the rocks. If this interpretation is correct, most of Earth's early greenhouse atmosphere lies buried in its rocks instead of concentrated in the atmosphere, as on Venus.

4-3 Was Methane Part of the Thermostat?

Decades ago, the astronomer Carl Sagan suggested that higher concentrations of methane (CH$_4$) warmed the early Earth. In the modern atmosphere, methane is a minor trace gas. After being emitted from stagnant carbon-rich wetlands, methane is broken down in a decade or less by chemical interaction with oxygen in the atmosphere. In the early geologic record, however, the absence of rocks with typical red-brown "rust" staining prior to 2.4 billion years ago indicates that Earth's early atmosphere held much less oxygen than it does today. As a result, methane could have stayed in the atmosphere longer, attained higher concentrations, and helped to warm the early Earth.

The subsequent long-term increase in atmospheric oxygen over several billion years would then have reduced the methane content of the atmosphere and weakened its greenhouse effects. Although this explanation provides a long-term cooling trend to counter the warmth from the strengthening Sun, it is not so obvious why methane would have acted like a thermostat by reacting to long-term climate changes. It seems to have acted more as an independent (and coincidentally opposing) climatic factor.

▶ Is Life the Ultimate Control on Earth's Thermostat?

Although chemical weathering provides a plausible thermostat-like mechanism to moderate Earth's climate, we have also seen that the processes involved are not strictly physical and that biological processes also take a part.

4-4 The Gaia Hypothesis

The biologists James Lovelock and Lynn Margulis proposed in the 1980s that life itself has been responsible for regulating Earth's climate. They called their idea the **Gaia hypothesis,** after the ancient Greek Earth goddess. A crude analogy of how their hypothesis works is the way the fur on an animal fluffs out to create an insulated layer and keep the creature warm when the weather turns cold. The animal in effect unconsciously regulates its own environment for its own good. The Gaia hypothesis holds that life regulates climate on Earth for its own good.

An extreme version of the Gaia hypothesis holds that all evolution on Earth has occurred for the greater good of the planet by producing the succession of life-forms needed to keep the planet habitable. This view is controversial: it goes far beyond Darwin's concept that evolution occurs to enhance the chance of reproductive survival of each species.

Still, modern biological processes are, without question, important components of the processes of chemical weathering and carbon cycling. Land plants photosynthesize carbon dioxide and transfer it to the soil in organic form as part of the vegetation litter, thereby forming carbonic acid that enhances chemical weathering. In addition, shell-bearing ocean plankton extract CO$_2$ from the ocean and store it in their CaCO$_3$ shells. These biologic processes are clearly part of the thermostat that moderates Earth's climate today. (In addition, a substantial amount of the carbon that moves through Earth's reservoirs does so in organic form, as part of a separate and smaller subcycle [Box 4-1]).

Earth's long history reveals a sequence of life-forms different from those that exist now (Figure 4-9). No record of life exists before 3.5 billion years ago, although it is possible that primitive life-forms did exist and have simply escaped detection because the rock record is so scarce and poorly preserved. By 3.5 billion years ago, primitive single-celled marine algae capable of photosynthesis had developed (Figure 4-10A). Over the next 3 billion years, slightly more complex organisms evolved: by 2.9 billion years ago, moundlike clumps of marine algae called stromatolites that attached to the seafloor; by 2.5 billion years ago, organisms that contained a cell nucleus; and by 2.1 billion years ago, a variety of multicelled algae.

Most complex forms of life did not arrive until later in Earth's history. Near 540 million years ago, hard shells of many kinds of organisms abruptly appear

Looking Deeper into Climate Science

Box 4-1

The Organic Carbon Subcycle

Nearly 20% of the carbon that cycles among Earth's carbon reservoirs today does so in organic form. Photosynthesis is critical to the organic carbon subcycle, mainly because land plants extract CO_2 from the atmosphere, and also because ocean plankton extract CO_2 from inorganic carbon dissolved in the surface ocean. Most of the organic carbon temporarily stored in land vegetation and ocean plankton is recycled and quickly returned to the ocean-atmosphere system by means of oxidation, which uses available oxygen in water or air to convert organic carbon back to inorganic form.

On land, oxidation consumes organic carbon just after the seasonal fall of leaves or dieback of green vegetation, and after the death of the woody tissue of trees. In the oceans, oxidation slowly consumes organic debris sinking out of the sunlit surface layers where photosynthesis occurs.

Only a small fraction of the organic carbon formed by these processes is buried in the geologic record, and carbon from the land and oceans contributes roughly equal amounts to this total. Burial of organic carbon is favored in water-saturated environments (marine or terrestrial) characterized by (1) low oxygen levels that minimize oxidation, and (2) rapid production of organic matter that consumes the remaining oxygen and allows the rest of the organic debris to escape oxidation. These conditions produce fine-grained carbon-rich muds that eventually turn into mudstones and then into harder rocks called shales.

The carbon buried in sediments and then rocks represents a net loss of CO_2 from the interactive carbon reservoirs in the ocean, atmosphere, soil, and vegetation. Once buried, organic carbon stays in the rocks until tectonic processes return it to the surface by slow-acting methods: (1) weathering (and oxidation) of carbon-bearing rocks at Earth's surface, and (2) thermal breakdown of organic carbon in rocks deep in Earth's interior, with release of liberated CO_2 through volcanoes.

Because this organic carbon subcycle carries one-fifth of the carbon moving between Earth's rocks and its surface reservoirs, it has the potential to have substantial effects on the global carbon balance and on atmospheric CO_2 over long (tectonic-scale) time intervals. Also, under conditions that cause the onset of high productivity and carbon burial in the ocean, large amounts of organic carbon can be quickly extracted from the atmosphere, causing rapid reductions of CO_2 levels and rapid climatic cooling.

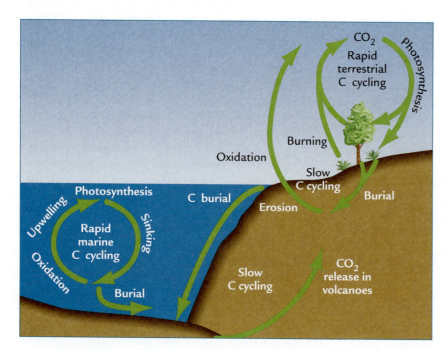

The organic carbon subcycle
About 20% of the carbon that shifts between Earth's surface reservoirs (air, water, and vegetation) and its deep rock reservoirs moves in the organic carbon subcycle. Photosynthesis on land and in the surface ocean turns inorganic carbon into organic carbon, most of which is quickly returned to the atmosphere or surface ocean. A small fraction of this organic carbon is buried in continental and oceanic sediments that slowly turn into rock. This carbon is eventually returned to the atmosphere as CO_2, either by erosion of continental rocks or by crustal melting and volcanic emissions.

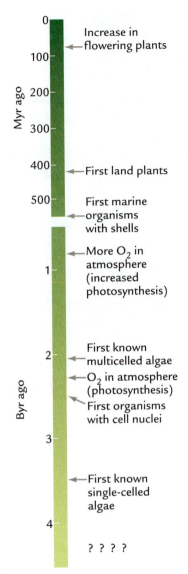

FIGURE 4-9

The Gaia hypothesis

Over time, life-forms gradually developed in complexity and played a progressively greater role in chemical weathering and its control of Earth's climate. In the extreme form of the Gaia hypothesis, life evolved for the purpose of regulating Earth's climate.

Critics of the Gaia hypothesis point out that many of the active roles played by organisms in the biosphere today are a relatively recent development in Earth's history, and that the role of life in the distant past was probably negligible. In this view, early life-forms were too primitive to have had much effect on chemical weathering, and the delicate climatic balance maintained through Earth's history must have been achieved primarily by physical-chemical means (the effects of temperature and precipitation on weathering rates), rather than by biological intervention.

Critics also note that the very late appearance of shell-bearing oceanic organisms near 540 million years ago means that life had played no obvious role in transferring the products of chemical weathering on land to the seafloor for the preceding 4 billion years. Instead, most CaCO$_3$ in the oceans was presumably deposited in warm shallow tropical seas where concentrations of dissolved ions increased to levels that permitted chemical precipitation, apparently with little or no biological intervention. Floating planktic plants capable of photosynthesis (coccolithophorida) evolved even later, in the last 250 million years.

Supporters of the Gaia hypothesis respond with several counterarguments. First, they claim that critics underestimate the role of primitive life-forms such as algae in the ocean and microbes on land in Earth's earlier history. They point to recent discoveries that modern bacteria with similarities to early primitive life-forms are now thought to play a greater role in the weathering process than has generally been recognized, and they suggest that these organisms must also have been more important than generally thought early in Earth's history, when they were the only life-forms present on land.

One indication that early life-forms were important at a global scale is the first development of an oxygen-rich atmosphere near 2.4 billion years ago, even before the first multicelled algae (see Figure 4-9). Evidence for this important event includes the first appearance of rocks that show red staining (rusting) of iron (Fe) minerals. The appearance of oxidized iron minerals at this time coincides roughly with the disappearance of previously widespread minerals such as FeS (pyrite, or "fool's gold"), which form only under reducing conditions (no oxygen). The only conceivable source of the oxygen that caused the widespread change to oxidized forms of iron is photosynthesis by marine organisms, implying an active global-scale role for these organisms far back in Earth's history. (Also, as noted above, the increase in oxygen would have reduced the amount of methane in the atmosphere.)

in the fossil record. Before that time, the only fossilized records of life consisted of ghost impressions left imprinted on the surfaces of soft sediment layers. The first primitive land plants did not evolve until near 430 million years ago (Figure 4-10B). These plants had acquired the ability to survive on land because their stems and roots delivered water from the ground. The first treelike plants appeared by 400 million years ago (Figure 4-10C). Trees and grasses are important in modern chemical weathering because they acidify groundwater by adding carbon to soils as litter.

A

B

C

FIGURE 4-10
Life-forms and weathering

Over Earth's 4.55-Byr history, plants evolved toward more complicated forms capable of playing a greater role in chemical weathering. Primitive organisms similar to the modern day bacteria *Oscillatoria* (A) existed by 3.5 Byr ago. The first simple land plants with roots and stems similar to those of the modern plant *Psilotum* (B) appeared by 430 Myr ago. Increasingly complex treelike plants similar to modern tropical cycads (C) appeared by 400 Myr ago and led to the modern diversity of trees and shrubs. (A: SINCLAIR STAMMERS/ SCIENCE PHOTO LIBRARY/SCIENCE SOURCE; B: BLICKWINKEL/ALAMY; C: GERALD CUBITT.)

Gaia hypothesis supporters also point out that the general path of biological evolution matches Earth's need for progressively greater chemical weathering through time. The more primitive organisms played a much smaller role in accelerating the process of chemical weathering during a time when it was to Earth's advantage to retain CO_2 in its atmosphere to counter the weakness of the faint young Sun. Then, as the Sun strengthened and provided more heat to Earth, more advanced organisms capable of accelerating the weathering process appeared, increased the rates of weathering, and pulled CO_2 out of the atmosphere to keep the climate system in approximate balance.

In Summary, the Gaia hypothesis is fascinating and still being argued. Scientists generally agree about the "minimum" form of Gaia: the idea that living organisms have played a significant role in the history of physical-chemical processes on Earth, including chemical weathering. Still, the "maximum" claim by proponents of the Gaia hypothesis—that individual life-forms regulate their own evolution for the greater benefit of all life on the planet—is not accepted by most scientists. Somewhere in between lies the answer to the role of life in determining the presence of life on Earth.

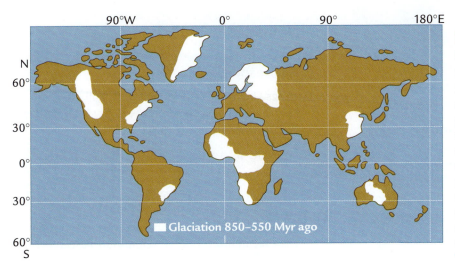

90°W   0°   90°   180°E

N
60°
30°
0°
30°
60°
S

■ Glaciation 850–550 Myr ago

FIGURE 4-11
Snowball Earth?

Evidence of several glaciations between 750 and 580 Myr ago exists in rocks on Earth's modern continents. If these glaciated regions were located in the tropics, Earth must have been much colder than today. (ADAPTED FROM L. A. FRAKES, *CLIMATES THROUGH GEOLOGIC TIME* [AMSTERDAM: ELSEVIER, 1979], AND FROM J. G. MEERT AND R. VAN DER VOO, "NEOPROTEROZOIC (1000–540 MYR) GLACIAL INTERVALS: NO MORE SNOWBALL EARTH," *EARTH AND PLANETARY SCIENCE LETTERS* 123 [1994]: 1–13.)

▶ Was There a Thermostat Malfunction? A Snowball Earth?

Ice sheets occur today at high latitudes, yet they coexist on the same planet with hot tropical regions where a strong overhead Sun heats the land and ocean. With the large pole-to-equator gradient in temperature, polar ice sheets can easily coexist on a planet with tropical heat.

For a continent-sized ice sheet to have extended to sea level near the equator, temperatures in the normally hot tropics would have to be near or below freezing through most of the year. Today's frigid polar climates would have to have extended all the way to the tropics to allow such ice sheets.

Climate scientists have found evidence that Earth came very close to freezing completely between about 750 and 580 million years ago. Sedimentary deposits from glaciers are found on several continents during this interval, providing evidence that ice sheets were present (Figure 4-11). These rocks contain ice-deposited mixtures of coarse boulders and cobbles along with fine silts and clays (see Chapter 3). Because these ancient deposits are difficult to date and correlate accurately, scientists have inferred that as few as two, or as many as four, major glacial eras occurred during this long interval.

A critical question is whether these ice sheets were located at high or low latitudes. For the glacial intervals between 715 and 640 million years ago (but not the ones before or after that interval), the geologic evidence suggests that at least some of the glaciated continents were in the tropics. This conclusion forms the basis for the novel idea that Earth was once nearly frozen—the **snowball Earth hypothesis.**

One obvious factor contributing to this much cooler interval was weaker solar heating from a Sun that was 6% below its modern luminosity (see Figure 4-2).

According to the thermostat concept, a cooler Earth would have reduced the rate of chemical weathering, kept CO₂ values higher, and moderated global temperature. In this case, however, climate models indicate that CO₂ concentrations would have to have been much lower than today to permit ice sheets to exist in tropical latitudes. The thermostat mechanism seems to have malfunctioned, at least for a while.

The reason for the thermostat malfunction remains unresolved. One explanation for the onset of a snowball Earth is that continents were clustered near the equator, where they would initially have been subject to heavy rainfall. Paradoxically, unusually heavy tropical precipitation could have driven unusually strong chemical weathering that greatly reduced CO₂ concentrations and chilled the planet.

A debate continues about how cold this world was. Although some scientists feel that Earth was frozen "hard," with sea ice extending right to the equator, evidence of water-deposited sedimentary rocks in several regions points to an incomplete freeze (called by some "slushball Earth"). Climate model simulations generally indicate an Earth with sea ice reaching into middle latitudes, but not as far as the tropics. Because the large amount of solar heat stored in the ocean at low latitudes tends to keep its surface free of ice, model simulations tend to fall short of a hard freeze.

Key Terms

greenhouse era (p. 81)

icehouse era (p. 81)

faint young Sun paradox (p. 84)

thermostat (p. 84)

silicate minerals (p. 87)

chemical weathering feedback (p. 89)

Gaia hypothesis (p. 91)

snowball Earth hypothesis (p. 95)

Review Questions

1. Why is Venus so much warmer than Earth today?
2. What factors explain why Earth is habitable today?
3. Why does the faint young Sun pose a paradox?
4. What evidence suggests that Earth has always had a long-term thermostat regulating its climate?
5. Why is volcanic input of CO_2 to Earth's atmosphere not a candidate for its thermostat?
6. What climate factors affect the removal of CO_2 from the atmosphere by chemical weathering?
7. Where did the extra CO_2 from Earth's early atmosphere go?
8. What arguments support and oppose the Gaia hypothesis that life is Earth's true thermostat?

Additional Resources

Basic Reading

Kastings, J. F., O. B. Toon, and J. B. Pollack. 1988. "How Climate Evolved on the Terrestrial Planets." *Scientific American* (February), 90–97.

Advanced Reading

Hoffman, P. F., and Schrag, D. P. 2007. "The Snowball Earth Hypothesis. Testing the Limits of Global Change." *Terra Nova* 14: 129–155.

Kasting, J. F. 2005. "Methane and Climate During the Precambrian Era." *Precambrian Research* 137: 119–129.

Lovelock, J. 1995. *The Ages of Gaia: A Biography of Our Living Earth*. New York: W. W. Norton.

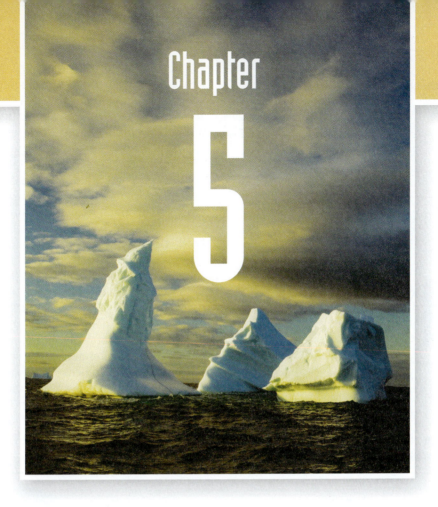

Plate Tectonics and Long-Term Climate

The last 550 million years of Earth's history are far better known than the first 4 billion years. From this time forward to the present, the locations of the continents and the shapes of the ocean basins become progressively clearer. Better-preserved sedimentary rock archives also hold more abundant evidence of past climates, including alternations between icehouse intervals with large ice sheets present and greenhouse intervals without ice on land (Figure 5-1). These fluctuations are the focus of this chapter.

First we examine how plate tectonic processes work. Next, we explore the possibility that icehouse intervals occur when plate tectonic motions shift continents across cold polar regions. Then we use climate models to investigate the factors that controlled climate 200 million years ago, a time when all landmasses on Earth existed as a single giant continent. These investigations reveal that changes in atmospheric CO_2 levels are needed to explain the sequence of changes from icehouse to greenhouse conditions over the last half-billion years. Finally, we evaluate two hypotheses that link changes in plate tectonic processes to changes in CO_2 levels.

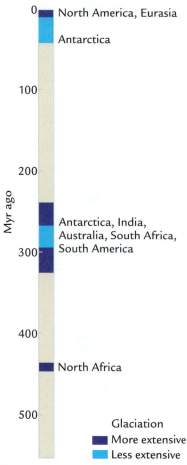

FIGURE 5-1
Icehouse intervals
Three intervals of major glaciation occurred during the last 500 Myr.

Plate Tectonics

In 1914, the German meteorologist Alfred Wegener proposed that continents have slowly moved across Earth's surface for hundreds of millions of years. He based his hypothesis in part on the obvious fact that continental margins such as those of eastern South America and western Africa fit together like pieces of a jigsaw puzzle. Research in the last half of the twentieth century showed that Wegener was correct in claiming that these continents were once joined and have since moved apart, but that he underestimated the mobility of Earth's outer surface. In fact, *all* of Earth's surface is on the move.

5-1 Structure and Composition of Tectonic Plates

Wegener's assumption that continents move in relation to ocean basins had a reasonable basis. The contrast

between the elevated continents and the submerged ocean basins is the most obvious division on Earth's surface. It also reflects the large difference in thickness and composition of the crustal layers that comprise the continents and ocean basins (Figure 5-2).

Continental crust is 30–70 kilometers thick, has an average composition like that of granite, and is low in density (2.7 g/cm³). This thick, low-density crust stands much higher than the floor of the ocean basins, which average near 4,000 meters below sea level. **Ocean crust** is 5–10 kilometers thick, has an average composition like that of basalt, and is higher in density (3.2 g/cm³). Below each of these crustal layers lies the **mantle**, which is richer in heavy elements like iron (Fe) and magnesium (Mg) and has an even higher density (> 3.6 g/cm³). The mantle extends 2,890 kilometers into Earth's interior, almost halfway to its center at a depth of 6,370 kilometers.

But these differences in elevation, crustal thickness, and composition are not the primary explanation for the fact that continents (and ocean basins) move. Instead, the critical reason for this mobility lies in the way different layers of rock behave.

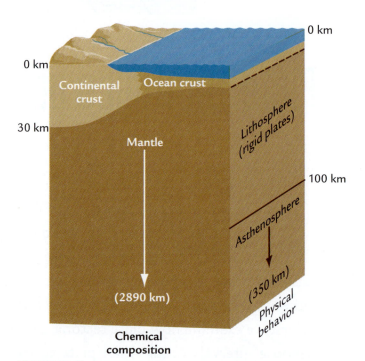

FIGURE 5-2
Earth's structure
Earth's outer layers can be subdivided in two ways. The basalts of ocean crust and the granites in continental crust differ from each other and from the underlying mantle in chemical composition. The other division is physical behavior: the lithosphere that forms the tectonic plates is hard and rigid, whereas the underlying asthenosphere is softer and capable of flowing slowly.

Two rock layers characterized by very different long-term behavior exist well below Earth's surface (see Figure 5-2). The outer layer, called the **lithosphere**, is 100 kilometers thick and generally behaves just the way the word "rock" implies: as a hard, rigid substance. The lithosphere encompasses not only the crustal layers (oceanic and continental) but also the upper part of the underlying mantle.

Below the lithosphere is a layer of partly molten yet mostly solid rock called the **asthenosphere**. This layer lies entirely within the upper section of Earth's mantle at depths of 100 to 350 kilometers. Compared to the rigid lithosphere, this deeper layer behaves like a soft, viscous fluid over long intervals of time, and flows more easily. It is the viscous behavior of this "softer" deeper layer that allows the overlying lithosphere to move.

The lithosphere consists of a dozen **tectonic plates**, each drifting slowly across Earth's surface (Figure 5-3). These plates move at rates ranging from less than 1 up to 10 centimeters per year and average about the same rate of growth as a fingernail. Over a time span of 100 million years, even slow plate motions of 5 centimeters per year add up to shifts of 5,000 kilometers, enough to create or destroy an entire ocean basin.

Most tectonic plates consist not just of continents or ocean basins but rather of combinations of the two. For example, the South American plate in Figure 5-3 consists of the continent of South America and the western half of the South Atlantic Ocean, all moving as one rigid unit.

These rigid tectonic plates have three basic types of edges, or margins. Most tectonic deformation on Earth (earthquakes, faulting, and volcanoes) occurs at these plate margins (Figure 5-4).

Plates move apart at **divergent margins**, the crests of ocean ridges like the one that runs down the middle of the Atlantic Ocean (see Figure 5-3). This motion allows new ocean crust to be created, and the new crust spreads away from the ridge. Plates diverging at ocean ridges carry not just the near-surface layer of ocean crust but also a much thicker layer of mantle lying underneath. Plates come together at **convergent margins** (see Figure 5-4 left). At these locations, the lithosphere (ocean crust and upper mantle) plunges deep into Earth's interior at ocean trenches in a process called **subduction**.

Some convergent margins occur along continent-ocean boundaries, such as the western coast of South

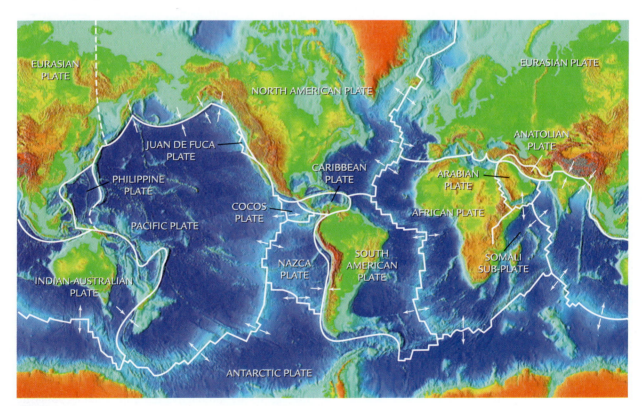

FIGURE 5-3

Tectonic plates

Earth's lithosphere is divided into a dozen major tectonic plates and several smaller plates, which move as rigid units in relation to one another, as the arrows indicate. (COURTESY NATIONAL GEOPHYSICAL DATA CENTER, NATIONAL OCEANIC AND ATMOSPHERIC ADMINISTRATION, U.S. DEPARTMENT OF COMMERCE.)

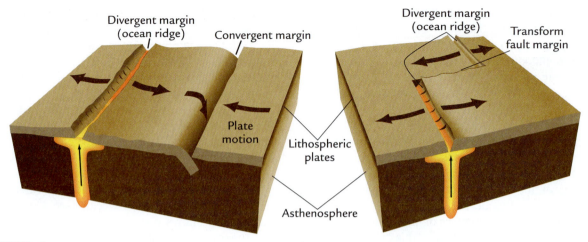

FIGURE 5-4

Plate margins

Earth's tectonic plates move apart at ocean ridges (divergent margins), slide past each other at faults (transform fault margins), and push together at convergent margins. (MODIFIED FROM F. PRESS AND R. SIEVER, *UNDERSTANDING EARTH*, 2ND ED., © 1998 BY W. H. FREEMAN AND COMPANY.)

America. In this case, narrow mountain chains such as the Andes form on the adjacent continents because of the compressive (squeezing) forces produced when two plates move together. Subduction can also occur within the ocean, where the ocean crust of one plate plunges under another and forms volcanic ocean islands, such as those in the western Pacific. A less common but important example of converging plates is the **continental collision** of landmasses such as India and Asia, which can create massive high-elevation regions such as the Tibetan Plateau.

Plates also can slide past each other at **transform fault margins** (see Figure 5-4 right), moving horizontally along faults such as the San Andreas Fault in western California. Sliding of plates at transform faults again involves the lithosphere—both the upper 30 kilometers of continental crust and the underlying 70 kilometers of upper mantle.

Even though geoscientists do not yet know the exact balance of forces that have caused past movements of plates and eventually produced their present distributions, they can accurately measure the way these processes have changed Earth's surface during the last several hundred million years. With this knowledge, the history of tectonic changes can be compared with changes in climate over the same interval in order to evaluate how tectonic changes may have influenced Earth's climate.

5-2 Evidence of Past Plate Motions

The past effects of plate tectonics in rearranging Earth's geography can be assessed from a broad range of evidence, the most important of which starts with the fact that Earth has a **magnetic field**. Molten fluids circulating in Earth's liquid iron core today create a magnetic field analogous to that of a bar magnet (Figure 5-5). Compass needles today point to magnetic north, which is located a few degrees of latitude away from the geographic North Pole, the axis of Earth's rotation.

This close link between "magnetic north" and Earth's north-polar axis of rotation is assumed to have held in the past. Some of Earth's once-molten rocks contain "fossil compasses" that record their past magnetic field. These natural compasses were frozen into the rocks shortly after they cooled from a molten state. Today, they give scientists studying **paleomagnetism** a way to reconstruct past positions of continents and ocean basins with respect to the pole of rotation.

The best rocks to use as ancient compasses are basalts, which are rich in highly magnetic iron. Basalts form the floors of ocean basins and are also found on land in actively tectonic regions. They form from molten lavas, which cool quickly after being extruded onto Earth's surface. As the molten material cools, its iron-rich components align with Earth's magnetic field like a compass. After the lava turns into basaltic rock (when its temperature drops below 1,200°C), continued cooling to temperatures near 600°C allows the "fossilized" magnetic compasses to become fixed in position in the rock. Also locked in the basalts are radioactive minerals such as potassium (K). Their slow decay (see Chapter 3) can date the time when the basalt layers cooled and acquired their magnetic compasses.

Paleomagnetism is used to reconstruct changes in the configuration of Earth's surface in two ways. (1) Back to about 500 million years ago (and in some cases earlier), paleomagnetic compasses recorded in

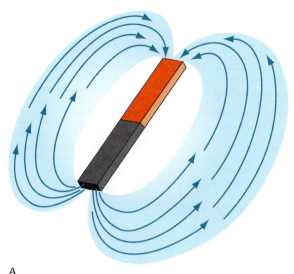

A

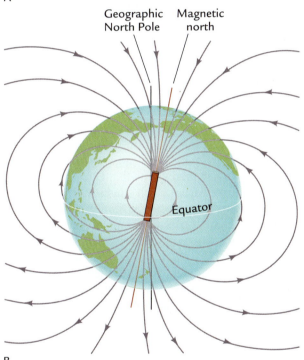

Geographic North Pole   Magnetic north

Equator

B

FIGURE 5-5

Earth's magnetic field

Like the magnetic field indicated by iron filings around a bar magnet (A), Earth has a magnetic field that determines the alignment of compass needles (B). Basaltic rocks contain iron minerals that align with Earth's prevailing magnetic field shortly after the molten magma cools to solid rock. (B: F. PRESS AND R. SIEVER, *UNDERSTANDING EARTH*, 2ND ED., © 1998 BY W. H. FREEMAN AND COMPANY.)

PALEOMAGNETIC DETERMINATION OF PAST LOCATIONS OF CONTINENTS Because ocean crust is constantly being destroyed at convergent plate boundaries, no ocean crust older than 175 million years survives except as highly altered fragments crumpled along continental margins. For older intervals, paleomagnetism must rely on basalts deposited on the continents. The orientations of the magnetic compasses frozen in these basalt layers are used to determine the past latitude of that rock (and of the portion of continental crust in which it is embedded) in relation to the magnetic poles.

In molten lavas that cool at high latitudes, the internal magnetic compasses point in a nearly vertical direction because Earth's magnetic field has that orientation at high latitudes (see Figure 5-5B). In contrast, lavas that cool near the equator have internal compasses oriented closer to horizontal, nearly parallel to Earth's surface. After they form, the basaltic rocks may be carried across Earth's surface by plate tectonic processes, but the angle of dip of their embedded magnetic compasses still records the latitude at which they formed. Rocks older than about 500 million years are less reliable for these studies because of the increasing likelihood that their magnetic compasses have been reset to the magnetic field of a later time by subsequent tectonic activity.

PALEOMAGNETIC DATING OF OCEAN CRUST Paleomagnetism is also used to trace the movement of the seafloor during the last 175 million years because of an entirely different aspect of Earth's magnetic field: the fact that it has repeatedly reversed direction. Compasses that today point to magnetic north in the present "normal" magnetic field would have pointed to magnetic south (a position very near the South Pole) during times when the field was in a "reversed" orientation.

Past changes in the magnetic field are recorded in fossil magnetic compasses in well-dated basaltic rocks from many regions. Because widely dispersed basaltic rocks have yielded the same sequence of reversals through time, the magnetic-reversal history they record must be a worldwide phenomenon. The reversals occur at irregular intervals ranging from as long as several million years to as short as a few thousand years.

Soon after this worldwide magnetic reversal sequence was established on land, marine geophysicists found stripe-like magnetic patterns called **magnetic lineations** on the ocean floor (Figure 5-6). Ships surveying the ocean towed instruments that measured Earth's regional magnetic field. To the surprise of most scientists, these magnetic lineations along the flanks of the mid-ocean ridges were found to be symmetrical around the ridge axis.

continental basalts can be used to track movements of landmasses with respect to latitude. (2) Over the last 175 million years, paleomagnetic changes recorded in basaltic oceanic crust can be used to reconstruct movements of plates and (over part of that interval) rates of spreading of the seafloor.

Even more surprisingly, the mapped pattern of horizontal highs and lows measured in the magnetic lineations at sea closely matched the pattern of normal and reversed intervals defined by the magnetic-reversal history from basalt sequences on land. Because of this match, scientists realized that the time framework that had been developed on land could be transferred directly to the lineations in the ocean. Based on this remarkable link, ocean crust could be dated in any region where ships measured the magnetic lineations.

This unexpected match of magnetic patterns on land with those in the ocean proved that new (zero-age) ocean crust is constantly being formed at the crests of the ocean ridges and that the ocean crust and underlying lithosphere then slowly move away from the ridge axis in both directions. As a result, the age of the ocean crust steadily increases with distance from the ridges (see Figure 5-6).

Scientists have now used this information about the age of existing ocean crust to evaluate causes of past climate changes in two ways. First, the dated magnetic lineations on the seafloor can be used to roll back the recent motions of the seafloor and restore the continents and oceans to their positions during much of the last 175 million years. Second, the lineations in ocean crust can be used to reconstruct the rate of **seafloor spreading**. Changes in the rate of spreading refer both to the rate at which new ocean crust and lithosphere are created at ocean ridges and the rate at which older ocean crust and underlying lithosphere are subducted at ocean trenches.

> **In Summary**, we can reconstruct the positions of continents on Earth's surface with good accuracy back to 300 million years ago, and less accurately back to 500 million years ago or earlier. For the last 100 million years, we can compile spreading rates over enough of the world's ocean to attempt to estimate the global mean rate of creation and destruction of ocean crust.

▶ The Polar Position Hypothesis

An early hypothesis of long-term climate change focused on latitudinal position as a likely cause of glaciation of continents. The **polar position hypothesis** made two key predictions that can best be tested over the younger part of Earth's history: (1) ice sheets should appear on continents that were located at polar or near-polar latitudes, but (2) no ice should appear at times when continents were located outside of polar regions. Rather than explaining icehouse intervals as caused by worldwide climate changes, this hypothesis simply calls on the movements of continents and tectonic plates across Earth's surface.

The fact that modern ice sheets occur on the polar continent of Antarctica and the near-polar landmass of Greenland makes this hypothesis seem plausible. Modern ice sheets exist at high latitudes for several reasons: cold temperatures caused by low angles of incident solar radiation, high albedos resulting from the prevalent cover of snow and sea ice that reflect most solar radiation, and sufficient moisture to replenish the ice despite melting along their lower margins (see Chapter 2).

5-3 Glaciations and Continental Positions since 500 Myr Ago

We can directly test the polar position hypothesis against evidence in the younger geologic record. Over the last 450 million years, seafloor spreading has slowly moved continents across Earth's surface between the warmer low latitudes and colder high latitudes (Table 5-1). If latitudinal position alone controls

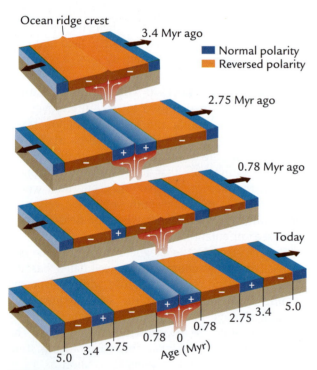

FIGURE 5-6
Magnetization of ocean crust

As molten lava erupts at the seafloor, cools, and solidifies, successive bands of ocean crust form and are magnetized in the normal or reversed polarity prevailing at the time. As the plates move apart, equal amounts of magnetized crust are carried away from the ridge axis in both directions and can be used to date the seafloor. (MODIFIED FROM F. PRESS AND R. SIEVER, *UNDERSTANDING EARTH*, 2ND ED., © 1998 BY W. H. FREEMAN AND COMPANY.)

Table 5-1 Evaluation of the Polar Position Hypothesis of Glaciation

Time (Myr ago)	Ice sheets present?	Continents in polar position?	Hypothesis supported?
445	Yes	Yes	Yes
425–325	No	Yes	No
325–240	Yes	Yes	Yes
240–125	No	No	Yes
125–35	No	Yes	No
35–0	Yes	Yes	Yes

climate, these movements should have produced predictable changes in glaciations over intervals of tens to hundreds of millions of years. For simplicity, we refer to the "south magnetic pole" as it exists in the Southern Hemisphere today, even though the magnetic poles were constantly reversing through time.

Since 450 million years ago, major (continent-sized) ice sheets have existed on Earth during three icehouse eras: a brief interval centered near 445 million years ago, a much longer interval from 325 to 240 million years ago, and the current icehouse era of the last 35 million years. During most of the long intervening intervals (430–325 Myr and 240–35 Myr ago), large ice sheets do not seem to have existed.

Near 420 million years ago, small landmasses that were later to form modern North America and the northern part of Eurasia lay scattered across a wide range of latitudes (Figure 5-7A). The other land areas, equivalent to modern Africa, Arabia, Antarctica, Australia, South America, and India, were combined in a much larger southern supercontinent called **Gondwana**. Gondwana was located on the opposite side of the globe from North America, but it had begun a long trip that would carry it

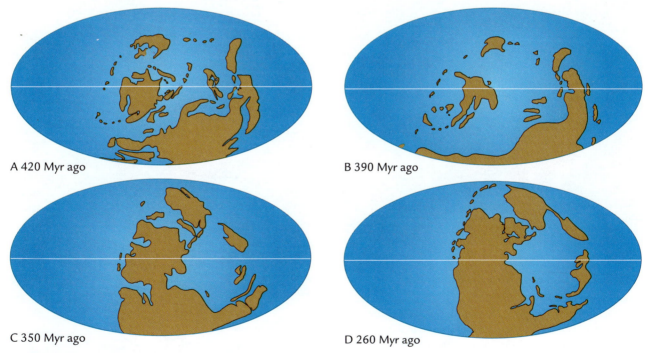

A 420 Myr ago

B 390 Myr ago

C 350 Myr ago

D 260 Myr ago

FIGURE 5-7
Moving continents

After 450 Myr ago, plate tectonic activity carried the southern continent of Gondwana across the South Pole on a path headed toward continents scattered across the Northern Hemisphere (A–C). Subsequent collisions formed the giant continent Pangaea (D). (ADAPTED FROM S. STANLEY, *EARTH SYSTEM HISTORY*, © 1999 BY W. H. FREEMAN AND COMPANY.)

across the South Pole and then northward to a collision with the northern landmasses, creating the giant supercontinent **Pangaea**, meaning "All Earth" (Figure 5-7B–D).

This motion is conveniently represented by plotting the changing position of the magnetic south pole in relation to the land (Figure 5-8). But this convention makes it look as if the south magnetic pole moved southward across Gondwana. Instead, the Gondwana continent was moving across the pole.

How well does the pattern shown in Figure 5-8 explain the intervals of glaciation and nonglaciation listed in Table 5-1? The position of the south magnetic pole 445 million years ago agrees with the evidence of glaciation in the area of the modern Sahara Desert. The weight of the ice pressing down on the loose rubble carried in its base left striations (grooves) cut into bedrock (Figure 5-9).

At first glance, this match is consistent with the polar position hypothesis, but closer inspection raises problems. One problem is that this glacial era near 445 million years ago was very brief in terms of geologic time. Although its duration was once thought

to be about 10 million years, new evidence suggests that large amounts of ice may only have been present for a million years or less. That brief a glaciation is not easily explained by the slow motion of Gondwana across the South Pole (Box 5-1).

A far more perplexing problem is the lack of glaciations between 425 and 325 million years ago, even though the Gondwana continent was still continuing its slow transit across the pole (see Figures 5-7 and 5-8). Somehow land existed at the South Pole for almost 100 million years without major ice sheets forming. This observation argues against the hypothesis that a polar position is the *only* requirement for large-scale glaciation.

From 325 to 240 million years ago, Gondwana continued its slow journey across the South Pole, and a huge region centered on the south-central part of the continent was glaciated (see Figure 5-8). These ice sheets were centered on modern Antarctica and South Africa, and they spread out into adjoining regions of South America, Australia, and India. Because of the correspondence between the area of Gondwana that was glaciated and its position at or near the south

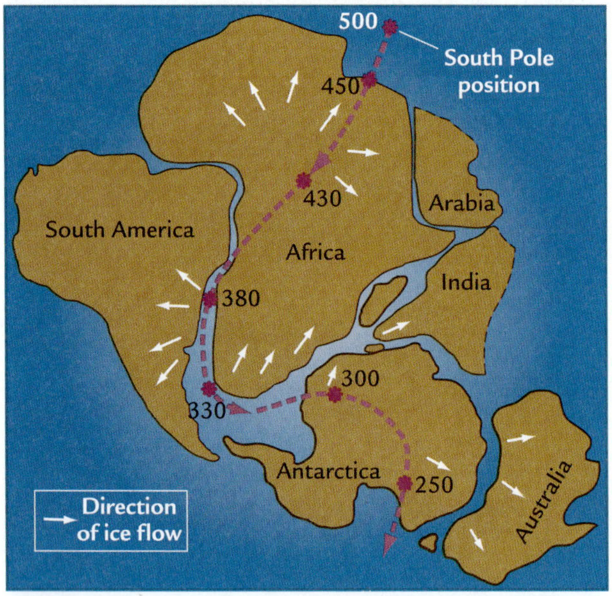

FIGURE 5-8
Gondwana glaciation and the South Pole

Changes in the position of the south magnetic pole in relation to the continent of Gondwana are largely the result of the slow movement of Gondwana. Glaciations occurred in the northern Sahara about 445 Myr ago and in southern Gondwana (South Africa, Antarctica, India, South America, and Australia) 325–240 Myr ago. The (shallow) water shown between the modern continental outlines was land during Pangaean times. (ADAPTED FROM T. J. CROWLEY ET AL., "GONDWANALAND'S SEASONAL CYCLE," *NATURE* 329 [1987]: 803–7, BASED ON P. MOREL AND E. IRVING, "TENTATIVE PALEOCONTINENTAL MAPS FOR THE EARLY PHANEROZOIC AND PROTEROZOIC," *JOURNAL OF GEOLOGY* 86 [1978]: 535–61.)

FIGURE 5-9
Glacial striations

Bedrock with grooves cut into the surface by glacial scouring and gouging. (LEONARD LEE RUE III/SCIENCE SOURCE.)

Looking Deeper into Climate Science

Box 5-1

Brief Glaciation 445 Myr Ago

An ice sheet comparable in size to that on modern Antarctica covered the North African part of the Gondwana continent near 445 million years ago. This glaciated interval has been thought to have lasted for 10 million years or more and has been attributed to a combination of factors: the general cooling effect from a Sun that was 4% weaker than today, the positioning of the North African part of Gondwana directly over the South Pole, and a reduction of atmospheric CO_2 values, caused by some combination of slower CO_2 input by volcanoes and faster chemical weathering. Faster weathering may have been caused by small continental collisions prior to the ones that later formed the supercontinent Pangaea, perhaps aided by the first appearance of vegetation on land and its effect in enhancing weathering (see Chapter 3).

More recent dating of the geologic record suggests that the peak expression of this glaciation may have lasted for only a million years—very brief in comparison with the 35 million years of the present glacial era and the glaciation that lasted from 325 to 240 million years ago. If this glaciation was indeed only a million years long, neither seafloor spreading nor chemical weathering seems likely to have changed the CO_2 concentration in the atmosphere fast enough to explain it. Volcanoes and chemical weathering rates can gradually drive CO_2 levels low enough to produce glaciation, but this episode appears to require a mechanism capable of dropping and then raising CO_2 values within a million years.

One mechanism under consideration is an abrupt increase in the rate of burial of organic carbon. The organic carbon subcycle (see Chapter 4, Box 4-1) meets several requirements for explaining a large but rapid climate cooling. Because it carries one-fifth of the total flow of carbon through the upper parts of Earth, this subcycle has the potential to alter the global carbon balance and atmospheric CO_2 levels. Also favoring this explanation is the fact that large amounts of organic carbon can be quickly buried in the sedimentary record, causing a rapid reduction of CO_2 levels.

Several kinds of changes can cause rapid burial of organic carbon: changes in wind direction that cause increased upwelling along coastal margins; an increase in the amount of organic carbon and nutrients delivered to the ocean; a change toward wetter climates on continental margins, where low relief naturally favors formation of vegetation-rich swamps; or the isolation of small ocean basins in regions of high rainfall that generates carbon-rich river runoff.

magnetic pole, this long interval of glaciation is consistent with the polar position hypothesis.

By 240 million years ago, Gondwana had moved farther northward and the glaciation had ended. The lack of ice after that time agrees with the positioning of major landmasses away from the South Pole. By that time, the northern part of Gondwana had begun to merge with the northern continents and form the even larger supercontinent Pangaea.

After 180 million years ago, Pangaea began to break up. Its southernmost part, which included the modern continents of Antarctica, India, and Australia, moved back over the South Pole by 125 million years ago, yet no major ice sheet developed. Antarctica remained directly over the pole but largely free of ice from 125 million years ago until 35 million years ago, when significant amounts of ice reappeared. Here again we face the mystery encountered earlier: How could a landmass centered on a pole have remained ice-free (or largely so) from 125 to 35 million years ago?

Clearly, the polar position hypothesis accounts for part of Earth's glaciation history during the last half-billion years (see Table 5-1). During that interval, ice sheets developed only on landmasses that were at polar or near-polar positions, consistent with the presence of ice at the South Pole today. This general correlation confirms that continents occupied polar positions when large-scale glaciations occurred, but it cannot explain the absence of ice during some intervals of the last 500 million years. The geologic record tells us that the presence of continents in a polar position is favorable to the formation of ice sheets, but does not guarantee that they actually will form.

In Summary, the polar position hypothesis is part of the explanation of Earth's glaciations, but not the entire story. Some other factor must also be at work, a factor that allows ice sheets to form over polar continents during some intervals and prohibits them from doing so during others.

Modeling Climate on the Supercontinent Pangaea

One fortunate aspect of studying the history of Earth's climate on tectonic time scales is the number of natural experiments Earth has run by altering its geography in major ways. Because the locations of continents are accurately known for the past 300 million years, climate scientists can use general circulation models (GCMs) to evaluate the impact of these geographic factors on climate. Here we examine a time near 200 million years ago when collisions of continents had formed the giant supercontinent Pangaea. Because this configuration differs considerably from the more dispersed continents today, Pangaea provides climate scientists with a very different and yet a real Earth for testing the performance of climate models.

5-4 Input to a Model Simulation of the Climate on Pangaea

Chapter 3 showed that GCM runs require the major physical aspects of a past world to be specified in advance as *boundary-condition inputs* in order to run simulations of past climates. The most basic physical constraint is the distribution of land and sea. Pangaea remained intact from the time it had formed (250 Myr ago) until it broke up after 180 million years ago. The focus here is on this long interval of relatively stable land-sea geometry. The only tectonic change of significance during this time was a very slow northward movement of Pangaea.

Around 200 million years ago, Pangaea stretched from high northern to high southern latitudes and was almost symmetrical around the equator (Figure 5-10A). The landmasses of Gondwana (Antarctica, Australia, Africa, Arabia, South America, and India) formed its larger southern part. The somewhat smaller northern part of Pangaea, sometimes referred to as Laurasia, consisted of North America, Europe, and north-central Asia. A wedge-shaped tropical seaway formed an indentation deep into the east coast of Pangaea, while the west coast had a smaller seaway in the north. This single landmass represented almost one-third of Earth's surface. It spanned 180° of longitude across its northern and southern limits near 70° latitude, and one-quarter of Earth's circumference (90°) at the equator.

Modelers have simplified this configuration for use as input to climate simulations by making the land distribution symmetrical around the equator (Figure 5-10B). This simplification requires relatively small changes in the way Pangaea is represented by the model grid boxes, which were large in size at the time this experiment was run (1994). A benefit of this

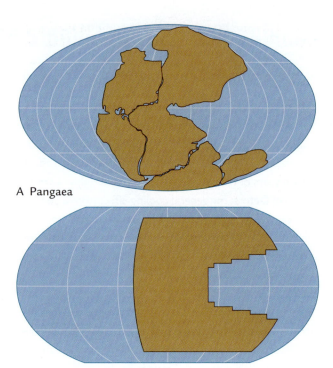

A Pangaea

B Pangaea in model grid

FIGURE 5-10

The supercontinent Pangaea

Geographic reconstructions of the interval around 200 Myr ago show all the continents joined in a single landmass called Pangaea (A). Climate modelers have simplified this configuration into an idealized continent symmetric around the equator (B). (A: ADAPTED FROM J. E. KUTZBACH AND R. G. GALLIMORE, "MEGAMONSOONS OF THE MEGACONTINENT," *JOURNAL OF GEOPHYSICAL RESEARCH* 94 [1989]: 3341–57; B: FROM J. E. KUTZBACH, "IDEALIZED PANGEAN CLIMATES: SENSITIVITY TO ORBITAL CHANGE," *GEOLOGICAL SOCIETY OF AMERICA SPECIAL PAPER* 288 [1994]: 51–55.)

simplification is that each seasonal model run in each hemisphere is the exact mirror image of the same season in the other hemisphere because the seasons simply switch back and forth between hemispheres. This symmetry effectively doubles the number of years the model simulates.

A second important decision on input to the model is global sea level. Evidence from rocks on Pangaea indicates that global sea level 200 million years ago was comparable to its present level. As a result, sea level was placed close to the structural edges of the continents, where it lies today.

A third important decision is the distribution of elevated topography on the continents, but this aspect of Pangaea is not as well known and has to be approximated. In the simulation examined here, all land in the interior of Pangaea was represented as a low-elevation plateau at a uniform height of 1,000 meters, with its edges sloping gradually down to sea level along the outer margins of the continents.

Another important boundary condition that needs to be specified is the CO_2 level in the atmosphere, but the CO_2 concentration for 200 million years ago is not certain. Although long-term CO_2 levels are determined by tectonic factors, they are also an integral part of the climate system. Because the choice of CO_2 concentration will have a direct impact on the climate simulated by the model, the danger of circular reasoning caused by choosing the wrong CO_2 level is present.

Fortunately, other considerations help climate modelers constrain the CO_2 level, if only by inference. Astronomers know that the Sun had not yet reached its present strength and was still about 1% weaker than it is today (see Chapter 4). By itself, this weaker Sun should have made Pangaea significantly colder than the modern world, with snow and ice closer to the equator than today.

Yet evidence from Pangaea refutes a colder world. No ice sheets existed on Pangaea 200 million years ago, even though its northern and southern limits lay within the Arctic and Antarctic circles (see Figure 5-10A). Today, landmasses at similar high latitudes are either permanently ice-covered (Greenland) or alternately ice-covered and ice-free through time (North America, Europe, and Asia). The absence of polar ice suggests that Pangaea's climate was somewhat warmer than today's climate.

Fossil evidence of vegetation on Pangaea supports this conclusion. Except for a few surviving tree types such as the ginkgo (Figure 5-11), Earth's vegetation has evolved to different forms since Pangaean times, and comparisons between plant types now and then have to be based on types with similar appearances rather than on actual species composition. Several kinds of palm-like vegetation that would have been killed by hard freezes existed on Pangaea to latitudes as high as 40°. This suggests that the equatorward limit of hard freezes on Pangaea was 40°, slightly higher than the modern limit (30° to 40°).

The most likely reason for a warmer Pangaea is that a higher CO_2 level 200 million years ago compensated for the weaker Sun. The model experiment examined here assumed a CO_2 level of 1,650 parts per million, almost six times the recent preindustrial value of 280 parts per million. As we will see, this choice not only produced temperature distributions consistent with the evidence from the lack of permanent ice and frost-sensitive vegetation, but it also simulated other climatic features that match independent evidence from the Pangaean geologic record. These other features, particularly precipitation and evaporation, are the main focus here.

With the critical boundary conditions specified, the model simulation is ready to run. After 15 years

FIGURE 5-11
Pangaean trees
Modern gingko trees are descended from similar forms that first evolved some 200 Myr ago. (COURTESY OF MICHAEL BOWERS, BLANDY FARM, BOYCE, VA.)

of simulated time to allow the model climate to come to a state of equilibrium, the results shown are based on the last 5 years of simulated seasonal changes.

5-5 Output from the Model Simulation of Climate on Pangaea

Because of its huge size and the reduction of the moderating influence of oceanic moisture, we might anticipate that the interior of Pangaea would have had an extremely dry continental climate. The climate model simulation confirms this expectation.

The model simulates widespread aridity at lower latitudes, especially in the Pangaean interior. Simulated mean annual precipitation and soil moisture levels are very low across vast expanses of interior and western Pangaea between 40°S and 40°N (Figure 5-12). Precipitation values of 1–2 millimeters per day in these regions are equivalent to annual totals of 15–25 inches (35–70 cm) per year, comparable to those in semi-arid grassland areas such as the western plains of the United States today.

This pervasive aridity reflects two factors: (1) the large amount of land at subtropical latitudes beneath the dry, downward-moving limb of the Hadley cell, and (2) the large amount of land in the tropics, causing trade winds to lose most of their ocean-derived water vapor before reaching the continental interior

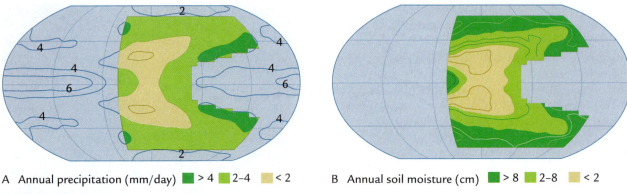

A   Annual precipitation (mm/day)   ■ > 4 ■ 2–4 ■ < 2   B   Annual soil moisture (cm)   ■ > 8 ■ 2–8 ■ < 2

FIGURE 5-12

Precipitation on Pangaea

Climate models simulate patterns of annual mean precipitation (A) and soil moisture (B) on Pangaea. Broad areas of the tropics and subtropics were very dry. (ADAPTED FROM J. E. KUTZBACH, "IDEALIZED PANGEAN CLIMATES: SENSITIVITY TO ORBITAL CHANGE," *GEOLOGICAL SOCIETY OF AMERICA SPECIAL PAPER 288* [1994]: 41–55.)

(see Chapter 2). In contrast, the simulated ocean around Pangaea received far more rainfall than the land, and considerably more than it does today.

Geologic evidence supports the model simulation of widespread Pangaean aridity. The clearest evidence is the presence of **evaporite** deposits, salts that precipitated out of water in lakes and coastal margin basins with limited connections to the ocean. Evaporite salts form only in arid regions where evaporation far exceeds precipitation. More evaporite salt was deposited during the time of Pangaea than at any time in the last several hundred million years (Figure 5-13). These evaporite deposits occurred in regions the model simulates as having been arid.

Because the moderating effects of ocean moisture failed to reach much of Pangaea's interior, the continent was left vulnerable to seasonal extremes of solar heating in summer and cooling during winter. As a result, the model simulates a large seasonal temperature response (Figure 5-14). In some mid-latitude regions, summer daily average temperatures of +25°C (77°F) alternated with winter daily temperatures of −15°C (+5°F).

This wide range of seasonal temperatures could explain the lack of ice sheets on Pangaea. The simulated winter temperatures were cold enough to provide the snowfall needed for ice sheets to grow, but the hot summers on Pangaea, even along the poleward margins of the sun-warmed landmass, caused rapid melting of the snow and thereby prevented glaciation. Ice sheets form more readily on smaller continents where summer temperatures are kept cooler by moist winds from the nearby ocean.

The model simulation also indicates that average daily land temperatures in winter would have reached the freezing point as far equatorward as 40° latitude (see Figure 5-14), close to the low-latitude

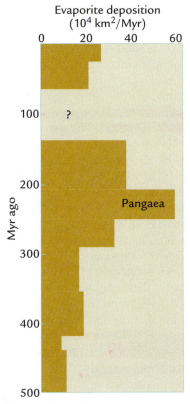

FIGURE 5-13

Pangaean evaporites

The volumes of rock salt deposits (evaporites) formed about 200 Myr ago were larger than at any other time in the last 500 Myr, indicating very dry conditions on Pangaea. (ADAPTED FROM W. A. GORDON, "DISTRIBUTION BY LATITUDE OF PHANEROZOIC EVAPORITES," *JOURNAL OF GEOLOGY* 83 [1975]: 671–84.)

limit of frost-sensitive vegetation on Pangaea indicated by geologic evidence. But with winter nights likely to have been considerably colder than daily mean

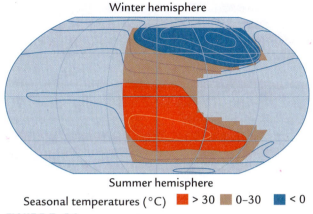

Winter hemisphere

Summer hemisphere

Seasonal temperatures (°C)   ■ > 30   ■ 0–30   ■ < 0

FIGURE 5-14
Temperature on Pangaea

Climate model simulations show very large seasonal temperature contrasts on Pangaea between the summer hemisphere, which was warmed by solar radiation, and the winter hemisphere, which lost heat by longwave back radiation to space. (ADAPTED FROM J. E. KUTZBACH, "IDEALIZED PANGEAN CLIMATES: SENSITIVITY TO ORBITAL CHANGE," *GEOLOGICAL SOCIETY OF AMERICA SPECIAL PAPER* 288 [1994]: 41–55.)

temperatures, the model results disagree to some extent with the ground-truth evidence. Despite the high CO_2 values used as input to the simulation, freezing occurs closer to the equator in the model simulation than the evidence from past vegetation indicates.

Another fundamental characteristic of the climate of Pangaea was the strong reversal between summer and winter monsoon circulations. Monsoon circulations are driven by the different rates of response of the land and the oceans to solar heating in summer and radiative heat loss in winter (recall Chapter 2). The large seasonal swings in land temperature and smaller seasonal changes in ocean temperature reflect these contrasting responses of land and ocean.

Strong solar heating over the part of Pangaea situated in the summer hemisphere (alternating between north and south) caused heated air to rise over the land. This upward motion of heated air created a strong low-pressure cell at the surface (Figure 5-15A), which produced a net inflow of moisture-bearing winds from the ocean. This moist inflow brought heavy rains to the east coast, especially in the subtropics (Figure 5-15B).

The situation in the winter hemisphere was the reverse. Weak seasonal heating from the Sun and strong heat loss by longwave back radiation caused cooling over the interior of Pangaea. This cooling caused air to sink toward the land surface, built up high pressures over the continent, and pushed cold, dry air out over the ocean. As a result, precipitation over the land was greatly reduced.

Note that the winds on the eastern margins of Pangaea from 0° to 45° latitude reversed direction

between the seasons: warm, wet, summer monsoon winds blew from the sea onto the land, but cold, dry, winter monsoon winds blew from the land out to sea. The subtropical margins of Pangaea were places of enormous contrasts in seasonal precipitation, alternating between very wet summers and dry winters.

Geologic evidence of seasonal moisture contrasts on Pangaea comes from the common occurrence of **red beds**, sandy or silty sedimentary rocks stained various shades of red by oxidation of iron minerals. Red-colored soils accumulate today in regions where the contrast in seasonal moisture is strong. Red beds were more widespread on Pangaea than during other geologic intervals, consistent with the model simulation of highly seasonal changes in moisture between wet summer monsoons and dry winter monsoons.

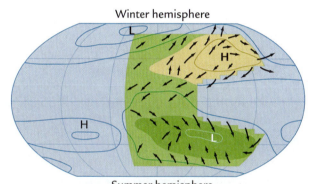

Winter hemisphere

Summer hemisphere
A Seasonal pressure and winds

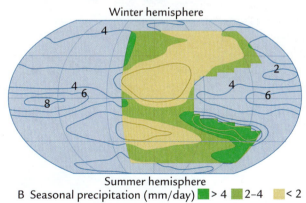

Winter hemisphere

Summer hemisphere
B Seasonal precipitation (mm/day) ■ > 4 ■ 2–4 ■ < 2

FIGURE 5-15
"Supermonsoons" on Pangaea

Climate models simulate very large seasonal changes in surface pressure and winds (A) and monsoonal precipitation (B) on Pangaea. Summer heating creates a low-pressure region (L) and draws in moist oceanic winds, which drop heavy precipitation along the subtropical east coast. Winter cooling creates a high-pressure cell (H) that sends dry air out from land to sea and reduces precipitation. (ADAPTED FROM J. E. KUTZBACH, "IDEALIZED PANGEAN CLIMATES: SENSITIVITY TO ORBITAL CHANGE," *GEOLOGICAL SOCIETY OF AMERICA SPECIAL PAPER* 288 [1994]: 41–55.)

Tectonic Control of CO_2 Input: The BLAG (Spreading Rate) Hypothesis

Our examinations of both the polar position hypothesis and the climate of Pangaea suggest that changes in Earth's geography alone cannot explain the variations between warm greenhouse climates and cold icehouse climates during the last 500 million years and that variations in the CO_2 concentration in the atmosphere played a role. In the remainder of this chapter, we examine two hypotheses that attempt to explain why CO_2 has changed over very long intervals of time. One hypothesis emphasizes changes in CO_2 input by volcanoes; the other focuses on changes in CO_2 removal by weathering.

5-6 Control of CO_2 Input by Seafloor Spreading

In 1983, a new hypothesis proposed that climate changes during the last several hundred million years were driven mainly by changes in the rate of CO_2 input to the atmosphere and ocean by plate tectonic processes. This hypothesis is named BLAG after its authors, the geochemists Robert *B*erner, Antonio *L*asaga, *A*nd Robert *G*arrels). It is also referred to as the **spreading rate hypothesis**.

In a world of active plate tectonic processes, carbon cycles constantly between Earth's interior and its surface (Figure 5-16). Most CO_2 is expelled to the atmosphere by volcanic activity at two kinds of locations: (1) margins of converging plates, where parts of the subducting lithosphere melt and form molten magmas that rise to the surface in mountain belt and island arc volcanoes, delivering CO_2 and other gases from Earth's interior; and (2) margins of divergent plates (ocean ridges), where hot magma carrying CO_2 erupts directly into ocean water.

Some volcanoes also emit CO_2 at sites distant from plate boundaries, where thin plumes of molten material rise from deep within the interior and reach the

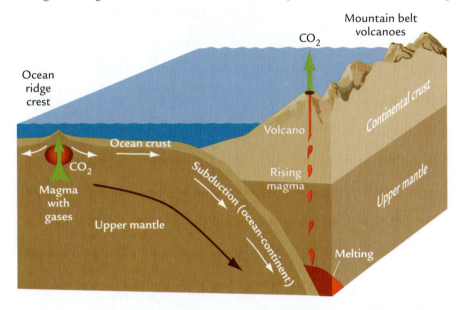

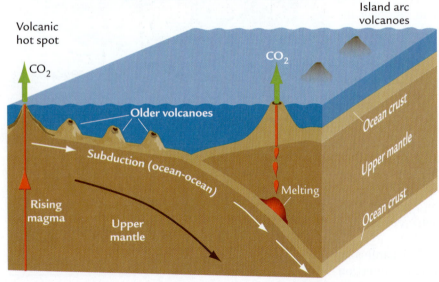

FIGURE 5-16
CO_2 input

CO_2 is transferred from Earth's interior to the atmosphere-ocean system primarily at ocean ridges (top left) and subduction zones (top and bottom right). Lesser emissions of CO_2 occur when volcanoes erupt at hot spots in the middle of plates (bottom left).

surface at volcanic **hot spots** (see Figure 5-16 bottom). Additional CO_2 is released to the atmosphere by the slow oxidation of old organic carbon in sedimentary rocks eroded at Earth's surface (recall Chapter 4).

The central feature of the BLAG hypothesis is the concept that changes in the average rate of seafloor spreading over millions of years have controlled the rate of delivery of CO_2 to the atmosphere from the large subsurface rock reservoir of carbon and that the resulting changes in atmospheric CO_2 concentrations have had an impact on Earth's climate.

Well-dated magnetic lineations show that the various ocean ridges that exist today have been spreading at different rates for millions of years (Figure 5-17). For example, the ridge in the southern Pacific Ocean spreads as much as ten times faster than the one in the Atlantic Ocean.

The BLAG hypothesis is based on the concept that the *globally averaged* rate of seafloor spreading has also changed over time. Changes in the mean rate of spreading should alter the transfer of CO_2 from Earth's rock reservoirs to its atmosphere at ocean ridges and subduction zone volcanoes because these plate margins are vital participants in the process of seafloor spreading (see Figure 5-16).

Faster rates of spreading at ridge crests create larger amounts of new ocean crust and more frequent releases of magma, which deliver greater amounts of CO_2 to the ocean (Figure 5-18). Faster spreading also causes more rapid subduction of crust and sediment in ocean trenches and delivers larger volumes of carbon-rich sediment and rock for subsequent melting and CO_2 release through volcanoes. Conversely, slower spreading reduces both kinds of CO_2 input to the atmosphere.

Although the BLAG hypothesis focuses on changes in spreading rates as a driver of long-term climate change, it also calls on chemical weathering for negative feedback to moderate these changes (see Chapter 4). Increased volcanic emissions caused by faster seafloor spreading lead to higher atmospheric CO_2 levels and a warmer climate (see Figure 5-18 top). This initial shift toward a greenhouse climate then activates the combined effects of higher temperature, greater precipitation, and more vegetation to speed up the rate of chemical weathering and draw CO_2 out of the atmosphere at a faster rate. The resulting CO_2 removal opposes and reduces some of the initial warming driven by faster spreading rates and higher CO_2 concentrations. Similarly, chemical weathering feedbacks work to offset some of the impact of cooling caused by slower volcanic input of CO_2 (Figure 5-18 bottom). In effect, the BLAG hypothesis relies on chemical weathering to moderate any fluctuations in climate driven by changes in volcanic CO_2 input.

The BLAG hypothesis further proposes that much of the cycling of carbon between the deeper Earth and the atmosphere occurs in a long, slow-acting loop (Figure 5-19). Carbon taken from the atmosphere during chemical weathering is initially stored in dissolved HCO_3^{-1} ions that are carried by rivers to the sea. As we have seen (recall Chapter 3), marine plankton use this dissolved carbon to form $CaCO_3$ shells, and the shells are deposited in ocean sediments when the organisms die. The movement of carbon through this surface part of the cycle is rapid, occurring in just a few years.

The $CaCO_3$-bearing sediments are then carried by seafloor spreading toward subduction zones at

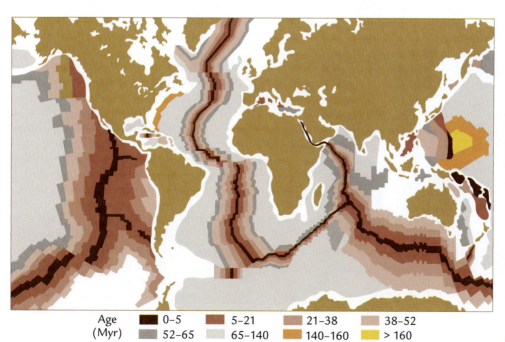

FIGURE 5-17
Age of the seafloor

Some ocean crust dates as far back as 175 Myr ago. Modern spreading rates are as much as ten times faster in the Pacific than in the Atlantic. (MODIFIED FROM S. STANLEY, *EARTH SYSTEM HISTORY,* © 1999 BY W. H. FREEMAN AND COMPANY, AFTER W. C. PITMAN ET AL., MAP AND CHART SERIES MC-6 [BOULDER, CO: GEOLOGICAL SOCIETY OF AMERICA, 1974].)

Age (Myr)		
0–5	5–21	21–38	38–52
52–65	65–140	140–160	> 160

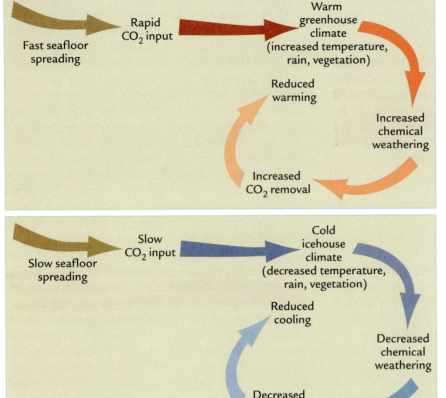

FIGURE 5-18
The BLAG (spreading rate) hypothesis

This hypothesis predicts that atmospheric CO_2 concentrations and global climate are driven by the global mean rate of seafloor spreading, which controls the rate of CO_2 input at ocean ridge crests and subduction zones. The spreading rate hypothesis also invokes chemical weathering as a negative feedback that partially counters changes in atmospheric CO_2 and global climate initiated by varying rates of seafloor spreading.

continental margins. Some sediment is scraped off at the ocean trenches, and some is carried downward in the subduction process (see Figure 5-19). This slow journey of carbon-bearing sediments across the ocean floor and down into the trenches and beyond takes tens of millions of years.

Most of the $CaCO_3$ (and other carbon) carried down into Earth's interior by subduction melts at the hot temperatures found at great depths or is transformed in other ways. These processes eventually return CO_2 to the atmosphere through volcanoes and complete the cycle. Almost none of the subducted carbon is carried deep into the mantle. Movement of carbon through this deep part of the cycle takes tens of millions of years.

The two chemical reactions that summarize the basic chemical changes involved at the beginning

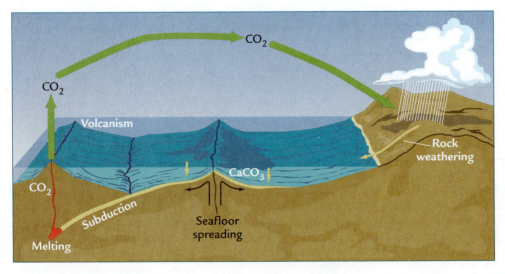

FIGURE 5-19
Carbon cycling

In the BLAG (spreading rate) hypothesis, carbon cycles continuously between rock reservoirs and the atmosphere: CO_2 is removed from the atmosphere by chemical weathering on land, deposited in the ocean, subducted, and returned to the atmosphere by volcanic activity. (ADAPTED FROM W. F. RUDDIMAN AND J. E. KUTZBACH, "PLATEAU UPLIFT AND CLIMATE CHANGE," *SCIENTIFIC AMERICAN* 264 [1991], 66–75.)

and end of this tectonic-scale carbon cycle are mirror opposites:

Chemical weathering on land

$$CaSiO_3 \quad + \quad CO_2 \quad \rightarrow \quad CaCO_3 \quad + \quad SiO_2$$

Silicate rock   Atmosphere   Plankton   Plankton

Melting and transformation in subduction zones

$$CaCO_3 \quad + \quad SiO_2 \quad \rightarrow \quad CaSiO_3 \quad + \quad CO_2$$

Ocean sediments   Silicate rock   Atmosphere

Together, the two reactions form a complete (closed) cycle with no net chemical change over tens of millions of years. The longest part of the cycle is caused by the slow spreading and subduction of the seafloor, the slow transformation of $CaCO_3$ in the lower crust and upper mantle, and the slow delivery of CO_2 to the surface through volcanic action. In contrast, changes in spreading rates can alter the rate of melting and CO_2 release to the atmosphere with little or no delay because carbon-bearing sediment is already "in the pipeline." At any interval in time, carbon-bearing sediments are in the process of being subducted into Earth's interior, and changes in the average rate of subduction will soon result in different rates of melting of this down-going material.

The BLAG hypothesis proposes that this cycling of carbon provides long-term stability to the climate system by moving a roughly constant amount of total carbon back and forth between the rocks and the atmosphere over long intervals of time. As a result, atmospheric CO_2 levels are constrained to vary only within moderate limits. But the long delays between carbon weathering and burial permit small imbalances to occur between the rate of carbon burial and the return of CO_2 to the atmosphere. These imbalances could drive climate changes over intervals of tens of millions of years.

5-7 Initial Evaluation of the BLAG (Spreading Rate) Hypothesis

Unfortunately, the predictions of the BLAG hypothesis cannot be directly tested over most of the geologic past because no undeformed ocean crust older than 175 million years exists to use for calculating past spreading rates. All older crust has been subducted

in ocean trenches. Half of the crust that formed 50 million years ago has already been destroyed by rapid subduction under western South America (see Figure 5-17) and by the total disappearance of ocean crust in the former tropical seaway (called Tethys) by subduction and collision along the southern coast of Asia.

Most reconstructions have inferred that the global mean spreading rate was faster near 100 million years ago than it is at present. If this conclusion holds up to future scrutiny, the BLAG hypothesis would predict that the rate of input of CO_2 to the atmosphere should have been higher 100 million years ago than it is today (Table 5-2). This prediction would agree with geologic evidence of a warmer climate 100 million years ago, and with the absence of large polar ice sheets. We will revisit this important issue in the next two chapters.

▶ Tectonic Control of CO₂ Removal: The Uplift Weathering Hypothesis

A second hypothesis that attempts to explain how plate tectonic processes control atmospheric CO_2 levels emerged from work by the marine geologist Maureen Raymo and colleagues in the late 1980s. Parts of this concept date back to work by the geologist T. C. Chamberlain a century ago. The **uplift weathering hypothesis** proposes that chemical weathering is an active driver of climate change, rather than just a passive negative feedback that moderates climate.

5-8 Rock Exposure and Chemical Weathering

The BLAG hypothesis emphasizes changes in CO_2 delivery to the atmosphere by seafloor spreading, and it assumes that removal of CO_2 by chemical weathering responds only to climate-related changes in temperature, precipitation, and vegetation. Although these factors certainly affect chemical weathering (see Chapter 4), they are not the only processes that do so.

The uplift weathering hypothesis starts from a different perspective. It asserts that the global mean

Time (Myr ago)	Ice sheets present?	Spreading rates	Hypothesis supported?
100	No	Faster (?)	Yes (?)
0	Yes	Slower (?)	Yes (?)

Table 5-2 Evaluation of the BLAG Spreading Rate (CO₂ Input) Hypothesis

rate of chemical weathering is mainly affected by the availability of fresh rock and mineral surfaces for weathering to attack, and it proposes that this exposure effect can override the combined effects of the climate-related factors (temperature, precipitation, and vegetation) on a global basis.

Figure 5-20 shows a simplified example of the importance of rock exposure. This simplification starts with a large boulder-sized cube of rock with six surfaces consisting of squares 1 meter across, each having a surface area of 1 m^2. This rock cube has a total surface area of 6 m^2, calculated from the total area of all six sides.

Now we slice this cube into halves along each of its three axes. This slicing creates eight smaller cubes, each 0.5m on a side, with each side having a surface area of 0.25m^2. The total surface area of these eight smaller cubes is 12m^2:

$$(8 \text{ cubes}) \times (6 \text{ sides each}) \times (0.25m^2 \text{ of surface area per side})$$

Note that simply cutting the large cube into smaller cubes has doubled the exposed surface area of the rock without changing its volume, at least for the imaginary (idealized) laser-sharp cut assumed in this example. The fragmentation has created more exposed surface area for weathering to attack.

This idealized process can be continued to progressively finer grain sizes, with similar results. Ten sequential halvings of the rock's dimensions will produce over 1 billion cubes each 1 mm on a side, about the same size as grains of sand on a beach. Together these tiny cubes have a total surface area 1,000 times larger than the original block, yet they still retain the same total volume. Fragmentation to even smaller sizes will expose still more surface area.

With over 1,000 times more surface area to act on, chemical weathering would increase by a factor of 1,000 or more. The huge increase of weathering in this hypothetical example far exceeds the combined changes estimated from changes in temperature, precipitation, and vegetation (see Chapter 4). Based on this analysis, climate-related factors are not the only processes to consider in evaluating chemical weathering.

5-9 Case Study: The Wind River Basin of Wyoming

Direct evidence of the importance of rock exposure in chemical weathering comes from studies of a drainage basin in the Wind River Mountains of Wyoming. Because all the bedrock in this basin consists of granite, the kind of silicate rock that is most typical of continental crust, this watershed is reasonably representative of the average response of continental rocks to weathering.

The Wind River Mountains have been glaciated repeatedly over the last several hundred thousand years, and each glaciation has left deposits of unsorted debris (glacial moraines) in the valley and foothills below. Because some older deposits have not been overridden by later glacial advances, undeformed moraines ranging in age from 200 to 130,000 years are present in the same valley.

These Wind River moraines provide an opportunity to quantify the amount of weathering of ground-up debris that is identical in composition but widely varying in age. The extent of weathering can be determined by analyzing the soils that have subsequently developed on the moraines. These soils gradually lose their major cations (Mg^{+2}, Na^{+1}, K^{+1}, Ca^{+2}, etc.) to the chemical

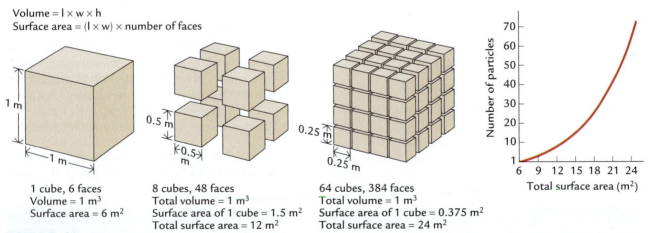

FIGURE 5-20

Fragmentation of rock

Each time a cube-shaped rock is sliced into smaller cubes (with each side half as long as before), the total surface area of rock doubles, even though the volume remains the same. (D. MERRITTS ET AL., *ENVIRONMENTAL GEOLOGY*, © 1997 BY W. H. FREEMAN AND COMPANY.)

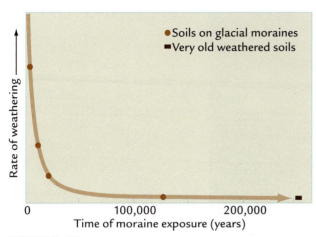

FIGURE 5-21

Weathering and exposure time

Glacially eroded and fragmented granite weathers quickly soon after deposition, but much more slowly 100,000 years later.

(ADAPTED FROM J. D. BLUM, "THE EFFECT OF LATE CENOZOIC GLACIATION AND TECTONIC UPLIFT ON SILICATE WEATHERING RATES," IN *TECTONIC UPLIFT AND CLIMATE CHANGE*, ED. W. F. RUDDIMAN [NEW YORK: PLENUM PRESS, 1997].)

weathering process. The cumulative amount of chemical weathering that has occurred since each moraine was deposited can be determined by measuring the total loss of these weathering-sensitive cations. Dividing the total amount of weathering since the moraine was deposited by the time elapsed yields the *average* rate of chemical weathering over that entire interval.

The Wind River deposits show a rapid (exponential) decrease in the mean rate of cation weathering with time of exposure (Figure 5-21). The younger moraines have average rates of weathering that are at least a factor of 100 faster than the older ones. The older moraines presumably also went through a phase of fast weathering during the tens of thousands of years soon after their deposition, and then weathered much more slowly later on. But why would younger glacial deposits weather so much faster?

The most likely explanation is that freshly ground rock has far more weatherable material—more of the kinds of fresh, unweathered silicate grains that are most vulnerable to the weathering process. As these vulnerable minerals are preferentially removed through time, only more resistant minerals are left, and rates of weathering slow.

Another part of the uplift weathering hypothesis relates to the effect of grain sizes on weathering (see Figure 5-20). Finer grain sizes expose more mineral surface area and cause faster weathering early in the process, but the finer sizes of eroded material disappear earlier as weathering consumes them. The coarser grain sizes that remain weather more slowly because they expose less surface area per unit of volume. Coarser fragments may also develop an outer

coating, or "rind," of weathering-resistant material that protects fresher material in their interiors and slows the weathering attack.

5-10 Uplift and Chemical Weathering

The uplift weathering hypothesis focuses on evidence that exposure of fragmented and unweathered rock is a key factor in the intensity of chemical weathering. It then links this evidence to the fact that freshly fragmented rock is exposed mainly in regions of tectonic uplift.

Several factors increase rates of exposure of fresh rock in uplifting areas. Mountains and plateaus have steep slopes both on their margins and in valleys between high peaks. Erosional processes known as **mass wasting** are unusually active on such slopes. Mass-wasting processes include rock slides and falls, flows of water-saturated debris, and a host of other processes that dislodge everything from huge slabs of rock to loose boulders, pebbles, and soil. Every event that removes overlying debris exposes fresh bedrock and unweathered material. Many high-mountain slopes consist almost entirely of fresh debris moving downslope (Figure 5-22).

Another important factor is earthquakes. Mountains and high plateaus are built by tectonic forces that push together and stack huge slivers of faulted rock at the margins of converging plates. This stacking process is accompanied by earthquakes that generate large amounts of energy, shake the ground, and dislodge debris. Even more fresh rock is exposed as a result.

A third important characteristic of steep slopes is that they are focal points for orographic precipitation (see Chapter 2). When warm air is forced up and over high terrain and then cooled, water vapor condenses and precipitation occurs. High but narrow mountain belts in the tropics and mid-latitudes capture much of the moisture carried by winds. In addition, large plateaus such as the Tibetan Plateau create their own wet (monsoonal) circulations by pulling moisture in from adjacent oceans. Major erosion events also occur in high-mountain areas when heavy rain falls on snow. Heavy precipitation favors chemical weathering.

Glacial ice also enhances chemical weathering in high terrain. Uplift can elevate rock surfaces to altitudes where temperatures are cold enough for mountain glaciers to form. Mountain glaciers pulverize blocks of underlying bedrock and deposit the debris in moraines at lower elevations. As we saw in the case study of the Wind River Range, glacial grinding greatly enhances rates of chemical weathering.

All these factors (steep slopes, mass wasting, earthquakes, heavy precipitation, and glaciers) are present in high mountains and plateaus. The uplift weathering

FIGURE 5-22
Debris on steep slopes
Steep slopes of actively eroding mountains consist of highly fragmented debris periodically dislodged downslope. (PHOTOSPHERE IMAGES/PICTURE QUEST.)

hypothesis proposes that uplift accelerates chemical weathering through the combined action of these processes (Figure 5-23). Faster weathering draws more CO_2 out of the atmosphere and cools global climate toward icehouse conditions. Conversely, during times when uplift is less prevalent, chemical weathering is slower, and CO_2 stays in the atmosphere and warms the climate, producing greenhouse conditions.

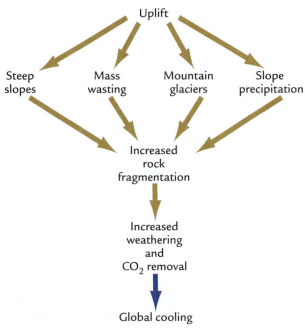

FIGURE 5-23
Uplift weathering hypothesis
Active tectonic uplift produces several tectonic and climatic effects that cause strong weathering of freshly fragmented rock. This process removes CO_2 from the atmosphere and cools global climate.

The two major kinds of plate tectonic processes that cause uplift have different implications for the uplift weathering hypothesis. The first process, subduction of ocean crust underneath continental margins, is an integral part of plate movements and a process that is continually active in many regions on Earth. Because subduction occurs relatively steadily over time, the total amount of high mountain terrain on Earth is likely to be relatively constant through time, even though the locations and heights of individual mountain ranges vary considerably.

The second process that creates high terrain is the collision of continents, and these events are far less common. Collision between India and Asia over the last 55 million years created the Tibetan Plateau in southern Asia, but no plateau-like feature remotely close in size to Tibet existed on Earth between 240 and 55 million years ago because no major continental collision of this kind occurred during that interval. Earlier, between 325 and 240 million years ago, the collisions between Gondwana and other continents that created the supercontinent Pangaea also formed a plateau of moderate size in east-central Europe, as well as high mountain ranges in eastern North America (the Appalachians) and in northwestern Africa.

The uplift weathering hypothesis focuses mainly on plateaus created by occasional collisions of continents, rather than on the ever-present mountain belts. As Table 5-3 indicates, times of continental collisions that created plateaus match times of glaciations over the last 325 million years. Like the BLAG hypothesis, the uplift weathering hypothesis is consistent with the icehouse-greenhouse-icehouse climatic sequence. But if recent discoveries prove correct, neither the uplift weathering hypothesis nor the BLAG hypothesis nor the polar position of Gondwana is a complete explanation for the

Table 5-3 Evaluation of the Uplift Weathering (CO_2 Removal) Hypothesis			
Time (Myr ago)	Ice sheets present?	Continents colliding?	Hypothesis supported?
325–240	Yes	Yes	Yes
240–35	No	No	Yes
35–0	Yes	Yes	Yes

short glaciation in the Sahara near 445 million years ago (see Box 5-1).

5-11 Case Study: Weathering in the Amazon Basin

The effect of uplift on chemical weathering can also be evaluated by examining the drainage basin of the Amazon River of South America (Figure 5-24). This region can be divided into two major units: (1) the low-lying Amazon Basin, where trade winds blowing westward from the Atlantic Ocean bring frequent precipitation to the rain forests of Brazil, and (2) the high-elevation eastern slopes of the Andes Mountains, which collect most of the rest of the incoming Atlantic precipitation.

Scientists have determined the regional distribution of chemical weathering in this drainage basin by sampling the amount of chemically weathered ions flowing down to the Amazon River in dissolved form. They found that the upper tributaries of the Amazon emerging from the foothills of the Andes carry almost 80% of the total dissolved chemical load discharged by the river when it enters the Atlantic. Despite its vast size and warm moist environment, the lower Amazon Basin adds only the remaining 20% of the total. Most of the chemical weathering in the Amazon drainage basin occurs in the Andes, at rates per unit area that are a factor of 40 higher than those in the lowlands.

How can this be so? This evidence seems at odds with what the eye actually sees in the two regions. In the lower Amazon rain forest, highly weathered clays that are the end products of intense chemical weathering dominate. In the Andes, rock debris produced by strong physical-mechanical weathering dominates, with little apparent evidence of intense chemical weathering.

The answer to this mystery is deceptively simple. The lower Amazon Basin is a place where chemical weathering does indeed dominate in percentage terms, but in which the fresh minerals have long since been "used up" in the weathering process. The only fresh, unweathered bedrock remaining in the lowlands lies buried hundreds of meters beneath a protective cover of highly weathered clays, out of reach of intense weathering processes. These older clays at and near the surface are the end products of slow bedrock weathering over many tens of millions of years, and they have little weatherable material left. As a result, the average rate of chemical weathering in this region is extremely low.

In contrast, the physical impacts of active uplift in the Andes (steep slopes, earthquakes, mass wasting, heavy precipitation, and glacial erosion) combine to generate a continual supply of fresh, finely ground

FIGURE 5-24

Weathering in the Amazon Basin

Almost 80% of the chemically weathered ions that reach the Atlantic Ocean from the Amazon River come from the small area of the eastern Andes, and just 20% from the extensive lowlands of the Amazon Basin.

rock debris for weathering. Some of this weathering occurs on the steep upper slopes of exposed high terrain even in the absence of much vegetation or soil cover. Much of it occurs in basins on the lower flanks of the mountains, where soils and vegetation have gained a foothold yet the supply of fresh unweathered rock debris from higher-elevation streams and rivers is continuous.

The lack of obvious visible chemical weathering in the Andes has two explanations. First, chemical weathering products such as clays are continually overwhelmed by the much larger supply of physically fragmented debris cascading down the steep slopes. Second, the fine clays and other products of weathering are soon removed from the steep slopes by fast-flowing streams and rivers and carried away to the Amazon lowlands or the ocean.

The Amazon Basin studies confirm that the rate of chemical weathering is rapid in the Andes, and presumably in many of Earth's other high-elevation regions as well, even though the visible effects of chemical weathering are not apparent. These studies also show that some warm, wet, vegetated regions may be places of surprisingly slow chemical weathering.

5-12 Weathering: Both a Climate Forcing and a Feedback?

The original uplift weathering hypothesis left an important issue unresolved. It did not specify a negative feedback that would act as a thermostat and moderate the climatic effects that uplift and weathering produce. Without such a thermostat, what would stop rapid uplift from accelerating chemical weathering to the point where Earth would freeze? As we have seen in Chapter 4, this nearly happened some 700 million years ago. And why wouldn't Earth overheat during times when uplift was minimal?

One possible mechanism that could moderate the degree of uplift-induced climate change is the total amount of fresh rock exposed at Earth's surface. Plate tectonic processes that cause uplift only affect the relatively small areas actively involved in subductions and collisions. The small size of these areas in turn limits both the amount of exposure of fresh rock at any one time and the intensity of cooling caused by uplift.

In addition, the effects of the uplift weathering processes are probably opposed by the chemical weathering thermostat. Uplift of geographically limited regions (perhaps 2% of Earth's total land area) could drive climatic cooling by promoting increased chemical weathering and CO_2 removal from the atmosphere, but chemical weathering on the other 98% of the continental area might well slow with the onset of cooler, drier climates and the reduction in vegetation cover. A slowing of the rate of CO_2 removal would leave more CO_2 in the atmosphere and moderate the overall cooling. In the end, the uplift-induced weathering increase would succeed in causing a net global cooling, but not nearly so large a cooling as would have occurred without the negative weathering feedback.

In Summary, both the BLAG (spreading rate) hypothesis and the uplift weathering hypothesis seem to provide plausible explanations of the icehouse-greenhouse changes of climate over the last 325 million years (see Tables 5-2, 5-3). In Chapter 7, we will revisit both hypotheses by examining in greater detail the sequence of changes that transformed the warm greenhouse climate of 100 million years ago to the later (and present) icehouse climate.

Key Terms

continental crust (p. 98)

ocean crust (p. 98)

mantle (p. 98)

lithosphere (p. 99)

asthenosphere (p. 99)

tectonic plates (p. 99)

divergent margins (p. 99)

convergent margins (-p. 99)

subduction (p. 99)

continental collision (p. 100)

transform fault margins (p. 100)

magnetic field (p. 100)

paleomagnetism (p. 100)

magnetic lineations (p. 101)

seafloor spreading (p. 102)

polar position hypothesis (p. 102)

Gondwana (p. 103)

Pangaea (p. 104)

evaporite (p. 108)

red beds (p. 109)

spreading rate hypothesis (p. 110)

hot spots (p. 111)

uplift weathering hypothesis (p. 113)

mass wasting (p. 115)

Review Questions

1. Does each lithospheric plate correspond to an individual continent or ocean basin?

2. What kind of physical behavior in Earth's deeper layers allows the plates to move?

3. Explain how paleomagnetism tells us about past latitudes of continents.

4. Explain how paleomagnetism tells us about rates of spreading at ocean ridges.

5. Do glaciations *always* occur when continents are located in polar positions?

6. What are the major characteristics of the climate of Pangaea?

7. What is the central concept behind the BLAG (spreading rate) hypothesis?

8. What role does chemical weathering play in the BLAG hypothesis?

9. Write a chemical reaction showing how weathering removes CO_2 from the atmosphere.

10. How soon after deposition does freshly fragmented debris undergo most chemical weathering?

11. Why is chemical weathering faster in the eastern Andes than in the Amazon lowlands?

12. How could chemical weathering be both the driver and the thermostat of Earth's climate?

13. Fast subduction in the modern Pacific Ocean carries down sediments with low amounts of $CaCO_3$, while almost no subduction occurs in the Atlantic Ocean, with its carbonate-rich sediments. What would happen if subduction suddenly began in the Atlantic and replaced an equal amount of subduction in the Pacific?

Additional Resources

Basic Reading

"Geology: Plate Tectonics." University of California Museum of Paleontoglogy, accessed January 3, 2013. http://www.ucmp.berkeley.edu/geology/tectonics.html.

Advanced Reading

Berner, R. A. 1999. "A New Look at the Long-Term Carbon Cycle." *GSA Today* 9:1–6.

Blum, J. D. 1997. "The Effect of Late Cenozoic Glaciation and Tectonic Uplift on Silicate Weathering Rates and the Marine $^{87}Sr/^{86}Sr$ Record." In *Tectonic Uplift and Climate Change*, ed. W. F. Ruddiman. New York: Plenum Press.

Chamberlain, T. C. 1899. "An Attempt to Frame a Working Hypothesis of the Cause of Glacial Periods on an Atmospheric Basis." *Journal of Geology* 7:545–84, 667–85, 751–87.

Kutzbach, J. E. 1994. "Idealized Pangaean Climates: Sensitivity to Orbital Parameters." *Geological Society of America Special Paper* 288:41–55.

Raymo, M. E., W. F. Ruddiman, and P. N. Froelich. 1986. "Influence of Late Cenozoic Mountain Building on Ocean Geochemical Cycles." *Geology* 16:649–53.

Stallard, R. F., and J. E. Edmond. 1983. "Geochemistry of the Amazon 2: The Influence of the Geology and Weathering Environments on the Dissolved Load." *Journal of Geophysical Research* 88: 9671–88.

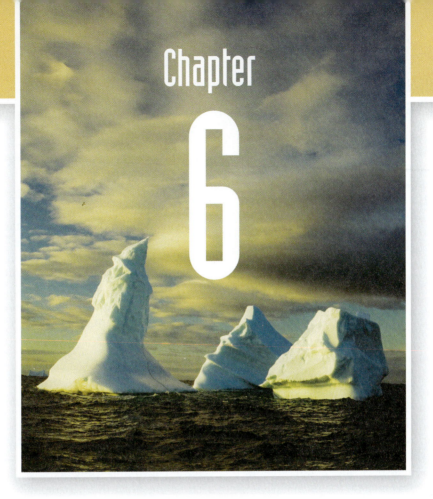

Greenhouse Climate

E arth's geography is much easier to reconstruct for 100 million years ago than for earlier times. The positions of all the continents are known, as are the locations of the edge of the sea along continental margins and the shapes of the ocean basins. Regional and global average rates of seafloor spreading are also better constrained. In addition, climate scientists know much more about Earth's climate 100 million years ago, including the fact that it was warm enough at the South Pole to keep large (or possibly any) ice from accumulating. With this array of evidence, scientists can ask an important question about this interval: Was this global warmth caused by a high level of atmospheric CO_2? The answer to this question potentially holds lessons about our climatic future in a world warmed by rising levels of the same gases. This chapter also explores the reasons why sea level 100 million years ago was higher than today and assesses the effect of this higher sea level on climate. We also investigate the climatic and other environmental effects of a giant asteroid impact 65 million years ago, during the later stages of warm greenhouse conditions. And we examine a brief interval of unusual warmth near 55 million years ago.

What Explains the Warmth 100 Million Years Ago?

Around 175 million years ago, the giant single continent of Pangaea began to break apart. By 100 million years ago, most of the continents had separated from one another, producing a very different-looking Earth consisting of a half-dozen smaller continents. In addition, global sea level stood at least 80 to 100 meters higher than it does today, and a shallow layer of ocean water flooded continental margins and low-lying interior areas.

The geographic effect of this flooding was to fragment the existing continents into even smaller land areas (Figure 6-1), making the geography of this greenhouse world even more unlike the single Pangaean landmass of 200 million years ago. Geologists call this interval the middle **Cretaceous**, a word meaning "abundance of chalk," because marine limestones deposited by the high seas during this interval are common around the world. This interval is important to climate scientists because geologic records contain no evidence of permanent ice anywhere on Earth, even on the parts of the Antarctic continent situated right over the South Pole (see Figure 6-1). Much of this interval seems to have been a warm greenhouse world.

In the Northern Hemisphere, evidence for unusual warmth comes from several fossil indicators, including warm-adapted vegetation such as broad-leafed evergreen trees that hold their leaves throughout the year except for a brief interval of leaf fall and regrowth. Other evidence includes warm-adapted animals such as dinosaurs, turtles, and crocodiles, all of which existed north of the Arctic Circle (Figure 6-2). This evidence contrasts markedly with the cold climatic conditions at these latitudes today.

The rest of Earth's surface was also warmer than today. Dinosaurs lived even on those parts of Australia and Antarctica lying south of the Antarctic Circle. At lower latitudes, coral reefs indicative of warm tropical ocean temperatures extended some 10° of latitude farther from the equator than they do today (40° vs. 30°).

6-1 Model Simulations of the Cretaceous Greenhouse

During the 1980s, the climate modeler Eric Barron and his colleagues pioneered the search for an explanation of the warmth 100 million years ago. They used constraints provided by plate tectonic reconstructions of Earth's geography to run general circulation model (GCM) simulations of the climate of the

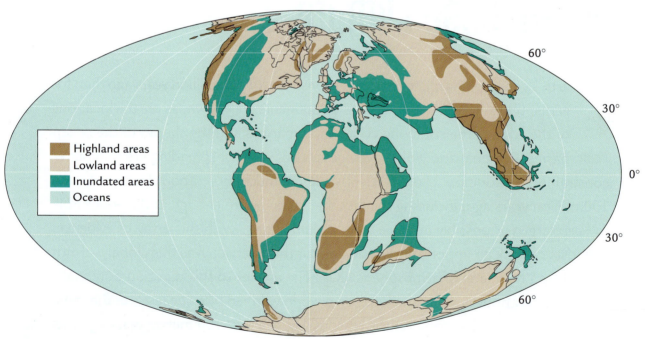

FIGURE 6-1

The world 100 Myr ago

By 100 Myr ago, plate tectonic processes had broken the giant Pangaean continent into separate smaller continents that were flooded by shallow seas. (D. MERRITTS ET AL., *ENVIRONMENTAL GEOLOGY,* © 1997 BY W. H. FREEMAN AND COMPANY.)

A   B

FIGURE 6-2

Evidence of greenhouse warmth 100 Myr ago

Vegetation and animals that appear to have been warm-adapted lived in both polar regions 100 Myr ago: fossils of breadfruit trees like those that are found today in the tropics (A) and dinosaurs (B), many species of which lived poleward of the Arctic and Antarctic circles.

(A: SWEDISH MUSEUM OF NATURAL HISTORY, YVONNE ARREMO, STOCKHOLM; B: CHRIS BUTLER/SCIENCE SOURCE.)

Cretaceous world. The surface temperature estimates from the resulting simulations were plotted as averages over lines of latitude (Figure 6-3). These plots represent the complex spatial pattern of temperature changes across Earth's surface in a shorthand form. The latitudes shown in this and subsequent figures are plotted in such a way as to correct for the much larger area of Earth's surface at low latitudes than in polar regions. These plots show all latitudes in proportion to the area they actually cover.

Initial reconstructions of Earth's present temperature trend are also shown in Figure 6-3 for comparison. Temperatures near the equator 100 million years ago were estimated to be a few degrees higher than

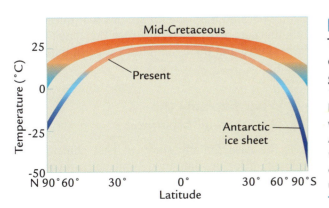

FIGURE 6-3

The Cretaceous target signal

Climate scientists used geologic data (faunal, floral, and geochemical) to compile an initial estimate of temperatures 100 Myr ago. Temperatures were warmer than they are today at all latitudes, especially in polar regions. (ADAPTED FROM E. J. BARRON AND W. M. WASHINGTON, "WARM CRETACEOUS CLIMATES: HIGH ATMOSPHERIC CO₂ AS A PLAUSIBLE MECHANISM," IN *THE CARBON CYCLE AND ATMOSPHERIC CO₂: NATURAL VARIATIONS, ARCHAEAN TO PRESENT*, ED. E. T. SUNDQUIST AND W. S. BROECKER, GEOPHYSICAL MONOGRAPH 32 [WASHINGTON, D.C.: AMERICAN GEOPHYSICAL UNION, 1985].)

those today, but those at the North Pole were as much as 20°C higher and at the South Pole, 40°C or more higher. The reason for the especially large temperature difference at the South Pole is the absence of the Antarctic ice sheet in the Cretaceous model simulation. Today, this ice sheet reaches elevations of 3 to 4 kilometers, where temperatures are 20° to 25°C colder than at sea level because of the lapse rate effect (see Chapter 2). Because the top of the modern ice sheet represents Earth's "surface" at the South Pole, the surface temperature for that area today is bitterly cold.

Geologic evidence indicates that this ice sheet did not exist during much of Cretaceous time. As noted in Chapter 3, GCMs can simulate only a few years or decades of elapsed time, and ice sheets cannot grow or melt that quickly. Because of this time limitation, the presence or absence of ice sheets is specified *in advance* as boundary-condition input to this kind of GCM simulation. With the geologic evidence indicating that no ice sheets existed for much of mid-Cretaceous time, the boundary condition specified for these simulations did not include an ice sheet. Without the Antarctic ice sheet, Earth's surface at the South Pole was close to sea level and the simulated polar temperatures were cold, but much less frigid than if an ice sheet had been present.

As part of this series of experiments, Barron and colleagues also defined a "target signal"—an independent estimate of Cretaceous climate based on the observations then available—to compare against the simulations. They compiled this target signal from estimates of surface temperatures 100 million years ago based on past distributions of temperature-sensitive vegetation, animals, and geochemical indicators. Every target signal estimate of past temperature includes some amount of uncertainty. The red and blue regions in Figure 6-4 show the range of possible temperature values they found.

Several experiments were then run as sensitivity tests to evaluate the significance of different factors (see Chapter 3). The first experiment used the Cretaceous geography (land-sea distribution and mountain elevations) as input to the model simulation. CO_2 levels were kept at modern (preindustrial) levels. For this simulation, the tropical temperatures simulated by the model fell within the range of estimated (target) temperatures, but model-simulated temperatures poleward of 40° latitude were considerably colder than the geologically based estimates (see Figure 6-4).

The second simulation retained the changes in geography but added another factor: a different CO_2 concentration. A range of indirect evidence suggests that atmospheric CO_2 concentrations 100 million years ago were somewhere between four and ten

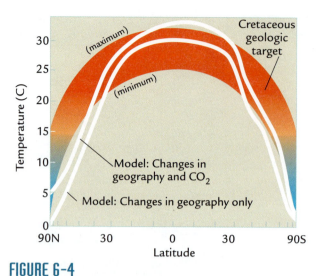

FIGURE 6-4
Data-model comparisons

Climate model simulations are matched against the temperature target signal for 100 Myr ago provided by geologic data. One simulation based on the altered geography of 100 Myr ago and another that also incorporates higher CO_2 values reproduce some aspects of the target signal but simulate a steeper pole-to-equator gradient. (ADAPTED FROM E. J. BARRON AND W. M. WASHINGTON, "WARM CRETACEOUS CLIMATES: HIGH ATMOSPHERIC CO_2 AS A PLAUSIBLE MECHANISM," IN *THE CARBON CYCLE AND ATMOSPHERIC CO_2: NATURAL VARIATIONS, ARCHAEAN TO PRESENT*, ED. E. T. SUNDQUIST AND W. S. BROECKER, GEOPHYSICAL MONOGRAPH 32 [WASHINGTON, D.C.: AMERICAN GEOPHYSICAL UNION, 1985].)

times higher than the recent preindustrial average, and Barron and colleagues chose a CO_2 value slightly more than four times the modern (preindustrial) level. The higher CO_2 level in this second simulation further warmed the planet and removed much of the mismatch with the target signal at high latitudes, but this time the equatorial temperatures became hotter than the range suggested by geologic evidence.

From one perspective, the general agreement between the two model simulations and the target signal in Figure 6-4 could be judged a success; the disagreements are not that large for a time so far in the past. Nevertheless, these mismatches subsequently became the focus of studies that questioned both the data used to reconstruct the target signal temperatures and the assumptions used in the model simulations.

6-2 What Explains the Data-Model Mismatch?

As is the case for any study of past climates, the explanation for the data-model mismatches could lie either in the geologic data, in the climate model simulation, or in some combination of the two.

PROBLEMS WITH THE MODELS? One possible explanation for the mismatch has focused on shortcomings in the climate models. At the time Barron and colleagues ran their initial experiments, the treatment of ocean circulation in O-GCMs (see Chapter 3) was still very crude. The process of upwelling of cool water along coastlines and near the equator was not included in global-scale ocean models, and deepwater circulation was handled with little success or not even attempted. In view of these shortcomings, it seemed possible that climate scientists were asking more from models than they could deliver at that point in their development.

An idea that initially intrigued many climate scientists was the possibility that the Cretaceous ocean operated in a fundamentally different way compared to the present ocean. Today, the surface ocean transports about half as much heat poleward as does the atmosphere. Some scientists suggested that if the ocean carried twice as much heat to the poles in the Cretaceous as it does today, the two major problems in the data-model mismatch might be resolved. In this **ocean heat transport hypothesis**, the poles would be warmed by the greater influx of ocean heat from low latitudes, and the tropics would not be heated as much because of their greater loss of heat to the poles.

A related idea was that deep water might have formed in the northern subtropics in regions of very high ocean salinity (>37‰), rather than in polar regions like today. The concept was that high salinities could have made surface waters dense enough to sink into the deep ocean as **warm saline bottom water**. A strong flow of this warm deep water from the tropics to the poles might then have contributed to the poleward heat flux needed to warm polar regions and resolve the data-model mismatch.

More recent experiments with improved ocean models have not supported the ocean heat transport hypothesis. The models do not show significant increases of poleward transport of heat by the ocean in a climate warmer than that today. To a first approximation, the combined transfer of heat by the ocean and atmosphere does not change much.

A second kind of data-model mismatch occurred in climate simulations for 100 million years ago, and this mismatch persists in model simulations of the warm climates that continued over the next several tens of millions of years. Data from warm-adapted vegetation (early palm-like trees) and from fossil reptiles suggest that continents at high and middle latitudes had moderate climates and did not freeze in winter. In contrast, all the GCM experiments described above (those with altered geography, higher CO_2 levels, and increased ocean heat transport, alone or in combination) simulated hard freezes in winter across the interiors of the northern hemisphere continents.

As a further test of the possible role of the ocean, an experiment was run in which warm waters were specified (imposed) in the Arctic Ocean as an initial boundary condition before running the simulation. One purpose of this experiment was to find out whether a very warm Arctic Ocean could explain the warm interiors of the northern continents. The simulation showed that even a warm Arctic Ocean failed to keep the interiors of the northern continents warm enough in winter to prevent freezing because the winter heat losses were too large.

PROBLEMS WITH THE DATA? A second path of investigation of the data-model mismatch focused on the possibility that the proxy data used to reconstruct past climate might have been giving an incorrect "target signal" for the models. Several studies have produced evidence in support of this explanation.

The tropics have been the main focus of these reassessments. Most of the data used to reconstruct Cretaceous temperatures at lower latitudes have come from geochemical analyses of ocean plankton using oxygen isotope ratios (Appendix 1). The shells of the plankton carry oxygen isotope ratios that reveal changes in the temperature of the ocean water in which they form. New evidence suggests that tropical temperatures were considerably warmer than the target signal values shown in Figure 6-4.

After the tropical plankton formed their shells in warm surface water, they died and fell to the seafloor, where water temperatures were 10°–15°C cooler than those at the surface. As the shells lay on the seafloor or in the uppermost sediments before being completely buried, they were bathed by cooler bottom waters. Some of these shells were partly recrystalized. Scientists found that the more pristine the shells they examined, the warmer the temperatures indicated by their oxygen isotope ratios, whereas the shells that were more heavily recrystalized yielded cooler temperature estimates. This difference indicated that chemical alteration on the seafloor had reset the isotopic temperatures toward cooler values. The best-preserved shells indicated that Cretaceous temperatures in the tropics were as warm as 35°C, or at least 5°C above the initial target signal shown in Figure 6-4. Several other kinds of ocean fossils from the tropics give a similar result: warmer tropical temperatures from the best-preserved shells and from those most resistant to alteration.

This explanation may resolve the data-model mismatch shown in Figure 6-4. The much warmer tropical target signal suggests that CO_2 levels were actually considerably higher than those used in the model simulation with altered geography and higher CO_2 levels. These higher CO_2 levels also warmed the higher latitudes enough to explain why the interiors of the mid-latitude continents didn't freeze in winter.

In Summary, CO_2 levels even higher than four times the modern level appear to be a major cause of the warmer climate 100 million years ago.

6-3 Relevance of Past Greenhouse Climate to the Future

The results from studies of Cretaceous climate fit into a larger picture of the effect of CO_2 on Earth's climate. Climate scientists have run a series of GCM sensitivity tests using Earth's present geography as a common boundary-condition input to all simulations, but allowing the level of atmospheric CO_2 to vary from as low as 100 ppm to as high as 1,000 ppm. The preindustrial ("modern") value of 280 ppm lies in the lower part of this range.

This set of simulations (Figure 6-5) shows that global average temperature rises with increasing CO_2 levels, but the relationship is not linear (directly proportional). Instead, Earth's temperature reacts strongly to CO_2 changes at the lower end of the range, but much less so to changes at the high end of the range.

A major reason for the shape of this curve is the positive feedback effect provided by snow and sea ice. At low (<200 ppm) CO_2 values, sea ice and snow advance well past their average limits today and cover a relatively large fraction of Earth's high and middle latitudes. These bright surfaces reflect incoming solar

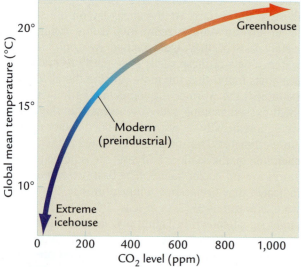

FIGURE 6-5

Effect of CO_2 on global temperature

Climate model simulations of the effects of changing atmospheric CO_2 levels on global temperature show greater warmth for higher CO_2 concentrations. (ADAPTED FROM R. J. OGLESBY AND B. SALTZMAN, "SENSITIVITY OF THE EQUILIBRIUM SURFACE TEMPERATURE OF A GCM TO CHANGES IN ATMOSPHERIC CARBON DIOXIDE," *GEOPHYSICAL RESEARCH LETTERS* 17 [1990]: 1089–92.)

radiation back to space, further cooling the planet (see Chapter 2). With so much of the planet covered by snow and ice (especially in winter), even small changes in CO_2 can have a relatively large impact in altering Earth's average temperature. CO_2 increases greatly reduce the extent of this bright reflective area, while CO_2 decreases enlarge it. This albedo-temperature feedback makes the climate system in an icehouse world react strongly to changes in CO_2.

In contrast, much higher CO_2 values reduce the average amount of snow and ice present at high latitudes. At CO_2 levels near 1,000 ppm, little or no sea ice is present in the Arctic in summer, and only the central Arctic has ice in winter. These small areas of snow and ice provide little positive feedback to changes in CO_2. As a result, the climate system in a greenhouse world is much less sensitive to changing CO_2 levels.

A second factor that accounts for the slower rise of temperature at higher CO_2 levels is **CO_2 saturation**. As CO_2 concentrations rise, the atmosphere gradually reaches the point at which further CO_2 increases have little effect in trapping additional back radiation from Earth's surface.

A third factor, but one acting in an opposite sense, is water vapor feedback (see Chapter 2). A warm atmosphere with CO_2 values of 1,000 ppm can hold much more water vapor than a cold atmosphere with values of 100 ppm. In this case, the positive feedback effect of water vapor on temperature grows stronger in warmer, higher CO_2 conditions, in contrast to the diminished effect of albedo-temperature feedback.

These sensitivity experiments, and others with different models, show that large ice sheets cannot exist anywhere on Earth when CO_2 concentrations exceed 1,000 ppm. At these higher CO_2 values, with global mean temperature well above 20°C, mean annual temperatures in polar regions exceed 0°C. Because ice sheets require mean annual temperatures well below freezing to persist through the summer melting season, they cannot survive in high CO_2 greenhouse warmth.

In the last 100 years, the CO_2 concentration in the atmosphere has risen from 280 ppm to nearly 400 ppm because of the industrial activities of humans. Projections for the next two centuries indicate increases to at least 550 ppm, and possibly 1,000 ppm or more, nearly as high as those in the Cretaceous. We are heading back into a greenhouse world, but this time at a far more rapid rate than anything found in the geologic record.

Why Were Sea Levels Higher 80 to 100 Million Years Ago?

Over tectonic time scales, the average level of the world ocean has risen and fallen by 200 meters or more against the margins of the continents. These changes, called marine **transgressions** and **regressions**, have been small in relation to the 4,000-meter or greater average depth of ocean basins, but they would have had significant climatic effects in the regions that were alternately flooded and exposed.

Many continental margins are flat, with changes of just 1 part in the vertical for every 1,000 parts in the horizontal. As a result, the vertical sea level changes of hundreds of meters in the past translate into horizontal movements of the coastline measured in hundreds of kilometers. During times of high sea level, the ocean flooded shallow low-lying interiors of continents and formed large seas. These long-term sea level changes also deposited and eroded thick sequences of coastal sediments.

When sea level is low, the coastline tends to be situated near the topographic break between the relatively flat continental shelf and the steeper continental slope (Figure 6-6A). At these times, erosion prevails on continental margins, and most of the eroded sediment is carried across the continental slope and dispersed down in the deeper ocean. When sea level is high, the ocean floods the low-gradient continental margin to depths of 100 meters or more (Figure 6-6B). At such times, sediment is deposited on the submerged continental shelf.

Local tectonic factors that cause uplift or subsidence of the land also affect the relative vertical position of the ocean margin against the land, even in the absence of changes in global sea level. These regional processes include mountain building, warping of Earth's surface because of deep-seated heating, and local depression and rebound of the land caused by the weight of ice sheets. Here, we ignore these local effects and focus on changes in **eustatic sea level**—changes that are global in scale.

The most persuasive evidence for higher global sea levels in the past comes from the presence of marine sediments deposited simultaneously on coastal margins and in shallow interiors of continents at levels well above present sea level. Deposition of marine sediments on several continents at the same time indicates that the changes in sea level are global in scale and not just local in extent.

In the Cretaceous world of 100 to 80 million years ago, the coastlines and interiors of most continents were flooded by an ocean that rose well above the modern level (see Figure 6-1). Areas flooded included much of southern Europe (Figure 6-7) and interior regions of North America penetrated by seaways linked to the Gulf of Mexico and the Arctic Ocean. Since that time, sea level has slowly fallen to its modern position, close to the lowest level on record (except for glacial intervals with more ice on land than now and sea levels more than 100 meters lower than today).

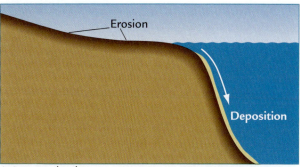

A Low sea level

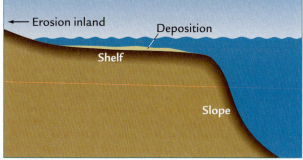

B High sea level

FIGURE 6-6
Sea level

When sea level is low (A), the coastline lies near the base of the continental shelf and sediment is deposited on the continental slope. When high sea levels flood the continental margin (B), more sediment is trapped and deposited on the submerged shelf.

The higher sea levels 80 to 100 million years ago have been attributed to two kinds of factors: (1) tectonically driven changes that altered the volume of the ocean basins and their capacity to hold water, and (2) changes in the volume of water in the ocean basins caused by variations in climate.

6-4 Changes in the Volume of the Ocean Basins

1. CHANGES IN THE VOLUME OF OCEAN RIDGES Ocean ridges owe their high elevations to unusual heating from molten material located below the surface of the ocean crust. Heating causes the rocks in the ridges to expand, and expansion of the rock causes the surface of the ocean crust to rise.

All ocean ridges today have crests that lie at an average depth of 2,500 meters below the sea surface. Away from the crest, the depth profiles of these ridges follow a simple equation:

$$\text{Ridge depth} = 2500 \text{ m} + 350 \text{ (crustal age)}^{1/2}$$
$$\text{(in meters)} \quad \text{(at 0 age)} \qquad \text{(in millions of years)}$$

FIGURE 6-7
Marine limestone exposed on land

Marine limestone deposits that today form the coasts of southern England and northern France are evidence of higher sea levels 100 Myr ago. (ANDREW WARD/LIFE FILE/PHOTODISC.)

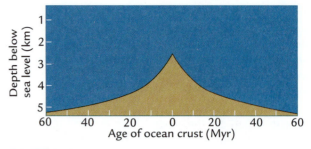

FIGURE 6-8
Subsidence of ocean ridges with time

All ocean ridges show the same average profile of age (time since formation) vs. depth. Heat elevates the ridge crests to a depth of 2,500 meters below the ocean surface, a level high above the rest of the seafloor. As the crust spreads away from the ridge crest and ages, it cools and contracts, rapidly at first and then more slowly. The crust eventually reaches a stable depth of more than 5,000 meters below sea level. (ADAPTED FROM J. G. SCLATER ET AL., "THE DEPTH OF THE OCEAN THROUGH THE NEOGENE," *GEOLOGICAL SOCIETY OF AMERICA MEMOIR* 163 [1985]: 1–19.)

This equation describes a ridge crest that starts at an initial depth of 2,500 meters below the sea surface and gradually deepens with age away from the ridge crest (Figure 6-8). Ridge crests that stand high above the rest of the seafloor because of heating from below initially subside rapidly because of rapid heat loss while moving away from the crest, but by 60 million years later, the crust and upper mantle have lost most of their excess heat, and the ridge elevations have reached a nearly stable depth of 5,500 meters. Local variations in depth of a few hundred meters around the values in this equation occur as a result of small-scale tectonic irregularities, but the mean values of ocean ridge depths follow the equation remarkably well throughout the world's oceans.

Paleomagnetic evidence from today's ocean shows that different ridges spread at different rates (see Chapter 5, Figure 5-17). Because all ridge depths are nearly constant with age (as shown by the preceding equation and Figure 6-8), crust of a given age (and a particular depth below sea level) will have been carried much farther from the ridge crest in a given amount of time in fast-spreading areas like the South Pacific than in slow-spreading ones like the North Atlantic. Fast spreading gives those ridges a wider elevation profile than the slow-spreading ones.

Through time, ridge profiles also vary on a *globally averaged* basis (Figure 6-9). At times like the present, when the globally averaged rate of seafloor spreading is relatively slow, mean ridge profiles are relatively narrow, and little water is displaced onto the continents. At times in the past when the average spreading rate was faster, mean ridge profiles would have been wider, pushing more ocean water up onto the land. Ridge profiles in the past can be reconstructed by resetting the ages of the past ridge crests to zero for the time being examined, using the dated magnetic lineations to recalculate the past ages of the ridge flanks relative to this adjusted "zero" age, and applying the equation shown above.

The size of the ridge-profile effect on sea levels 80 to 100 million years ago is not well constrained. One reason for this uncertainty is that the rate of spreading prior to 80 million years ago is not known for the seaway that existed in the tropics but has now been destroyed. In addition, estimated spreading rates for the few preserved areas of ocean crust in the northwest Pacific that formed during that interval have recently been revised downward.

2. COLLISION OF CONTINENTS Most plate tectonic movements do not change the net area of either

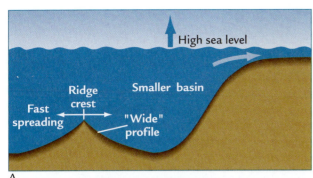

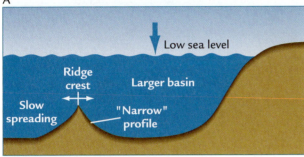

FIGURE 6-9

Spreading rates and sea level

Rates of spreading at ocean ridges vary widely. Fast spreading creates wider ridge profiles that reduce the volume of the ocean basins and displace more water onto the continents (A). Slow spreading produces narrower profiles that create larger ocean basins that can hold more seawater (B).

the oceans or the continents: creation of new ocean crust at ocean ridge crests is balanced by destruction of ocean crust subducting into trenches, leaving the area of the ocean basins constant. However, collision of continents does alter the area of the ocean basins and in that way affects sea level.

Because continental crust is low in density, colliding continents tend to float near Earth's surface rather than being pushed or pulled down into Earth's mantle (Figure 6-10). In the region where they collide, continental crust thickens from its normal value of 30 kilometers to about twice that amount. This process builds a high plateau that rises well above sea level and at the same time thickens the subsurface low-density "root" of the plateau as far as 60 to 70 kilometers below Earth's surface. In the upper 15 kilometers of Earth's crust, the thickening that creates the plateau occurs by movements along faults that cause thin slivers of crust to shear off and stack up on top of each other. Below a depth of 15 kilometers, thickening occurs when slow flow causes rock layers to be squeezed and folded.

Because collision drives continents together and forms plateaus with a double-thick crust, this thickening must result in a net loss in the area of continental crust. To a first approximation, the area of the plateau across which the crust doubles in thickness should equal the net loss of area of continental crust. This decrease in areal extent of the continents requires an equivalent increase in area of the ocean basins (see Figure 6-10). With a larger area of ocean to fill, the water level in the ocean should drop.

As noted in Chapter 5, continental collisions have occurred only sporadically through geologic time. One major collision began when northern India first made contact with southern Asia some 55 million years ago. This collision, still in progress, has increased the area of the ocean by some 2 million km^2 over the last 55 million years. As seawater has flowed in to fill this new area of ocean basin, estimates are that global sea level has fallen by at least 10 meters to as much as several tens of meters below the level of 100 million

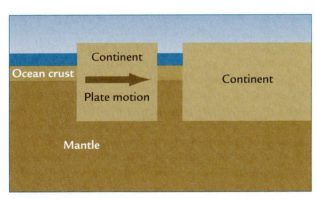

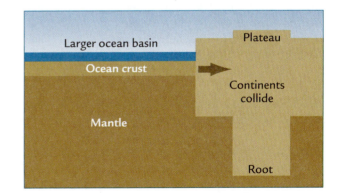

FIGURE 6-10

Continental collisions and sea level

When continents collide, the continental crust doubles in thickness and creates a high plateau with a thick low-density crustal "root." This thickening reduces the original area of the continents and increases the area of the ocean basins. The increased area and volume of the ocean basins causes sea level to fall.

years ago, a time when no collisions were occurring. Viewed in the reverse sense, the absence of continental collisions 80 to 100 million years ago would have kept sea level tens of meters higher than now.

3. TRANSFER OF CONTINENTAL SEDIMENTS TO THE OCEAN Along many ocean margins, sediments eroded from the continents are delivered to ocean trenches where they are subducted. But along those margins without subduction zones, ocean sediments from major rivers pile up on the seafloor and displace ocean water by reducing the volume of the ocean basins. The Bengal Fan in the northern Indian Ocean is an enormous sediment pile built by eroded sediments from the Himalaya Mountains to the north during the last 55 million years. Other large deposits include debris dumped into the Atlantic Ocean by the Amazon River from South America and by rivers like the Congo flowing from Africa. By some estimates, sediment filling of ocean basins by these eroded deposits can cause sea level increases of as much as 50 meters.

6-5 Climatic Factors

4. WATER STORED IN ICE SHEETS Continent-sized ice sheets several kilometers in thickness and thousands of kilometers in lateral extent can extract enormous volumes of water from the ocean and store it on land (Figure 6-11). Because no major ice sheets existed between 100 and 80 million years ago, little or no water was stored on land as ice. Today, the Antarctic ice sheet holds the equivalent of almost 60 meters of global sea level, and the Greenland ice sheet another 6 meters. The Antarctic ice sheet has come into existence and grown to its present size within the last 35 million years, and the Greenland ice sheet has done the same within the last few million years. Together, these ice sheets have extracted a volume of ocean water equivalent to 65 meters of global sea level. Their absence 80 to 100 million years ago would have made sea level 65 meters higher than today.

5. THERMAL EXPANSION AND CONTRACTION OF SEAWATER Ocean water has the capacity to expand and contract with temperature changes. The **thermal expansion coefficient** of water (the fractional change in its volume per degree of change in temperature) averages about 1 part in 7,000 for each 1°C of temperature change. Because of this behavior, even a constant amount of seawater would have lost volume during the cooling of the last 80 to 100 million years. Low-latitude surface waters have cooled by about 5°C over that interval, while the high-latitude surface ocean and the deep ocean have both cooled by 10°–15°C. Thermal expansion of the warmer ocean

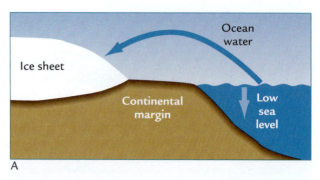

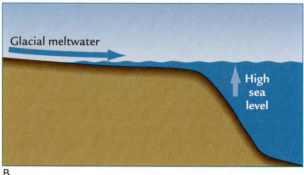

FIGURE 6-11
Ice sheets and sea level

Ice sheets covering large parts of continents hold volumes of water equivalent to tens of meters of global ocean level. Sea level falls when ice sheets are present on the land (A) and rises when they melt (B).

that existed 80 to 100 million years ago would have resulted in a global sea level about 7 meters higher than today.

The quantitative effects of several of the above factors on global sea level have to be adjusted for further complications (Box 6-1). One problem is the fact that water moving into (or out of) the ocean basins represents a large weight added to (or removed from) the underlying ocean crust, which sags (or rebounds) accordingly. This response of the ocean crust decreases the net magnitude of the change in sea level. The other complication has to do with translating a change in the volume of ocean water or in the volume of the ocean basins into actual movement of sea level against the complex shapes of the world's continental margins.

6-6 Assessment of Higher Cretaceous Sea Levels

Estimates of the highest Cretaceous sea level have ranged widely (Figure 6-12). During the 1980s, research scientists at the Exxon Oil Company estimated sea levels of 250 meters or more above present values. Subsequent estimates that have taken into account important processes that occurred after

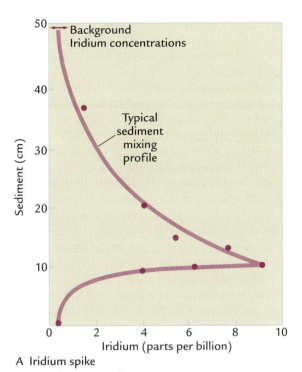

A Iridium spike

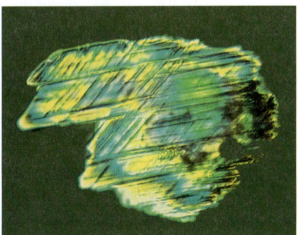

B Quartz grain subject to shock waves

FIGURE 6-13
Evidence of an asteroid impact

Ocean sediments containing a layer enriched in the element iridium (A) are evidence of a large asteroid impact 65 Myr ago. Sediments deposited in Montana 65 Myr ago (B) contain grains of quartz crisscrossed by multiple lineations produced by high-pressure shock waves from an asteroid impact. (A: ADAPTED FROM W. ALVAREZ ET AL., "EXTRATERRESTRIAL CAUSE FOR THE CRETACEOUS-TERTIARY EXTINCTION," SCIENCE 280 [1980]: 1095–1108; B: GLEN IZETT, WILLIAMSBURG, VA.)

Other supporting evidence for an impact event includes small grains of quartz with distinctive textures called "shock lamellae" that are formed by the shock wave of sudden pressures much larger than those found on Earth, even in highly explosive volcanoes (Figure 6-13B). The best candidate for the site

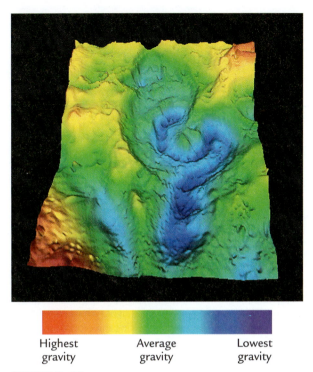

| Highest gravity | Average gravity | Lowest gravity |

FIGURE 6-14
A 65-Myr-old impact crater?

Mexico's Yucatán Peninsula has a circular area more than 200 km in diameter that is a good candidate for the site of the asteroid impact 65 Myr ago. The pattern shown is based on measurements of Earth's gravity that can detect low-density pulverized rock (in blue) and higher-density rock (in green and yellow). (B. SHARPTON, LUNAR AND PLANETARY INSTITUTE.)

of the impact 65 million years ago is a crater in eastern Mexico on the Yucatán Peninsula, between the Caribbean Sea and the Gulf of Mexico (Figure 6-14).

Climate scientists have concluded that asteroid impacts could have affected Earth on time scales ranging in duration from instantaneous to as long as a few hundred or even a few thousand years (Figure 6-15). The instantaneous effects were caused when the asteroid blasted a hole through Earth's atmosphere. The speed of the incoming asteroid, 20 kilometers per second, created a shock wave that moved outward, flattening objects for hundreds of miles around the impact site and heating Earth's atmosphere. Seismic waves sent through Earth's interior are thought to have been equivalent to those caused by an earthquake that would have measured 11 on the Richter scale, 100 to 1,000 times stronger than the strongest earthquakes in recorded human history.

Some of the water and rock in the vicinity of the impact were instantly vaporized by the heat of the impact and blasted back out into space through the hole in the atmosphere created by the incoming asteroid. The rest of the hot debris remained in Earth's

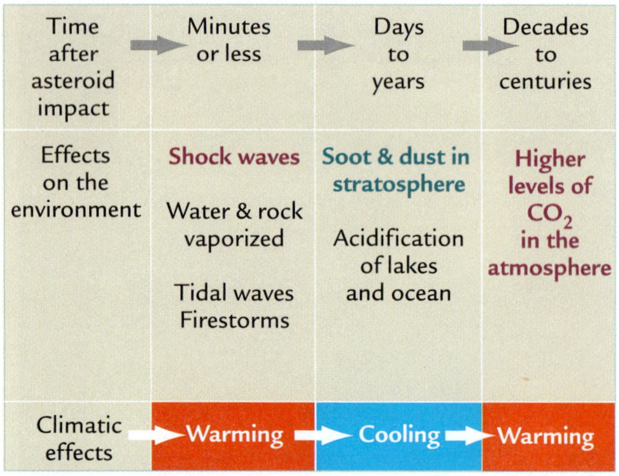

Time after asteroid impact	Minutes or less	Days to years	Decades to centuries
Effects on the environment	**Shock waves** Water & rock vaporized Tidal waves Firestorms	**Soot & dust in stratosphere** Acidification of lakes and ocean	**Higher levels of CO_2 in the atmosphere**
Climatic effects	Warming	Cooling	Warming

FIGURE 6-15

Climatic and environmental effects of asteroid impacts

The asteroid impact 65 Myr ago is thought to have had major effects on Earth's environment, including the extinction of over two-thirds of the species then alive. The likely climatic effects vary with the amount of elapsed time after the initial impact and appear to have been restricted to a few centuries.

atmosphere, heating it still further. The combined heating caused large-scale (possibly global) wildfires that ignited much of the aboveground vegetation and sent a thick layer of soot into the atmosphere.

Over the slightly longer term of days to years, the dust and soot that remained within the atmosphere spread around the planet, blocking most incoming solar radiation (see Figure 6-15). The debris injected into the lower atmosphere (the troposphere) would probably have been removed over a period of days to at most weeks, because rainfall quickly clears debris from that part of the atmosphere. In contrast, it would have taken months or years for the dust and soot injected into the stratosphere to settle out. The only means of removing debris at those altitudes is the slow pull of gravity on small particles. As a result, stratospheric particles (particularly dark soot) would have blocked significant amounts of sunlight for a year or more and cooled Earth's climate. Another likely effect over the course of a few years would have been the partial acidification of the oceans due to the creation of nitric acid (a component of acid rain) from atmospheric nitrogen, oxygen, and water vapor from the heat of the impact.

On the longer term of decades to centuries, the initial injection of carbon biomass into the atmosphere by burning should have produced higher CO_2 levels (see Figure 6-15). Those plants that recovered from the firestorm and began to grow would have pulled some of the excess CO_2 out of the atmosphere, but full recovery and the development of

new forms of vegetation to replace the ones that became extinct took longer. The warming induced by higher atmospheric CO_2 may have lasted for centuries or longer.

By any standard used, this impact event was an enormous short-term environmental catastrophe, but what were the long-range effects on Earth's climate? Surprisingly, little effect is evident. Earth's climate was in a warm greenhouse state at the time of the impact, and it remained there afterward.

Several problems make it difficult to determine the climatic effects of such a brief impact event. One is that rapid changes in sedimentary records are blurred and smeared by burrowing animals on land and by burrowers and bottom currents in the oceans (see Chapter 3). Distinctive impact-related features such as the iridium layer can still be detected despite this blurring (see Figure 6-13A) because they contrast clearly with the material in which they are deposited. But for most kinds of sedimentary archives, the signal from a single year (or even a decade or century) of climate that is warmer or cooler than normal will be mixed into and combined with the signals left by "normal" years and blurred beyond recognition.

A second problem with detecting the climatic effect of the impacts of even huge asteroids is that climate change was already occurring for other reasons before the impact. Records from marine sediments show a large (3°–4°C) warming and then cooling during the 500,000 years before the impact, but little or no additional change after it. Some land vegetation records suggest a long-term warming for hundreds of thousands of years after the impact, but it has not been demonstrated that this warming was related to the impact event. The increase in CO_2 levels caused by the immediate effects of the event seems unlikely to have persisted for that long.

In Summary, impact events such as the one at the Cretaceous-Tertiary boundary have clearly had apocalypse-like effects on the environment, including mass extinctions of organisms, transforming life on Earth. Despite this environmental apocalypse, the background state of the climate system 65 million years ago seems to have been changed little or not at all.

Greenhouse Episode 55 Million Years Ago: Another Thermostat Malfunction?

The warm greenhouse world was still in existence when a relatively brief episode of even warmer climate began near 55.5 million years ago. Because this warm episode falls near the boundary between the geologic epochs known as the Paleocene and the Eocene, it is called the PETM (Paleocene-Eocene Thermal Maximum). Within less than 20,000 years, ocean and terrestrial climate warmed by 5°C at low latitudes and 9°C at the poles. The excess warmth then slowly faded away over the next 150,000 years. Compared to the slow scale of tectonic changes, this thermal maximum was a brief event.

Evidence for the warming comes from the spread of warm-adapted plants and mammals into high latitudes and from the trend toward more negative values of the oxygen-isotope "paleothermometer" explained in Appendix 1 (Figure 6-16). Mg/Ca (magnesium/calcium) ratios in marine fossil shells also confirm a large warming.

Dramatic changes in the deep ocean accompanied the warming. Deep-ocean waters became far more acidic, dissolved $CaCO_3$ sediment on the seafloor, and left behind a clay-rich sediment layer. During this same interval, between a third and a half of the species of benthic foraminifera that had lived on the seafloor went extinct.

Another characteristic signature of this interval was a large (2–3‰) shift to more negative carbon isotope values in the shells of marine microfossils. This change requires a major release of carbon enriched in the ^{12}C isotope compared to the heavier ^{13}C form (see Appendix 2). Estimates of the amount of extra carbon needed to explain this negative excursion range from 2,500 to almost 7,000 billion tons.

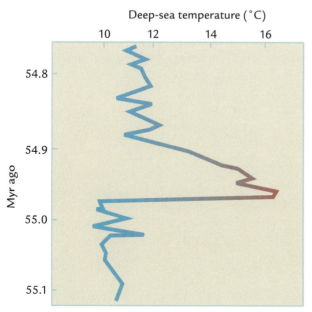

Deep-sea temperature (°C)

FIGURE 6-16

Unusual warmth 55 Myr ago

An episode of unusual warmth that developed near 55 Myr ago and persisted for tens of thousands of years warmed the ocean by several degrees centigrade.

The combined evidence of a warmer climate and a major carbon release points to major increases in atmospheric CO_2, CH_4, or both. Methane (CH_4) exists in the atmosphere as a greenhouse gas (see Chapter 2), but it is also present as a slushy mixture called **methane clathrate** not far below the ocean floor. If the subsurface ocean were to warm significantly, this ^{12}C-rich slush could be converted to a gas and released to the ocean and then to the atmosphere. Carbon enriched in ^{12}C is also present on the continents, both in forest vegetation and peat bog deposits.

The source of the excess carbon remains unclear. Also uncertain at this point is the initial trigger for the carbon releases. Proposed causes include volcanic eruptions on the seafloor, comet impacts, burning of carbon-rich peat deposits on land, and other tectonic instabilities such as slumps that affected methane clathrate deposits on the seafloor.

In any case, the addition of large amounts of CO_2 and/or CH_4 left the climate warmer for almost 200,000 years. Plausible mechanisms for removing the carbon excess from the atmosphere include increased chemical weathering and burial of the excess carbon in ocean sediments, and higher ocean productivity that transferred more organic carbon to the deep ocean.

This episode is potentially important to our future because it is a possible analog to the changes we humans will produce by our industrial releases of fossil-fuel carbon into the climate system. During the two centuries since the start of the Industrial Revolution, we have added roughly 300 billion tons of carbon to the atmosphere and ocean combined. As a result, climate has begun to warm and the oceans have become slightly more acidic, the same kinds of changes that occurred at the PETM.

The total amount of carbon that remains stored in economically exploitable fossil-fuel reserves (coal, oil, and natural gas) is thought to be about 800 billion tons, less than the amount emitted at the PETM. But the total amount of carbon estimated for all fossil-fuel resources (most of which are not presently commercially viable) is ten times as large (8,000 billion tons), slightly above the upper range of estimates for the PETM. As we start to extract and burn a substantial fraction of those resources, we could add enough extra carbon to the atmosphere to rival the amounts estimated for the PETM, potentially with similar effects on the environment.

The current rate of fossil-fuel carbon emissions is already five to ten times higher than it was during the early phase of the PETM, and the rate has been steadily increasing for decades. If this trend continues to grow, major environmental disruptions caused by burning fossil fuels will arrive within at most a century or two, rather than being spread out over many thousands of years as it was in the PETM.

Key Terms

Cretaceous (p. 122)

ocean heat transport hypothesis (p. 125)

warm saline bottom water (p. 125)

CO_2 saturation (p. 126)

transgressions (p. 127)

regressions (p. 127)

eustatic sea level (p. 127)

thermal expansion coefficient (p. 130)

hypsometric curve (p. 131)

methane clathrate (p. 135)

Review Questions

1. What evidence shows that the world was warmer 100 Myr ago than today?

2. Why do higher CO_2 concentrations make sense as an explanation for this greater warmth?

3. How well do model simulations capture the distribution of temperatures 100 Myr ago?

4. What are the possible causes of mismatches between the models and geologic observations?

5. Why do some climate scientists believe that increased ocean heat transport is required?

6. Which regions of the continents were flooded by high seas 100 Myr ago?

7. What were the major factors that explain higher sea level 100 Myr ago?

8. How did higher sea levels affect global climate?

9. Do sea level changes explain past glaciations?

10. Did asteroid impacts have long-lasting effects on climate?

Additional Resources

Basic Reading

Alvarez, L. W., W. Alvarez, F. Asaro, and H. V. Michel. 1980. "Extraterrestrial Cause for the Cretaceous-Tertiary Extinction." *Science* 208: 1095–1108.

Advanced Reading

Barron, E. J., S. L. Thompson, and S. H. Schneider. 1981. "An Ice-Free Cretaceous? Results from a Model Simulation." *Science* 212: 501–8.

Miller, K., et al. 2005. "The Phanerozoic Record of Global Sea-Level Change." *Science* 310: 1293–98.

Oglesby, R. J., and B. Saltzman. 1992. "Equilibrium Climate Statistics of a General Circulation Model as a Function of Atmospheric Carbon Dioxide. Part I: Geographic Distribution of Primary Variables." *Journal of Climate* 5: 66–92.

Zachos, J. C. et al. 2005. "Rapid Acidification of the Ocean During the Paleocene-Eocene Thermal Maximum." *Science* 308: 1611–1615.

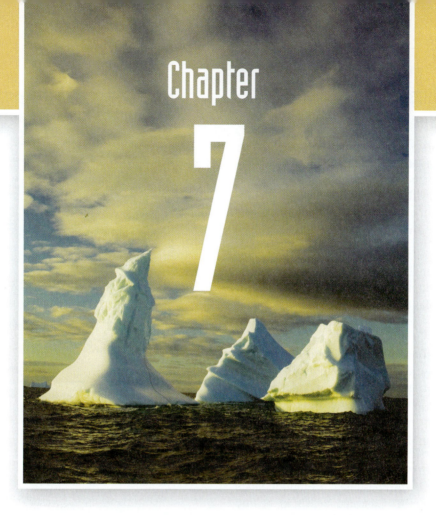

Chapter 7

From Greenhouse to Icehouse: The Last 50 Million Years

Our recent icehouse climate, marked by large fluctuations of ice sheets and sea ice in both hemispheres, is the result of a gradual cooling over many tens of millions of years. Earth's record of this transition is rich in information, including the initial appearance of mountain glaciers and continental ice sheets, the replacement of warm-adapted vegetation by cold-adapted forms, and a range of evidence indicating progressively cooler ocean temperatures. In this chapter, we first examine evidence showing when this greenhouse-to-icehouse cooling occurred. Then we explore three possible explanations for the cooling: changes in ocean heat transport tied to opening or closing of key interoceanic "gateways," the BLAG spreading rate hypothesis, and the uplift weathering hypothesis.

▶ Global Cooling Since 50 Million Years Ago

Earth has undergone a profound cooling at both poles and across the lower latitudes of both hemispheres during the last 50 million years. Both ice and vegetation have left abundant evidence of this cooling (Figure 7-1).

7-1 Evidence from Ice and Vegetation

As climate cools, two kinds of glacial ice form on land (see Chapter 2). Small mountain glaciers appear on the tops of high mountains, along with ice caps on low-lying Arctic islands, and large ice sheets cover much larger areas of the continents. Because temperatures vary widely from region to region and with altitude, the conditions that permit ice to persist year-round do not appear at the same time in all areas.

In the Southern Hemisphere, no evidence exists for any substantial amount of ice on Antarctica until 35 million years ago, when ice-rafted debris was first deposited in ocean sediments on the nearby continental margin. Since that time, the size of the Antarctic ice sheet has increased irregularly toward the present, with a major growth phase near 13 million years ago. Greater amounts of ice-rafted debris in nearby ocean sediments suggest additional increases in Antarctic ice during the last 10 million years. Coring by the ANDRILL Program along the Antarctic margin has shown that the Ross Ice Shelf and its upstream ice sources fluctuated widely from 13 to 2 million years ago, although the size of the central Antarctic ice sheet is not constrained. Today, even in a warm

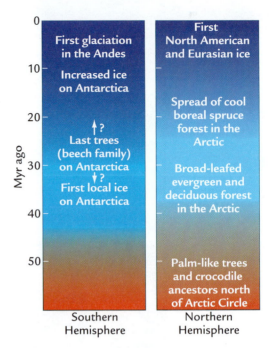

FIGURE 7-1
Global cooling for 50 Myr

Gradual cooling during the last 50 Myr is demonstrated by the first appearance of mountain glaciers and continental-scale ice sheets, and by a progressive trend toward cold-adapted vegetation in both hemispheres.

interglacial interval, more than 97% of Antarctica is buried under ice (Figure 7-2 left). Elsewhere, in the lower-middle latitudes of the Southern Hemisphere, the earliest evidence of mountain glaciers in the high Andes is dated to between 7 and 4 million years ago.

FIGURE 7-2
Cooling in Antarctica

Today, an ice sheet up to 4 km thick covers most of Antarctica, although mountains protrude through the thinner cover around the margins (left). Until 30 Myr ago, *Nothofagus* trees, members of the beech family like those living today at the southern tip of South America (right), still existed in parts of Antarctica. (LEFT: KEVIN SCHAFER/ALAMY; RIGHT: MICHAEL GIANNECHINI/SCIENCE SOURCE.)

FIGURE 7-3
Cooling in the Arctic
Warm-adapted breadfruit trees lived above the Arctic Circle in Canada until 60 Myr ago (left), but land around the Arctic Ocean is now covered by scrubby tundra vegetation grazed by caribou (right). (LEFT: SWEDISH MUSEUM OF NATURAL HISTORY, PHOTO BY YVONNE ARREMO, STOCKHOLM; RIGHT: REED KAESTNER/CORBIS.)

In the Northern Hemisphere, significant glacial ice first developed on Greenland sometime between 7 and 3 million years ago, although small mountain glaciers likely existed around the North Atlantic Ocean before that time (see Figure 7-1). The first evidence of glaciers in the high coastal mountains of southern Alaska dates to about 5 million years ago. The first continental ice sheets of significant size appeared 2.75 million years ago. These ice sheets grew and melted in repeated cycles, and their maximum size in North America and Eurasia increased after 0.9 million years ago. Although these northern ice sheets developed more than 30 million years later than the ones in Antarctica, they are a response to the same overall global cooling trend.

Fossil remains of vegetation in both hemispheres also indicate a progressive cooling over the last 50 million years. A form of beech tree called *Nothofagus* (Figure 7-2 right) lived on Antarctica before 40 million years ago, along with several types of ferns. This vegetation disappeared as climate became more frigid and ice spread across the continent. Today, the only vegetation on Antarctica is lichen and algae found in summer meltwater ponds in ice-free regions of a few coastal valleys.

A similar long-term cooling trend is evident in north polar regions. Palm-like and other broad-leaf evergreen vegetation existed in the Canadian Arctic at 80°N near 55 million years ago (Figure 7-3 left), as did the ancestors of modern alligators that would presumably have been ill-adapted to extreme cold. Sea ice was apparently absent, even along the coastal Arctic margins.

Gradually the warm conditions in the Arctic gave way to colder climates. The appearance of conifer forests of spruce and larch by 20 million years ago indicates cooling, and the ring of tundra that has encircled the Arctic Ocean in the last few million years indicates deepening cold. Tundra is scrubby grasslike or shrublike vegetation that lives on thawed layers lying above **permafrost**, ground that is frozen each winter by intense cold (Figure 7-3 right). The appearance of tundra and permafrost is thought to be linked to the onset of extremely frigid winters brought on by expanding sea ice.

The shapes of tree leaves have also been used to reconstruct past climate (Figure 7-4). Leaves of trees living today in the warm tropics tend to have smoothly rounded margins, while leaves of trees in cooler climates generally have irregular edges, usually jagged or serrated in outline. The reason for this trend is not known, but the correlation with temperature in the modern vegetation is strong enough that climate scientists have used this relationship to estimate past

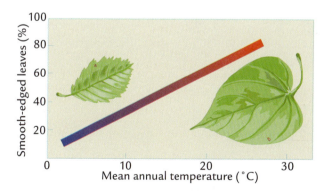

FIGURE 7-4
Leaf outlines indicate temperature
Trees with smooth-edged leaves flourish today in the tropics, while trees with more jagged-edged leaves grow in colder climates. (ADAPTED FROM S. STANLEY, *EARTH SYSTEM HISTORY*, © 1999 BY W. H. FREEMAN AND COMPANY, AFTER J. A. WOLFE, "A PALEOBOTANICAL INTERPRETATION OF TERTIARY CLIMATES IN THE NORTHERN HEMISPHERE," *AMERICAN SCIENTIST* 66 [1978]: 994–1003.)

temperatures from assemblages of fossil leaves preserved in sedimentary rocks.

One low-resolution record derived from leaf-margin evidence in western North America shows an ongoing cooling over the last 55 million years (Figure 7-5). Although interrupted by small warm intervals, the longer-term trend toward lower temperatures persists.

7-2 Evidence from Oxygen Isotope Measurements

Evidence of climate change on the continents over the last 50 million years is incomplete. The first occurrences of ice sheets on land are difficult to define and their subsequent fluctuations in size even more so. In addition, lakes that accumulate remains of past continental vegetation in their muddy sediments rarely persist for millions of years. In contrast, parts of the deep ocean have accumulated a continuous climatic record with quantitative information about climate change across the entire 50-million-year interval.

The most important climatic record in the ocean is the oxygen isotope signal (see Appendix A for a full summary). Most of the oxygen in nature occurs either as the very abundant ^{16}O isotope or the less abundant ^{18}O isotope. Scientists refer to changes in the relative amounts of these two isotopes as variations in $\delta^{18}O$, measured as changes in parts per thousand (‰).

Typical modern $\delta^{18}O$ values are 0 to $-2‰$ for the surface ocean, $+3$ to $+4‰$ for the deep ocean, and -30 to $-55‰$ for ice sheets (Figure 7-6). Changes in the $\delta^{18}O$ values of water in the ocean and of ice in glaciers and ice sheets occur through time. In this chapter, the main focus is on changes in $\delta^{18}O$ values in the ocean. Foraminifera living both in surface waters and on the seafloor use HCO_3^{-1} ions dissolved in seawater as the source of carbon and oxygen for their $CaCO_3$ shells. Because the oxygen in HCO_3^{-1} comes directly from seawater, it gives climate scientists information on past variations in the two isotopes of oxygen in the ocean, as quantified by the $\delta^{18}O$ signal.

Two climatic factors are primarily responsible for $\delta^{18}O$ variations in the past: (1) changes in the temperature of ocean water, and (2) changes in the size of ice sheets on the continents. Changes in $\delta^{18}O$ values measured in the foraminiferal shells decrease by 1‰ for each 4.2°C increase in the temperature of the ocean water in which the foraminifera live. The same relationship holds if seawater cools, but in the reverse sense ($\delta^{18}O$ values become heavier).

Changes in size of the ice sheets also alter $\delta^{18}O$ values in foraminiferal shells. Ice sheets are formed from water vapor evaporated from the oceans and later precipitated as snow (see Appendix A). Because the snow and ice are enriched in the lighter ^{16}O isotope, more of the heavier ^{18}O isotope is left behind in the oceans. This enrichment process is called **fractionation**. As a result, the $\delta^{18}O$ value of ocean water becomes more positive as ice sheets grow and preferentially store the ^{16}O isotope.

The opposite is true if ice melts. For example, if all the ice present on Antarctica and Greenland today

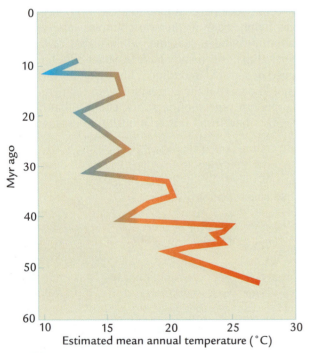

FIGURE 7-5
Cooling in western North America

Temperature trends estimated from the outline shapes of fossil leaves indicate an erratic but progressive cooling of northern middle latitudes during the last 55 Myr. (ADAPTED FROM J. A. WOLFE, "TERTIARY CLIMATIC CHANGES IN WESTERN NORTH AMERICA," PALAEOGEOGRAPHY, PALAEOCLIMATOLOGY, PALAEOECOLOGY 108 [1994]: 195–205.)

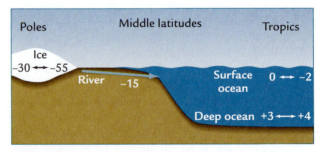

FIGURE 7-6
$\delta^{18}O$ values in the modern world

$\delta^{18}O$ values in the ocean vary from 0 to $-2‰$ in warm tropical surface waters today, to as much as $+3$ to $+4‰$ in cold deep ocean waters. In present ice sheets, typical $\delta^{18}O$ values are $-30‰$ in Greenland and $-55‰$ in Antarctica.

melted and flowed back into the ocean, it would deliver a large volume of ^{16}O-rich meltwater that would shift the ocean's average δ^{18}O value ($\delta^{18}O_w$) from its present value of 0‰ to a value of −1‰.

A simple equation that summarizes these processes is

$$\Delta\delta^{18}O_c = \Delta\delta^{18}O_w \times 0.23\Delta T$$

where Δ means "change in." This equation indicates that measured changes in δ^{18}O values in the shells of foraminifera ($\Delta\delta^{18}O_c$) result from changes in the mean δ^{18}O value of the oceans ($\Delta\delta^{18}O_w$) caused by storage of ^{16}O in ice sheets, and from variations in the temperature of the water in which the shell forms (ΔT). The value 0.23 results from inverting the 4.2°C value noted earlier (1/4.2 = 0.23).

A record of $\delta^{18}O_c$ over the last 70 million years has been compiled from benthic foraminifera living on the ocean floor (Figure 7-7). Although the trend is shown as a single line, it is actually derived from hundreds of individual analyses that scatter in a wide band because of local temperature conditions specific to each site and because of short-term changes. This signal begins to trend erratically toward more positive values near 50 million years ago, and intervals of fastest change occur near 35 million years ago, 13 million years ago, and in the last 3 million years.

These changes toward more positive $\delta^{18}O_c$ values are caused by a combination of (1) cooling of the deep ocean, and (2) growth of ice sheets on land. Both of these factors are critical aspects of the transition from a greenhouse to an icehouse climate.

We can disentangle the effects of temperature and ice volume in this signal to some extent. No evidence exists of significant amounts of ice on Antarctica or anywhere else on Earth prior to 35 million years ago. During the interval between 50 and 35 million years ago, the $\delta^{18}O_c$ values increased from −0.75‰ to +0.75‰, a net change of +1.5‰. The temperature/ $\delta^{18}O_c$ relationship tells us that deep waters must have cooled by more than 6°C (1.5‰ × 4.2°C/‰) during this interval before major ice sheets appeared.

Between 35 million years ago and today, the deep-ocean $\delta^{18}O_c$ values increased from about +0.75‰ to +3.5‰, a further increase of 2.75‰ (see Figure 7-7). Both long-term cooling and the growth of ice sheets contributed to this trend. Following early glaciation of Antarctica near 35 million years ago, the first substantial ice appeared just after 3 million years ago in Greenland. Together, the modern Antarctic and Greenland ice sheets make the δ^{18}O value of ocean water about 1‰ heavier than it would otherwise be. Consequently, 1‰ of the 2.75‰ $\delta^{18}O_c$ increase between 35 million years ago and today can be

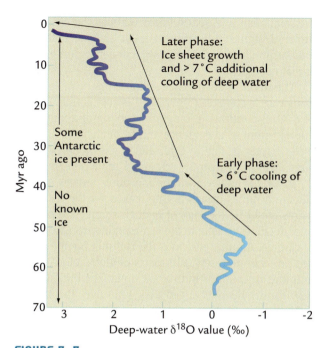

FIGURE 7-7

Long-term δ^{18}O trend

Measurements of δ^{18}O in benthic foraminifera show an erratic long-term trend toward more positive values. From 50 to 35 Myr ago, the increase in δ^{18}O was caused by cooling of the deep ocean. After 35 Myr ago, it reflects further ocean cooling and formation of ice sheets on Antarctica. (ADAPTED FROM K. G. MILLER ET AL., "TERTIARY OXYGEN ISOTOPE SYNTHESIS: SEA LEVEL HISTORY AND CONTINENTAL MARGIN EROSION," *PALEOCEANOGRAPHY* 2 [1987]: 1–19.)

explained by the growth of ice sheets in that interval. The remainder of the $\delta^{18}O_c$ increase—1.75‰—must have been caused by an additional cooling of the deep ocean by more than 7°C (1.75‰ × 4.2°C/‰). This additional cooling brings the total deep-ocean cooling since 50 million years ago to almost 14°C.

Because the temperature of today's deep ocean averages about 2°C and has cooled by about 14°C over the last 50 million years, the average deep-ocean temperature must have been near 16°C before 50 million years ago. If deep water formed mainly in high latitudes as it does today, the polar climates that sent such warm water into the deep ocean must have been much warmer than they are today (see Chapter 6).

7-3 Evidence from Mg/Ca Measurements

Another valuable index of the climatic response of the ocean comes from analyzing the ratio of the elements magnesium (Mg) and calcium (Ca) in the shells of foraminifera. The process by which Mg substitutes for Ca in the foraminiferal shells is dependent on the temperature of the waters in which the shells form. Across the cold temperature range of the deep ocean, the relationship is nearly linear. As was the case with the δ¹⁸O signal, adjustments must also be made for long-term changes in the Mg concentration of the global ocean.

The trend in deep-water temperature reconstructed from Mg/Ca changes (Figure 7-8) is very similar to that of δ¹⁸O. It confirms a long-term cooling of about 14°C inferred from δ¹⁸O evidence and from changes on land (see Chapter 6, Figure 6-1), and it shows particularly large steps at or near 35, 13, and 3 million years ago.

In detail, however, these and other indices of past climate at times disagree. One reason for the disagreements could be the lack of enough samples to resolve finer detail, especially in the leaf-margin reconstructions of climate. Another reason is the fact that no one signal can possibly represent all aspects of global climate. For example, δ¹⁸O trends from tropical planktic foraminifera (not shown here) changed very little over the last 15 million years, indicating only a small cooling of low-latitude surface waters. Yet ice sheets were growing and both high-latitude and deep-ocean temperatures were cooling markedly during the last 15 million years.

A wide and convincing array of evidence documents the progressive cooling of both poles and of mid-latitude areas during the last 50 million years. Several hypotheses have been put forward to explain this gradual greenhouse-to-icehouse transition.

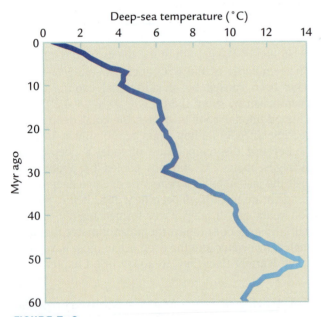

FIGURE 7-8
Long-term Mg/Ca trend

Measurements of Mg/Ca ratios in benthic foraminifera indicate a progressive cooling of deep water over the last 50 Myr. (ADAPTED FROM LEAR ET AL., "CENOZOIC DEEP-SEA TEMPERATURES AND GLOBAL ICE VOLUMES FROM MG/CA IN BENTHIC FORAMINIFERAL CALCITE," *SCIENCE* 287 [2000]: 269–272.)

▶ Do Changes in Geography Explain the Cooling?

As noted in Chapter 5, the polar position hypothesis cannot explain this slow cooling over the last 50 million years. Antarctica was located at the South Pole during the greenhouse climate of 100 million years ago, and it is still at the pole during the icehouse climate of today. The largest subsequent latitudinal shift of the continents during the last 50 million years has been the northward movement of India and Australia into tropical latitudes (Figure 7-9). These movements away from the poles are unlikely to have produced the onset or intensification of glaciation.

7-4 Evaluation of the Gateway Hypothesis

Some climate scientists have called on the opening or closing of **ocean gateways** to explain the onset of both southern and northern glaciation during the last 50 million years. These hypotheses have focused on narrow passages that allow or impede exchanges of ocean water between ocean basins. They have proposed that changes in key gateways caused glaciations by altering the poleward transport of heat or salt.

CASE STUDY 1: ANTARCTICA During the last 50 million years, the last of the Gondwana continents connected to Antarctica split off and moved north, leaving Antarctica isolated and surrounded by a circumpolar ocean (Figure 7-10). In the late 1970s, the marine geologist James Kennett proposed that this breakup caused the onset of glaciation on Antarctica. Before the continents separated, oceanic flow around Antarctica had been impeded by the land connections with South America and Australia. Kennett hypothesized that these barriers had diverted warm ocean currents poleward from lower latitudes and had delivered enough heat to Antarctica to prevent glaciation.

After the continents separated, a strong, unimpeded west-to-east flow developed around Antarctica. Kennett proposed that the loss of the warm poleward flow of heat isolated Antarctica, causing it to cool and glaciation to begin. Subsequent investigations have concluded that Australia separated from Antarctica around 37 to 33 million years ago, the same interval as the first glaciation on Antarctica. The timing of the opening of Drake's Passage between South America and Antarctica (see Figure 7-10) is more uncertain, dated by some studies to as early as 48 to 34 million years ago and by others to as late as 22 to 17 million years ago.

Climate scientists have used sensitivity tests with atmosphere/ocean general circulation models to

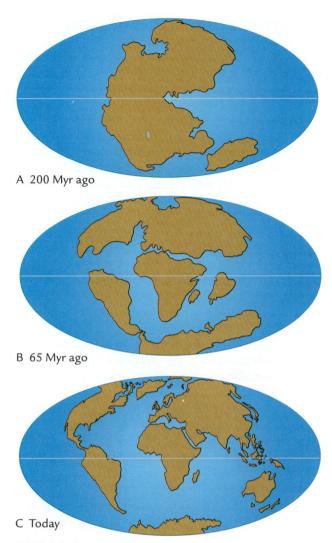

FIGURE 7-9
Continental movements since 200 Myr ago

Since the time of Pangaea, 200 Myr ago (A), the Atlantic Ocean has widened, the Pacific Ocean has narrowed, and India and Australia have separated from Antarctica and moved northward to lower latitudes (B, C). (MODIFIED FROM F. PRESS AND R. SIEVER, *UNDERSTANDING EARTH*, 2ND ED., © 1998 BY W. H. FREEMAN AND COMPANY.)

FIGURE 7-10
Opening of Drake's Passage

Opening of an ocean gap between South America and Antarctica that is estimated to have occurred near 25 to 20 Myr ago allowed a strong Antarctic circumpolar current (arrows) to flow uninterrupted around the Antarctic continent. The passageway between Australia and Antarctica had opened 10 Myr earlier. (ADAPTED FROM E. J. BARRON ET AL., "PALEOGEOGRAPHY: 180 MILLION YEARS AGO TO THE PRESENT," *ECOLOGAE GEOLOGICAE HELVATIAE* 74 [1981]: 443–70.)

evaluate this hypothesis. Drake's Passage was closed in one simulation and left open in a companion simulation, while all other features of Earth's geography were kept the same. The model results suggested that opening Drake's Passage did not significantly alter ocean (or atmospheric) temperatures near Antarctica. Instead, the models simulated a frigid climate over Antarctica regardless of the pattern of ocean flow. The combined heat transport by the ocean and the atmosphere remained about the same in both experiments. Apparently, the opening of the circum-Antarctic gateway was not a critical factor in the onset and development of Antarctic glaciation.

CASE STUDY 2: THE CENTRAL AMERICAN SEAWAY

During the last 10 million years, uplift in Central America gradually closed a deep ocean passage that had previously separated North and South America in the region of Panama. The last stages of uplift created the Central American part of the cordilleran mountain chain. Final closure of this open passage occurred just before 4 million years ago with the emergence of the Isthmus of Panama, and the first large-scale glaciation of North America followed at 2.75 million years ago.

Several climate scientists have suggested that closure of this passage and the start of northern glaciation are linked. They hypothesized that construction of the Isthmus of Panama blocked the strong flow of warm, salty tropical water that had previously been driven westward out of the tropical Atlantic Ocean and into the eastern Pacific by trade winds. The newly formed isthmus should have redirected this flow into the Gulf Stream and toward the high latitudes of the Atlantic. They further hypothesized that this strengthened northward flow of warm, salty water would have suppressed the formation of sea ice in north polar regions because saltier waters resist freezing better than fresh waters (see Chapter 2). According to this hypothesis, the reduced cover of sea ice would have made more moisture from the ocean available to nearby landmasses and triggered the growth of ice sheets.

Climate modelers have used atmosphere/ocean general circulation models to test this hypothesis by running pairs of experiments with the Panama region configured both as an open gateway passage and as a closed-off isthmus. These two configurations roughly correspond to the end points of the tectonic changes that occurred between about 10 and 4 million years ago. These simulations confirmed the prediction that warm, salty water would have been retained in the Atlantic Ocean and redirected toward northern latitudes (Figure 7-11). Closing of the Panamanian isthmus also cut off a return flow of low-salinity Pacific water into the Atlantic. Blockage of this low-salinity return flow by the isthmus further increased the salinity of the northward-flowing Atlantic water.

In a critical respect, however, the model simulations contradicted the gateway hypothesis. The stronger northward flow of salty water in the Atlantic and the resulting reduction of sea ice caused by closing the Isthmus of Panama did not greatly alter precipitation patterns around the high-latitude North Atlantic. As a result, the increase in moisture hypothesized as necessary to grow ice sheets did not occur. Instead, the stronger northward flow of water from the tropics and subtropics transferred a large amount of heat to the atmosphere and warmed the regions where ice sheets were eventually to form. This warming increased summer melting of snow and opposed glacial inception, contrary to the gateway hypothesis.

Another problem with the gateway hypothesis is that northern hemisphere climate continued to cool, and ice sheets grew larger in size, millions of years after the Panamanian isthmus had emerged and closed off all flow between the Atlantic and Pacific oceans.

ASSESSMENT OF GATEWAY CHANGES These two case examples invoke very different (actually opposed) roles for the ocean in glacial inception. For Antarctica, a *reduced* poleward flow of warm ocean water was proposed to have caused a cooling and subsequent glaciation. For the Isthmus of Panama, an *increased* poleward flow of warm ocean water was invoked as the cause of an increase in moisture flux that promoted glaciation.

These differing assumptions reflect past disagreements among climate scientists about how the ocean affects ice sheets. Some climate scientists have suggested that a warmer ocean will release more *latent* heat (water vapor) to the atmosphere and thereby supply more moisture (snow) to aid ice growth. Most climate scientists, however, emphasize the fact that a warmer ocean will release more *sensible* heat to the overlying atmosphere and thereby potentially melt more ice (see Chapter 2).

In any case, neither set of modeling experiments supports the hypothesis that changes in poleward flow of warm ocean water tied to gateway changes would have had a large enough effect on climate to initiate the growth of ice sheets. This criticism needs to be tempered by the realization that the oceanic components of general circulation models are still at a relatively early stage of development and that they

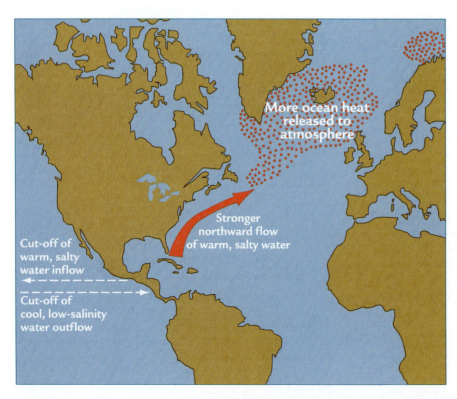

FIGURE 7-11
Closing of the Isthmus of Panama

Simulations with ocean models indicate that gradual closing of the Central American isthmus between 10 and 4 Myr ago redirected warm salty water northward into the Atlantic Ocean, reduced the extent of high-latitude sea ice, and handed off additional heat to the atmosphere. (ADAPTED FROM E. MAIER-REIMER ET AL., "OCEAN GENERAL CIRCULATION MODEL SENSITIVITY EXPERIMENT WITH AN OPEN CENTRAL AMERICAN ISTHMUS," *PALEOCEANOGRAPHY* 5 [1990]: 349–66.)

simulate precipitation patterns rather poorly. A second criticism of the gateway hypothesis is that each opening or closing of a gateway occurs over perhaps 10 million years, while the cooling has lasted far longer.

In Summary, it seems unlikely that such discontinuous gateway episodes could have driven a progressive global-scale climatic cooling for 50 million years.

Whether or not gateway changes affect climate on a global scale, they certainly have the potential to alter the flow of deep and bottom water. Major gateway changes redistribute heat and salt at the high-latitude sites where deep waters form, and these surface-water changes may affect formation of deep water. For example, the model experiments indicated that closing the Isthmus of Panama increased the formation of deep water in the high latitudes of the North Atlantic. Formation of deep water increased in this simulation because of the increased salinity of northward-flowing Gulf Stream waters (Figure 7-11). Higher-salinity surface waters promote stronger deep-water formation because the water is already dense when it encounters cold winter air masses at high latitudes. The results from this experiment agree with independent evidence that indicates increased formation of deep water in the North Atlantic between 10 and 4 million years ago, the interval over which the Central American seaway was gradually reduced and finally closed off by the emergence of the Central American isthmus.

▶ Hypotheses That Invoke Changes in CO₂

The evidence summarized in sections 7-1 through 7-3 requires a mechanism that cooled climate at both poles and middle to lower latitudes more or less continuously for 50 million years. Most climate scientists regard falling CO_2 levels as the cause, whether by slower CO_2 input (BLAG spreading rate) or by faster removal (uplift weathering). Simulations with climate models support the importance of CO_2.

The climate modelers Rob DeConto and David Pollard and colleagues used a general circulation model to run a series of sensitivity tests with CO_2 concentrations specified at values ranging from 280 to 1,400 ppm. These simulations showed that major ice sheets formed on the pole-centered Antarctic continent when CO_2 values fell below 750 ppm, but did not form in Greenland or elsewhere in the Northern Hemisphere until the concentrations reached 280 ppm or lower. This offset sequence of glacial onsets is broadly consistent with geologic evidence for earlier appearance of ice on Antarctic and later appearance in the Northern Hemisphere (see Figure 7-1).

How well do these results match the reconstructed history of CO_2 changes? As we will see in Chapter 11, actual measurements of past CO_2 (from air bubbles in ice cores) only extend back to 800,000 years ago, long after ice sheets had already appeared in both polar regions. As a result, scientists have turned to numerous proxy indicators to try to estimate changes in atmospheric CO_2 concentrations during the last 50 million years.

These proxy reconstructions suggest a major decrease in CO_2 levels during the interval between 50 and 30–25 million years ago (Figure 7-12). This trend is broadly consistent with the first appearance of substantial ice on Antarctica by 34 million years ago and supports the idea that decreasing CO_2 levels drove the cooling around that time.

But the reconstructed CO_2 concentrations for later intervals pose a major problem. By 25 to 20 million years ago, the proxy CO_2 estimates suggest values around 250 ppm. Values that low should have caused northern ice sheets to appear by 20 million years ago, but they did not.

Even more perplexingly, no further decrease in CO_2 occurred between 20 million years ago and today. Yet evidence from both poles and lower latitudes clearly shows that the planet continued to cool during this interval (see Figures 7-1, 7-7, and 7-8).

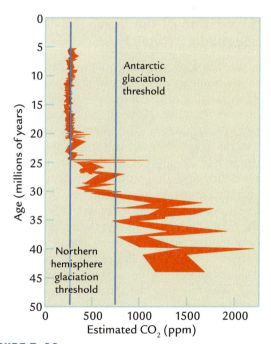

FIGURE 7-12

Did decreasing CO₂ cause glaciation on Antarctica?

Climate model experiments indicate that an ice sheet would develop on Antarctica if atmospheric CO_2 concentrations fell below 750 ppm, as indicated by some estimates based on CO_2 proxies. (ADAPTED FROM M. PAGANI ET AL., "MARKED DECLINE IN ATMOSPHERIC CARBON DIOXIDE CONCENTRATIONS DURING THE PALEOGENE," *SCIENCE* 309 [2005]: 600–603; AND R. M. DECONTO ET AL., "THRESHOLDS FOR CENOZOIC BIPOLAR GLACIATION," *NATURE* 455 [2008]: 652–656, DOI:10.1038/NATURE07337.)

If the proxy CO_2 estimates are accurate, CO_2 could not have been the cause of the cooling after 20 million years ago, but this would leave climate scientists with no widely accepted explanation for that cooling. The other (seemingly more plausible) alternative is that the CO_2 proxy estimates are not valid. Some support for this idea comes from the fact that multiple proxies spanning the same intervals can disagree by very large amounts. In any case, the operating assumption used for the rest of this chapter is that CO_2 concentrations fell throughout the last 50 million years, not just during the time prior to 20 million years ago.

Beginning after 5 to 4 million years ago, many areas of the ocean show a marked trend toward colder sea surface temperatures, often with a particularly prominent step near 3 to 2.5 million years ago. Examples include the eastern tropical Pacific and Atlantic oceans, the high-latitude North Pacific Ocean, and the Southern Ocean. Some scientists have proposed that cooling in one or another of these regions triggered the onset of northern hemisphere glaciation by 2.75 million years ago, but cause and effect are difficult to disentangle when changes occur nearly synchronously. Other scientists have interpreted the nearly global expression of these changes as evidence of a common driving factor, with decreasing CO_2 the cause most often invoked.

7-5 Evaluation of the BLAG (Spreading Rate) Hypothesis

To explain the global cooling of the last 50 million years, the spreading rate hypothesis must pass one critical test. A slowing of global mean spreading and subduction rates must have occurred through that interval, leading to slower rates of CO_2 input to the atmosphere.

It might seem initially that the spreading rate hypothesis has already passed this test. We saw in Chapter 6 that spreading rates 100 million years ago are generally thought to have been faster than they are today, and slower spreading today would be consistent with the colder modern climate. But when we look more closely at the last 50 million years, this explanation is not completely satisfactory. Before 15 million years ago, global mean spreading (and subduction) rates show a decrease consistent with the spreading rate hypothesis (Figure 7-13). But since 15 million years ago, the mean rates of spreading and subduction have leveled out or even slightly *increased* to a present value comparable to the one that existed almost 30 million years ago.

The slight increase in spreading rates since 15 million years ago should have put more CO_2 into the atmosphere and warmed global climate. Instead, climate continued to cool, with a substantial increase in size of the Antarctic ice sheet, the first appearance of mountain glaciers in the Andes and coastal Alaska, and the onset of northern hemisphere ice sheets (see Figure 7-1).

Could additional volcanic input of CO_2 at sites away from ocean ridges and subduction zones explain this discrepancy? During some intervals in the past, extra CO_2 was released at oceanic and continental hot spots located well away from plate margins. Some of these episodes of volcanism have been radiometrically dated, and the estimated volume of volcanic rock that was produced can be added to the amount produced by spreading and subduction (see Figure 7-13). This adjustment does not change the basic picture. Inferred rates of volcanism and CO_2 input still increase during the last 15 million years, and in this case the modern rate of CO_2 addition appears comparable to that 40 million years ago, even though most of the greenhouse-to-icehouse cooling occurred during the last 40 million years.

In Summary, the evidence indicates that the spreading rate hypothesis may have been a cause of global cooling before 15 million years ago, and particularly before 30 or 40 million years ago. But it predicts a warming during the last 15 million years, when in fact a substantial cooling has actually occurred.

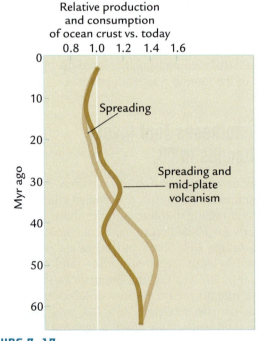

FIGURE 7-13
Changes in spreading rates

The average rate of seafloor spreading slowed until 15 Myr ago, but later increased slightly. Adding in the effects of generation of new crust by volcanism at hot spots away from plate margins does not change this basic trend. (ADAPTED FROM L. R. KUMP AND M. A. ARTHUR, "GLOBAL CHEMICAL EROSION DURING THE CENOZOIC," IN *TECTONIC UPLIFT AND CLIMATE CHANGE*, ED. W. F. RUDDIMAN [NEW YORK: PLENUM PRESS, 1997].)

An alternative possibility has been informally considered by some scientists, but not yet posed as a full hypothesis. The amount of carbon carried down into ocean trenches may have varied in the past because of changes in the *kind* of sediments being subducted, even in the absence of changes in spreading rates. Today, most subduction occurs on the margins of the Pacific Ocean, where sediments are $CaCO_3$-poor because of strong dissolution on the seafloor by corrosive deep waters (see Chapter 3). In contrast, carbonate-rich sediment is now being deposited more abundantly on the Atlantic seafloor where dissolution is less intense. If a future change in the plate tectonic regime were to initiate subduction in the Atlantic Ocean, an enormous amount of carbonate would begin to be subducted in trenches for later melting and eventual release to the atmosphere through volcanoes. As a result, atmospheric CO_2 values would increase, even in the absence of changes in spreading rates. At this point, however, this idea has not yet been tested.

7-6 Evaluation of the Uplift Weathering Hypothesis

To demonstrate that the uplift weathering hypothesis explains global cooling during the last 50 million years, three main requirements must be met: (1) the amount of high-elevation terrain in existence today must be unusually large in comparison with earlier intervals; (2) this high terrain must be causing unusual amounts of rock fragmentation; and (3) the exposure of fresh debris must be causing unusually high rates of chemical weathering.

To determine whether or not these requirements are met, we have to compare the present with some interval in the past. The last half of the Cretaceous interval, from 100 to 65 million years ago, is a useful basis for comparison, for two reasons: (1) abundant evidence of this interval is still left in the geologic record; and (2) it was an interval of full greenhouse climate (see Chapter 6).

PREDICTION 1: EXTENSIVE HIGH TERRAIN At first glance, it might seem obvious that uplift has been unusually active in most mountain ranges during the last few tens of millions of years. Marine sediments deposited at or below sea level 100 to 65 million years ago are now found at high elevations in Tibet and the Himalayas of Asia, the South American Andes, the North American Rocky Mountains, and the European Alps (Figure 7-14). These sediments have been uplifted from sea level to their present heights in the last 70 million years or less.

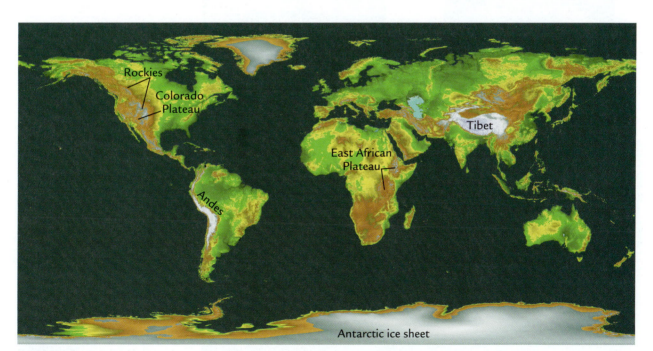

FIGURE 7-14
Earth's high topography
Earth today has only a few regions where broad areas of land stand more than 1 km high (shown in brown, blue, and white). Except for the high ice domes on Antarctica and Greenland, the highest bedrock surfaces are the Tibetan Plateau and other high terrain in southern Asia, the Andes of South America, the Rocky Mountains and Colorado Plateau of North America, and the volcanic plateaus of eastern and southern Africa. (COURTESY NOAA, NATIONAL GEOPHYSICAL DATA CENTER.)

Although much of today's high topography is geologically youthful, this evidence alone does not prove that the modern elevations are *uniquely* high or extensive. Plate tectonic processes continually cause uplift in many regions throughout geologic time, while erosion continually attacks the highest topography and wears it down. As a result, the highest topography during any interval of geologic time is always recent in origin, just as it is today.

The strongest evidence that the amount of high terrain is indeed more massive today than in earlier geologic eras is the presence of the Tibetan Plateau, with an area of some 2.5 million square kilometers (1 million square miles) at an average elevation above 5 kilometers. This plateau has slowly risen since the initial collision of India and Asia 55 million years ago (Figure 7-15A and B).

In contrast, no major continental collisions occurred from 100 to 65 million years ago, and no massive plateaus existed then, or for the preceding 150 million years. The presence of the Tibetan Plateau and Himalayan complex (Figure 7-15B and C) provides a conclusive argument that an unusually massive amount of high topography exists at the present.

Most other high-elevation regions on Earth (see Figure 7-14) have been formed by subduction of ocean crust beneath continental margins. Because subduction is an ongoing process, mountain terrain has existed continuously through time, in contrast to plateaus produced by sporadic continental collisions. The modern Andes and narrow central plateau called the Altiplano are the result of subduction along the west coast of South America. Because subduction has been under way there for more than 100 million years,

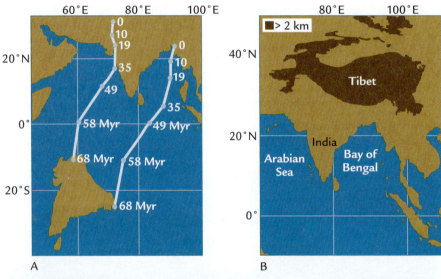

A

B

C

FIGURE 7-15

India-Asia collision and Tibet

Collision of India and Asia (A) produced the Tibetan Plateau (B), the largest high-elevation rock feature on Earth today. The Himalaya Mountains tower over the Indian subcontinent (in the foreground) to the south (C). Behind the Himalayas, to the north, lies the vast Tibetan Plateau at an average elevation above 5,000 m. (A AND B: ADAPTED FROM P. MOLNAR ET AL., "MANTLE DYNAMICS, THE UPLIFT OF THE TIBETAN PLATEAU, AND THE INDIAN MONSOON," *REVIEW OF GEOPHYSICS* 31 [1993]: 357–96; C: EMIL MUENCH/SCIENCE SOURCE.)

FIGURE 7-16
Sediments suspended in rivers
The annual yield of suspended sediments is highest in two regions: the Himalayas of southeast Asia and the Andes of South America. (ADAPTED FROM D. E. WALLING AND B. W. WEBB, "PATTERNS OF SEDIMENT YIELD," IN *BACKGROUND TO PALEOHYDROLOGY*, ED. K. J. GREGORY [NEW YORK: JOHN WILEY, 1983].)

a mountain range has long existed where the western Andes currently sit, along the western coast of South America, although several studies indicate that much of the high topography of the central Altiplano and the eastern Andes was created in the last 15 million years.

Another kind of high terrain is the extensive low plateau in eastern and southern Africa at an elevation of 1 km (see Figure 7-14). This plateau results from deep-seated heating that causes a broad upward doming and outpouring of volcanic lava. Much of the East African plateau was built in the last 30 million years, but similar amounts of plateau construction may have occurred on Africa near 100 million years ago as the giant continent of Pangaea broke up. Subduction has also occurred along western North America for some 200 million years, but scientists disagree about the history of uplift in this region over the last 50 million years (Box 7-1).

> **In Summary**, the existence of the massive Tibetan Plateau alone makes modern topography unusual, consistent with the uplift weathering hypothesis. Other regions of recently elevated terrain also exist in the South American Andes Mountains and in East Africa.

PREDICTION 2: UNUSUAL PHYSICAL WEATHERING
The second test of the uplift weathering hypothesis is whether or not today's high topography has caused higher rates of physical weathering and rock fragmentation than intervals farther in the past. The concentrations of suspended particles carried southward and eastward from the Himalayas and Tibet by Asian rivers are larger than any on Earth (Figure 7-16). The youthful topography of the eastern Andes drained by the Amazon River is another region of high particle concentrations. These measurements clearly indicate intense physical weathering at the present in the two regions with highest terrain.

The best record of rates of erosion lies in sediments deposited in ocean basins by rivers. By far the largest mass of young sediment in the ocean today is found on the Indian Ocean seafloor south of the Himalayas. Deposition of this pile of sediment began by 40 million years ago, increased near 25 million years ago, and accelerated rapidly near 10 million years ago (Figure 7-17).

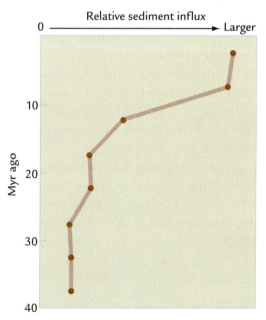

FIGURE 7-17
Himalayan sediments in the Indian Ocean
The rate of influx of sediments from the Himalayas and Tibet to the deep Indian Ocean has increased almost tenfold since 40 Myr ago. (ADAPTED FROM D. K. REA, "DELIVERY OF HIMALAYAN SEDIMENT TO THE NORTHERN INDIAN OCEAN AND ITS RELATION TO GLOBAL CLIMATE, SEA LEVEL, UPLIFT, AND SEAWATER STRONTIUM," IN *SYNTHESIS OF RESULTS FROM SCIENTIFIC DRILLING OF THE INDIAN OCEAN*, ED. R. A. DUNCAN ET AL. [WASHINGTON, DC: AMERICAN GEOPHYSICAL UNION, 1992].)

Climate Debate

The Timing of Uplift in Western North America

Scientists disagree about the timing of uplift of high terrain across western North America. All agree that mountains of some size have existed in Nevada and eastern California for 200 million years or more because ongoing subduction occurred along the west coast until about 30 million years ago. They also agree that the Rocky Mountain West (the modern High Plains, Colorado Plateau, and both the U.S. and Canadian Rocky Mountains) was flooded by an inland sea from 100 to 70 million years ago and has since been uplifted to its present height. But just about everything else about the timing and amount of uplift in this region is in question.

One group emphasizes recent uplift. In their view, the earlier mountain terrain near Nevada was a series of discontinuous low-elevation peaks, not a major topographic feature. This group infers that broad, large-scale upwarping of the *entire* West from the Sierra Nevadas of California to the High Plains of Colorado and Wyoming began about 20 million years ago because of some kind of deep-seated heating process. Visitors to parks in the American West will notice that this view is widely promoted by the U.S. National Park Service.

The other group interprets major uplift as occurring at an earlier time, followed by a more recent *loss* of elevation in many regions. In their view, the mountain belt that existed in the Far West before 100 million years ago was a continuous feature at 3–4 km elevation like the modern Andes and has subsequently dropped to its modern height of 1–2 km. This group acknowledges large-scale uplift of the Rocky Mountain West since 70 million years ago, but they infer that most of the uplift actually occurred between 70 and 45 million years ago during an interval

of heightened tectonic activity, with most of the Rocky Mountain area losing elevation once that activity ceased.

The geophysicist Peter Molnar has offered an explanation for this surprising difference in opinion. He proposed that the global cooling trend has increased the extent of glaciation on high-elevation mountains and plateaus by lowering the freezing line onto the high topography. As a result, glacial erosion has caused increases in rates of erosion and major incision of mountain valleys. Both the appearance of glaciers and the increase of erosion can be (and has been) mistaken for evidence of active uplift in mountain belts that are actually tectonically "dead." More recent investigations have shown that lowering of the winter snowline due to long-term cooling could have dramatically increased the amount of weathering caused by "snowmelt flooding," with 8 or 9 months of snow accumulation removed during 3 or 4 warm summer months.

Temperature reconstructions based on leaf-margin types (see Figure 7-4) have been used to test these opposing views. Temperatures are estimated from deposits of the same age both in higher-elevation mountain areas and in nearby coastal regions. Because temperature decreases with elevation in a known way, the estimated temperature *differences* between coastal and high-elevation regions can be converted to estimates of past elevation. This method suggests a small decrease in elevation of most parts of the Rocky Mountain West during the last 40 or 50 million years. One exception during the last 10 or 20 million years is the Yellowstone hot spot area, which has been domed upward by shallow subsurface heating. The northern Canadian Rockies have also risen recently because subduction is still underway along the Pacific coast.

Climate model experiments indicate that this influx of sediment to the Indian Ocean is a result of two factors: (1) creation of steep terrain along the southern Himalayan margin of the Tibetan Plateau, and (2) the fact that a plateau the size of Tibet in effect creates its own weather, including the powerful South Asian monsoons (Figure 7-18).

Monsoons result from different rates of heating of continents and oceans due to the different heat capacities of land and water (see Chapters 2 and 5). In summer, solar heating warms the continents

(including the surface of Tibet) more rapidly than the nearby oceans. The heated air over the land rises and pulls in moist ocean air. The high Himalayas on the southern margins of the plateau form an obstacle to the incoming ocean air, forcing it to rise and its water vapor to condense in cooler temperatures at high altitudes. As a result, these steep slopes become a natural focal point for strong summer monsoon rains, which release latent heat and fuel even more powerful monsoons. The powerful South Asian monsoons owe part of their existence to the rise of the Tibetan Plateau.

Box 7-1

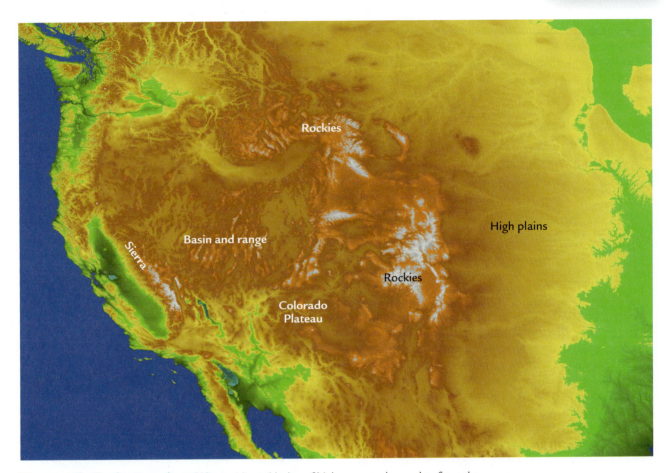

Topography in the American West A broad bulge of high topography reaches from the Sierra Nevadas in the far western United States to the Rocky Mountains and High Plains farther east, with the Colorado Plateau and Basin and Range lying in between. Low-elevation regions a few hundred meters above sea level are shown in green, with progressively higher elevations in yellow, brown, pale orange, and white. (COURTESY NOAA, NATIONAL GEOPHYSICAL DATA CENTER.)

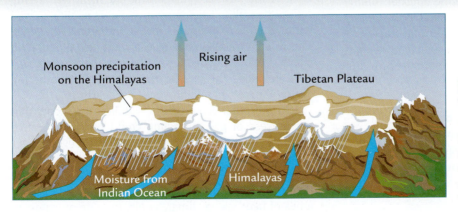

FIGURE 7-18
Tibet and the monsoon

Heating of the Tibetan Plateau draws in moisture from the Indian Ocean and enhances the intensity of the warm, moist summer monsoon on its southern (Himalayan) margin.

Unfortunately, ocean sediments cannot give us a definitive estimate of *global* rates of physical weathering in the past. Much of the sediment eroded from coastal mountain ranges and deposited in the nearby ocean is soon subducted into nearby trenches. The large amount of sediment lost in this way cannot be quantified. In addition, some sediments deposited on the seafloor are eroded and redeposited, and this reworking skews compilations of sediment deposition rates through time toward younger ages. Rapid deposition of huge amounts of Himalayan sediment thus supports the hypothesis that physical weathering is stronger today on a global basis than in earlier times, but this conclusion remains tentative because of the sediment lost to subduction.

PREDICTION 3: UNUSUAL CHEMICAL WEATHERING

The final test of the uplift weathering hypothesis is whether or not the global average rate of chemical weathering is higher today than it was in the past. Unfortunately, chemical weathering rates are difficult to determine even on regional scales, much less for the entire Earth.

Climate scientists quantify modern rates of chemical weathering on a regional basis by measuring the total amount of ions dissolved and transported in rivers. This measure reflects the amount of chemical weathering in the watershed drained by each river, but disturbances of natural weathering processes by humans complicate such studies. In addition, it is difficult to distinguish between the ions provided by slow weathering of silicate rocks (hydrolysis) and those resulting from rapid dissolution of carbonate rocks. Only hydrolysis affects the CO_2 balance in the atmosphere (see Chapter 4). Another limitation is strategic: it is virtually impossible to study enough rivers to reach an accurate estimate of the global weathering rate because too many rivers contribute significantly to the global total.

It is even more difficult to reconstruct rates of chemical weathering during earlier intervals. As a result, the case for unusual chemical weathering during the last tens of millions of years rests on a plausibility argument based on several observations: the unusual height and extent of the Tibetan-Himalayan complex, the unusual strength of the monsoon rains, and the unusual volume of physically eroded sediments deposited in the nearby ocean. By inference, this combination of favorable factors should promote unusually rapid chemical weathering and cause CO_2 removal from the atmosphere. But inference is not proof.

In Summary, the cause of global cooling during the last 50 million years remains uncertain. The most likely culprit is a decrease in atmospheric CO_2, driven by some combination of decreased input from sea-floor spreading processes and increased removal tied to chemical weathering.

Future Climate Change at Tectonic Time Scales

Despite the wealth of evidence that climate has cooled over the last 50 million years, it is impossible to predict the changes in climate that will be caused by tectonic processes in the future. Changes in these processes are inherently unpredictable because the driving forces are not sufficiently understood. If we cannot accurately predict how plate tectonic processes will operate in the future, we cannot predict the resulting climatic effects. The safest prediction—that long-term (tectonic-scale) climate will continue to cool—is not really a prediction at all, just a forward projection of past trends.

The tectonic-scale climatic trends of the distant future will also be influenced by positive and negative feedback processes. We have already seen that negative feedback from chemical weathering is an integral part of the BLAG hypothesis (see Chapter 5), but a rough calculation shows that it could also have acted to offset much of the increase in chemical weathering driven by uplift (Figure 7-19). In this calculation, we assume that uplift has affected an area equivalent to 1% of the total area of continental crust, approximately the size of the Himalayas and the southern parts of the Tibetan Plateau. We also assume that uplift increased the rate of exposure of fresh bedrock in this small region by a factor of 50, broadly consistent with the relative differences in dissolved fluxes between the Andes Mountains and the lowlands in the Amazon Basin. A 50-fold increase in weathering over 1% of Earth's land surface would increase global chemical weathering by 50%.

This localized increase in chemical weathering in the uplifted region would then be opposed by a decrease in weathering across the rest of Earth's land surface. As climate cools, the negative feedback role of weathering should begin to moderate climate across the globe because chemical weathering is slower in cooler, drier, less vegetated conditions (see Chapter 4). A global temperature decrease of 3° to 4°C would be enough to reduce weathering rates by 50% over the remaining 99% of the land surface and thereby offset most of the localized increase in chemical weathering caused by uplift. This amount of global cooling is well within the range estimated for the last 50 million years. This rough calculation suggests that the moderating effects of the chemical weathering thermostat may have balanced the effects of uplift-driven weathering during the last 50 million years (or nearly so).

On the other hand, the possibility also exists that the process of global cooling could produce positive feedbacks that keep driving climate toward even colder conditions. These feedbacks result from the increased amounts of freshly fragmented rock generated by ice (Figure 7-20A). If the mountain glaciers grind large volumes of bedrock as climate cools, the rate of exposure

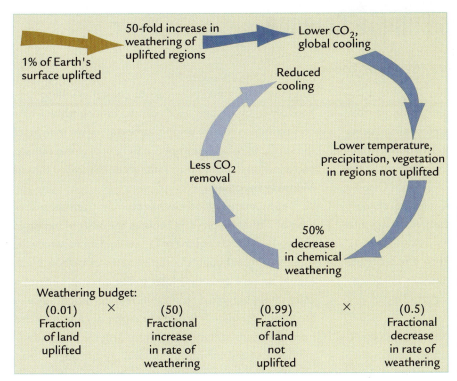

Weathering budget:

(0.01) Fraction of land uplifted	×	(50) Fractional increase in rate of weathering	(0.99) Fraction of land not uplifted	×	(0.5) Fractional decrease in rate of weathering

FIGURE 7-19

Negative feedback from chemical weathering?

If increased chemical weathering in localized regions of uplift causes global climate cooling, the resulting reduction of chemical weathering in other regions may act as a negative feedback that moderates much of the cooling driven by uplift.

of freshly fragmented rock should increase, thereby speeding up rates of chemical weathering, pulling more CO_2 out of the atmosphere, and causing additional cooling (Figure 7-20B). This kind of feedback can happen both in regions of active uplift and in high terrain that is no longer being uplifted (see Box 7-1).

An increase in weathering is particularly likely if the lower limits of the mountain glaciers move up and down the sides of mountains, alternately grinding fresh rock and then exposing it to the atmosphere for chemical weathering (see Figure 7-20B). These kinds of fluctuations result from the orbital-scale cycles of warming and cooling we will examine in Part III.

A similar positive feedback may occur in connection with large, continental-scale ice sheets (Figure 7-20C). As global climate cools and vast ice sheets appear,

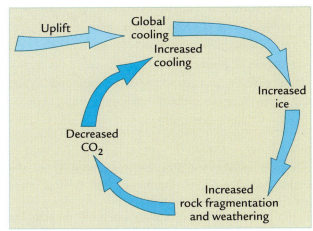

A Positive weathering feedback

FIGURE 7-20

Positive feedback from ice?

Global cooling produces more ice on Earth (A), and the ice increases rock fragmentation in high mountain terrain (B) and near ice sheets (C). Chemical weathering of this fragmented debris may cause further cooling by positive feedback.

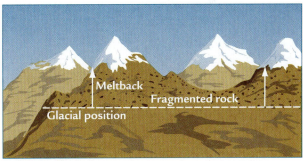

B Mountain glaciers

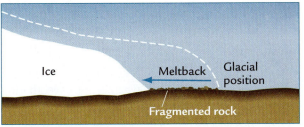

C Continental ice sheets

Looking Deeper into Climate Science

Box 7-2

Organic Carbon: The Monterey Hypothesis

The long-term cooling that produced the present icehouse climate was somewhat erratic, with interruptions by shorter intervals of warming and cooling lasting a few million years. One possible source of shorter-term climate changes at tectonic time scales is variations in the rate of burial and exposure of organic carbon. Organic carbon is a plausible driver of these climate changes because it accounts for 20% of the carbon cycling into and out of Earth's sediments and rocks (see Chapter 4). Organic carbon also has the potential to affect climate relatively rapidly because large amounts can be quickly buried in the sedimentary record, causing rapid reductions of atmospheric CO_2 levels.

Several kinds of climatic and tectonic changes could favor rapid increases in burial of organic carbon (see Chapter 2): changes in wind direction along coastal regions that cause increased upwelling and carbon production; an increase in the total amount of organic carbon and nutrients delivered to the ocean; or a change toward a wetter climate on continental margins, where flat topography naturally favors development of swamps and deposition of organic matter.

An increase in the rate of burial of organic carbon has been proposed as the cause of the cooling trend near 13 million years ago. The large increase in deep-ocean $\delta^{18}O$ values at this time indicates some combination of deep-water cooling and increase in size of the ice sheet on Antarctica. These changes followed an interval when carbon-rich sediments were deposited in shallow waters around the margins of the Pacific Ocean, including the Monterey coast of California. The marine geologists Edith Vincent and Wolfgang Berger

suggested that a major increase in coastal upwelling, perhaps driven by stronger winds caused by long-term climate cooling, buried enough organic carbon along the margins of the Pacific to reduce atmospheric CO_2, cool global climate, and allow ice to build up on Antarctica. They called this the **Monterey hypothesis**.

The Monterey hypothesis has been criticized because of a lag of 2 to 3 million years between the onset of increased carbon burial and the time of fastest cooling shown by the $\delta^{18}O$ record, although the fastest rates of carbon burial appear to have occurred closer to the cooling. Other scientists have suggested that the increased carbon burial in the Pacific could be linked to the supply of carbon eroded from older sedimentary rocks on land (in the Himalayas).

Burial of organic carbon on shallow continental margins also tends to produce its own negative feedback. Carbon-rich sediments deposited in shallow areas are later re-exposed to the atmosphere if sea level falls because ice sheets grow. Exposure of this buried organic carbon allows it to be oxidized back to CO_2 and returned to the atmosphere. The return of CO_2 then causes the climate to warm.

Some climate scientists speculate that changes in rates of weathering on land and burial of organic carbon in the ocean could also be an important cause of longer-term cooling over tens of millions of years. If imbalances between these rates persist for many millions of years, the result could be increases or decreases in the total amount of carbon in the ocean-atmosphere system and of the level of CO_2 in the atmosphere. Scientists are investigating whether the organic carbon subcycle has been adding or removing carbon (and CO_2) from the ocean-atmosphere system over long intervals.

they erode the preexisting cover of already weathered soils, expose fresh underlying bedrock, and grind the rock down to finer particles. This debris is then deposited in extensive moraine ridges at the ice margins.

If the ice sheets simply remain at or near their maximum extents for millions of years (as the Antarctic ice sheet seems to have done for the last several million years), they would likely slow the overall rate of chemical weathering. Weathering may increase along the ice margins where piles of fragmented debris are exposed, but these localized

increases would probably be overwhelmed by the larger decrease in weathering caused by ice cover across much of the continent.

If, however, the ice sheets fluctuate in size (as they did initially in Antarctica and have done for the last 2.75 million years in the Northern Hemisphere), these variations should cause a net increase in chemical weathering. When shorter-term climatic variations cause the ice sheets to melt back from their maximum extents or even disappear (as has happened in North America and Europe since 20,000 years ago), vast areas of finely ground debris are exposed (see Figure 7-20C).

This fresh debris can then be rapidly weathered during the warmer climates between glaciations. The net result should be an increase in chemical weathering.

Further research is obviously needed to assess the role of tectonic forcing and internal feedbacks, both positive and negative. In any case, projections of past trends suggest that Earth's long-term "forecast" over tectonic time scales would entail colder temperatures and more ice, assuming the current plate tectonic regime does not change. But in the meantime, other factors operating on shorter time scales will drive climate changes that are far more relevant to our immediate concerns, including the orbital-scale changes explored in Parts III and IV and especially the human-induced changes examined in Part V.

Key Terms

permafrost (p. 139)

$\delta^{18}O$ (p. 140)

fractionation (p. 140)

ocean gateways (p. 142)

Monterey hypothesis (p. 154)

Review Questions

1. What kinds of changes in vegetation and ice show that Earth has cooled in the last 50 Myr?

2. What two things do changes in $\delta^{18}O$ values in the last 50 Myr tell us about climate changes?

3. How well does the BLAG (spreading rate) hypothesis explain the last 50 Myr of cooling?

4. How well does the uplift weathering hypothesis account for the last 50 Myr of cooling?

5. Explain how chemical weathering could either moderate or deepen the long-term cooling.

6. The volume of water in the world ocean is 48.5 times larger than the amount stored in the two largest ice sheets. The average $\delta^{18}O$ value of the ocean is near zero, while the mean $\delta^{18}O$ value of ice on Antarctica and Greenland is $-50‰$. Show a calculation indicating how much the mean $\delta^{18}O$ value of ocean water would decrease if the two ice sheets melted.

7. Over the last 15 Myr, ice volume has increased, $\delta^{18}O_c$ values measured in deep-ocean foraminifera have increased, and long-term $\delta^{18}O_c$ values in planktonic foraminifera from the tropical Pacific Ocean have remained almost constant. What explains the difference between the two $\delta^{18}O_c$ trends?

Additional Resources

Basic Reading

1996. *Cracking the Ice Ages.* NOVA Video. Boston: WGBH.

Ruddiman, W. F., and J. E. Kutzbach. 1991. "Plateau Uplift and Climatic Change." *Scientific American* (March), 66–75.

Advanced Reading

Berner, R. A. 1999. "A New Look at the Long-Term Carbon Cycle." *GSA Today* 9: 1–6.

DeConto, R. M., and D. Pollard, 2003. "Rapid Cenozoic Glaciation of Antarctica Induced by Declining Atmospheric CO_2." *Nature* 421: 245–249.

Kennett, J. P. 1977. "Cenozoic Evolution of Antarctic Glaciation, the Circum-Antarctic Ocean, and Their Impact on Global Paleoceanography." *Journal of Geophysical Research* 82: 3843–60.

Mikolajewicz, U., T. Maier-Reimer, T. J. Crowley, and K.-Y. Kim. 1993. "Effect of Drake and Panamanian Gateways on the Circulation of an Ocean Model." *Paleoceanography* 8: 409–26.

Miller, K. G., R. A. Fairbanks, and G. S. Mountain. 1987. "Tertiary Oxygen Isotope Synthesis, Sea-Level History, and Continental Margin Erosion." *Paleoceanography* 2: 1–19.

Molnar, P., and P. England. 1990. "Late Cenozoic Uplift of Mountain Ranges and Global Climate Change: Chicken or Egg?" *Nature* 346: 29–34.

Raymo, M. E., and W. F. Ruddiman. 1992. "Tectonic Forcing of Late Cenozoic Climate." *Nature* 359: 117–22.

Zachos, J. C., et al. 2001. "Trends, Rhythms and Aberrations in the Global Climate 65 Ma to Present." *Science* 292: 686–691.

Orbital-Scale Climate Change

In Part III, we move from tectonic-scale climate changes over tens to hundreds of millions of years to orbital-scale changes during the last several million years, a time when the continents and oceans were at or very near their present positions. During this interval, changes in Earth's orbit have been the major driver of climate by altering the amount of solar radiation received by season and by latitude (Chapter 8). Three aspects of Earth's orbit have varied over cycles ranging in length from roughly 20,000 to 400,000 years: the tilt of its axis, the shape of its yearly path of revolution around the Sun, and the slowly changing positions of the seasons along that path.

part **III**

Orbital-scale changes have occurred throughout Earth's history, but our focus here is on the last 3 million years in part because of the availability of well-dated climate records from many sites. The resulting increase in regional coverage provides considerable insight into the operation of the climate system. Most climate records that document orbital changes over this time span come from ocean sediments, which are dated by radiometric methods and by orbital "tuning." Records covering the last 800,000 years also come from ice cores, which are dated by counting annual layers in favorable locations and elsewhere by ice flow models and orbital tuning. These techniques make it possible to resolve time to within a few thousand years in many records. Because ocean sediments and ice cores are multichannel recorders that carry several kinds of climate signals side by side, scientists can also determine the relative timing of climatic responses in the oceans and on land, including the ice sheets.

Major orbital cycles have been detected in records of several important climatic responses on Earth: the strength of the low-latitude monsoons on Asia, Africa, and South America (Chapter 9), the size of north polar ice sheets (Chapter 10), and the concentrations of important greenhouse gases (CO_2 and CH_4) through time (Chapter 11). Because climate scientists now have in hand accurate records of both the orbital forcing and many of the internal climate system responses, the mechanisms of orbital-scale changes in Earth's climate are gradually becoming clearer (Chapter 12).

The major questions addressed in this Part are:

▷ How do orbital variations drive the strength of tropical monsoons?

▷ How do changes in Earth's orbit affect the size of northern hemisphere ice sheets?

▷ What controls orbital-scale fluctuations of atmospheric greenhouse gases?

▷ What explains the 41,000-year ice-age cycles between 2.75 and 0.9 million years ago and the variations centered near 100,000 years during the last 0.9 million years?

Orbital changes All aspects of Earth's present orbit have changed with time: the tilt of its axis, the shape of its path around the Sun, and the positions of the seasons on this path. These changes in orbit have driven climatic changes on Earth. (ADAPTED FROM F. K. LUTGENS AND E. J. TARBUCK, THE ATMOSPHERE [ENGLEWOOD CLIFFS, NJ: PRENTICE-HALL, 1992].)

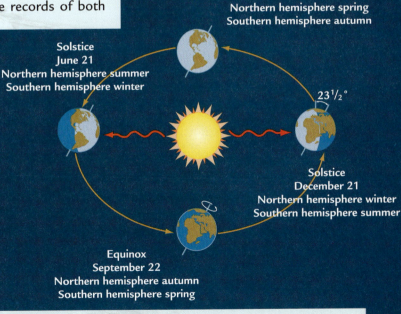

Equinox
March 20
Northern hemisphere spring
Southern hemisphere autumn

Solstice
June 21
Northern hemisphere summer
Southern hemisphere winter

$23\frac{1}{2}°$

Solstice
December 21
Northern hemisphere winter
Southern hemisphere summer

Equinox
September 22
Northern hemisphere autumn
Southern hemisphere spring

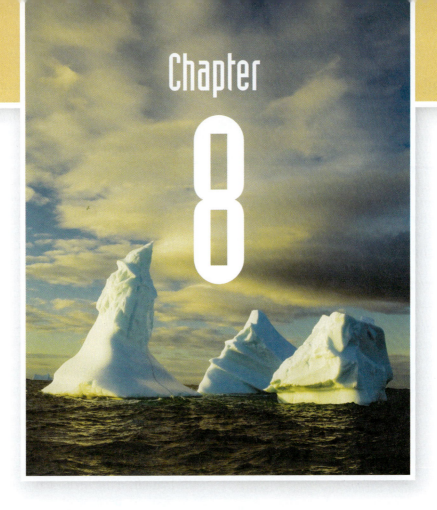

Chapter

8

Astronomical Control of Solar Radiation

Each year, we feel the effects of Earth's orbit around the Sun through seasonal changes in the angle of the Sun's rays and their effects on temperature and other responses. We experience seasonal changes because Earth is tilted as it orbits the Sun—toward the Sun in summer and away from it in winter. The seasonal cycle is by far the largest climate-related signal people experience in a lifetime.

In this chapter, we examine much longer-term changes in Earth's orbit that are equally important to the climate system: changes in the angle of tilt of Earth's axis of rotation, in the shape of its orbit as it revolves around the Sun, and in the timing of the seasons in relation to its noncircular orbit. These longer-term variations in Earth's orbit occur at cycles ranging from ~20,000 to ~400,000 years in length, and they cause cyclic variations in the amount of solar radiation received at the top of the atmosphere by latitude and by season. These changes in incoming radiation drive the climatic changes explored in subsequent chapters of Part III.

▶ # Earth's Orbit Today

The geometry of Earth's present solar orbit is the starting point for understanding past changes in Earth-Sun geometry. Much of our knowledge of Earth's orbit dates back to investigations in the seventeenth century by the astronomer Johannes Kepler. The larger frame of reference for understanding Earth's present orbit is the plane in which it moves around the Sun, the **plane of the ecliptic** (Figure 8-1).

8-1 Earth's Tilted Axis of Rotation and the Seasons

Two fundamental motions describe the present orbit. First, Earth spins on its axis once a day. One result is the daily "rising and setting" of the Sun, but of course that description is inaccurate. Days and nights are caused by Earth's rotational spin, which every 24 hours carries different regions into and out of the portion of Earth's surface that receives the Sun's radiation.

Earth rotates around an axis (or line) that passes through its poles (see Figure 8-1). This axis is tilted at an angle of 23.5°, called Earth's "obliquity," or tilt. This tilt angle can be visualized in either of two ways: (1) as the angle Earth's axis of rotation makes with a line perpendicular to the plane of the ecliptic, or (2) as the angle that a plane passing through Earth's equator makes with the plane of the ecliptic.

The second basic motion in Earth's present orbit is Earth's once-a-year revolution around the Sun. This motion results in seasonal shifts between long summer days, when the Sun rises high in the sky and delivers stronger radiation, and short winter days, when the Sun stays low in the sky and delivers weaker radiation. These seasonal differences culminate at the summer and winter **solstices**, which mark the longest and shortest days of the year (June 21 and December 21,

respectively, in the Northern Hemisphere, and the reverse in the Southern Hemisphere).

Moving outside our Earthbound perspective, we find that the cause of the seasons, the solstices, the changes in length of day, and the angle of incoming solar radiation actually lies in the changing *position* of the tilted Earth with respect to the Sun. During each yearly revolution around the Sun, Earth maintains a constant *angle* of tilt (23.5°) and a constant *direction* of this tilt in space. The seasonal changes we experience arise only from Earth's position in space. When the Northern or Southern Hemisphere arrives at the position in its orbit where it is tilted directly toward the Sun, it receives the more direct radiation of summer. When it tilts directly away from the Sun, it receives the less direct radiation of winter. But at both times, and all times of year, it keeps the same 23.5° tilt.

If we now switch back to our Earthbound perspective, we see the overhead Sun appearing to move back and forth through the year between the north tropic (Cancer) at 23.5°N and the south tropic (Capricorn) at 23.5°S. But again, this apparent motion actually results from Earth's revolution around the Sun with a constant 23.5° tilt. Earth's 23.5° tilt also defines the 66.5° latitude of the Arctic and Antarctic circles: 90° − 23.5° = 66.5°. Because of the 23.5° tilt away from the Sun in winter, no sunlight reaches latitudes poleward of 66.5° on the shortest winter day (winter solstice).

Midway between the extremes of the winter and summer solstices, during intermediate positions in Earth's revolution around the Sun, the lengths of night and day become equal in each hemisphere at the March and September **equinoxes** (which means "equal nights"—that is, nights equal in length to days). Again, Earth's tilt angle remains at 23.5° during the equinoxes, and its direction of tilt in space stays the same. The only factor that changes is Earth's position in respect to the Sun. The two equinoxes and two solstices are handy reference points for describing distinctive features of its orbit.

8-2 Earth's Eccentric Orbit: Distance Between Earth and Sun

Up to this point, everything that has been described would be true whether Earth's orbit was perfectly circular or not. But Earth's actual orbit (Figure 8-2) is not a perfect circle: it has a slightly eccentric or elliptical shape. The noncircular shape of Earth's orbit is the result of the gravitational pull of other planets on Earth as it moves through space.

Geometry tells us that ellipses have two focal points, rather than the single focus (center) of a circle. In Earth's case, the Sun lies at one of the two focal points in its elliptical orbit, as required by the

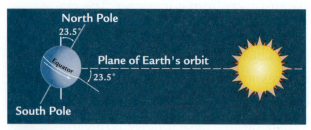

FIGURE 8-1
Earth's tilt

Earth's rotational (spin) axis is currently tilted at an angle of 23.5° away from a line perpendicular to the plane of its orbit around the Sun.

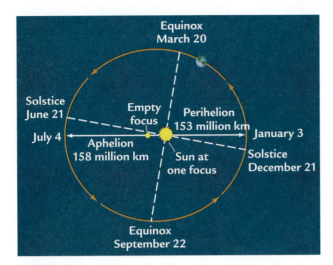

FIGURE 8-2
Earth's eccentric orbit
Earth's orbit around the Sun is slightly elliptical. Earth is most distant from the sun at aphelion, on July 4, just after the June 21 solstice, and closest to the Sun at perihelion, on January 3, just after the December 21 solstice. (MODIFIED FROM J. IMBRIE AND K. P. IMBRIE, *ICE AGES: SOLVING THE MYSTERY* [SHORT HILLS, NJ: ENSLOW, 1979].)

physical laws of gravitation. The other focus is empty (see Figure 8-2).

Earth's distance from the Sun changes according to its position in this elliptical orbit. Not surprisingly, these changes in Earth-Sun distance affect the amount of solar radiation Earth receives, especially at two extreme positions in the orbit. The position in which Earth is closest to the Sun is called **perihelion** (the "close pass" position, from the Greek meaning "near the Sun"), while the position farthest from the Sun is called **aphelion** (the "distant pass" position, from the Greek meaning "away from the Sun"). On average, Earth lies 155.5 million km from the Sun, but the distance ranges between 153 million km at perihelion and 158 million km at aphelion. This difference is equivalent to a total range of variation of slightly more than 3% around the mean value.

Earth is presently in the perihelion position (closest to the Sun) on January 3, near the time of the December 21 winter solstice in the Northern Hemisphere and summer solstice in the Southern Hemisphere (see Figure 8-2). The fact that this close pass position occurs in January causes winter radiation in the Northern Hemisphere and summer radiation in the Southern Hemisphere to be slightly stronger than they would be in a perfectly circular orbit.

Conversely, Earth lies farthest from the Sun on July 4, near the time of the June 21 summer solstice in the Northern Hemisphere and winter solstice in the Southern Hemisphere. The occurrence of this present distant pass position in July makes summer radiation in the Northern Hemisphere and winter radiation in the Southern Hemisphere slightly weaker than they would be in a circular orbit.

The effect of Earth's elliptical orbit on its seasons is small, enhancing or reducing the intensity of radiation received by just a few percent. Remember that the main cause of the seasons is the direction of tilt of Earth's axis in its orbit around the Sun (see Figure 8-1).

Another feature of Earth's eccentric orbit is that the time intervals between the two equinoxes are not equal: there are seven more days in the long part of the orbit, between the March 20 equinox and the September 22 equinox, than in the short part of the orbit, between September 22 and March 20. The greater length of the interval from March 20 to September 22 tends to compensate for the fact that Earth is farther from the Sun in this part of the orbit and is thus receiving less solar radiation.

▶ Long-Term Changes in Earth's Orbit

Astronomers have known for centuries that Earth's orbit around the Sun is not fixed over long intervals of time. Instead, it varies in a regular (cyclic) way because of the mass gravitational attractions among Earth, its moon, the Sun, and the other planets and their moons. These changing gravitational attractions cause cyclic variations in Earth's angle of tilt, its eccentricity of orbit, and the relative position of the solstices and equinoxes around its elliptical orbit (Box 8-1).

8-3 Changes in Earth's Axial Tilt Through Time

If we assume for simplicity that Earth has a perfectly circular orbit around the Sun, we can examine two hypothetical cases that show the most extreme differences in tilt. For both cases, we look at the summer and winter solstices, the two seasonal extremes in Earth's orbit.

For the first case, Earth's axis is not tilted at all (Figure 8-3A). Incoming solar radiation is directed straight at the equator throughout the year, and it always passes by the poles at a 90° angle. Without any tilt, no seasonal changes occur in the amount of solar radiation received at any latitude. As a result, solstices and equinoxes do not even exist because every day has the same length. This configuration shows that a tilted axis is necessary for Earth to have seasons.

Next, consider the opposite extreme with a maximum tilt of 90° (Figure 8-3B). Solar radiation is directed straight at the summer-season pole, while the winter-season pole lies in complete darkness. Six months later, the two poles have reversed position.

Tools of Climate Science

Box 8-1

Cycles and Modulation

Slow changes in Earth's orbit around the Sun occur in a cyclic or rhythmic way, as do the changes in amount of incoming solar radiation they produce. The science of wave physics provides the terminology to describe these changes. The length of a cycle is referred to as its **wavelength**. Expressed in units of time, the wavelength of a cycle is called the **period**, the time between successive pairs of peaks or valleys.

The opposite (or inverse) of the period of a cycle is its **frequency,** the number of cycles (or in this case fractions of one cycle) that occur in one year. If a cycle has a period of 10,000 years, its frequency is 0.0001 cycle per year (one cycle every 10,000 years). In this book, we will refer to cycles in terms of their periods.

Another important aspect of cycles is their **amplitude,** a measure of the amount by which they vary around their long-term average. Low-amplitude cycles barely depart from the long-term mean trend; high-amplitude cycles fluctuate more widely.

Not all cycles are perfectly regular. More commonly the sizes of peaks and valleys oscillate irregularly around the long-term mean value through time. Behavior in which the amplitude of peaks and valleys changes in a repetitive or cyclic way is called **modulation,** a concept that lies behind the principle of AM (amplitude modulation) radio. Modulation creates an envelope that encompasses the changing amplitudes that occur at a specific cycle. Note that *modulation of a cycle is not in itself a cycle;* it simply adds amplitude variations to an actual cycle.

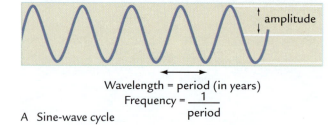

Wavelength = period (in years)

$$\text{Frequency} = \frac{1}{\text{period}}$$

A   Sine-wave cycle

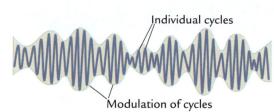

Individual cycles

Modulation of cycles

B   Amplitude modulation

Description of wave behavior   Perfectly cyclic behavior can be represented by a sine wave with a particular period and amplitude (A). Cycles may show regular variations in amplitude, or modulation (B).

If variations in a particular signal are regular *both* in period and in amplitude, it is appropriate to use the term "cycle." For the case of perfect cyclicity, this behavior is described as "sinusoidal" or **sine waves**. If the variations are irregular in period, the term "cycle" is technically incorrect, and "quasi-cyclic" or "quasi-periodic" is preferable. In the case of orbital-scale changes, we will informally use the term "cyclic" or "periodic" for climatic signals that are nearly regular but vary slightly in wavelength or amplitude.

A   No tilt

B   90° tilt

FIGURE 8-3
Extremes of tilt

If Earth's orbit were circular and its axis had no tilt (A), solar radiation would not change through the year and there would be no seasons. For a 90° tilt (B), the poles would alternate seasonally between conditions of day-long darkness and day-long overhead Sun. (ADAPTED FROM J. IMBRIE AND K. P. IMBRIE, *ICE AGES: SOLVING THE MYSTERY* [SHORT HILLS, NJ: ENSLOW, 1979].)

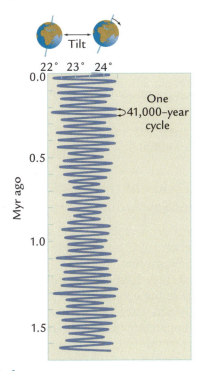

FIGURE 8-4

Long-term changes in tilt

Changes in the tilt of Earth's axis have occurred in a regular 41,000-year cycle.

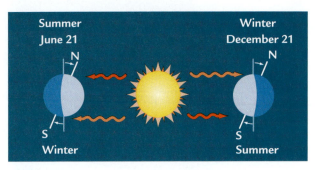

FIGURE 8-5

Effects of increased tilt on polar regions

Increased tilt brings more solar radiation to the poles in the summer season and less radiation to the poles in the winter season.

Decreases in tilt have the opposite effect: they diminish the amplitude of seasonal differences. Smaller tilt angles put the Earth slightly closer to the configuration shown in Figure 8-3A, which has no seasonal differences at all.

In Summary, changes in tilt mainly amplify or suppress the seasons, particularly at the poles.

The difference between these two extreme configurations shows that tilt is an important control on the amount of solar radiation at polar latitudes.

The angle of Earth's tilt has varied through time in a narrow range, between values as small as 22.2° and as large as almost 24.5° (Figure 8-4). The French astronomer Urbain le Verrier discovered these variations in the 1840s. The present tilt (23.5°) is near the middle of this range, and the angle is currently decreasing at a slow rate. Cyclic changes in tilt angle occur mainly at a period of 41,000 years, the interval between successive peaks or successive valleys (see Box 8-1). The 41,000-year cycle is fairly regular, both in period (wavelength) and in amplitude.

Changes in tilt amplify or suppress the strength of the seasons, especially at high latitudes (Figure 8-5). Larger tilt angles turn the poles more directly toward the Sun in the summer and increase the amount of solar radiation received. The increase in tilt that turns the North Pole more directly toward the Sun at its summer solstice on June 21 also turns the South Pole more directly toward the Sun at its summer solstice six months later (December 21). On the other hand, the increased angle of tilt that turns a particular polar region more directly toward the Sun in summer also turns the pole away from the Sun in winter in the other hemisphere.

8-4 Changes in Earth's Eccentric Orbit Through Time

The shape of Earth's orbit around the Sun has also varied in the past, at times becoming more circular and at other times more elliptical (or "eccentric") than today. Urbain le Verrier also discovered these variations in the 1840s. The shape of an ellipse can be described by reference to its two main axes: the "major" (or longer) axis and the "minor" (or shorter) axis (Figure 8-6). The degree of departure from a perfectly circular orbit can be described by:

$$\varepsilon = \frac{\sqrt{a^2 - b^2}}{a}$$

where ε is the **eccentricity** of the ellipse and a and b are half of the lengths of the major and minor axes (the "semimajor" and "semiminor" axes).

The eccentricity of the elliptical orbit increases as these two axes become more unequal in length. At the extreme where the two axes become exactly equal ($a = b$), the eccentricity drops to zero because the orbit is circular ($a^2 - b^2 = 0$). Eccentricity (ε) has varied over time between values of 0.005 and 0.0607 (Figure 8-7). The present value (0.0167) lies well toward the lower (more circular) end of the range.

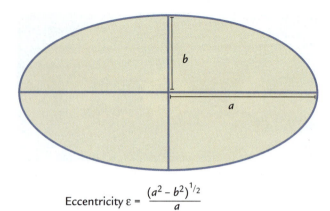

$$\text{Eccentricity } \varepsilon = \frac{\left(a^2 - b^2\right)^{1/2}}{a}$$

FIGURE 8-6
Eccentricity of an ellipse
The eccentricity of an ellipse is related to half of the lengths of its longer (major) and shorter (minor) axes as shown in the equation above.

Changes in orbital eccentricity are concentrated mainly at two periods. One eccentricity cycle shows up as variations at intervals near 100,000 years (see Figure 8-7). This cycle actually consists of four cycles of nearly equal strength and periods ranging between 95,000 and 131,000 years, but these cycles blend into a cycle near 100,000 years.

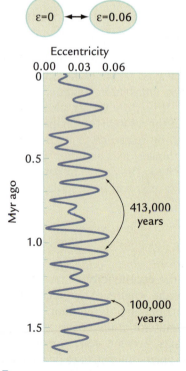

FIGURE 8-7
Long-term changes in eccentricity
The eccentricity (ε) of Earth's orbit varies at periods of 100,000 and 413,000 years.

The second eccentricity cycle has a wavelength of 413,000 years. This longer cycle is not quite as obvious, but it shows up as alternations of the 100,000-year cycles between larger and smaller peak values. Larger amplitudes can be seen ~200,000, ~600,000, ~1,000,000, and ~1,400,000 years ago (Figure 8-7). A third eccentricity cycle has a wavelength of 2,100,000 years, but this cycle is much weaker in amplitude.

8-5 Precession of the Solstices and Equinoxes Around Earth's Orbit

The positions of the solstices and equinoxes in relation to the eccentric orbit have not always been fixed at their present locations (shown in Figure 8-2). Instead, they have slowly shifted through time with respect to the eccentric orbit and the perihelion (close pass) and aphelion (distant pass) positions. Although Hipparchus in ancient Greece first noticed these changes, the eighteenth-century French mathematician, scientist, and philosopher Jean Le Rond d'Alembert was the first to understand them.

The cause of these changes lies in a long-term wobbling motion similar to that of a spinning top. Tops typically move with three superimposed movements (Figure 8-8). They spin very rapidly (rotate) around a tilted axis. They also revolve with a slower near-circular motion across the surface on which they spin, with many spins (rotations) for each complete revolution. Finally, tops wobble, gradually leaning in different directions through time. The wobbling motion referred to here is not caused by changes in

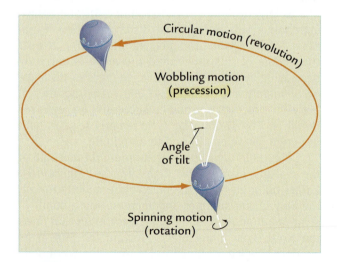

FIGURE 8-8
Earth's wobble
In addition to its rapid (daily) rotational spin and its slower (yearly) revolution around the Sun, Earth wobbles slowly like a top, with one full wobble every 25,700 years.

the *amount* of lean of the top (its angle of tilt), but rather by changes in the *direction* in which it leans.

Earth's wobbling motion, called **axial precession**, is caused by the gravitational pull of the Sun and Moon on the slight bulge in Earth's diameter at the equator. Axial precession can also be visualized as a slow turning of Earth's axis of rotation through a circular path, with one full turn every 25,700 years. Today, Earth rotates around an axis that points to the North Star (Polaris), but over time the wobbling motion causes the axis of rotation to point to other celestial reference points (Figure 8-9). Earth wobbles very slowly; it revolves 25,700 times around the Sun and rotates almost 10 million times on its axis during the time it takes to complete just a single wobble (25,700 × 365 = 9,380,000).

A second kind of precessional motion is known as **precession of the ellipse**. In this case, the entire elliptically shaped orbit of the Earth rotates, with the long and short axes of the ellipse turning slowly in space (Figure 8-10). This motion is even slower than the wobbling motion of axial precession.

The combined effects of these two precessional motions (wobbling of the axis and turning of the

FIGURE 8-10
Precession of the ellipse

The elliptical shape of Earth's orbit slowly precesses in space so that the major and minor axes of the ellipse slowly shift through time. (ADAPTED FROM N. G. PISIAS AND J. IMBRIE, "ORBITAL GEOMETRY, CO$_2$, AND PLEISTOCENE CLIMATE," *OCEANUS* 29 [1986]: 43–49.)

ellipse) cause the solstices and equinoxes to move around Earth's orbit, with one full orbit around the Sun completed approximately every 22,000 years (Figure 8-11). This combined movement, called the **precession of the equinoxes**, describes the absolute motion of the equinoxes and solstices in the larger reference frame of the universe. It consists of a strong cycle near 23,000 years and a weaker one near 19,000 years, with an average of one cycle every 21,700 years. For the rest of this book, we will concentrate mainly on the strong precession cycle near 23,000 years.

The precession of the equinoxes involves complicated angular motions in three-dimensional space, but these motions need to be reduced to a simple, easy-to-use mathematical form that can be plotted against time in the same way as the changes in tilt shown in Figure 8-4. To accomplish this goal, we make use of two basic geometric characteristics of precessional motion.

The first characteristic has to do with the angular form of Earth's motion with respect to the Sun. We define ω (omega) as the angle between two imaginary lines (Figure 8-12A): (1) a line connecting the Sun to Earth's position at perihelion (its closest pass to the Sun), and (2) a line connecting the Sun to Earth's

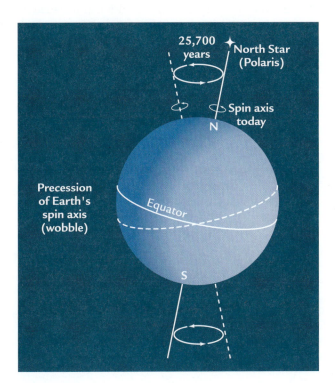

FIGURE 8-9
Precession of Earth's axis

Earth's slow wobbling motion causes its rotational axis to point in different directions through time, sometimes (as today) toward the North Star, Polaris, but at other times toward other stars. (ADAPTED FROM J. IMBRIE AND K. P. IMBRIE, *ICE AGES: SOLVING THE MYSTERY* [SHORT HILLS, NJ: ENSLOW, 1979].)

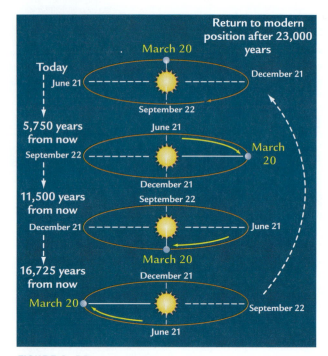

FIGURE 8-11

Precession of the equinoxes

Earth's wobble and the slow turning of its elliptical orbit combine to produce the precession of the equinoxes. Both the solstices and equinoxes move slowly around the eccentric orbit in cycles of 23,000 years. (ADAPTED FROM J. IMBRIE AND K. P. IMBRIE, *ICE AGES: SOLVING THE MYSTERY* [SHORT HILLS, NJ: ENSLOW, 1979].)

The changing angle ω slowly sweeps out a 360° arc, starting at 0° (where the March 20 equinox coincides with the perihelion position), increasing to 90°, then to 180° (where the March 20 equinox occurs on the other side of the orbit, coincident with the aphelion position), later to 270°, and finally to 360°, at which point the cycle is complete and the angle returns to 0° (Figure 8-12B).

This complicated angular motion can be represented in a simplified mathematical form by using basic geometry and trigonometry to convert the angular motions in Figure 8-12 to a rectangular coordinate system. Box 8-2 shows how the mathematical sine wave function projects the motion of a radius vector sweeping around a circle onto a vertical coordinate. This conversion allows the circular motion to be represented as an oscillating sine wave on a simple *x-y* plot. The amplitude of sinω moves from a value of +1 to −1 and back again over each 23,000-year precession cycle.

The second aspect of Earth's orbital motion to be considered is eccentricity. If Earth's orbit were perfectly circular, the slow movements of the solstices and equinoxes caused by precession would not alter the amount of sunlight received on Earth because its distance from the Sun would remain constant through time. Because the orbit is not circular, however, movements of the solstices and equinoxes (see Figure 8-11) cause long-term changes in the amount of solar radiation Earth receives.

These gradual movements of precession bring the solstices and equinoxes into orbital positions that vary in distance from the Sun. Consider the two extreme positions of the solstices in the eccentric orbit shown in Figure 8-13. As noted earlier, in the present orbit, the position of the June 21 solstice (northern hemisphere summer and southern hemisphere winter)

position at the March 20 equinox. The first line is firmly tied to the elliptical shape of Earth's orbit and the second to the varying positions of the seasons in the orbit. As a result, the slow change in the angle ω is a measure of Earth's wobbling motion—the very slow changes in the positions of the seasons with respect to the elliptical orbit.

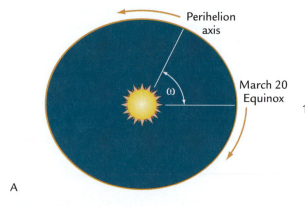

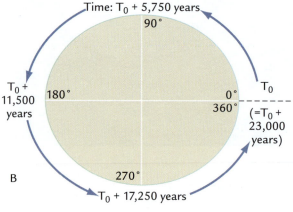

FIGURE 8-12

Precession and the angle ω

The angle between lines marking Earth's perihelion axis and the vernal equinox (March 20) is called ω (A). The angle ω increases from 0° to 360° with each full 23,000-year cycle of precession (B).

Looking Deeper into Climate Science

Box 8-2

Earth's Precession as a Sine Wave

For a right-angle triangle (shown in A), the sine of the angle ω is defined as the length of the *opposite* side over the length of its *hypotenuse* (the longest side). Shown in B is a circle whose radius is a vector r that sweeps around a 360° arc in an angular motion measured by the changing angle ω. Note that the circular motion described by the angle ω is analogous to the nearly circular long-term changes in Earth-Sun geometry.

The angular motion of the radius vector r around the circle can be converted into changes in the dimensions of a triangle lying in the circle, such that the sides of this triangle can be measured in a rectangular (horizontal and vertical) coordinate system (C). In this conversion, the hypotenuse of the triangle is also the radius vector r of the circle.

The sweeping motion of the radius vector r around the circle causes the shape of the internal triangle to change. The radius vector (r) *always* has a value of +1 because its length stays at that same value and its sign is defined as a positive value in the angular coordinate system.

But the length of the opposite side of the triangle (y) is defined in the rectangular coordinate system, and it can change both in amplitude and sign (positive or negative). As the radius vector r sweeps around the circle, y increases and decreases along the vertical scale, cycling back and forth between values of +1 and −1. When r lies in the top half of the circle, y has values greater than 0. When it lies in the lower half, y is negative.

The angular motion of r can be converted to the more convenient linear mathematical form by plotting changes in sinω as the radius vector r sweeps out a full 360° circle, with the angle ω increasing from 0° to 90°, 180°, 270°, and back to 360° (= 0°). As before, sinω is defined as the ratio of the length of the opposite side y over the hypotenuse r (which always retains a value of +1).

The mathematical function sinω cycles smoothly from +1 to −1 and then back to +1 for each complete revolution of the radius vector r. At the starting point (ω = 0°), the length of the opposite side y is 0 and the radius is +1, so the value of sinω is 0/1, or 0. As the angle ω increases, the length of the opposite side of the triangle (y) increases in relation to the constant +1 value of the radius of the circle. When ω reaches 90°, sinω = +1 because the lengths of the opposite side and the radius (hypotenuse) are identical (1/1). At 180°, sinω returns to 0 because the length of the opposite side (y) is again 0. For angles greater than 180°, the sinω values become negative because the opposite side of the triangle y now falls in negative rectangular coordinates (values below 0 on the vertical axis). Sinω values reach a minimum value of −1 at ω = 270° (−1/1). After that, sinω again begins to increase, returning to a value of 0 at ω = 360° (= 0°).

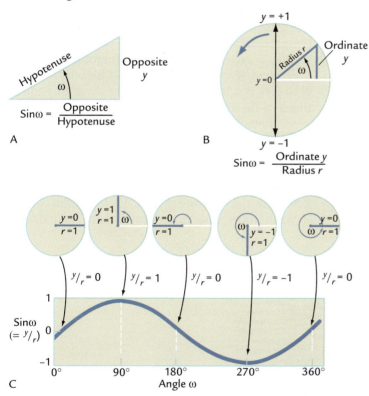

Converting angular motion to a sine wave The sine of an angle is the length of the opposite side of a triangle over its hypotenuse (A). This concept can be applied to a circle where the hypotenuse is the radius (amplitude = 1) and the length of the opposite side of the triangle varies from +1 to −1 along a vertical coordinate axis (B). As the radius vector sweeps out a full circle and ω increases from 0° to 360°, the sine of ω changes from +1 to −1 and back to +1, producing a sine wave representation of circular motion and of Earth's precessional motion (C).

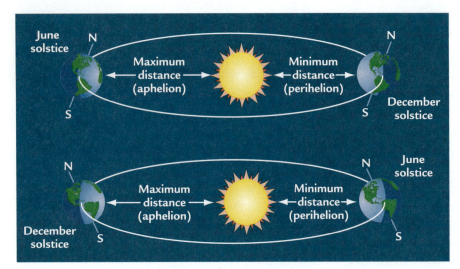

FIGURE 8-13
Extreme solstice positions

Slow precessional changes in the attitude (direction) of Earth's spin axis produce changes in the distance between Earth and Sun as the summer and winter solstices move into the extreme (perihelion and aphelion) positions in Earth's eccentric orbit. (MODIFIED FROM W. F. RUDDIMAN AND A. MCINTYRE, "OCEANIC MECHANISMS FOR AMPLIFICATION OF THE 23,000-YEAR ICE-VOLUME CYCLE," *SCIENCE* 212 [1981]: 617–27.)

occurs very near aphelion, the most distant pass from the Sun (Figure 8-13 top). This greater Earth-Sun distance on June 21 slightly reduces the amount of solar radiation received during those seasons. Conversely, with the December 21 solstice (northern hemisphere winter and southern hemisphere summer) currently occurring near perihelion, the closest pass to the Sun, solar radiation is higher during those seasons than it would be in a perfectly circular orbit. But approximately 11,000 years ago, half of a precession cycle before now, this configuration was reversed (Figure 8-13 bottom). The June 21 solstice occurred at perihelion, and the December 21 solstice occurred at aphelion.

The solstice positions shown in Figure 8-13 are extreme points in a continuously changing orbit. Precession also moves the solstices through orbital positions with intermediate Earth-Sun distances like those shown in Figure 8-11. In the next 11,000 years, the solstices will move from their present positions back to those shown at the bottom of Figure 8-13.

Eccentricity plays an important role in the effect of precession on the amount of solar radiation received on Earth. The full expression for this impact is $\varepsilon\sin\omega$, the **precessional index** (Figure 8-14). The $\sin\omega$ part of this term is the sine wave representation of the movement of the equinoxes and solstices around the orbit (see Box 8-2). The eccentricity (ε) acts as a multiplier of the $\sin\omega$ term.

As noted earlier, the present value of ε is 0.0167. If this value remained constant through time, the $\varepsilon\sin\omega$ index would cycle smoothly between values of $+0.0167$ and -0.0167 over each precession cycle of ~23,000 years. As shown earlier in Section 8-4, however, the eccentricity of Earth's orbit varies through time, ranging between 0.005 and 0.0607

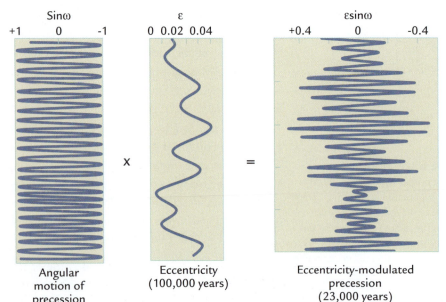

Sinω
+1   0   -1

ε
0 0.02 0.04

εsinω
+0.4   0   -0.4

Angular motion of precession (23,000 years)

×

Eccentricity (100,000 years)

=

Eccentricity-modulated precession (23,000 years)

FIGURE 8-14
The precessional index

The precessional index, $\varepsilon\sin\omega$, is the product of the sine wave function ($\sin\omega$) caused by precessional motion and the eccentricity (ε) of Earth's orbit.

(see Figure 8-7). These changes in ε cause the εsinω term to vary in amplitude (see Figure 8-14).

Long-term variations in the precessional index have two major characteristics (Figure 8-15). First, they occur at a cycle with a period near 23,000 years because of the regular angular motion of precession at that cycle (see Figure 8-14). Second, individual cycles vary widely in amplitude because changes in eccentricity modulate the 23,000-year signal (see Box 8-1). At times, the 23,000-year cycle swings back and forth between extreme maxima and minima, while at other times the amplitude of the changes is small.

The changing values of εsinω affect the extreme perihelion and aphelion positions shown in Figure 8-13 by altering the distance between Earth and the Sun. With greater eccentricity, the differences in distance between a close pass and a distant pass are magnified. With a nearly circular orbit, differences in distance almost vanish.

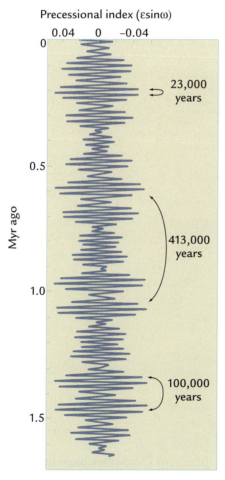

FIGURE 8-15

Long-term changes in precession

The precessional index (εsinω) changes mainly at a cycle of 23,000 years. The amplitude of this cycle is modulated at the eccentricity periods of 100,000 and 413,000 years.

In Summary, changes in eccentricity magnify or suppress contrasts in Earth-Sun distance around the orbit at the 23,000-year precession cycle. These changes in distance from the Sun in turn alter the amount of solar radiation received on Earth (more radiation at the perihelion close pass position, less at the aphelion distant pass position).

The modulation of the εsinω signal by eccentricity is not a real cycle (see Box 8-1), although this statement probably goes against your intuition. You have learned that eccentricity varies at cycles of 100,000 and 413,000 years (see Figures 8-7 and 8-14), and you can see that the upper and lower envelopes of the εsinω signal vary at these periods (see Figure 8-15). But the offsetting effects of the upper and lower envelopes cancel each other out.

For example, at times when the 23,000-year cycle is varying between large minima and large maxima, the adjacent minima and maxima are approximately equal in size. Over the longer (100,000-year) wavelengths of the eccentricity variations, the amplitudes of these opposing shorter-term (23,000-year) oscillations cancel each other out, leaving a negligible amount of net variation. Similarly, short-term variations between small amplitude maxima and minima at other times also offset each other. The importance of this point will become obvious in Chapters 10 and 12.

In Summary, the combined effects of eccentricity and precession cause the distance from Earth to the Sun to vary by season, primarily at a cycle of 23,000 years. Times of high eccentricity produce the largest contrasts in Earth-Sun distance in the orbit, and times of low eccentricity produce the smallest contrasts. As Earth precesses in its orbit, the changes in Earth-Sun distance are registered as seasonal changes in arriving radiation.

Changes in Insolation Received on Earth

Changes in Earth's orbit alter the amount of solar radiation received by latitude and by season. Climate scientists refer to the radiation arriving at the top of Earth's atmosphere as **insolation**. Some of this incoming insolation does not arrive at Earth's surface because clouds and other components of the climate system alter the amount that actually penetrates the atmosphere (recall Chapter 2). Still, these calculations of insolation are the best guide to the effects of orbital changes on Earth's climate.

8-6 Insolation Changes by Month and Season

The long-term trends of tilt (see Figure 8-4) and $\varepsilon\sin\omega$ (see Figure 8-15) contain the information needed to calculate the amount of insolation arriving at any latitude and season. By convention, climate scientists usually show the amount of insolation (or the departures of insolation from a long-term average) during the solstice months of June and December in units of W/m². Some studies use an alternate form: calories per cm² per second.

June and December insolation values over the last 300,000 years show a strong dominance of the 23,000-year precession cycle at lower and middle latitudes, and also at higher latitudes during the summer season (Figure 8-16). Just like the $\varepsilon\sin\omega$ precessional index, individual insolation cycles at lower latitudes occur at

wavelengths near 23,000 years, but their amplitudes are modulated at periods of 100,000 and 413,000 years. The June and December monthly insolation curves at each latitude in Figure 8-16 are also opposite in sign. Both can vary by as much as 12% (40 W/m²) around the long-term mean value for each latitude.

The 41,000-year cycle of tilt (obliquity) is not evident at lower latitudes but is visible in the low-amplitude variations of winter-season insolation at higher mid-latitudes (northern hemisphere January and southern hemisphere June at 60°). For example, two precession cycles evident near 50,000 years ago in the June insolation signal for latitude 20°N gradually blend and merge into a single tilt cycle at latitude 80°N (see Figure 8-16).

Changes in annual mean insolation at the 41,000-year tilt signal at high latitudes have the same sign as the summer insolation anomalies, but are lower in

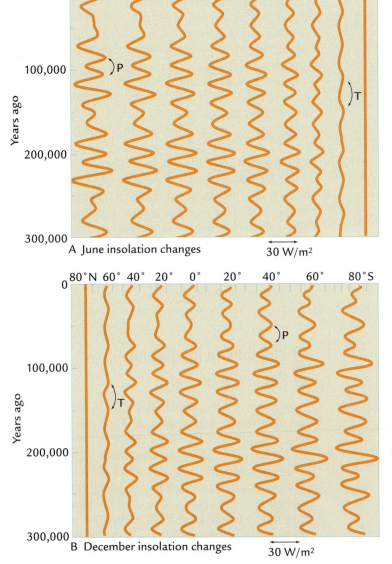

A June insolation changes   30 W/m²

B December insolation changes   30 W/m²

FIGURE 8-16

June and December insolation variations

June and December monthly insolation values show the prevalence of precessional changes at low and middle latitudes and the presence of tilt changes at higher latitudes. Cycles of tilt and precession are indicated by *T* and *P*. The double arrows indicate variations of 30 W/m² for these signals.

amplitude. The lesser significance of winter season changes in tilt at the highest polar latitudes results from the fact that no insolation at all arrives during long stretches of polar winter.

> **In Summary**, monthly seasonal insolation changes are dominated by precession at low and middle latitudes, with the effects of tilt evident only at higher latitudes.

As noted earlier, cycles of insolation change at 100,000 or 413,000 years are not evident in these signals because eccentricity is *not* a source of seasonal insolation changes. Actually, very small variations in received insolation do occur in connection with Earth's eccentric orbit around the Sun, but these appear only as changes in the total energy received by the entire Earth, not as seasonal variations. These changes are governed by the term $(1 - \varepsilon^2)^{1/2}$. We have already seen that ε varies through time between 0.005 and 0.0607. Substituting these values for ε in the term above reveals that changes in total insolation received because of changes in eccentricity have varied by at most 0.002 (0.2%) around the long-term mean. Compared to changes in seasonal insolation of 10% or more at the tilt and precession cycles, these annual eccentricity changes are negligible (smaller by a factor of about 50).

The patterns of insolation changes for tilt and precession can be compared by season and by hemisphere (northern versus southern). Insolation variations at high latitudes caused by changes in tilt are *in phase* between the hemispheres from a seasonal perspective: tilt maxima in the northern winter solstice of December match tilt maxima in the southern winter solstice of June. At high tilt angles (Figure 8-17A), summer (June) insolation maxima in the Northern Hemisphere occur at the same time in the 41,000-year cycle as summer (December) insolation maxima in the Southern Hemisphere on the opposite side of the orbit. Higher tilt produces more insolation at both poles in their respective summers because both poles are turned more directly toward the Sun. For the same reason, more pronounced insolation minima also occur at both poles in winter as a result of a higher tilt: the two winter poles are tilted away from the Sun during the same orbit.

If we compare the North Pole with the South Pole *during a particular month* in the orbit, however, the two hemispheres are exactly out of phase (Figure 8-17A). The increased tilt angle that turns north polar regions more directly toward the Sun in northern hemisphere summer also tilts the southern polar regions farther away from the Sun at that same place in the orbit (southern hemisphere winter). As

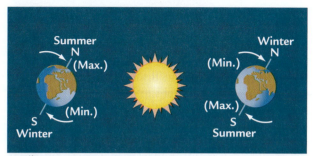

FIGURE 8-17
Phasing of insolation maxima and minima

Tilt causes in-phase changes for polar regions of both hemispheres in their respective summer and winter seasons (A). Precession causes out-of-phase changes between hemispheres for their summer and winter seasons (B).

a result, tilt causes opposite insolation effects at the North and South poles for a given point in the orbit.

For precession, the relative sense of phasing between seasons and hemispheres differs from that of tilt (see Figure 8-17B). Because Earth-Sun distance is the major control on these changes in insolation, a position close to the Sun (at perihelion) produces higher insolation than normal over *all* of Earth's surface. A precessional-cycle insolation maximum occurring at June 21 (or December 21) will be simultaneous everywhere on Earth. Distant pass positions (at aphelion) will simultaneously diminish insolation everywhere on Earth.

An important fact to remember about precession is that the seasons are reversed across the equator. As a result, an insolation maximum at June 21 is a *summer* insolation maximum in the Northern Hemisphere, but it is a *winter* insolation maximum in the Southern Hemisphere, where June 21 is the winter solstice. As a result of the seasonal reversal at the equator, insolation signals considered in terms of the season of the year are *out of phase* between the hemispheres for precession. This pattern is opposite in sense to the in-phase pattern for tilt at high latitudes of both hemispheres.

Another way of looking at the relative phasing of precessional insolation is to track changes between seasons in a single hemisphere. The orbital position on the left in Figure 8-17B, which produces minimum summer (June 21) insolation in the Northern Hemisphere because it occurs at a distant position from the Sun (aphelion), must six months later cause a maximum in winter (December 21) insolation in the same hemisphere when Earth revolves around to the perihelion position (Figure 8-17B right). As a result, precessional variations in insolation at any one location always move in opposite directions for the summer versus winter seasons.

Precessional changes in insolation have an additional characteristic not found in changes caused by tilt: an entire family of insolation curves exists for each season and month (and even day) of the year. As a matter of convention, insolation changes are typically shown only for the extreme solstice months of June and December, but in fact every season and month precesses into parts of the eccentric orbit that are alternately farther from the Sun and closer to the Sun at the same 23,000-year cycle.

As a result, each season and month experiences the same 23,000-year cycle of increasing and decreasing insolation values relative to the long-term mean, but the anomalies (departures from the mean) are offset in time from those of the preceding month or season. These offsets produce a family of monthly (and seasonal) insolation curves (Figure 8-18). Each successive month passes through perihelion (or aphelion) roughly 1,916 years later than the previous month did (1/12 × 23,000 = 1,916).

8-7 Insolation Changes by Caloric Seasons

Calculations of monthly insolation are complicated by an additional factor related to the eccentricity of Earth's orbit. Although Earth gradually moves through

a 360° arc in its orbit around the Sun, its rate of angular motion in space is not constant. Instead, Earth speeds up as it nears the extreme perihelion position and slows down near aphelion. As a result, as the solstices move slowly around the eccentric orbit, they gradually pass through regions of faster or slower movement in space.

These changes in speed cause changes in the lengths of the months and seasons in relation to a year as determined by "calendar time" (day of the year). The net effect is that changes in the amplitude of insolation variations in the monthly signals tend to be canceled by opposing changes in the lengths of the seasons. For example, times of unusually high summer insolation values at the perihelion position are also times of shorter summers. It is not obvious how to balance these two offsetting factors.

One way of minimizing these complications is to calculate the changes in insolation received on Earth in the framework of caloric seasons (the **caloric insolation season**). The summer caloric half-year is defined as the 182 days of the year when the incoming insolation exceeds the amount received during the other 182 days. Caloric seasons are not fixed in relation to the calendar because the insolation variations caused by orbital changes are added to or subtracted from different parts of the calendar year (see Figure 8-18). As a result, the caloric summer half-year falls during the part of the year we think of as summer, but it is not precisely centered on the June 21 summer solstice.

Changes in insolation viewed in reference to the half-year caloric seasons put a somewhat different emphasis on the relative importance of tilt and precession. Although low-latitude insolation anomalies are still dominated in both seasons by the 23,000-year precession signal, the 41,000-year tilt rhythm is much more obvious in high-latitude anomalies during the summer caloric half-year (Figure 8-19) than it is in the monthly insolation curves (see Figure 8-16). Another aspect of caloric season calculations is that the insolation values vary by a maximum of only ~5% around the mean, compared to variations as large as 12% for the monthly insolation changes.

Searching for Orbital-Scale Changes in Climatic Records

In the next four chapters, we will explore evidence showing that orbital-scale cycles are recorded in many of Earth's climate records. Most such records contain two or even three superimposed orbital-scale cycles, and it can often be difficult to disentangle them visually.

For example, consider the three cycles shown in Figure 8-20A, with periods of 100,000 years, 41,000 years, and 23,000 years. These three cycles are equivalent to the three most prominent cycles of orbital

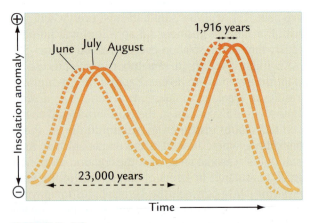

FIGURE 8-18

Family of monthly precession curves

Because all seasons change position (precess) around Earth's orbit, each season (and month) has its own insolation trend through time. Monthly insolation curves are offset by slightly less than 2,000 years (23,000 years divided by 12 months).

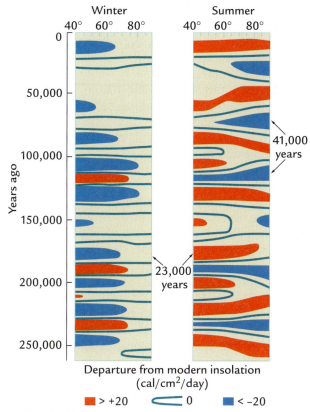

FIGURE 8-19
Caloric season insolation anomalies

Plots of insolation anomalies for the summer and winter caloric half-year show a larger influence of tilt in relation to precession at higher latitudes than do the monthly anomalies. (ADAPTED FROM W. F. RUDDIMAN AND A. MCINTYRE, "OCEANIC MECHANISMS FOR AMPLIFICATION OF THE 23,000-YEAR ICE-VOLUME CYCLE," *SCIENCE* 212 [1981]: 617–27.)

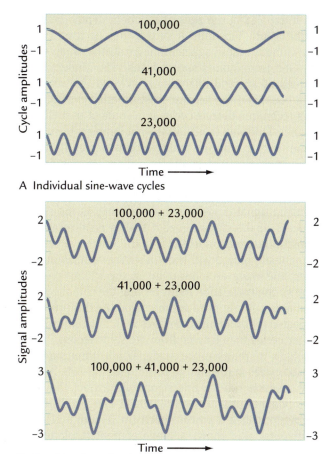

A Individual sine-wave cycles

B Combination of cycles

FIGURE 8-20
Complications from overlapping cycles

If perfect sine wave cycles with periods of 100,000 years, 41,000 years, and 23,000 years are added together so that they are superimposed on top of one another, the original cycles are almost impossible to detect by eye in the combined signal.

change, but for simplicity they are shown here as perfect sine waves rather than the more complex forms of the actual variations (because of amplitude modulations).

We can combine these three cycles by adding them together in various ways (Figure 8-20B). When the 23,000-year and 100,000-year cycles are combined, the resulting signal is obviously a simple addition of the two separate cycles. The two cycles are easy to distinguish because they differ in period by a large amount (4.3, or 100,000 divided by 23,000).

It becomes more difficult to detect the two original signals when only the 23,000-year and 41,000-year cycles are combined, as in the middle plot of Figure 8-20B. Because the periods of these two cycles are more similar, they reinforce and cancel each other in somewhat complicated ways. The task becomes even more difficult when all three cycles are combined, as in the bottom plot. It is not at all obvious to the eye that this signal is a simple addition of three perfect sine waves.

In the case of Earth's actual climate records, the situation is even more complex because the three cycles are not only superimposed on each other but also change in amplitude through time (Figures 8-4 and 8-15). Obviously, it will be impossible to disentangle all this information simply by eye.

8-8 Time Series Analysis

To simplify analyses of cyclic variations in climate changes, scientists use **time series analysis**. The term "time series" refers to records plotted against age (time). These techniques extract rhythmic cycles embedded in records of climate.

The first step in time series analysis is to convert climatic records to a time framework. After individual measurements of a climatic indicator have been made (for example, across an interval in a sediment core), all available sources of dating are used to define the ages of particular levels in the sediment sequence. A complete time scale for the entire sequence is then created by interpolating the ages of sediment depths between dated levels. This time scale can then be used to plot the climatic record against time for further analysis.

One technique is **spectral analysis**. Modern techniques of spectral analysis are beyond the scope of this book, but we need at least a basic sense of how this technique detects cycles in records of past climate change. One way to visualize what happens in spectral analysis is to imagine taking a climate record plotted on a time axis and gradually sliding a series of sine waves of different periods across it. As this is done, the correlation between each sine wave and the full climatic signal is measured at each step in the sliding process. If the climate record that is being examined contains a strong cycle at one of the sine wave periods, the climate record will show a strong correlation with that sine wave at some point in the sliding process. The strong correlation indicates that the climatic signal contains a strong cycle at that period. As this process is repeated for different sine waves with different periods, other cycles may emerge from the same record.

Now we return to the example of the three superimposed cycles in the bottom plot of Figure 8-20B. A spectral analysis run on this signal will extract the three component (orbital) cycles, which can be displayed on a plot called a **power spectrum** (Figure 8-21). The horizontal axis shows a range of periods plotted on a log scale, with the shorter periods to the right. The vertical axis represents the amplitude of the cycles, also known as their "power." The height of the lines plotted on the power spectrum is related to the square of the amplitude of the cycle at that period.

For the example shown in Figure 8-21, all three cycles detected by spectral analysis plot as narrow

"line spectra," with their power concentrated entirely at the periods shown by the solid vertical lines. In this idealized example, no power occurs anywhere else in the spectrum except at those three cycles.

In actual studies of climate, however, power spectra are never this simple. One reason is that even the most regular looking orbital cycles such as the tilt changes in Figure 8-4 are not perfect sine waves but instead vary over a small range of periods. In addition, errors in dating records of climate change or in measuring their amplitude also have the effect of spreading power over a broader range of periods than would be the case for perfectly measured and dated signals. As a result of these complications, the total amount of power associated with each cycle looks more like the area under the dashed curves in Figure 8-21.

Still another reason that real-world spectra are more complicated is that random noise exists in the climate system, consisting of irregular climatic responses not concentrated at orbital or other cycles. In most records, the effect of noise is spread out over a range of periods in the spectrum. In general, the amount of power tends to be larger at longer periods. As a result, spectra from real-world climatic signals tend to look like the thick curved line in Figure 8-21. The spectral peaks that rise farthest above the baseline of the trend are the most significant (believable) ones in a statistical sense.

A second useful time series analysis technique is called **filtering**. This technique extracts individual cycles at a specific period (or narrow range of periods) from the complexity of the total signal. This process is often referred to as "bandpass filtering" (filtering of a narrow band or range of the many periods present in a given signal). Filtering is analogous to using glasses with colored lenses to block out all colors of the light spectrum except the one color (wavelength) we wish to see.

Filters are constructed directly from well-defined peaks in power spectra like those in Figure 8-21. The highest point on the spectral peak defines the central period of the filter, and the sloping sides of the spectral peak define the shape of the rest of the filter.

To understand the importance of filtering, consider again the three hypothetical sine waves in Figure 8-20. We can create filters for these three cycles based on the peaks in the power spectrum shown in Figure 8-21. If we pass these filters across the combined signal at the bottom of Figure 8-20B, the filters will extract the original form of all three of the individual cycles (at 23,000 years, 41,000 years, and 100,000 years). In effect, the filtering operation extracts the varying shapes of individual cycles in specific bands of time embedded in the complexities of actual climate records.

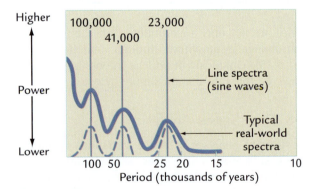

FIGURE 8-21
Spectral analysis

Spectral analysis reveals the presence of cycles in complex climate signals. In this example, the original sine wave cycles from Figure 8-20A form line spectra (vertical bars) whose heights indicate their amplitudes. Actual climate records have peaks that are spread over a broader range of periods (dashed lines and curving solid line).

8-9 Effects of Undersampling Climate Records

The technique of spectral analysis can be used only for a specific range of cycles in any climate record. Confident identification of a cycle by time series analysis requires that the cycle be repeated at least four times in the original record (the record must be at least four times longer than the cycle analyzed). At the other extreme (for the shortest cycles in a record), at least two samples per cycle are needed to verify that a given cycle is present, although many more are needed to define its amplitude accurately. With fewer than two samples per cycle, time series analysis runs into the problem of **aliasing**, a term that refers to false trends generated by undersampling the true complexity in a signal.

Consider the hypothetical case of a climatic signal that has the form of the 23,000-year cycle of orbital precession, with the wide range of amplitude variation typical of such signals (Figure 8-22). Assume that three scientists sample a record containing this underlying signal, with all three sampling the record at an average spacing of 23,000 years, but with each beginning the sampling process at a different place in the record. If one scientist happened to start sampling exactly at a maximum in the signal, he or she would end up measuring only a record of successive maxima, but if another scientist happened to start at a minimum, the record would only show successive minima. These sampling attempts give completely different results because they are persistently biased toward different sides of a highly modulated cycle. A third scientist who happened to start sampling exactly at a crossover point between minima and maxima might extract a record suggesting that no signal exists at all.

These differences show the danger of aliasing. Although this example is obviously chosen to show the worst possible effect of aliasing, undersampling is a common problem in climate records.

8-10 Tectonic-Scale Changes in Earth's Orbit

Over time scales of hundreds of millions of years, some of Earth's orbital characteristics slowly evolved, as shown by evidence in ancient corals. Corals are made of banded $CaCO_3$ layers caused by changes in environmental conditions. The primary annual banding reflects seasonal changes in sunlight and water temperature (see Chapter 3). A secondary banding follows the tidal cycles created by the Moon and Sun. The tidal cycles also affect water depth and other factors in the reef environment that influence coral growth.

Corals from 440 million years ago show 11% more tidal cycles per year than modern corals do, implying that Earth spun on its rotational axis 11% more times per year than at present. As a result, each year had 11% more days. Gradually over the last 440 million years, the spin rate and number of days decreased to their current levels. This gradual slowing in Earth's rate of rotation was caused by the frictional effect of the tides.

Other changes in Earth's orbit that can be inferred from this kind of information, such as changes in Earth-Moon distance, are thought to have affected the wavelengths of tilt and precession over tectonic-scale intervals. One estimate of the slow, long-term increases in the periods of tilt and precession toward their present values is shown in Figure 8-23.

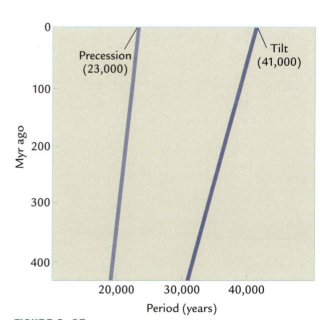

FIGURE 8-23
Tectonic-scale orbital changes
Gradual changes in Earth's orbit over long tectonic time scales have caused a slow increase in the periods of the tilt and precession cycles. (ADAPTED FROM A. BERGER ET AL., "PRE-QUATERNARY MILANKOVITCH FREQUENCIES," *NATURE* 342 [1989]: 133-34.)

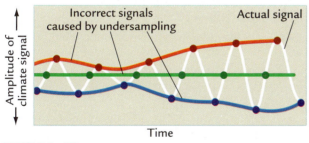

FIGURE 8-22
Aliasing (undersampling) of climate signals
Undersampling of a climate signal (in this case one that is a direct response to changes in orbital precession) can produce aliased climate signals completely unlike the actual one.

Key Terms

plane of the ecliptic (p. 160)

tilt (p. 160)

solstices (p. 160)

equinoxes (p. 160)

perihelion (p. 161)

aphelion (p. 161)

wavelength (p. 162)

period (p. 162)

frequency (p. 162)

amplitude (p. 162)

modulation (p. 162)

sine waves (p. 162)

eccentricity (p. 163)

axial precession (p. 164)

precession of the ellipse (p. 165)

precession of the equinoxes (p. 165)

precessional index (p. 168)

insolation (p. 169)

caloric insolation seasons (p. 172)

time series analysis (p. 173)

spectral analysis (p. 174)

power spectrum (p. 174)

filtering (p. 174)

aliasing (p. 175)

Review Questions

1. Why does Earth have seasons?

2. When is Earth closest to the Sun in its present orbit? How does this "close pass" position affect the amount of radiation received on Earth?

3. Describe in your own words the concept of modulation of a cycle.

4. Earth's tilt is slowly decreasing today. As it does so, are the polar regions receiving more or less solar radiation in summer? In winter?

5. How is axial precession different from precession of the ellipse?

6. How does eccentricity combine with precession to control a key aspect of the amount of insolation Earth receives?

7. Do insolation changes during summer and winter have the same or opposite timing at any single location on Earth? Why or why not?

8. Do the following changes occur at the same time (same year) in Earth's orbital cycles?

   (a) Summer insolation maxima at both poles caused by changes in tilt?

   (b) Summer insolation maxima in the tropics of both hemispheres caused by precession?

Additional Resources

Basic Reading

Imbrie, J., and K. P. Imbrie, 1979. *Ice Ages: Solving the Mystery.* Short Hills, NJ: Enslow.

Advanced Reading

Berger, A. L., 1978. "Long-Term Variations of Caloric Insolation Resulting from the Earth's Orbital Elements." *Quaternary Research* 9: 139–167.

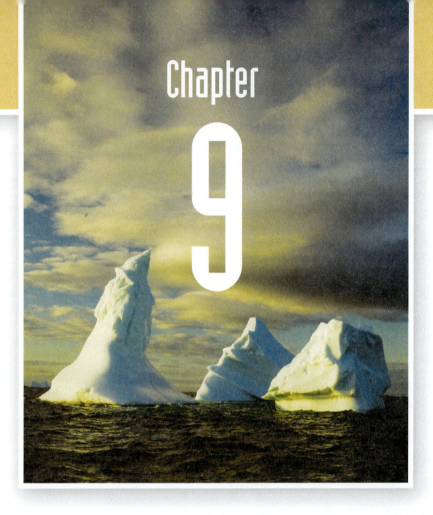

Chapter 9

Insolation Control of Monsoons

Monsoonal circulations exist on Earth today because the land responds to seasonal changes in solar radiation much more quickly than the ocean does. The reason for these differing rates of response is that the land has a far smaller capacity to store heat than the ocean. In this chapter, we will examine evidence showing that changes in insolation have been the primary driver of summer monsoons over orbital time scales. Variations in insolation produced large changes in heating of tropical landmasses and in the strength of summer monsoons at the 23,000-year cycle of precession. Just 11,000 years ago, strong summer insolation drove a vigorous summer monsoon circulation in North Africa, and the southern part of the present Sahara Desert was dotted with lakes, seasonally flowing rivers, and grassland vegetation. Similar changes have occurred across the northern tropics for millions of years. Precession-driven changes in monsoon strength have also occurred in the Southern Hemisphere, but with opposite timing. Especially large changes occurred 200 million years ago across the northern tropics of the supercontinent Pangaea.

Monsoonal Circulations

In summer, strong solar radiation causes a rapid and large warming of the land, but a slower and much less intense warming of the ocean. Rapid heating over the continents causes air to warm, expand, and rise, and the upward movement of air creates an area of low pressure at the surface (see Chapter 2). The air flowing toward this low-pressure region also warms and rises (Figure 9-1A). The air arriving from nearby oceans carries water vapor that condenses and contributes to monsoonal rainfall. Continental regions far from the ocean or protected behind intervening mountain ranges may lie beyond the reach of oceanic moisture and, therefore, may bake under the strong summer radiation.

During winter, when solar radiation is weaker, air over the land cools off rapidly, becomes denser than the air over the still-warm ocean, and sinks from higher levels in the atmosphere. This downward movement creates a region of high pressure over the land, in contrast to the lower pressure over the still-warm oceans. The overall atmospheric flow in winter is a downward-and-outward movement of cold, dry air from the land to the sea (Figure 9-1B).

Most strong summer monsoons occur in the Northern Hemisphere because landmasses there are large (Asia and North Africa) and elevations are high (especially in the Tibetan Plateau and Himalayan Mountain region of southern Asia). Monsoons are weaker in the Southern Hemisphere, where landmasses at tropical and subtropical latitudes are smaller, and high topography is more limited in extent.

Here the initial focus is on past variations of the North African monsoon for two reasons. First, North Africa lies far from the high-latitude ice sheets that might complicate the direct response of land surfaces to solar heating. In addition, the nearby seas and oceans yield a rich variety of climate records that document monsoon-related signals on North Africa.

Africa is a deceptively large landmass compared to its appearance on Mercator maps. It stretches from 37°N to 35°S, with far more of its land area north of

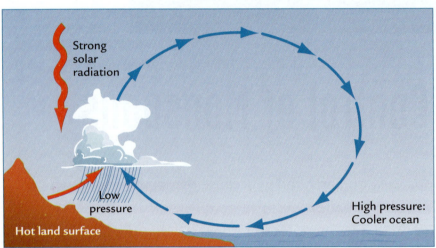

A Summer monsoon

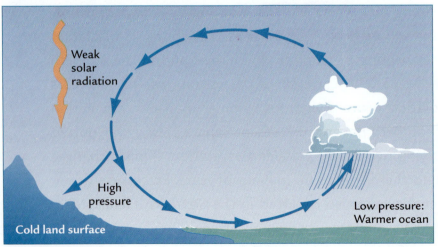

B Winter monsoon

FIGURE 9-1

Seasonal monsoon circulations

Seasonal changes in the strength of solar radiation affect the surface of the land more than the ocean. In summer, intense solar heating of the land causes an in-and-up circulation of moist air from the ocean (A). In winter, weak solar radiation allows the land to cool off and creates a down-and-out circulation of cold dry air (B).

the equator (in fact an area almost twice that of the U.S. mainland). Because the huge North African land surface is situated at tropical and lower subtropical latitudes, it is strongly influenced by the overhead Sun.

As a result of strong solar heating during northern hemisphere summer, a low-pressure region develops over west-central North Africa and draws in moisture-bearing winds from the tropical Atlantic Ocean (Figure 9-2A). During typical summers, this monsoonal rainfall penetrates northward to 17°N latitude (the southern edge of the Sahara Desert) before retreating southward later in the year.

During northern hemisphere winter, the overhead Sun moves to the Southern Hemisphere, and solar radiation over North Africa is weaker. Cooling of the North African land surface by back radiation causes sinking of air from above, and a high-pressure cell develops at the surface over the northwestern Sahara Desert (Figure 9-2B). Strong and persistent trade winds associated with this high-pressure cell and with similar circulation over the adjacent North Atlantic blow southwestward from North Africa across the tropical Atlantic.

Because the trade winds of the winter monsoon carry little moisture, winter precipitation is rare in North Africa. Only two areas receive much rain during this season: (1) the northernmost Mediterranean margin, where storms occasionally form over the nearby ocean, and (2) the southwest equatorial coast (the Ivory Coast), where the moist intertropical convergence zone (ITCZ) remains over the land.

Because most of the rainfall in North Africa occurs in association with the summer monsoons, the distribution of major vegetation types reflects the monsoonal delivery of precipitation from the south (Figure 9-3). Rain forest in the year-round wet climate near the equator gives way northward to a sequence of progressively drier vegetation, first the trees and grasses of the savannas of the Sahel region and then the desert scrub of the arid Sahara.

9-1 Orbital-Scale Control of Summer Monsoons

The idea that changes in insolation control the strength of monsoons over orbital time scales was proposed by the atmospheric scientist John Kutzbach in the early 1980s, having been anticipated in part by the astronomer Rudolf Spitaler late in the nineteenth century. This concept is called the **orbital monsoon hypothesis**.

The orbital monsoon hypothesis is a direct logical extension of factors at work in modern monsoonal circulations (Figure 9-4). Because seasonal monsoon circulations are driven in the modern world by changes in solar radiation, orbital-scale changes in summer and winter insolation (see Chapter 8) should have produced a similar response. When summer insolation was higher in the past than today, the summer monsoon circulation should have been stronger, with greater heating of the land, stronger rising motion, more inflow of moist ocean air, and more rainfall (Figure 9-4B). Conversely, summer insolation levels lower than those today should have driven a weaker summer monsoon in the past.

The same kind of reasoning applies to the winter monsoon. Winter insolation minima weaker than the one occurring today should have enhanced the cooling of the land surface, which should have driven a stronger down-and-out flow of dry air from land to sea (Figure 9-4C).

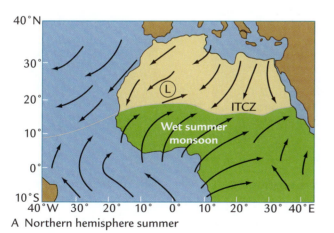

A Northern hemisphere summer

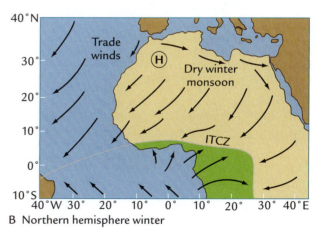

B Northern hemisphere winter

FIGURE 9-2

Monsoonal circulations over North Africa

Seasonal changes cause a moist inflow of monsoonal air toward a low-pressure center over North Africa in summer (A), and a dry monsoonal outflow from a high-pressure center over the land in winter (B). (ADAPTED FROM J. F. GRIFFITHS, *CLIMATES OF AFRICA* [AMSTERDAM: ELSEVIER, 1972].)

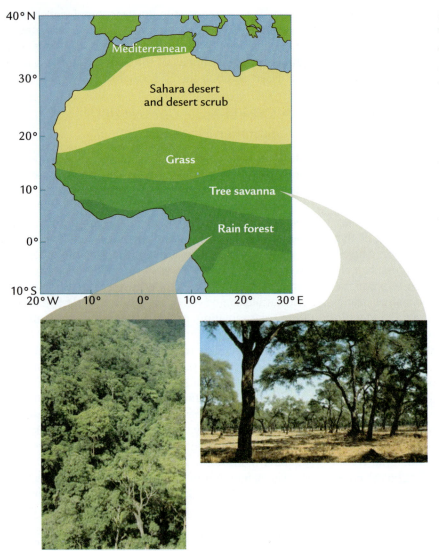

FIGURE 9-3
Vegetation in North Africa

Vegetation across the northern part of Africa ranges from rain forest near the equator to savanna and grassland in the Sahel to desert scrub vegetation in the Sahara. This pattern reflects the diminishing northward reach of summer monsoon moisture from the tropical Atlantic. (ADAPTED FROM J. F. GRIFFITHS, *CLIMATES OF AFRICA* [AMSTERDAM: ELSEVIER, 1972]. INSET PHOTOS: COURTESY OF TOM SMITH, UNIVERSITY OF VIRGINIA.)

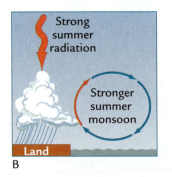

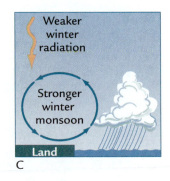

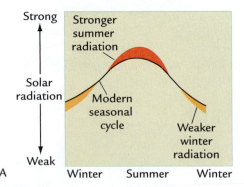

FIGURE 9-4
The orbital monsoon hypothesis

Departures from the modern seasonal cycle of solar radiation have driven stronger monsoonal circulations in the past (A). Greater summer radiation intensified the wet summer monsoon (B), while decreased winter insolation intensified the dry winter monsoon (C). (ADAPTED FROM J. E. KUTZBACH AND T. WEBB III, "LATE QUATERNARY CLIMATIC AND VEGETATIONAL CHANGE IN EASTERN NORTH AMERICA: CONCEPTS, MODELS, AND DATA," IN *QUATERNARY LANDSCAPES*, ED. L. C. K. SHANE AND E. J. CUSHING [MINNEAPOLIS: UNIVERSITY OF MINNESOTA PRESS, 1991].)

Recall from Chapter 8 that more intense summer insolation maxima and deeper winter insolation minima occur together at any one location. As a result, stronger in-and-up monsoonal flows in summer should occur at the same times in the past as stronger down-and-out monsoonal flows in winter.

At first, it might seem that the climatic effects of these opposed insolation trends in the two seasons might cancel each other, but this is not the case for annual precipitation. Monsoonal winters are always dry, regardless of the amount of insolation, because the air descending from higher in the atmosphere holds very little moisture (Figure 9-4C). As a result, orbital-scale changes in winter insolation have little or no effect on annual rainfall. In contrast, because summer monsoon winds coming in from the ocean carry abundant moisture, orbital-scale changes in summer insolation have the dominant impact on annual rainfall.

This imbalance is an example of a **nonlinear response** of the climate system to insolation: the amount of rainfall is highly sensitive to insolation change in one season (summer) but largely insensitive to changes in the other (winter). As a result, the annual response of the system has a strong summer signature even though the equal-but-opposite insolation trends in the two seasons might have been expected to cancel each other.

Orbital-Scale Changes in North African Summer Monsoons

The orbital monsoon hypothesis can be tested against a wide range of evidence, including changes in lake levels across arid regions like North Africa. At low and middle latitudes, changes in the amount of incoming solar insolation follow the 23,000-year rhythm of orbital precession (recall Chapter 8). A June insolation curve from latitude 30°N covering the last 140,000 years clearly shows this 23,000-year tempo (Figure 9-5). Note that today's June insolation level is well below the longer-term average. The orbital monsoon hypothesis predicts that stronger summer monsoons should have occurred at those times in the past when summer insolation values were significantly larger than the modern value.

The most recent instance when summer insolation values were substantially higher than today occurred near 11,000 years ago. Evidence we will examine in Part IV of this book shows that lakes across tropical and subtropical North Africa were at much higher levels during this insolation maximum, and for some time afterward. In fact, many lakes that were filled to high levels at that time are completely dry today. This evidence indicates that a strengthened summer monsoon circulation reached much farther northward into North Africa 11,000 years ago than it does today, in agreement with the orbital monsoon hypothesis.

But this interval was only one brief period in a long history of rainfall changes in North Africa. What about the much longer-term behavior of the summer monsoon? We can use the June 30°N insolation curve in Figure 9-5 to construct a simple conceptual model that predicts how the summer monsoon could have varied with time if the orbital monsoon hypothesis is valid. The predicted monsoon response is based on three assumptions.

First, we assume that a **threshold level** of insolation exists, below which the monsoon response will be too weak to leave any evidence in the geologic record (see Figure 9-5). A good example is the level of lakes in arid regions. The orbital monsoon hypothesis predicts that the higher the level of summer insolation, the stronger the summer monsoon rainfall, and the higher the lake levels. But if the insolation value falls below a certain level, the weakened summer monsoon may produce so little rain that the lakes dry up

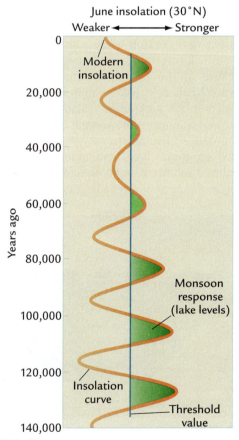

FIGURE 9-5
Conceptual model of monsoon response to summer insolation

Increases in summer insolation heating above a critical threshold value drive a strong monsoon response at the 23,000-year tempo of orbital precession. The amplitude of this strong monsoon response is related to the size of the increase in summer insolation forcing.

completely. A dry lake can no longer register changes toward even drier climates because it has reached the threshold limit of its ability to record climate.

This assumption has a good basis in fact. Many lakes in North Africa that existed 11,000 years ago during the strong monsoon interval are dry today, even though a weak summer monsoon still occurs at present in response to today's low levels of summer radiation. Apparently, it takes a threshold insolation value well above the present level to bring most North African lakes into existence.

Second, we assume that the strength of the monsoon response (such as the water level of the North African lakes) is directly proportional to the amount by which summer insolation exceeds the threshold value. This assumption has a reasonable physical basis: stronger insolation should drive stronger monsoons and fill lakes to higher levels.

Third, we assume that the strength of the monsoon in the past, as recorded in lake level records, is a composite of the average monsoon strength over many individual summers. Actually, this is more observed fact than assumption: the wet monsoonal circulations that develop every summer today inevitably cease during the following winter. The lakes fill because of the integrated effect of many wet summers. When scientists sample records of these changes, they are looking at year-by-year responses blended into a longer-term average over hundreds or even thousands of summers.

Based on these three assumptions, we arrive at the predicted monsoon response shown by the green shading in Figure 9-5. Insolation maxima above the

threshold value produce a series of pulse-like monsoon (and lake) maxima at regular 23,000-year intervals. These pulses vary in strength according to the amount by which summer insolation exceeds the threshold value. Insolation levels below the threshold value leave no monsoonal evidence in the geologic record.

The strongest predicted monsoon peaks in Figure 9-5 occur between 85,000 and 130,000 years ago, when the summer insolation curve reached large maxima because of modulation of the precession signal by orbital eccentricity (see Chapter 8). In contrast, the weaker insolation maxima near 35,000 and 60,000 years ago should have produced less powerful monsoons. All long-term summer insolation minima (such as the one we are in today) fall below the critical threshold for major lake filling.

We can examine several climate records for evidence of this predicted monsoon response. Because most of North Africa is arid, and because erosion of its sediments is much more prevalent than deposition, its climate history is sparse and difficult to date. Fortunately, the nearby seas and oceans contain continuous and well-dated records.

9-2 "Stinky Muds" in the Mediterranean

The water that fills the Mediterranean Sea today has a high oxygen content. Near-surface waters are well oxygenated because they exchange oxygen-rich air with the atmosphere and because photosynthesis by marine organisms produces O_2. The high oxygen content of waters deeper in the basin results from sinking of oxygen-rich surface water during winter (Figure 9-6A).

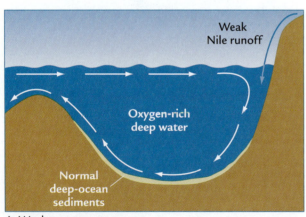

A Weak summer monsoon

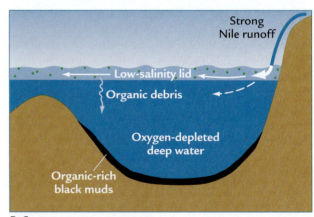

B Strong summer monsoon

FIGURE 9-6

Mediterranean circulation and monsoons

In today's Mediterranean circulation, salty surface water chilled by cold air in winter sinks and carries dissolved oxygen to deeper layers (A). At intervals in the past, strong summer monsoons in tropical Africa caused an increased discharge of Nile River fresh water into the eastern Mediterranean, creating a low-density surface-water lid that inhibited sinking of surface water and caused the deep ocean to lose its oxygen and deposit organic-rich black muds (B).

This sinking motion is a result of two factors (see Chapter 2): (1) the high salt content of the Mediterranean Sea, caused by the excess of summer evaporation over precipitation, and (2) winter chilling of salty water along the northern margins of the Mediterranean Sea during incursions of cold air from the north. These two factors make surface waters dense enough to sink to great depths in winter. The dense waters that sink into the deep Mediterranean Sea eventually exit westward into the Atlantic Ocean. As a result of this large-scale flow, the floor of the present Mediterranean Sea is covered by sediments typical of well-oxygenated ocean basins: light tan silty mud containing shells of plankton that once lived at the sea surface and benthic foraminifera that once lived on the seafloor.

Mediterranean sediments also contain distinct layers of black organic-rich muds, called **sapropels**. Their high organic carbon content indicates that they formed at times when the waters at the seafloor were **anoxic**: they lacked the oxygen needed to convert (oxidize) organic carbon to inorganic form. The lack of oxygen led to stagnation of the deep waters and deposition of iron sulfides, giving the sediments a "stinky" (rotten egg) odor. The lack of oxygen also kept benthic foraminifera and other creatures from living on the seafloor.

The paleoecologist Martine Rossignol-Strick proposed that these sinky muds mark times when the deep Mediterranean Basin was deprived of oxygen because sinking of oxygen-rich surface waters was prevented by a cap of low-density freshwater brought in by river water (see Figure 9-6B). Even though the surface waters were still chilled by cold winter air masses, the low-salinity lid kept them from becoming dense enough to sink deep into the basin. As a result, the deep Mediterranean Basin lost its supply of oxygen.

At the same time, production of planktic organisms continued at the surface, and probably even increased as the stronger river inflow delivered extra nutrients (food) to the Mediterranean. The high productivity at the surface continued to send organic-rich remains of dead plankton toward the seafloor. Sinking and oxidation of this organic carbon continually depleted the oxygen levels in the deep Mediterranean and produced the stinky muds on the seafloor.

The most recent sapropel in the eastern Mediterranean dates from 10,000 to 8,000 years ago, an interval when summer insolation levels were higher than today, the African summer monsoon was stronger, and African lakes were at higher levels. Earlier layers of organic-rich mud deeper in Mediterranean sediment cores occur at regular 23,000-year intervals during times when summer insolation was higher than it is today. The sapropels were best developed (thickest and most carbon-rich) near the time of the strongest summer insolation maxima, but poorly developed during weaker insolation maxima, and absent the rest of the

time. This history of sapropel deposition matches very well the conceptual pattern predicted by the orbital monsoon hypothesis (see Figure 9-5). This close match indicates a connection of some kind to the low-latitude monsoon over North Africa.

Initially, some climate scientists questioned this explanation. The Mediterranean Sea lies at high subtropical latitudes (30°–40°N), beyond the greatest northward expansions of past summer monsoons indicated by lake-level evidence across North Africa. If climate in the confines of the Mediterranean region never became truly monsoonal, how could the stinky muds deposited in that basin be a response to the North African monsoon?

The critical link turned out to be the Nile River (Figure 9-7), which gathers most of its water from the highlands of eastern North Africa at tropical latitudes much farther south. Even today these highlands

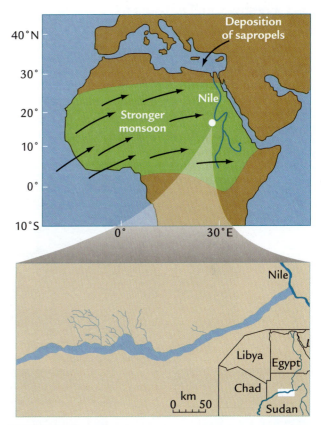

FIGURE 9-7
Monsoons and the Nile River

Strong summer monsoons in tropical North Africa periodically produced large discharges of Nile freshwater into the Mediterranean Sea. Satellite sensors have detected riverbed sediments deposited during strong monsoons but now buried beneath sheets of sand in the hyperarid eastern Sahara Desert (inset). (INSET: ADAPTED FROM H.-J. PACHUR AND S. KROPLEIN, "WADI HOWAR: PALEOCLIMATIC EVIDENCE FROM AN EXTINCT RIVER SYSTEM IN THE SOUTHEASTERN SAHARA," *SCIENCE* 237 [1987]: 298–300.)

receive summer rains during the relatively weak tropical monsoon, and the Nile delivers the water to the Mediterranean Sea far to the north. At times when summer insolation was much stronger than it is today, the strengthened summer monsoon expanded northward and eastward, bringing much heavier rainfall to these high-elevation regions. In effect, rainfall in the North African tropics exerts a remote control on the salinity of the subtropical Mediterranean Sea via the Nile River connection.

Satellite sensors have detected the buried remnants of streams and rivers that once flowed across Sudan (see Figure 9-7) but are now covered by sheets of sand blowing across the hyperarid southeastern Sahara Desert. The fact that these streams once flowed eastward and joined the Nile River indicates that lower-elevation regions also contributed to the Nile's stronger flow during major monsoons.

9-3 Freshwater Diatoms in the Tropical Atlantic

Evidence that North African lakes fluctuate at the 23,000-year tempo of orbital precession can also be found in sediment cores from the north tropical Atlantic Ocean (Figure 9-8). These sediments contain layers with high concentrations of the opaline ($SiO_2 \cdot H_2O$) shells of the freshwater diatom *Aulacoseira granulata*. Because

these diatoms could not have lived in the ocean, and because these ocean cores are hundreds to thousands of kilometers away from the land, the only explanation for the presence of the diatoms is wind transport. In arid and semiarid regions, winds scoop out ("deflate") sediment from the beds of dry lakes and blow the fine debris far away, some of it to the nearby oceans.

The only plausible source of these lake diatoms in these tropical Atlantic sediment cores is the one lying directly upwind in the prevailing northeasterly flow of winter trade winds—arid and semiarid North Africa. The intervals in the Atlantic cores containing freshwater diatoms mark times in the past when North African lakes were drying out and their muddy lakebeds were becoming exposed to, and eroded by, strong winter trade winds.

Records from the Atlantic sediment cores show that lake diatoms were delivered in distinct pulses separated by 23,000 years. As was the case for the Mediterranean sapropels, this 23,000-year tempo in diatom influxes is a direct indication of a connection to the tropical monsoon fluctuations in North Africa. In this case, however, each diatom pulse occurs later than the summer insolation maxima by as much as 5,000 to 6,000 years (Figure 9-9).

This delay makes sense if seen as a part of the sequence of likely events during a typical monsoon cycle. Lakes in North Africa filled to maximum size during the summer insolation maxima that drove the strong monsoons. These high lake levels deposited lakebeds rich in diatoms. Then, as summer insolation began to decrease toward the next insolation minimum, the monsoon weakened, summers became drier, and the lake levels began to drop. The fall in lake levels exposed the diatom-bearing silts and clays to winter winds, which scooped them up and blew them out to the ocean. Once the lakes had dried out completely and most of the diatom-bearing sediments had been blown away, transport of diatoms to the

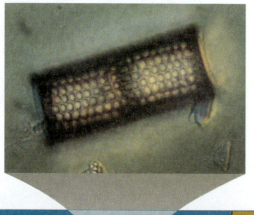

FIGURE 9-8
Drying of monsoonal lakes

North African lakes filled by strong monsoonal rains later dried out and were exposed to erosion by winds. Lake muds containing the freshwater diatom *Aulacoseira granulata* (inset) were carried by winds to the tropical Atlantic. (INSET: COURTESY OF BJORG STABELL, UNIVERSITY OF OSLO.)

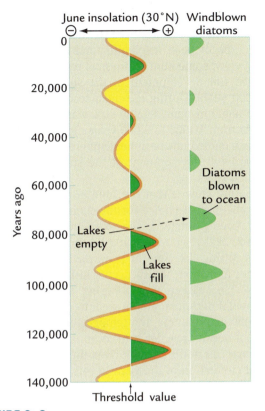

FIGURE 9-9

Delayed diatom deposition in the Atlantic

Diatoms from North African lakes were deposited in the tropical Atlantic Ocean several thousand years after the intervals of strongest monsoons, as the lakes dried out. (ADAPTED FROM W. F. RUDDIMAN, "TROPICAL ATLANTIC TERRIGENOUS FLUXES SINCE 25,000 YEARS B. P.," *MARINE GEOLOGY* 136 [1997]: 189–207; BASED ON E. M. POKRAS AND A. C. MIX, "EARTH'S PRECESSION CYCLE AND QUATERNARY CLIMATIC CHANGES IN TROPICAL AFRICA," *NATURE* 326 [1987]: 486–7.)

diatom pulses sent to the ocean lagged well behind the summer monsoon maxima because of the time needed to dry the lakes, but they preceded the subsequent summer insolation minima because many of the lakebeds had already been fully exposed just partway into the drying trend.

Another indication of a link to the North African summer monsoon comes from the amplitude of the diatom peaks. Each 23,000-year diatom pulse has the same relative strength as the immediately preceding summer insolation maximum (see Figure 9-9). This pattern is consistent with a scenario in which stronger insolation maxima drove stronger summer monsoon maxima, which created bigger lakes, which provided larger sources of diatom-bearing sediments for subsequent transport to the ocean.

9-4 Upwelling in the Equatorial Atlantic

Atlantic Ocean sediments contain additional evidence consistent with the hypothesis that the North African monsoon fluctuates at the 23,000-year tempo of orbital precession. Cores in the eastern Atlantic just south of the equator show that the structure of the upper water layers has varied with a prominent 23,000-year rhythm. Part of the reason for this response is that the North African summer monsoon imposes an atmospheric circulation pattern that overrides the local circulation.

When the North African summer monsoon is relatively weak (as it is today), trade winds along the equator have a strong east-to-west flow (Figure 9-10A). The strongest trade winds occur in southern hemisphere winter (July and August) and blow from the South Atlantic toward the equator. Part of this flow crosses the equator, turns to the northwest, and enters North Africa in the summer monsoon flow, but this part of the flow is not strong when the monsoon is weak, as it is today. Instead, strong trade winds at times like today blow mainly toward the west and drive warm surface

ocean slowed or stopped, even though the monsoon continued to weaken as summer insolation continued to fall toward the next minimum. As a result, the

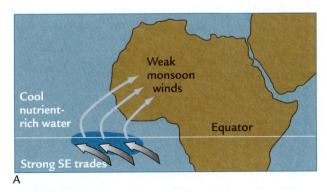 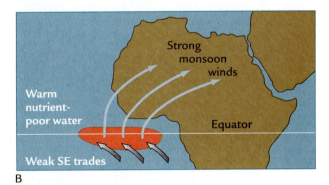

FIGURE 9-10

Effect of monsoons on southeast trade winds

When monsoonal circulation over North Africa is weak, strong southeasterly trade winds in the eastern tropical Atlantic cause cool nutrient-rich waters to rise close to the surface (A). When a strong monsoonal circulation over North Africa weakens the trade winds, tropical waters become warm and depleted in nutrients (B).

waters southward away from the equator (recall Chapter 2). This upper-ocean flow causes a shallowing of the seasonal thermocline, a subsurface region of steep temperature gradients between the warm surface waters and much cooler temperatures below. As the thermocline shallows, cooler waters rich in nutrients rise toward the sea surface just south of the equator.

In contrast, at times when summer insolation was higher than it is today, such as 11,000 years ago, the stronger summer monsoon flow overrode this circulation pattern (Figure 9-10B). A much larger portion of the southern trade-wind flow crossed the equator, turned to the northeast, and was drawn into North Africa in the monsoonal circulation. This strengthening of the monsoon flow into North Africa weakened the westward trade-wind flow along the equator, and the weaker trade winds reduced the upwelling of cold waters, leaving the surface waters poorer in nutrients from below.

Changes between these two circulation patterns over time can be measured by examining variations in the relative amounts of planktic organisms that inhabit near-surface waters and leave shells in the sediments below. In equatorial Atlantic sediments, planktic foraminifera and coccoliths are the most common shelled organisms (see Chapter 3). Different species of these two kinds of plankton prefer different environmental conditions near the sea surface—either warmer waters with fewer nutrients or cooler waters rich in nutrients. Sediment cores from the Atlantic Ocean just south of the equator show 23,000-year cycles of alternating abundances in these two types of plankton, still another indication of the effects of the North African summer monsoon.

9-5 The Phasing of Summer Monsoons

The idealized monsoon model presented so far in this chapter has suggested that peak development of past summer monsoons at the 23,000-year cycle occurred as a direct response to strong insolation forcing at the timing of the June 21 summer solstice. In fact, however, the strongest monsoons have occurred about 2,000 years later. For example, the most recent low-latitude insolation maximum was about 11,000 years ago, but the most recent Mediterranean sapropel didn't occur until 9,000 years ago. This offset has been interpreted in two ways.

Recall that Earth's precessional motion produces a family of monthly insolation curves, each offset from the preceding month by one-twelfth of a 23,000-year cycle, or slightly less than 2,000 years (see Chapter 8, Figure 8-18). This entire family of 23,000-year insolation curves is available to boost the strength of the summer monsoon, but the problem is to provide a specific physical justification for the one chosen.

One interpretation starts with the assumption that June 21 is the best choice because the summer solstice is the time of peak insolation forcing (Figure 9-11). In this view, some kind of physical process must have retarded full development of the summer monsoon by 2,000 years. During the most recent deglaciation, an ice sheet of considerable size was still present in Canada from 11,000 to 9,000 years ago, so perhaps its presence cooled northern hemisphere climate enough to retard full development of the summer monsoon for 2,000 years.

Another view is that the June 21 summer solstice is not the correct choice for the time of critical insolation forcing. Instead, the extra insolation forcing at the 23,000-year cycle could be more effective if it occurs during the time of peak development of the summer monsoon, which occurs in late July to early August (see Figure 9-11). During this month, the continental interiors are heated to their maximum summer temperatures, the temperature contrast with the cool oceans reaches a maximum, and the resulting temperature difference drives the strongest monsoonal circulation of the year. In this view, insolation changes at the 23,000-year precession cycle that are aligned with this most intense midsummer heating should have the greatest impact in boosting continental temperatures still further and driving even stronger

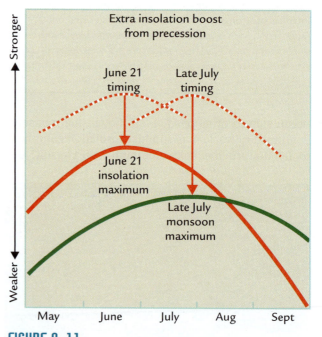

FIGURE 9-11
Alignment of insolation and summer monsoon
Peak summer monsoon development in the Northern Hemisphere occurs in late July to early August. If the increment of extra insolation caused by Earth's precession falls in late July, the summer monsoon will be stronger.

monsoons. Choosing late July or early August as the time of critical insolation forcing (rather than June 21) would eliminate the apparent 2,000-year lag in monsoon responses.

▶ Orbital Monsoon Hypothesis: Regional Assessment

The evidence from North Africa and its surrounding seas supports the orbital monsoon hypothesis, but strong monsoonal circulations are also found on Asia and South America (Figure 9-12). The South Asian monsoon is the most powerful monsoon on Earth because of the size of the landmass and the extent of the high topography of Tibet (see Chapter 7). The only strong monsoon system in the Southern Hemisphere is the air mass flow over the Amazon Basin in South America.

In recent years, studies of cave deposits (speleothems) have become a very important source of information on monsoon changes over orbital time scales. Groundwater dripping through soils and bedrock and into caves deposits stalactites and stalagmites constructed of calcite ($CaCO_3$). These deposits build up layer by layer over thousands to tens of

thousands of years, and they can be very accurately dated by radiometric analysis of small amounts of thorium and uranium (see Chapter 2). The relative amount of the two isotopes of oxygen (^{16}O and ^{18}O) in the calcite layers varies though time because of several factors, the most important of which is changes in the amount of surface precipitation that feeds the flow of groundwater into the caves. As a result, changes in overlying monsoonal air masses can be accurately reconstructed from variations in the $\delta^{18}O$ signal of cave calcite.

Records from caves in southern China and in South America both show $\delta^{18}O$ variations so large that they can only be explained by air mass changes linked to the changes in monsoon strength in those regions (Figure 9-13). Layers with highly negative $\delta^{18}O$ values indicate a very strong monsoon flow from the ocean and greater fractionation of the oxygen isotopes, and layers with less negative values indicate weaker monsoon flow from the ocean and lesser fractionation.

In China, these $\delta^{18}O$ changes correlate closely with changes in midsummer (late July) insolation at 25°N, especially during the interval prior to 100,000 years ago when both the insolation and $\delta^{18}O$ changes were largest. This record provides unambiguous evidence in support of the orbital monsoon hypothesis:

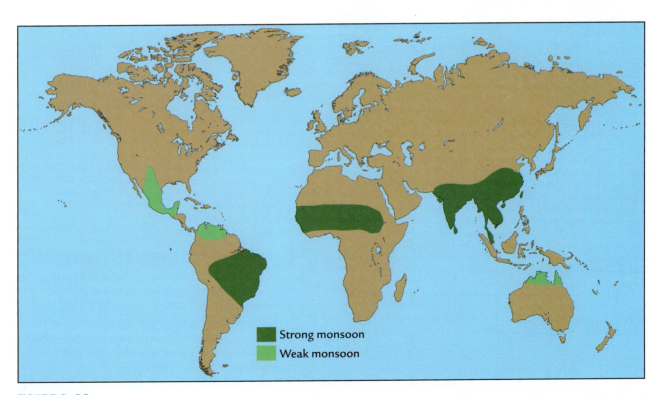

Strong monsoon
Weak monsoon

FIGURE 9-12
Major monsoon systems
Strong summer monsoons occur in North Africa, Asia, and South America.

the Asian monsoon has varied at the 23,000-year cycle and the changes had a midsummer (late July) phase. Other changes are also evident in the $\delta^{18}O$ signal over intervals shorter than the orbital-scale oscillations.

The $\delta^{18}O$ record in cave calcite from southeastern Brazil also shows a close link to summer insolation forcing, with a strong 23,000-year cycle (Figure 9-13). Note, however, that this 23,000-year cycle correlates to insolation during February rather than July. This result is an elegant confirmation of the orbital-monsoon hypothesis. Because late February is midsummer in the Southern Hemisphere, monsoon variations on southern continents should have this phase at the precession cycle, and they do.

In Summary, evidence from several continents fully supports John Kutzbach's orbital monsoon hypothesis. At this point, the hypothesis has passed so many tests that it merits the higher status of a theory (see Chapter 1).

▶ Monsoon Forcing Earlier in Earth's History

The concept of insolation forcing of summer monsoons can be used to investigate the more distant geologic past. Orbital-scale changes in summer insolation at the precession cycle have driven changes

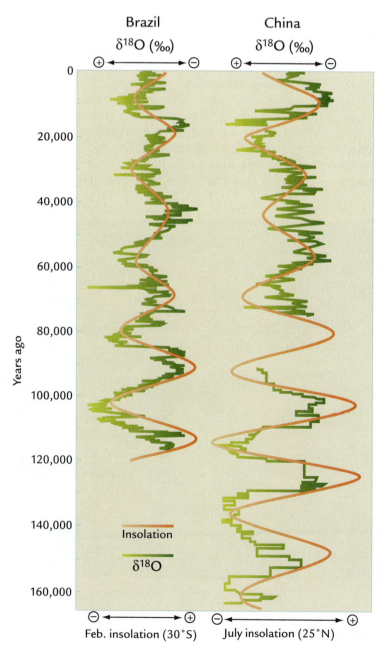

FIGURE 9-13
Monsoon $\delta^{18}O$ signals from caves

Calcite from cave deposits in China and Brazil shows $\delta^{18}O$ changes produced by variations in the strength of monsoonal air masses at the 23,000-year precession cycle. The $\delta^{18}O$ variations have the phase of midsummer insolation in each hemisphere. (ADAPTED FROM D. YUAN ET AL., "TIMING, DURATION, AND TRANSITIONS OF THE LAST INTERGLACIAL ASIAN MONSOON," *SCIENCE* 304 [2004]: 575–578; AND FROM F. W. CRUZ ET AL., "INSOLATION-DRIVEN CHANGES IN ATMOSPHERIC CIRCULATION OVER THE PAST 116,000 YEARS IN SUBTROPICAL BRAZIL," *NATURE* 434 [2005]: 63–66.)

in monsoonal precipitation, and processes linked to precipitation have left evidence in ancient climate records, such as pulses of sediment-laden runoff and changes in lake depth. As a result, many ancient sedimentary rocks contain valuable information about varying monsoon strength.

In cases where high-quality time control is available for ancient deposits, we can look for evidence of the kind of monsoon signature shown in Figure 9-5. Because many of these deposits extend over millions of years, we can expect to see records that look like those in Figure 9-14. A wide range of sediment indicators linked to precipitation, erosion, runoff, transport, and deposition may have this appearance. In some cases, this relationship can even be used to refine ("tune") time scales initially set by radiometric dating (Box 9-1).

As in the case of North African lakes, we expect the monsoon signature to show clusters of two or three strong maxima separated by clusters of two or three weaker maxima, with these clusters repeating in the record at intervals of about 100,000 years because of control of the amplitude of precession by orbital eccentricity (see Chapter 8). In this case, however,

because we are looking at much longer records, we should also see clusters of monsoon-driven maxima at the longer eccentricity period near 400,000 years (see Figure 9-14). The truncation of the summer monsoon response pattern at a critical threshold value is called a **clipped response**. As a result of this truncation, many monsoon responses register only one side of each 23,000-year precession cycle, with modulation of the amplitude of this one-sided response at 100,000 and 400,000 years.

9-6 Monsoons on Pangaea 200 Million Years Ago

Just before 200 million years ago, a chain of basins formed in a region that is now the eastern United States but at that time was deep in the interior of the giant supercontinent Pangaea (Figure 9-15). These deep depressions in a region of generally high terrain

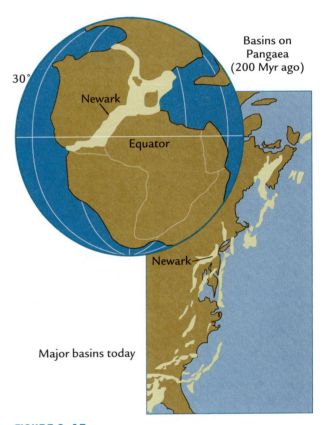

FIGURE 9-15
Mid-Pangaean basins

In the middle of the Pangaean supercontinent 200 million years ago (top left), the Newark Basin developed in what is now New Jersey as one of a chain of basins of equivalent age (bottom right). (ADAPTED FROM P. E. OLSEN AND D. V. KENT, "MILANKOVITCH CLIMATE FORCING IN THE TROPICS OF PANGAEA DURING THE LATE TRIASSIC," *PALAEOGEOGRAPHY, PALAEOCLIMATOLOGY, PALAEOECOLOGY* 122 [1996]: 1–26.)

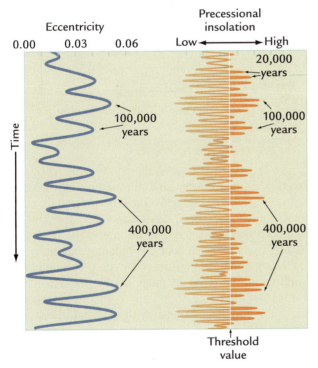

FIGURE 9-14
Monsoon signals recorded in sediments

Monsoonal influences can be detected in older sediment sequences. High orbital eccentricity values (left) should amplify individual 23,000-year precession cycles approximately every 100,000 and 400,000 years (right). The monsoon signal in the sediments could resemble the red-shaded area to the right of the threshold insolation value.

Looking Deeper into Climate Science

Insolation-Driven Monsoon Responses: A Chronometer for Tuning

The clearly demonstrated link between summer insolation forcing and monsoon responses at low latitudes has become part of the basis for a new way of dating sedimentary records on land and in the oceans. This method, **orbital tuning**, can provide even better time resolution than radiometric methods.

The tuning method is based on the relationship between the insolation signal (the forcing) and the summer monsoon changes (the response). The timing of orbital insolation changes is known with great accuracy from astronomical calculations, and a range of monsoon responses can be measured in sediments. By making the simple assumption that the insolation driver and the monsoon responses have kept the same relationship in the past, the monsoon responses in the sediments can be dated with nearly the same accuracy as the orbital forcing.

Tuning sediment sequences to orbital variations If lavas or magnetic reversal boundaries provide radiometrically dated levels in terrestrial or marine sediments, the age of intervening sediment intervals can be determined by tuning monsoon-driven sediment responses to insolation changes at the 23,000-year precession cycle.

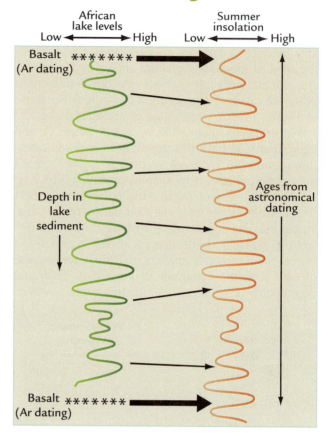

were formed by precursors of the forces that would eventually pull the Pangaean continent apart and create the Atlantic Ocean. But this part of Pangaea would not break up until tens of millions of years later (see Chapter 7).

Sediments deposited in one of these depressions, the Newark Basin in modern New Jersey, have been extensively investigated. The fossil compasses provided by magnetic evidence from volcanic rocks indicate that the Newark Basin was located in the tropics 200 million years ago, about 10° of latitude north of the equator (see Figure 9-15). Because of its tropical location, the Newark Basin was situated in a regime dominated by precessional insolation changes, similar to those in modern North Africa and southern Asia. Because the basin was far from the ocean, its climate was relatively arid, but enough moisture arrived to create a lake that varied greatly in size over time.

Evidence preserved in a thick (greater than 7,000 m) sequence of lake sediments shows that the size of this lake fluctuated at a tempo near 20,000 years. Several

layers of molten magma that intruded into the lakebed sequence and quickly cooled have been dated by radiometric methods. These dates show that the lakebed sequence was deposited over an interval of at least 20 million years centered near 200 million years ago.

This estimate is confirmed by the presence of fine laminations (varves) in parts of the sequence. The varves are tiny (0.2−0.3 mm) couplets of alternating light and dark layers, with one light/dark pair deposited each year. Darker organic-rich layers were deposited in summer and lighter mineral-rich layers in winter. Dissolved oxygen concentrations must have been low or zero in the deeper levels of the lake when the organic-rich layers accumulated to prevent destruction of the delicate varves by animals moving across and in the sediments. Use of these varves as an internal chronometer to count elapsed time (see Chapter 3) confirms that the total time of lake-sediment deposition was about 20 million years.

The types of sediment deposited in the Newark Basin varied widely in response to changes in lake depth.

Box 9-1

This method is most easily applied in ocean sediments because deposition in the ocean tends to be continuous. Assume that a sediment core contains two magnetic reversal boundaries that have been dated by correlation to the global magnetic stratigraphy established by dating basalt layers on land (see Chapter 5). The ages of these reversals constrain the intervening sequence of sediments to a particular interval of time. Also assume that this sediment sequence contains a record that is directly tied to the strength of the tropical monsoons, such as a sequence of sapropel layers.

In many cases, the monsoon-related response measured in the sediments will show an obvious correlation to the summer insolation forcing. Both the insolation changes and the monsoon responses will show cycles near 23,000 years and obvious modulation of these cycles at eccentricity periods near 100,000 and 400,000 years. The tuning process (matching maxima and minima in the monsoon response to correlative features in the insolation signal) allows the ages in the sedimentary sequence to be assigned to specific precession cycles in the past at a finer resolution than the length of each cycle.

This method has also been applied to sediments deposited in long-lived lakes on continents. In regions like East Africa, volcanic eruptions deposit beds of basalt (lava) or volcanic ash that can be radiometrically dated by K/Ar methods. The volcanic deposits provide the initial time framework for the tuning process, analogous to the use of magnetic reversal boundaries in marine sediments. The monsoon-driven variations in lake size during the sequence lying between the basalt layers can then be tuned to the astronomically dated record of summer insolation. Because records on land are generally much more difficult to date than ocean sediments, tuning of these lake sequences provides an enormous improvement over other dating methods.

For sedimentary records that contain climatic responses at the cycles of both precession and obliquity, the tuning method can be tested even more rigorously. In this case, the "tuned" time scale must match not just the amplitude-modulated precession cycle at 23,000 years, but also the tilt cycle at 41,000 years. This added requirement makes the tuning process an even more demanding exercise, but also one that is even more likely to yield a uniquely accurate time scale.

When the lake was deep (100 m or more), the sediments tended to be gray or black muds with large amounts of organic carbon. These sediments contain finely laminated varves and are rich in well-preserved remains of fish. Sediments deposited when the lake was shallower or entirely dried out tend to be red or purple because they were oxidized (rusted) by sporadic contact with air, and they often contain mud cracks due to exposure to dry air. Dinosaur footprints and the remains of plant roots are also common in sediments from the dried-out, vegetated parts of the lakebeds (Figure 9-16).

FIGURE 9-16
Evidence of changing lake levels
Dinosaur footprints in lake muds that have since hardened into rock show that the Pangaean lakes occasionally dried out completely. These footprints are from a basin in Connecticut formed at the same time as the Newark Basin in New Jersey. (DINOSAUR STATE PARK, ROCKY HILL, CT.)

The thick sequences preserved in the Newark Basin repeatedly fluctuate between sediments typical of deep lakes and those that indicate a shallowing or complete drying up of the lakes. Individual layers in these sequences are continuous over large areas, indicating that the wet-dry variations in climate affected the entire basin.

Extensive investigations show that these fluctuations in lake depth over millions of years were cyclic (Figure 9-17). The shortest cycles occur over rock thicknesses averaging 4 to 5 meters, equivalent to about 20,000 years in time based on the average thickness of each annual varve (0.2–0.3 mm). These cycles were driven by precession. Monsoons filled and emptied these Pangaean lakes (see Chapter 5) in response to orbital precession in the same way that North African lakes have filled and emptied during much more recent times. Because we are looking much farther back in time, the periods of the orbital cycles were slightly shorter than they are today (see Chapter 8).

Two larger-scale groupings of cycle peaks are also evident. The amplitude of individual 20,000-year peaks in lake depth rises and falls roughly every five or six cycles separated by 20 to 25 meters of sediment, or a little less than 100,000 years. An even larger-scale change in amplitude of the monsoon-cycle peaks occurs between approximately 530 and 620 meters depth in the core (Figure 9-17), equivalent to a time interval of about 400,000 years.

These two longer-term patterns match the expected monsoon signature shown in Figure 9-14 remarkably well. They reflect a modulation of the strength of the 20,000-year precession cycles by eccentricity changes at intervals of about 100,000 and 400,000 years. The full imprint of ancient monsoons is amazingly clear in the sediments of this basin despite the passage of 200 million years.

9-7 Joint Tectonic and Orbital Control of Monsoons

We saw in Chapter 5 that tectonic changes affect the intensity of monsoonal circulations. Large landmasses such as Pangaea intensify monsoons by offering a larger area for the Sun to heat. Positioning of landmasses at lower latitudes is important because solar radiation is more direct and albedos are much lower than at higher, snow-covered latitudes. Topography is a key control over monsoon strength at tectonic time scales because high-elevation regions focus strong monsoonal rains on their margins.

The processes that control monsoon intensity over tectonic time scales interact with those at orbital scales. Tectonic-scale processes alter the average

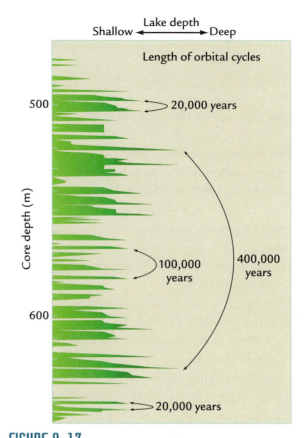

FIGURE 9-17
Fluctuations of Pangaean lakes

Newark Basin lake sediments varied in depth from very shallow to over 100 m deep at three tempos. Individual cycles in lake depth every 4–5 m occur at a period of 20,000 years, clusters of larger deep-lake maxima every 20–25 m occur at intervals near 100,000 years, and unusually large deep-lake clusters at intervals of 90–100 m occur every 400,000 years. (ADAPTED FROM P. E. OLSEN AND D. V. KENT, "MILANKOVITCH CLIMATE FORCING IN THE TROPICS OF PANGAEA DURING THE LATE TRIASSIC," *PALAEOGEOGRAPHY, PALAEOCLIMATOLOGY, PALAEOECOLOGY* 122 [1996]: 1–26.)

strength of the monsoon over millions of years, while the orbital-scale insolation changes drive shorter-term monsoon strength at a cycle near 20,000 years. One way the tectonic and orbital factors might interact is suggested in Figure 9-18. On the left is a schematic version of a low-latitude summer insolation curve, with individual maxima and minima at the 20,000-year precessional cycle and modulation of this cycle over intervals of 100,000 and 400,000 years. The smooth curve in the center represents gradually changing tectonic-scale processes, such as the slow uplift that gradually intensifies the average strength of the monsoon over millions of years. This slow tectonic-scale increase in monsoon strength combines with the orbital-scale monsoon cycles to produce the response shown on the right—a *slow increase in the*

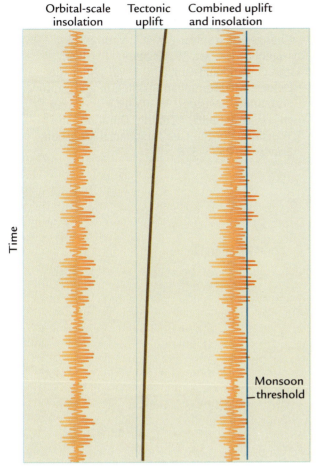

Orbital-scale insolation | Tectonic uplift | Combined uplift and insolation

Time

Monsoon threshold

FIGURE 9-18
Combined tectonic and orbital forcing of monsoons

Monsoons are driven by orbital-scale variations in insolation (left) and by slower-acting tectonic factors such as plateau uplift (center). The combined tectonic and orbital forcing causes the amplitude of orbital-scale monsoon responses to increase and gradually exceed critical thresholds (right).

amplitude of the orbital-scale cycles caused by tectonic amplification.

We can hypothesize the existence of a threshold value above which key climatic indices record monsoon responses but below which no response is registered (as in Figure 9-5). In the changes shown on the right in Figure 9-18, the tectonic influence may have been weak enough during the earlier intervals that the orbital-scale monsoon cycles never exceeded this threshold. Later, as tectonic processes created conditions more favorable to monsoons, peaks in summer insolation would have driven monsoons that began to exceed the threshold by small amounts, and then later by steadily increasing amounts.

Something like this kind of evolving climatic response is thought to have occurred in Southeast Asia over the last 30 or 40 million years. A long-term tectonic increase in monsoon intensity due to uplift progressively intensified the amplitude of orbitally driven monsoon cycles in this region. Simulations run with general circulation models indicate that the combined effects of orbital-scale insolation and uplift are not additive in a simple linear way. Instead, it appears that plateau uplift sensitizes the monsoon system to insolation forcing in such a way that the combined monsoon response to uplift and insolation is stronger than a simple linear combination of the two effects.

Key Terms

orbital monsoon hypothesis (p. 179)
nonlinear response (p. 181)
threshold level (p. 181)

sapropels (p. 183)
anoxic (p. 183)
clipped response (p. 189)
orbital tuning (p. 190)

Review Questions

1. In what way is the orbital monsoon hypothesis an extension of processes driving modern monsoons?

2. Why does the intensity of 23,000-year monsoon peaks vary at intervals of 100,000 and 413,000 years?

3. How did the Mediterranean Sea acquire a freshwater lid during times when very little precipitation was falling in that region?

4. Explain how the opposed July/February timing of past monsoon changes in China and Brazil lends strong support to the orbital monsoon hypothesis.

5. Does peak monsoon strength lag behind summer insolation forcing?

6. What similarities exist between monsoon changes in Pangaea 200 million years ago and those in North Africa during the last several hundred thousand years?

7. How do tectonic uplift and orbital variations combine to affect the long-term intensity of monsoons?

Additional Resources

Basic Reading

Kutzbach, J. E. 1981. "Monsoon Climate of the Early Holocene: Climate Experiment with Earth's Orbital Parameters for 9000 Years Ago." *Science* 214:59–61.

Ruddiman, W. F. 2006. "What Is the Timing of the Orbital-Scale Monsoon Changes?" *Quaternary Science Reviews* 25:657–58.

Advanced Reading

Olsen, P. E. 1986. "A 40-Million-Year Lake Record of Early Mesozoic Orbital Climatic Forcing." *Science* 234:842–48.

Pokras, E. M., and A. C. Mix. 1987. "Earth's Precession Cycle and Quaternary Climatic Changes in Tropical Africa." *Nature* 326:486–87.

Rossignol-Strick, M., W. Nesteroff, P. Olive, and C. Vergnaud-Grazzini. 1982. "After the Deluge: Mediterranean Stagnation and Sapropel Formation." *Nature* 295:105–10.

Yuan, D., et al. 2004. "Timing, Duration, and Transitions of the Last Interglacial Monsoon." *Science* 304:575–578.

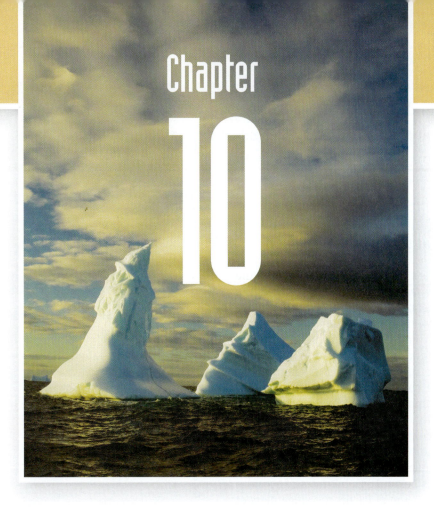

Insolation Control of Ice Sheets

ce sheets covered northern North America and Europe 20,000 years ago. The present locations of Toronto, New York, Chicago, Seattle, and London were buried under hundreds of meters of ice. Later, the ice melted, and the last remnants disappeared by 6,000 years ago, near the time human civilizations came into existence. The fact that ice sheets first appeared in the Northern Hemisphere in the last 3 million years can be explained by very slow tectonic-scale cooling (see Part II), but the evidence that ice sheets grew and melted over much shorter intervals of time requires a different explanation.

Orbital changes in insolation are the initial driver of these shorter-term variations in the amount of ice. In this chapter, we investigate how changes in summer insolation control the size of ice sheets by determining the rate of ice melting or accumulation. We explore two lags that are important to understanding the ice response: the lag of slow-responding ice sheets behind the insolation changes, and the delayed depression of bedrock beneath the weight of the overlying ice. Both of these lags are thousands of years in length. Then we examine past changes in ice volume based on evidence from oxygen isotopes and coral reefs. Finally, we analyze how the ice sheet history over the last 3 million years compares with predictions from the theory of orbital control of ice volume.

What Controls the Size of Ice Sheets?

Continental ice sheets exist in regions where the overall rate of snow and ice accumulation across the entire mass of ice equals or exceeds the overall rate of ice loss or ablation (see Chapter 2). Snow accumulates and turns into ice at high latitudes and altitudes where temperatures are cold enough to permit frozen precipitation and to suppress melting in summer. For continent-sized ice sheets that extend to sea level, temperatures cold enough to sustain ice occur today only at high latitudes.

Rates of ice accumulation and ablation vary with temperature, but the two relationships differ in a critical way. Ice accumulates (initially as snow) at mean annual temperatures below 10°C, but rates of accumulation remain below 0.5 m of ice per year regardless of temperature (Figure 10-1A). At higher temperatures, ice accumulation is limited by the fact that more of the precipitation falls as rain. At extremely low temperatures, all of the precipitation is snow, but frigid air carries so little water vapor that rates of ice accumulation are low.

In contrast, ablation of ice accelerates rapidly when temperatures warm. Melting begins at mean annual temperatures above −10°C, equivalent to summer temperatures above 0°C, and can reach rates equivalent to several meters of ice per year, much larger than the maximum rates of accumulation.

Ice ablation can occur as a result of incoming solar radiation, by uptake of sensible or latent heat delivered by warm air masses (and rain) and by shedding (calving) icebergs to the ocean or to lakes. Calving differs from other ablation processes because icebergs leave the main ice mass and drift elsewhere to melt, often ending up in a warmer environment than the one near the ice sheet.

The net balance between accumulation and ablation over an entire ice sheet is called the ice mass balance (Figure 10-1B). The mass balance at very cold temperatures (below −20°C) is positive but small because so little snow falls. The mass balance at mean annual temperatures near −15° to −10°C is more positive because snow accumulation rates are more rapid but ablation is not strong. The mass balance turns sharply negative at temperatures above

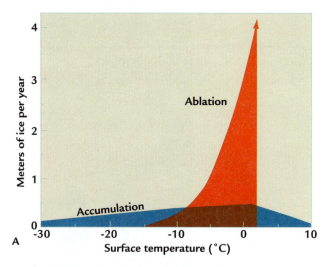

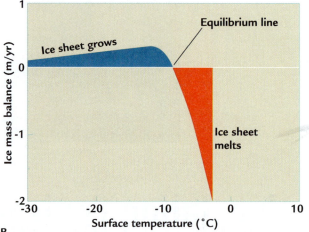

FIGURE 10-1
Temperature and ice mass balance

Temperature is the main factor that determines whether ice sheets are in a regime of net ablation (negative mass balance) or accumulation (positive mass balance). Ablation increases sharply at higher temperatures. (MODIFIED FROM J. OERLEMANS, "THE ROLE OF ICE SHEETS IN THE PLEISTOCENE CLIMATE," *NORSK GEOLOGISK TIDSSKRIFT* 71 [1991]: 155–61.)

−10°C because ablation accelerates and overwhelms accumulation. The boundary between positive and negative mass balance is called the **equilibrium line**.

If net accumulation and ablation are in balance over an entire ice sheet, the ice sheet is said to be in a condition of stable equilibrium. Net accumulation high on the ice sheet is exactly balanced by ablation at lower elevations, and no net change in total ice volume occurs. Ice flows in the ice sheet from areas of accumulation to areas of ablation, but the total mass of the ice sheet remains unchanged.

Here, our focus is not on ice sheets that remain in equilibrium. We seek to understand what makes ice sheets grow and shrink.

10-1 Orbital Control of Ice Sheets: The Milankovitch Theory

Beginning with the Belgian mathematician Joseph Adhemar in the 1840s, scientists suspected that orbitally driven changes in solar insolation might be linked in some way to the growth and melting of continent-sized ice sheets. Because orbital changes alter the amount of insolation received on Earth in all seasons (see Chapter 8), scientists were immediately faced with an important question: Which season is critical in controlling the size of ice sheets?

Winter would seem to be the obvious choice because snow falls mainly during that time of year. Scientists initially thought that colder winters caused by lower solar radiation at that time of year would help to accumulate larger amounts of snow and promote glaciation. But this seemingly reasonable idea turned out to be wrong. One problem is that ice sheets grow at high latitudes where temperatures are *always* cold in winter, even during intervals of relatively warm climate like the one we live in today. In addition, the Sun at these latitudes always lies low in the winter sky, regardless of ongoing orbital changes, and incoming solar radiation in winter is never strong. Winter is not the critical season.

The opposite possibility—*summer insolation control of ice sheets*—was proposed by several scientists working in the late nineteenth and early twentieth centuries, including Rudolf Spitaler (who was also the first to realize that summer insolation changes might drive monsoons), Wladimir Köppen, and Alfred Wegener (who also proposed the theory that continents drift). Their reasoning was simple: no matter how much snow falls during winter, it can all be easily melted if the following summer is warm and ablation is rapid (see Figure 10-1).

These scientists reasoned that low summer insolation is critical in producing summers cool enough for snow and ice to persist from one winter to the next. This idea gained popularity during the early and mid-twentieth century from work by the Serbian astronomer Milutin Milankovitch, who first calculated in a systematic way the impact of astronomical changes on insolation received on Earth at different latitudes and in different seasons. This explanation is now known as the **Milankovitch theory.**

Milankovitch proposed that ice growth in the Northern Hemisphere occurs during times when summer insolation is reduced. Low summer insolation occurs when Earth's orbital tilt is small and its poles are pointed less directly at the sun (Figure 10-2A). Low insolation also occurs when the northern summer solstice occurs with Earth farthest from the Sun (in the aphelion or distant pass position) and when

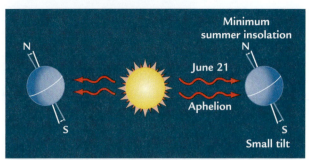

A Northern hemisphere ice growth

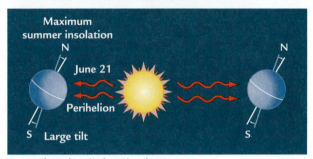

B Northern hemisphere ice decay

FIGURE 10-2
Orbital changes and ice sheets

According to the Milankovitch theory, ice sheets grow in the Northern Hemisphere at times when summer insolation is low because tilt is small and Earth lies in the aphelion position farthest from the Sun (A). Ice melts when summer insolation is high because tilt is larger and Earth lies in the perihelion position closest to the Sun (B). (ADAPTED FROM W. F. RUDDIMAN AND A. MCINTYRE, "OCEANIC MECHANISMS FOR AMPLIFICATION OF THE 23,000-YEAR ICE-VOLUME CYCLE," *SCIENCE* 212 [1981]: 617–27.)

the orbit is highly eccentric (further increasing the Earth-Sun distance). Milankovitch reasoned that the most sensitive latitude for low insolation values is 65°N, the latitude at which ice sheets first accumulate and last melt. He also proposed that ice melts during the stronger summer insolation resulting from the opposite orbital configuration (Figure 10-2B).

The amount of summer insolation arriving at the top of Earth's atmosphere at 65°N has varied by as much as ±12% around the long-term mean value (see Chapter 8). We have no way of knowing how much of this incoming solar radiation has actually made it through the atmosphere to Earth's high-latitude ice sheets because of the complicating effects of regional changes in atmospheric circulation, clouds, and water vapor. Milankovitch noted these complications but assumed that the amount of radiation penetrating to Earth's surface is closely related to the amount arriving at the top of the atmosphere.

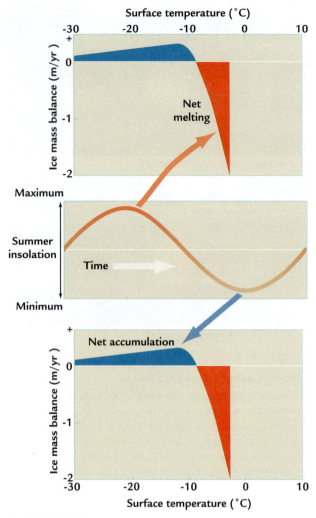

FIGURE 10-3
The Milankovitch theory

According to the Milankovitch theory, high summer insolation heats the land and results in greater ice ablation (top), while low summer insolation allows the land to cool and ice sheets to form (bottom). (MODIFIED FROM J. OERLEMANS, "THE ROLE OF ICE SHEETS IN THE PLEISTOCENE CLIMATE," *NORSK GEOLOGISK TIDSSKRIFT* 71 [1991]: 155–61.)

In Summary, the Milankovitch theory proposes that when summer insolation is strong, more radiation is absorbed at Earth's surface at high latitudes, making the climate in those regions warmer. Warming accelerates ablation, melts more snow and ice, and either prevents glaciation or shrinks existing ice sheets (Figure 10-3 top). Conversely, when summer insolation is weak, less radiation is delivered to high latitudes, and the reduction in radiation cools the regional climate. This cooling reduces the rate of summer ablation and allows snow to accumulate and ice sheets to grow (Figure 10-3 bottom).

Modeling the Behavior of Ice Sheets

To gain more insight into summer insolation control of ice sheets in the Northern Hemisphere, climate scientists have developed numerical models based on an idealized (simplified) representation of the ice sheets surrounding the Arctic Ocean. One kind of model portrays the changes in ice sheets across a single (idealized) high-latitude continent near the Arctic, but it ignores possible differences between various sectors of the Arctic (Europe versus North America). This simplification is reasonably well justified because ice sheets grew on all the continents around the Arctic, at least during the last glacial maximum 20,000 years ago (Figure 10-4).

10-2 Insolation Control of Ice Sheet Size

The mechanism by which changes in summer insolation control the size of ice sheets on northern landmasses in these models follows directly from Milankovitch's theory. Changes in summer insolation

FIGURE 10-4
Ice sheets around the Arctic Ocean

At the last glacial maximum, 20,000 years ago, ice sheets surrounded much of the Arctic Ocean. (MODIFIED FROM G. DENTON AND T. HUGHES, *THE LAST GREAT ICE SHEETS* [NEW YORK: JOHN WILEY, 1981].)

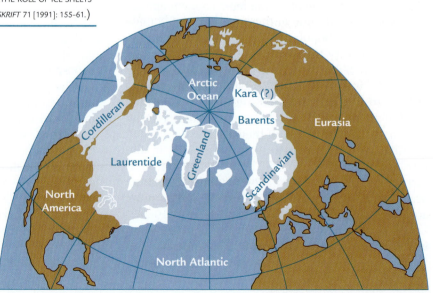

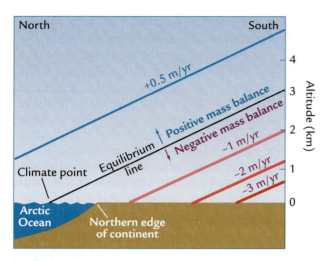

FIGURE 10-5
Ice sheet models

Two-dimensional models represent northern hemisphere ice sheets along a north-south line. The equilibrium line separates northern (and higher-elevation) regions of net accumulation from southern (and lower-elevation) regions of net ablation, and it intersects Earth's surface at the climate point. (ADAPTED FROM J. OERLEMANS, "MODEL EXPERIMENTS OF THE 100,000-YEAR GLACIAL CYCLE," *NATURE* 287 [1987]: 430-32.)

drive regional temperature responses that alter both melting rates and ice mass balance.

One kind of model represents this relationship as changes in mass balance along a north-south line (Figure 10-5). These transects have just two dimensions, one in a vertical direction (altitude) and the other in a north-south direction (latitude). Changes in the other horizontal dimension (longitude) are ignored to allow the models to simulate changes over longer intervals of time than would be possible with full three-dimensional models, which are computationally more demanding.

The equilibrium line in these models, the boundary between areas of net ice ablation and accumulation, slopes upward into the atmosphere toward the south at a low angle. This slope is consistent with conditions today: temperatures are colder toward higher latitudes and altitudes, and warmer toward lower latitudes and altitudes. As a result, subfreezing temperatures occur today only at high latitudes and altitudes. Long-distance air travel commonly occurs at these subfreezing altitudes.

Parallel to this moving equilibrium line are lines of ice mass balance. These lines show the thickness (in meters) of ice that accumulates or melts each year (as before, snowfall is converted to an equivalent thickness of ice). Ice accumulates above the equilibrium line and in the north because of the colder temperatures, and it melts in the warmer temperatures below the equilibrium line and toward the south. The rates of ice melting

are more closely spaced in the warmer areas because of the rapid melting rates shown in Figure 10-1.

The equilibrium line intercepts Earth's surface in the higher latitudes at a location called the **climate point** (see Figure 10-5). Ice sheet models use orbital-scale changes in summer insolation to move this climate point (and the equilibrium line attached to it) north and south across the landmasses (Figure 10-6). The amount of north-south shift of the equilibrium line is set proportional to the amount of change in summer insolation. These shifts can cover 10° to 15° of latitude.

Gradual changes in the amount of summer radiation received at high latitudes cause the areas of net snow accumulation and net ice ablation to shift back and forth across the land. Strong summer insolation warms the high-latitude landmasses in summer,

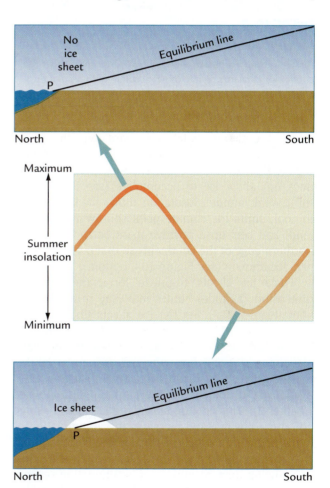

FIGURE 10-6
Insolation changes displace the equilibrium line

When the equilibrium line is driven north by high values of summer insolation, the continents lie in a regime of net ablation and no ice can accumulate (top). When it is driven south by summer insolation minima, the northern landmasses lie in a regime of net accumulation and ice sheets can grow (bottom). (P = climate point.) (MODIFIED FROM J. OERLEMANS, "THE ROLE OF ICE SHEETS IN THE PLEISTOCENE CLIMATE," *NORSK GEOLOGISK TIDSSKRIFT* 71 [1991]: 155-61.)

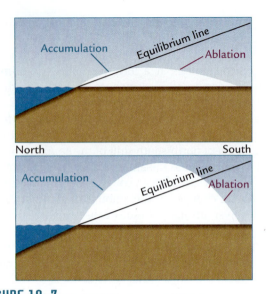

FIGURE 10-7
Ice elevation feedback

As ice sheets grow higher, more of their surface lies above the equilibrium line in a regime of net accumulation, even if the equilibrium line does not move.

moves the climate point northward over the Arctic Ocean, and puts northern landmasses in an ablation regime that melts all winter snow each summer and does not allow ice to accumulate (see Figure 10-6 top). Weak summer insolation allows the landmasses to cool, shifts the climate point southward over the land, and sets up a positive mass balance over the northern edge of the continents so that permanent ice can accumulate (see Figure 10-6 bottom).

Once ice sheets begin to form, their vertical dimension (altitude) comes into play in a powerful way (Figure 10-7). As the ice sheets thicken, their upper surfaces reach altitudes where temperatures are much colder. The tops of the ice sheets can reach elevations of 2 to 3 km, where temperatures are 12° to 19°C cooler than those at sea level, using an average lapse rate cooling of 6.5°C per km of altitude (see Chapter 2). This cooling makes the ice mass balance still more positive and increases the accumulation of snow and ice. In effect, once ice sheets start to grow, they contribute to their own positive mass balance by growing upward into a net accumulation regime. Eventually, the tops of the ice sheets reach elevations at which the rate of snowfall is much lower because the frigid air contains very little water vapor. At these elevations, this positive feedback effect weakens.

10-3 Ice Sheets Lag Behind Summer Insolation Forcing

The response of ice sheets to changes in summer insolation is far from immediate. Imagine what would happen

if climate suddenly cooled enough to permit snow to fall throughout the year and accumulate rapidly over Canada. Based on the mass balance values plotted in Figure 10-1, about 0.3 meters of ice might accumulate each year. But even under this unrealistically favorable assumption, a full-sized ice sheet that is 3,000 meters thick would take 10,000 years to form. In reality, the ice would take much longer to thicken that much because the initial cooling would not be instantaneous.

The geologist John Imbrie and his colleagues have led the exploration of the link between insolation and ice volume. They use as an analogy a conceptual model based on variations through time in the intensity of the flame in a Bunsen burner shown in Figure 10-8 (top; also see Chapter 1). Because water has a high heat capacity and reacts slowly to changes in the heat applied, changes in water temperature lag behind changes in intensity of the heat source.

Ice sheets have the same lagged response to summer insolation but on a much longer time scale (Figure 10-8 bottom). As summer insolation declines from a maximum value, ice begins to accumulate. The *rate* at which ice volume grows reaches a maximum when summer insolation has fallen to its lowest value because summer ablation is at a minimum at that time. Later, as summer insolation begins to increase, the ice sheet will continue to grow for a while longer because insolation values are still relatively low. The ice sheet does not reach its maximum *size* until the point that insolation crosses the value that initiates net ice ablation. As a result, the maximum in ice volume lags thousands of years behind the minimum in summer insolation.

As summer insolation continues to increase, the rate of ablation increases (see Figure 10-8 bottom). When insolation reaches its maximum value, the *rate* of ablation also reaches its maximum. Again, however, the minimum in ice volume does not occur at the insolation maximum because falling insolation values still remain high enough to melt even more ice for thousands of years. Later, about halfway through the drop in insolation, ice volume reaches its maximum value and then starts to decrease in a regime of increasing ablation. This relationship can be described by the equation shown in Box 10-1.

The persistent delay in ice volume relative to summer insolation shown in Figure 10-8 (bottom) is the **phase lag** between the two cycles. Remember that the period of a cycle is the interval of time separating successive peaks or successive valleys. In this example, the phase lag of ice volume behind summer insolation (Figure 10-8 bottom) represents one-quarter of the cycle length.

This same relationship can be applied to the separate ice volume responses driven by the insolation cycles of orbital tilt and precession. At the orbital

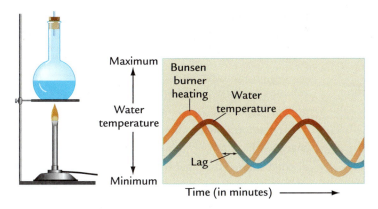

FIGURE 10-8
Ice volume lags insolation

As the flame of a Bunsen burner is alternately turned higher and lower, the water heats and cools, but with a short time lag behind the changes in heating (top). Similarly, long-term increases and decreases in summer insolation heating cause ice sheets to melt and grow, but with lags of thousands of years (bottom). (ADAPTED FROM J. IMBRIE, "A THEORETICAL FRAMEWORK FOR THE PLEISTOCENE ICE AGES," *JOURNAL OF THE GEOLOGICAL SOCIETY* (LONDON) 142 [1985]: 417-32.)

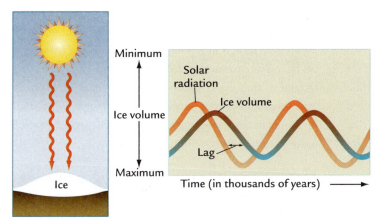

tilt cycle, the ice volume response would have the same regular sine wave shape as the summer insolation signal but would lag behind it by one-quarter of the 41,000-year wavelength or just over 10,000 years (Figure 10-9A). At the precession cycle, the ice volume response would again lag insolation by one-quarter of its 23,000-year length, or just under 6,000 years. The ice volume response at this cycle also shows the same amplitude modulation as the precession insolation signal (Figure 10-9B). As we will see later, the actual lags of the ice sheets behind the solar forcing are large, but not a full quarter wavelength.

10-4 Delayed Bedrock Response Beneath Ice Sheets

As ice sheets grow, so does the pressure of their weight on the underlying bedrock. Although the density of solid ice (just under 1 g/cm³) is much lower than that of the underlying rock (about 3.3 g/cm³), ice sheets can reach thicknesses in excess of 3,000 meters, with their weight equivalent to 1,000 meters of rock. This load is enough to depress the underlying bedrock far beneath the level it would be if no ice sheets existed.

One way to visualize this process is a thought experiment in which an ice sheet 3.3 kilometers thick is instantaneously loaded onto bedrock (Figure 10-10A). In time, the 3.3-km ice sheet would

eventually depress the underlying bedrock by 1 km. To put this bedrock change in a climatic context, 1 km of elevation change is equivalent to a 6.5°C

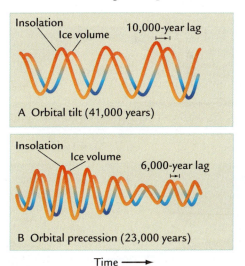

FIGURE 10-9
Ice volume lags tilt and precession

At the 41,000-year cycle of orbital tilt, the lag of ice sheet size behind changes in summer insolation approaches one-quarter wavelength, or 10,000 years (A). At the 23,000-year cycle of orbital precession, ice sheets lag roughly one-quarter wavelength (6,000 years) behind changes in summer insolation and show the same modulation of amplitude (B).

Looking Deeper into Climate Science

Box 10-1

Ice Volume Response to Insolation

The dependence of ice volume on summer insolation can be expressed by this equation:

$$\frac{d(I)}{d(t)} = \frac{1}{T}(S - I)$$

where I is ice volume, $\dfrac{d(I)}{dt}$ is the rate of change of ice volume per unit of time (t), T is the response time of the ice sheet (in years), and S is the summer insolation signal.

This equation specifies that the rate of change of ice volume with time is a function of two factors. One factor is T, the response time of the ice sheets, measured in thousands of years. The larger the time constant T of ice response is, the slower will be the resulting rate of ice

volume change, which depends on the inverse of the value of T, or $1/T$.

The second factor controlling the rate of ice volume change is the term $S - I$, which is a measure of the degree of disequilibrium (offset) between the summer insolation forcing signal (S) and the ice volume response (I). Conceptually, this disequilibrium can be thought of in this way: the ice volume response (I) is constantly chasing after the insolation forcing signal (S), but it never catches up to it. For example, when insolation is at a minimum, ice volume is growing at its fastest rate toward its maximum value, but it has not yet gotten to that value. When ice volume finally does reach its maximum size, the insolation curve has already turned and risen halfway toward its next maximum, causing the ice volume curve to reverse direction and shrink.

---

change in temperature at Earth's mean prevailing lapse rate. For this reason, these large changes in bedrock elevation can translate into significant effects on temperature and mass balance at the surface of the overlying ice sheet.

Bedrock responds to the ice load in two phases (Figure 10-10B). The initial reaction is a quick sagging beneath the weight of the ice. This **elastic response** represents about 30% of the total vertical change in the bedrock. Then, over the next several thousand

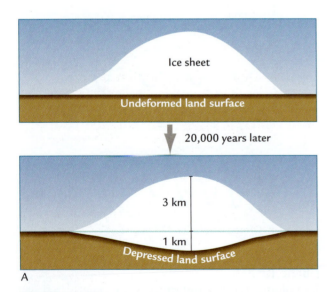

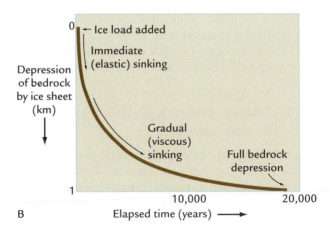

A   B

FIGURE 10-10

Bedrock sinking

If an ice sheet 3.3 km thick were suddenly placed on the land, the bedrock would sink almost 1 km under the load (A). The initial sinking would be elastic and immediate, but the later response would be viscous and slower, with about half of the remaining sinking occurring every 3,000 years (B).

years, the bedrock continues to sink in a much slower (and larger) **viscous response** caused by the extremely slow flow of rock in a relatively "soft" layer of the upper mantle at a depth of between 100 and 350 km (see Chapter 5).

This viscous response slows progressively as the bedrock adjustment moves toward a final state of equilibrium. Viscous behavior has a response time (see Chapter 1) of about 3,000 years: that is, about half of the remaining response needed to reach final equilibrium is achieved every 3,000 years. The rate of change of the curve gradually slows through time because each successive 3,000-year response time eliminates half of the remaining (unrealized) response ($1 > 1/2 > 1/4 > 1/8$, and so on). After six response times of 3,000 years each (totaling 18,000 years), only a tiny fraction ($\sim$1–2%) of the eventual bedrock depression remains unrealized.

Bedrock behavior works in the same sense but in the opposite direction if the ice load is abruptly removed. The rock surface will rebound toward the level that is in equilibrium with the absence of an ice load. The initial rapid elastic rebound will be followed by a slow viscous rebound lasting many thousands of years. Today, parts of Canada (in the Hudson Bay region) and Scandinavia (around the Baltic Sea) are still undergoing a slow viscous rebound in response to ice melting that occurred more than 10,000 years ago.

Actual ice sheets grow and melt much more slowly than these idealized (instantaneous) examples. When ice begins to accumulate on the land, the immediate (elastic) sagging of the bedrock depresses the land and promotes ice sheet melting. But the slower (and larger) viscous bedrock response keeps the growing ice sheet at higher elevations, where temperatures are colder, ablation is slower, and the ice mass balance is more positive. Overall, the delay in bedrock sinking provides a positive feedback to the growing ice sheet (Figure 10-11 top).

Bedrock plays the same overall positive feedback role when the ice is melting (Figure 10-11 bottom). The weight of a large ice sheet that exists for thousands of years creates a deep depression in the underlying bedrock. As the ice begins to melt, the (smaller) elastic part of the rebound quickly lifts the bedrock and eliminates part of the depression. But the (larger) viscous part of the rebound leaves the ice sheet at lower elevations in the depression it created and the warmer air at these levels causes further ice melting.

10-5 A Full Cycle of Ice Growth and Decay

In this section, we explore the interactions of insolation, ice volume, and bedrock responses during a typical cycle of ice sheet growth and decay. Because

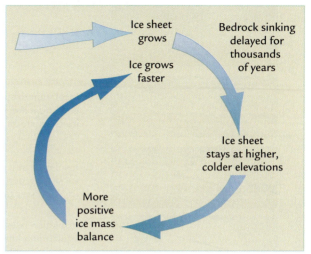

A Low summer insolation

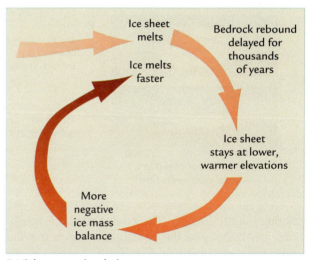

B High summer insolation

FIGURE 10-11

Bedrock feedback to ice growth and melting
Delayed bedrock sinking during ice accumulation (top) and delayed rebound during ice melting (bottom) provide positive feedback to the growth and decay of ice sheets.

of the long lags inherent in these responses, these factors interact in intricate ways to create and destroy ice sheets.

We start with an interglacial maximum like the one today, with the climate point P located in the Arctic Ocean and with no ice sheet present on the northern continent (Figure 10-12A). As summer insolation begins to decrease from a previous maximum, the equilibrium line shifts to the south and the climate point gradually moves onto the land. Some snow survives summer ablation on the far northern part of the continent, and a small ice sheet begins to form (Figure 10-12B).

As this ice sheet slowly grows, it reaches higher, colder elevations where accumulation dominates over

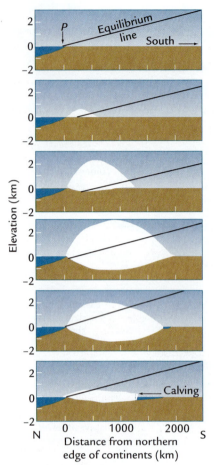

A No ice sheet
   (interglacial)

B Insolation drops,
   eqilibrium line shifts south,
   ice sheet starts to grow

C Insolation at a minimum,
   ice sheet grows rapidly,
   ice depresses bedrock

D Insolation rises,
   equilibrium line moves north,
   ice sheet at maximum size,
   bedrock depression increases

E Insolation at a maximum,
   equilibrium line far to north,
   ice melts rapidly,
   bedrock starts to rise

F Insolation starts to drop,
   last ice remnants melt,
   bedrock rises rapidly

FIGURE 10-12
Cycle of ice sheet growth and decay

Typical cycles of ice sheet growth and decay are affected by both the delayed response of the ice sheets to summer insolation forcing and the delayed response of the bedrock to the ice sheet loading and unloading. (ADAPTED FROM J. OERLEMANS, "THE ROLE OF ICE SHEETS IN THE PLEISTOCENE CLIMATE," *NORSK GEOLOGISK TIDSSKRIFT* 71 [1991]: 155–61.)

ablation (Figure 10-12C). The ice also advances southward, partly because the equilibrium line is moving south and partly because flow from the area of ice accumulation in the north carries ice to the south. The thickening ice sheet slowly begins to weigh down the bedrock, but most of the bedrock depression lags several thousand years behind ice accumulation. This delay in bedrock sagging helps to keep the surface of the ice sheet at higher and colder elevations where accumulation exceeds ablation.

The highest rate of ice accumulation occurs when summer insolation reaches a minimum value and the equilibrium line is displaced farthest south (see Figure 10-12C). At this point, the ice sheet has not yet reached maximum size because of the lag of ice volume behind the insolation driver. The rapid growth of new ice continues to weigh down the bedrock even more, with a lag of thousands of years for each new increment of ice.

Summer insolation then begins to increase and shift the equilibrium line slowly back to the north, but the ice sheet continues to grow to its maximum size for several thousand more years (Figure 10-12D). Ice growth continues because insolation levels are still relatively low and because most of the surface of the ice sheet lies above the equilibrium line, protected from the slowly increasing levels of ablation.

At some point, the combined effects of the ongoing northward shift of the equilibrium line along with the increasing amount of bedrock depression bring the southern end of the ice sheet below the equilibrium line (Figure 10-12E). With a larger surface area of the ice sheet now undergoing ablation, the overall mass balance turns negative, and ice volume begins to decrease. Melting is aided by the delayed rebound of bedrock from the weight of the earlier (larger) ice sheet. In effect, the southern edge of the ice sheet has now sagged into its own bedrock hole, where temperatures are relatively warm.

Eventually, rising summer insolation drives the equilibrium line far enough north to move the climate point back over the Arctic Ocean (Figure 10-12F). Most of the remaining ice now lies in an ablation regime, and the last remnants may disappear several thousand years later. But if a small amount of ice survives intact through a summer insolation maximum, it can serve as a nucleus from which the next ice sheet can grow.

10-6 Ice Slipping and Calving

Ice is transferred in the body of ice sheets by slow flow from colder, higher regions of net accumulation to lower, warmer regions of net ablation (see Chapter 2).

In two-dimensional ice models, this flow is usually represented in a simplified way as a slow diffusion (spreading) of ice from higher to lower elevations.

Several other types of ice behavior that are usually omitted from these models because they are inherently less predictable may also be important. One such process is **basal slip**. Slipping occurs because meltwater at the base of the ice sheet saturates soft sediments and creates a lubricated layer across which the ice can slide. This process is usually not included in models because of the difficulty in predicting when and where it will occur.

Iceberg calving, which occurs along the ocean margins of ice sheets, is another unpredictable process. The ice sheets that existed 20,000 years ago in North America and Scandinavia had large borders along the Atlantic Ocean (see Figure 10-4). These margins lost a substantial fraction of their mass by calving icebergs to the sea, but this loss is also difficult to quantify in models because it is unpredictable.

One method for modeling the long-term evolution of ice sheets is to couple a two-dimensional (altitude/latitude) ice sheet model like the one shown in Figure 10-5 to a simplified two-dimensional physical model of the atmosphere and ocean. Similar to the ice sheet models, the 2-D atmosphere-ocean models have one vertical dimension (altitude in the atmosphere, depth in the ocean) and one horizontal dimension (latitude), with changes in the other horizontal dimension (longitude) omitted. The goal of these coupled models is to simulate the linked changes in ice sheets and the atmosphere-ocean system.

Northern Hemisphere Ice Sheet History

The history of glaciation in the Northern Hemisphere has been reconstructed during the last four decades. The two most definitive kinds of evidence of this history have come from the ocean.

10-7 Ice Sheet History: $\delta^{18}O$ Evidence

At first thought, it might seem that the best records of past glaciations would be found on the continents where the ice sheets existed. Ice erodes underlying sediments and bedrock and deposits long moraine ridges containing unsorted sediment called till (see Chapter 3). Unfortunately, these deposits are of little use in reconstructing long-term glacial history because each successive glaciation erodes and destroys most of the sediment left by the previous ones. The few undisturbed deposits that remain are isolated fragments generally beyond the 50,000-year reach of radiocarbon dating.

Continuous records of glacial history come from ocean basins where sediment deposition is uninterrupted. Ocean sediments contain two key indicators of past glaciations: (1) ice-rafted debris, a mixture of coarse and fine sediments delivered to the ocean by melting icebergs that calve from ice sheet margins; and (2) $\delta^{18}O$ records from the shells of foraminifera, which provide a quantitative measure of the combined effects of changes in ice volume and in the temperature of ocean water (see Appendix A). These signals accumulate layer by layer in sediments on the ocean floor.

Decades ago, the marine scientists Cesare Emiliani and Nick Shackleton pioneered the use of oxygen isotope ratios recorded in the shells of marine foraminifera to study past climates. In the 1950s and 1960s, Emiliani analyzed $\delta^{18}O$ records extending back a few hundred thousand years and interpreted the $\delta^{18}O$ variations primarily as a record of past temperature changes. In the late 1960s, Shackleton proposed instead that the $\delta^{18}O$ signals are mostly a record of changing global ice volume, with only a small overprint from temperature changes. (Current thinking is that the effect of ice volume on $\delta^{18}O$ signals lies somewhere between these two views but is larger than the temperature effect in most regions.) Using mass spectrometers capable of analyzing very small samples, Shackleton published detailed $\delta^{18}O$ signals based on bottom-dwelling (benthic) foraminifera and extended our knowledge of glacial history much farther into the past.

In 1976, James Hays and John Imbrie joined with Shackleton on a landmark paper that conclusively linked changes in $\delta^{18}O$ to changes in orbital insolation. They found that orbital periods were clearly present in $\delta^{18}O$ changes over the last 300,000 years and that the $\delta^{18}O$ changes lagged behind changes in summer insolation forcing by several thousand years, consistent with the lag of ice volume behind summer insolation that Milankovitch had predicted.

The first continuous and detailed $\delta^{18}O$ record of the entire 2.75 million years of northern hemisphere glacial history was compiled in the late 1980s by isotopic analysis of benthic foraminifera from the North Atlantic Ocean (Figure 10-13). This long $\delta^{18}O$ record shows two trends: (1) a very gradual drift toward more positive values, and (2) numerous cyclic-looking oscillations between positive and negative values. Both of these features reflect some combination of changes in temperature and fluctuations in ice volume (see Appendix A). Changes toward more positive $\delta^{18}O$ values indicate more ice on the land and/or cooler deep-ocean temperatures. More negative $\delta^{18}O$ values indicate smaller ice sheets and/or warmer deep-ocean temperatures. This record also contained debris eroded from the continents by the ice sheets and carried to the ocean on ice "rafts."

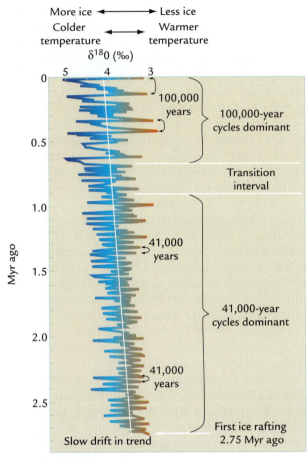

FIGURE 10-13
Evidence of ice sheet evolution: δ¹⁸O

A sediment core from the North Atlantic Ocean reveals a 3-million-year δ¹⁸O record that reflects changes in ice volume and deep-water temperature. No major ice sheets existed before 2.75 million years ago, after which small ice sheets grew and melted mainly at a cycle of 41,000 years until 0.9 million years ago. Since that time, large ice sheets grew and melted at intervals near 100,000 years. The diagonal white line shows a gradual long-term δ¹⁸O trend toward more ice and colder temperature. (ADAPTED FROM M. E. RAYMO, "THE INITIATION OF NORTHERN HEMISPHERE GLACIATION," *ANNUAL REVIEWS OF EARTH AND PLANETARY SCIENCES* 22 [1994]: 353–83.)

Prior to 2.75 million years ago, the δ¹⁸O values were relatively negative (3.5‰ or less) and no ice-rafted debris was present. During this interval, northern hemisphere ice sheets either did not exist or never reached the size needed to send large numbers of icebergs to the North Atlantic south of Iceland. The smaller variations in δ¹⁸O during this interval probably reflect temperature changes in the deep waters or changes in the size of the Antarctic ice sheet.

Beginning 2.75 million years ago, significant amounts of ice-rafted debris appeared in the record, an indication that ice sheets of considerable size were now present at least sporadically. This debris

accumulated during intervals of relatively positive δ¹⁸O values, which occurred mainly at a regular cycle of 41,000 years (see Figure 10-13). This part of the record suggests that ice sheets were now forming during intervals of low summer insolation, but that all or most of the ice probably disappeared during the subsequent summer insolation maxima.

This regime of 41,000-year cycles persisted for the first two-thirds of the interval of northern hemisphere glaciation from 2.75 to 0.9 million years ago. The forty or more δ¹⁸O oscillations that can be detected during this interval indicate at least forty episodes of glaciation. The slow background shift of the δ¹⁸O signal toward more positive values during this interval also indicates a gradual underlying drift into a colder world.

Beginning near 0.9 million years ago and becoming more obvious after 0.6 million years ago, the character of the δ¹⁸O record changes (see Figure 10-13). Maximum δ¹⁸O values reach larger amplitudes but are spaced farther apart, indicating that ice sheets persisted for longer intervals of time and grew larger in a colder world. These glacial intervals come to an end during abrupt δ¹⁸O decreases that indicate rapid ice melting and ocean warming. Over the last 0.6 million years, there have been six of these large δ¹⁸O maxima, each followed by an abrupt deglaciation (called a **termination**) at an average spacing near 100,000 years.

Almost hidden in the highly compressed record shown in Figure 10-13 are smaller 41,000-year and 23,000-year δ¹⁸O oscillations that persist during the last 0.9 million years as secondary cycles superimposed on the larger oscillations near 100,000 years. To get a clearer sense of the character of these later cycles, we zoom in on the most recent part of a record that begins during the major glaciation near 150,000 years ago (Figure 10-14). Near 130,000 years ago, an abrupt shift occurred into an interglacial interval that lasted until 120,000 years ago. Like the modern interglaciation, this interval had no ice-rafted debris in North Atlantic sediments, because northern ice sheets were not present except on Greenland.

Between 125,000 and 80,000 years ago, the δ¹⁸O signal oscillated several times between values that indicate more or less ice and colder or warmer temperatures. The spacing of these oscillations at approximately 23,000 years confirms the presence of the orbital precession signal in this record. The two later glacial maxima near 63,000 and 21,000 years ago are separated by about 42,000 years, an indication that the 41,000-year orbital tilt signal is also present in this record.

The rapid transition between 17,000 and 10,000 years ago marks another abrupt deglaciation, the first one since 130,000 years ago. These terminations are the most prominent marker of the longer-period oscillations near a period of 100,000 years.

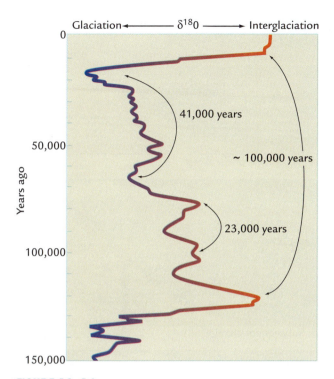

FIGURE 10-14
Ice sheet $\delta^{18}O$ changes over the last 150,000 years

A multicore combined $\delta^{18}O$ record covering the last 150,000 years shows 23,000-year and 41,000-year oscillations in addition to the larger oscillation near 100,000 years. (ADAPTED FROM D. MARTINSON ET AL., "AGE DATING AND THE ORBITAL THEORY OF THE ICE AGES: DEVELOPMENT OF A HIGH-RESOLUTION 0 TO 300,000-YEAR CHRONOSTRATIGRAPHY," *QUATERNARY RESEARCH* 27 [1987]: 1–29.)

In Summary, evidence from $\delta^{18}O$ signals indicates that ice sheets have fluctuated at orbital cycles of approximately 23,000 years, 41,000 years, and 100,000 years during the long history of northern hemisphere glaciation. Unfortunately, these $\delta^{18}O$ signals also contain a temperature overprint that makes it difficult to constrain the actual size of the ice sheets. For this reason, independent evidence is needed to confirm that $\delta^{18}O$ signals provide a reasonably accurate history of ice volume.

10-8 Confirming Ice Volume Changes: Coral Reefs and Sea Level

The oceans provide a second independent measure of ice volume—the fossil remains of coral reefs. Today, most coral reefs grow in warm tropical ocean water and prefer clear water near small islands (Figure 10-15) rather than in water muddied by river runoff from large continents. The coral reef species most useful to climate scientists (such as *Acropora palmata*)

FIGURE 10-15
Coral reefs

Coral reefs form in clear, shallow waters in warm tropical seas. (IAN CARTWRIGHT/PHOTODISC.)

grow just below sea level. Reefs have strong structural frameworks that remain intact long after individual coral organisms have died and that preserve records of past sea level positions.

As sea level rises and falls, coral reefs follow by migrating upslope and downslope. In effect, ancient reefs function as dipsticks that measure the past water level in the world ocean. Over orbital cycles of tens to hundreds of thousands of years, fluctuations in sea level result mainly from changes in the amount of water extracted from the ocean and stored in ice sheets on land. As a result, the sea level histories recorded by coral reef dipsticks are also a record of ice volume.

Old coral reefs can be dated by radiometric decay methods. Their skeletons contain small amounts of ^{234}U, which slowly decays to ^{230}Th (see Chapter 3). This dating technique is well suited for use during the last several hundred thousand years. The sea level record from dated coral reefs can be compared with changes in the δ^{18}O signal covering the last 150,000 years or more (see Figure 10-14).

Ocean islands in tectonically stable regions like Bermuda have a prominent fossil coral reef that lies about 6 m above modern sea level and dates to 125,000 years ago. These reefs are the only indication of a sea level higher than today during the last 150,000 years. This evidence agrees with the marine δ^{18}O record in Figure 10-14: the only δ^{18}O minimum comparable to the modern value in the last 150,000 years dates to between 130,000 and 120,000 years ago. Both types of evidence agree that this interval was the only time in the last 150,000 years when the amount of ice on Earth was as small as it is now. In fact, the amount was slightly smaller; the "extra" 6 m of higher sea level requires that some amount of ice on Greenland, or Antarctica, or both had melted at that time but is back in place today.

All other coral reefs that grew during the last 150,000 years formed when sea level was below its modern position because of the greater amount of seawater tied up in ice sheets. Today, these other reefs lie submerged on the underwater slopes of tectonically stable islands and beyond easy reach. To circumvent this problem, marine scientists have turned to ocean islands in a different tectonic setting: areas where slow uplift has raised older reefs that formed below modern sea level and exposed them above sea level (Figure 10-16).

The two most intensively studied of these islands are Barbados in the eastern Caribbean Sea, and New Guinea in the western Pacific Ocean. Both islands are rimmed by prominent coral reef terraces that have been lifted out of the ocean since they formed (Figure 10-17). Two major reefs—one exposed on Barbados and one on New Guinea—date from 125,000 years ago, the same age as the last interglacial reefs on tectonically stable islands. Two other reefs—one on each island—date from about 104,000 and 82,000 years ago. The ages of the two younger reefs on Barbados and New Guinea match the ages of two prominent minima in the δ^{18}O signal in Figure 10-14. These reefs record higher sea levels caused by ice melting, but the melting was not as complete as today and sea levels were not as high.

Drilling of deeply submerged reefs has shown that the lowest global sea level reached during the last 125,000 years was close to 120 meters below the modern level near 20,000 years ago. The timing of this sea level minimum correlates with the largest δ^{18}O maximum in Figure 10-14, again confirming that δ^{18}O is a good index of ice volume.

The sea level minimum from 20,000 years ago and the +6-meter maximum level from 125,000 years ago serve as anchor points for removing the effect of uplift on these tectonically active islands and calculating

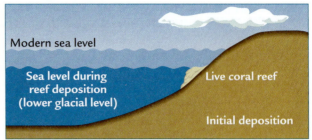

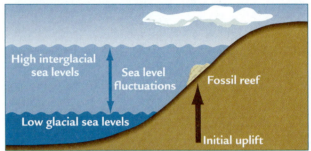

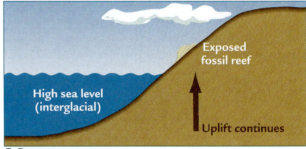

FIGURE 10-16
Gradual uplift of coral reefs

Coral reefs may initially form on the edge of an uplifting island at times when sea level lies below its modern position (A). As time passes, uplift steadily raises the island and the fossil reef toward higher elevations, while sea level moves up and down against the island in response to changes in ice volume (B). Today, old fossil reefs can lie well above sea level as a result of ongoing uplift (C).

changes in sea level during the intervening interval (Box 10-2). This method reveals that the reefs at 82,000 and 104,000 years ago were formed when sea level was lower than today by an estimated 17 meters. This estimate falls about 15% of the way from full interglacial to full glacial sea levels, about the same relative position as the proportional change in δ^{18}O between the minimum value 125,000 years ago and the maximum value 20,000 years ago (see Figure 10-14). This agreement provides even more confirmation that δ^{18}O is a good index of ice volume, despite the temperature overprint known to be present. Each 10-meter change in global sea level results in an isotopic (δ^{18}O) change of 0.8–1.1‰.

FIGURE 10-17
Uplifted coral reef terraces

Terraces formed by erosion-resistant coral reefs lie well above sea level on the island of Papua New Guinea, in the western Pacific Ocean. (COURTESY OF ARTHUR L. BLOOM, CORNELL UNIVERSITY.)

In Summary, coral reefs that formed during the last 150,000 years confirm that the $\delta^{18}O$ signal is a reasonable proxy for ice sheet size. The ages of the most prominent $\delta^{18}O$ minima correspond to the ages of coral reefs formed during high stands of sea level caused by ice sheet melting, and the amplitudes of the sea level changes estimated from the reefs correspond to the relative changes in ice volume inferred from the $\delta^{18}O$ signal.

Is Milankovitch's Theory the Full Answer?

As noted earlier, the 2.75-million-year history of northern hemisphere glaciation shown in Figure 10-13 has two defining characteristics. One feature is the long sequence of $\delta^{18}O$ oscillations, each of which lasted for tens of thousands of years. This aspect of the signal is linked to orbital forcing of glacial cycles. Milutin Milankovitch, who died in 1958, did not live to see these remarkable records of glacial history. During his lifetime, the only records available were those from glacial moraines on land, and most of those were deposited during or after the most recent glaciation, 20,000 years ago. Until well into the 1960s, most glacial geologists thought that only four or five glaciations had occurred, but the marine $\delta^{18}O$ record in Figure 10-13 reveals some fifty glacial maxima. The other obvious feature in this $\delta^{18}O$ record is the slow persistent increase in $\delta^{18}O$ values. This trend is part of the gradual cooling that has been underway for 50 million years and that led to northern hemisphere glaciation (see Chapter 7).

The interaction of these longer-term tectonic and shorter-term orbital controls on ice sheets can be combined to interact in a conceptual model of the history of northern hemisphere glaciation (Figure 10-18). The first component of this model is an oscillating signal representative of changes in summer insolation at high northern latitudes. This curve incorporates the combined influence of the cycles at 23,000 years (precession) and at 41,000 years (tilt). Because insolation

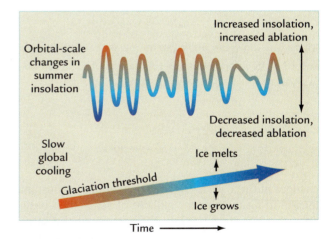

FIGURE 10-18
Factors in long-term evolution of ice sheets

The long-term evolution of ice sheets reflects the interaction of two factors: cyclic changes in summer insolation that drive shorter-term changes in ice sheet mass balance (top), and a much more gradual global cooling represented by a slowly changing glaciation threshold (bottom). Ice sheets accumulate when summer insolation falls below the critical glaciation threshold and melt when it rises above it.

Looking Deeper into Climate Science

Sea Level on Uplifting Islands

Coral reefs on islands that are slowly being lifted out of the ocean by tectonic processes can be used as sea level dipsticks if the overprint caused by uplift can be removed. Many ancient coral reefs now sit on the emerged flanks of islands at elevations above modern sea level. Their present elevation is the result of two factors: (1) the position of global sea level at the time the reef formed (the sea level difference from the modern position), and (2) the amount of uplift of the island since the reef formed.

The amount of subsequent uplift of each reef depends on its age (older reefs have undergone greater uplift because more time has passed) and on the rate of uplift of the island (reefs of a given age have undergone more uplift on fast-rising islands than on slow-rising ones). Although the present elevations of these older reefs are referenced to present sea level, modern sea level is just one position in a pattern of continual rises and falls through time.

First, the average rate of uplift (U) is calculated for the island under study by using the last interglacial reef on the island as a reference point. Both the age of this reef (125,000 years) and the sea level at the time it formed (6 m higher than today) are known. The calculation is:

$$U = \frac{(Ht - 6)}{125}$$

where U is the mean uplift rate in meters per 1,000 years and Ht is the *present* elevation (in meters) of the 125,000-year reef on the island under study.

This calculation removes the 6-meter difference in sea level between today and 125,000 years ago to isolate the average effect of the gradual uplift between the two times. With the mean uplift rate determined (and assumed to have remained constant over the last 125,000 years), the amount of uplift that has occurred since a reef of *any* intermediate age formed can be corrected in order to derive an estimate of sea level at the time that it formed:

$$S = h - (U)(t)$$

where S is the relative sea level at the time the older reef formed (in meters), h is the present elevation of the older reef (in meters), U is the mean uplift rate in meters per 1,000 years, and t is the time elapsed since the reef formed (in thousands of years).

With these equations, the differing effects of uplift that occurred on any island can be removed. For example, New Guinea is being uplifted at a rate close to 2 meters per 1,000 years, while Barbados is rising at a rate of about 0.3 meters per 1,000 years. A coral reef dated to 82,000 years ago will have been uplifted by 150 meters on New Guinea since it formed, but by only 25 meters on Barbados. With the correction for local uplift rates, the 82,000-year reef formed on Barbados must have formed when global sea level was about 17 meters lower than it is today. The same kind of calculation shows that the 104,000-year reef formed at roughly the same relative sea level on New Guinea.

---

changes at these cycles have varied around a constant long-term mean value for millions of years, their basic character has stayed nearly the same, and the same curve is repeated throughout the following description.

The second component of the model is an assumed threshold temperature for ice sheet formation. This threshold takes the form of an equilibrium line that separates temperatures cold enough to permit net ice accumulation (glaciation) from those warm enough to cause net melting of snow and ice. The position of this threshold line relative to the insolation curve changes very slowly through time as the gradual cooling of the last several million years proceeds.

The ice sheet response results from the interaction between the equilibrium line threshold and the summer insolation signal. When insolation values fall below the equilibrium line threshold existing at that time, ice sheets grow. When they rise above it, ice sheets melt. Both the growth and melting of ice sheets lag several thousand years behind the insolation forcing.

With this conceptual model, we can characterize the major stages in the development of northern hemisphere glaciation (Figure 10-19). During each stage, the threshold line is shown as flat, although it is actually changing very slowly with respect to the

Box 10-2

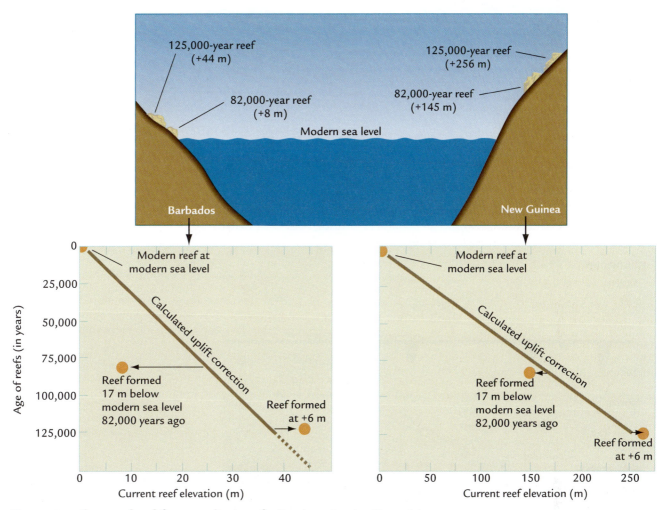

Reconstructing sea level from ancient reefs By subtracting the effects of slow tectonic uplift of the islands of Barbados and New Guinea, scientists can reconstruct sea level at earlier times when coral reefs formed.

insolation signal (as in Figure 10-18). For each glacial stage, changes in the position of the equilibrium line during intervals of maximum and minimum summer insolation are also shown on a north-south profile.

PRE-GLACIATION PHASE For millions of years prior to 2.75 million years ago, there were no ice sheets of significant size on North America or Eurasia. The small amounts of ice formed on Greenland are ignored here. Even the deepest summer insolation minima failed to reach the critical threshold necessary for glaciation to develop over the major continents (Figure 10-19A), and the climate point remained over

the Arctic Ocean at all times (Figure 10-19B). North America and Eurasia remained too warm for ice sheets to form.

SMALL GLACIATION PHASE (2.75–0.9 MILLION YEARS AGO) By 2.75 million years ago, global cooling had altered the position of the equilibrium line threshold enough that it began to intercept and interact with the summer insolation curve. At intervals of 41,000 or 23,000 years, as summer insolation minima crossed the equilibrium line threshold (Figure 10-19C), the climate point moved southward over the continents (Figure 10-19D), and ice sheets began to grow. But

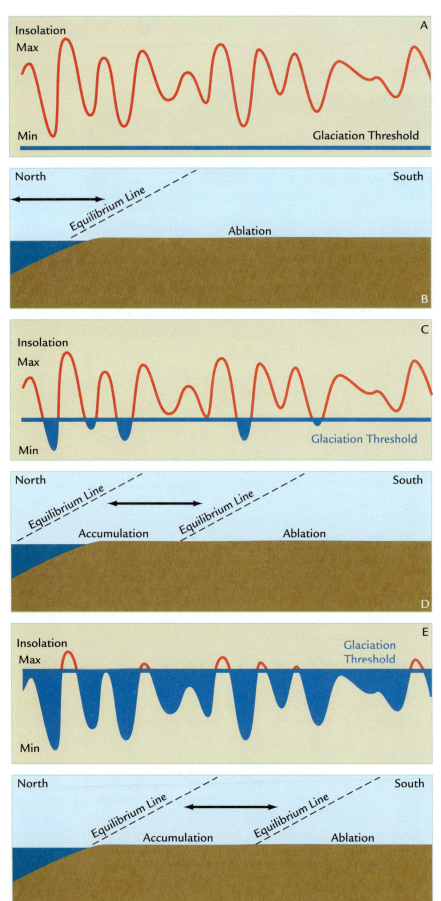

FIGURE 10-19

Conceptual phases of ice sheet evolution

The long-term evolution of northern hemisphere ice sheets passes through three phases. In the preglaciation phase (A, B), no ice accumulates. In the small glaciation phase (C, D), ice accumulates during summer insolation minima but melts entirely during the next insolation maximum. In the large glaciation phase (E, F), ice persists through weak summer insolation maxima and melts only during larger maxima.

before these ice sheets could grow very large, summer insolation had already reversed direction and begun to increase toward the next maximum, and the equilibrium line had moved back northward off the continents. As a result, the ice sheets melted away.

To some degree, this small glaciation phase fits the Milankovitch theory, which predicted discrete intervals of glaciation, each lagging just behind individual summer insolation minima and ending during subsequent summer insolation maxima. On closer inspection, however, the $\delta^{18}O$ oscillations during this period are not fully consistent with his theory. Insolation changes at high northern latitudes show large variations at the 23,000-year precession cycle, both in monthly changes and in the caloric summer index used by Milankovitch (see Chapter 8). Yet the $\delta^{18}O$ signal during the small glaciation phase is dominated by strong 41,000-year variations, with much weaker changes at 23,000 years (see Figure 10-13). A plot of the relative amplitude of these orbital cycle variations using the power spectrum method (see Chapter 8) highlights the mismatch between the rhythms present in the insolation signal and those found in the $\delta^{18}O$ (ice volume) response (Figure 10-20 top and middle).

LARGE GLACIATION PHASE (0.9 MILLION YEARS AGO TO THE PRESENT) As global cooling slowly continued, the glaciation threshold line again shifted relative to the insolation curve. In this new regime, conditions favorable for ice accumulation rather than ablation prevailed most of the time (see Figure 10-19E). Through much of this interval, the climate point remained over the land and allowed ice growth. It retreated to the Arctic Ocean only during unusually strong insolation maxima (see Figure 10-19F). Because the land remained in a regime of ice accumulation during many of the smaller insolation maxima, ice sheets did not disappear but persisted until a sufficiently strong maximum caused complete melting. With a longer time span in which to grow, these ice sheets became larger than the ones during the small glaciation phase.

This large glaciation phase corresponds in a general way with the interval of $\delta^{18}O$ changes after 0.9 million years ago (see Figure 10-13). During the last 900,000 years, larger ice sheets have grown over longer intervals of time. Milankovitch would probably have been surprised by this part of the $\delta^{18}O$ record. Although the continued presence of 23,000-year and 41,000-year glacial cycles during this interval is consistent with his theory, he did not anticipate the much larger oscillations at a period near 100,000 years (see Figure 10-20 bottom). He probably would also have been surprised by the sawtoothed character of these oscillations, with relatively gradual buildup of ice followed by rapid melting (see Figure 10-14).

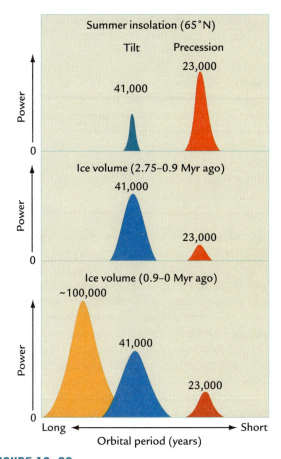

FIGURE 10-20

Spectral analysis: insolation and ice volume

Seasonal summer insolation changes at 65°N are strong at 23,000 years, weaker at 41,000 years, and negligible at 100,000 years (top). In contrast, the $\delta^{18}O$ signal of northern hemisphere ice volume (and deep-ocean temperature) between 2.75 and 0.9 million years ago was very strong at 41,000 years but much weaker at 23,000 years (middle). Since 0.9 million years ago, the $\delta^{18}O$ signal has been strongest at or near 100,000 years (bottom).

You might think that these large oscillations in ice volume at a cycle near 100,000 years could be easily explained by changes in orbital eccentricity at that period. But remember that changes in orbital eccentricity act only indirectly as a multiplier effect on the size of the precession cycle, not as a direct driver of ice volume cycles at 100,000 years. Incoming insolation is the mechanism that drives ice sheets, and the 100,000-year variations in the insolation signals are negligible. The large oscillations in ice volume at or near a period of 100,000 years do not seem to be explained by the Milankovitch theory.

If the natural global cooling trend of the last 50 million years were to persist for tens of millions of years into the future, it is possible that the Northern Hemisphere would enter a different phase in which the equilibrium line remained permanently on the continents and ice sheets never melted. During the last 900,000 years, we have only avoided this state during relatively brief interglaciations of 10,000 years or less that represent less than 10% of that entire time span. In this sense, we have moved closer to a state in which ice sheets are present permanently on the northern continents. With continued cooling, the northern continents (particularly North America) might begin to more closely resemble Antarctica, which has been in a state of permanent glaciation for many millions of years.

In Summary, the Milankovitch theory is a useful starting point for understanding the history of northern hemisphere glaciation. It explains the presence of ice volume responses at 41,000 and 23,000 years, and why they lag several thousand years behind the summer insolation forcing. But it fails to explain the dominance of the 41,000-year cycle over the 23,000-year signal for almost 2 million years, and it does not account for the emergence of large oscillations near 100,000 years in the last million years. Something more complicated must be happening to make ice sheets grow and melt at these rhythms. We will return to this problem in Chapter 12.

Key Terms

equilibrium line (p. 196)

Milankovitch theory (p. 197)

climate point (p. 199)

phase lag (p. 200)

elastic response (p. 202)

viscous response (p. 203)

basal slip (p. 205)

termination (p. 206)

Review Questions

1. What is the equilibrium line and why is it important?

2. Why are northern ice sheets likely to be more responsive to insolation changes than ice in Antarctica?

3. Why does the size of a growing or melting ice sheet lag well behind changes in insolation?

4. How does the delay in bedrock response to ice loading or unloading act as a positive feedback on ice volume?

5. If the average amplitude of a 41,000-year $\delta^{18}O$ cycle in the deep North Atlantic prior to 0.9 million years ago was 1.0‰, and if changes in ice volume account for half of that total, how large was the average changes in deep-water temperature?

6. How do corals provide an indication of the volume of water tied up in ice sheets on land?

7. What has been the average rate of uplift of a coral reef formed during the last interglaciation 125,000 years ago and now lying 131 meters above sea level?

8. Which part of the Milankovitch theory has proven accurate? Which parts are insufficient?

9. If a reef on New Guinea that has been dated to 104,000 years formed at 17 m below modern sea level, and if the average uplift rate on that island has been 2 m per thousand years, what is the current elevation of that reef?

Additional Resources

Basic Reading

Imbrie, J., and K. P. Imbrie. 1979. *Ice Ages: Solving the Mystery.* Short Hills, NJ: Enslow.

Advanced Reading

Hays, J. D., J. Imbrie, and N. J. Shackleton. 1976. "Variations in the Earth's Orbit: Pacemaker of the Ice Ages." *Science* 194:1121–32.

Imbrie, J. 1982. "Astronomical Theory of the Ice Ages: A Brief Historical Review." *Icarus* 50:408–22.

———. 1985. "A Theoretical Framework for the Ice Ages." *Journal of the Geological Society* (London) 142:417–32.

Mesolella, K. J., R. K. Matthews, W. S. Broecker, and D. L. Thurber. 1969. "The Astronomical Theory of Climate Change: Barbados Data." *Journal of Geology* 77:250–74.

Oerlemans, J. 1991. "The Role of Ice Sheets in the Pleistocene Climate." *Norsk Geologisk Tidsskrift* 71:155–61

Weertman, J. 1964. "Rate of Growth or Shrinkage of Non-equilibrium Ice Sheets." *Journal of Glaciology* 5:145–58.

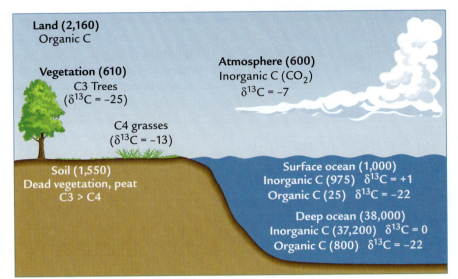

FIGURE 11-5

Carbon reservoir δ^{13}C values

The major reservoirs of carbon on Earth have varying amounts of organic and inorganic carbon (shown in parentheses as billions of tons of carbon), and each type of carbon has characteristic carbon isotope (δ^{13}C) values.

form. Terrestrial plants use atmospheric inorganic carbon (CO_2) with a δ^{13}C value near -7‰ and convert it to organic carbon with values that range between -11‰ and -28‰ for different kinds of plants. This photosynthetic fractionation of carbon by living systems is different from the physical fractionation of oxygen isotopes by evaporation and condensation (see Chapter 7 and Appendix A).

Both the organic and inorganic forms of carbon are preserved in the geologic record. $CaCO_3$ shells of marine foraminifera are formed from inorganic carbon dissolved in seawater, typically with δ^{13}C values near 0‰. Many kinds of geologic deposits on land and in the ocean also contain small amounts of organic carbon with more negative δ^{13}C values, typically averaging -25‰. These deposits are useful for tracking past transfers of carbon among the surface reservoirs over orbital (and shorter) time scales.

Orbital-Scale Changes in CO₂

Records of past CO_2 changes come from sequences of cores drilled into the Antarctic ice sheet. The longest of these sequences, from the Dome C site, shows oscillations between values as high as almost 300 ppm and as low as 180 ppm over the last 800,000 years (Figure 11-6). These CO_2 oscillations line up well with variations in marine δ^{18}O, which primarily indicate changes in ice volume, along with a smaller temperature effect. When the ice sheets were large, CO_2 concentrations were low. When ice sheets were small or absent in the Northern Hemisphere, CO_2 concentrations were high.

11-4 Where Did the Missing Carbon Go?

The observation of a general match between CO_2 concentrations and ice volume for the last 800,000 years tells us that some kind of cause-and-effect relationship

exists between these two components of the climate system. One way to begin exploring this problem is to find out the fate of the carbon that was removed from the atmosphere during glacial times. The ~90-ppm reductions in CO_2 values indicate that almost one-third of the carbon in the atmosphere during interglacial

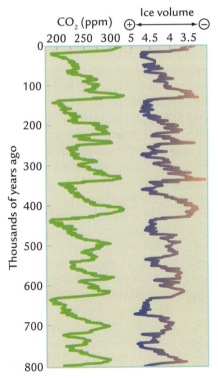

FIGURE 11-6

Long-term CO₂ changes

An 800,000-year record of CO_2 from Dome C in Antarctica shows eight large-scale cycles near a period of 100,000 years, similar to those in the marine δ^{18}O (ice volume) record. (ADAPTED FROM L. E. LISIECKI, "A BENTHIC ^{13}C-BASED PROXY FOR ATMOSPHERIC PCO$_2$ OVER THE LAST 1.5 MYR," *GEOPHYSICAL RESEARCH LETTERS* 37 [2010]: L21708, DOI:10.1029/2010GL045109.)

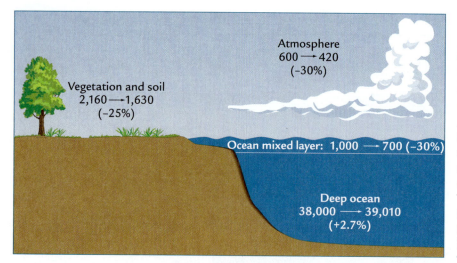

FIGURE 11-7

Interglacial-to-glacial changes in carbon reservoirs

During times like the glacial maximum 20,000 years ago, large reductions in carbon biomass occurred in the atmosphere, in vegetation and soils on land, and in the surface ocean compared to warm interglacial intervals. The total amount of carbon removed from these reservoirs (more than 1,000 billion tons) was added to the much larger reservoir in the deep ocean.

times moved elsewhere during glaciations. Converted from ppm concentrations to tons of carbon, this transfer amounted to ~180 billion tons (Figure 11-7).

One place to look for the "missing" CO_2 carbon is the vegetation-soil reservoir. Scientists have information on the amount of carbon stored on land in vegetation and soils during the last glacial maximum, 20,000 years ago. The evidence shows that the continents had less net vegetation cover and held less carbon during glaciations than they did during warm interglacial intervals like today.

The main reason for the decrease in glacial carbon was the expansion of ice sheets across large areas of North America and Eurasia that had been covered by forests of conifers and deciduous trees during interglacial times like today. In addition, other regions that were forested during warm interglacial times became covered by steppe and grassland with lower amounts of carbon during glaciations. Taken as a whole, the forested regions that survived during full glaciations were smaller than those today.

Continental lakes with sediments containing pollen can also tell us about past changes of nearby vegetation. The picture that emerges from these records is that most regions on Earth were drier and less vegetated during maximum glaciations than they are today. Even in the tropics, rain forests were less extensive. One region where a significant increase in glacial vegetation may have occurred was north of Australia. There, the fall of sea level caused by storage of water in glacial ice sheets exposed large expanses of now-submerged continental shelf, and these regions were probably covered by tropical rain forests. But this regional increase was not enough to offset losses elsewhere.

Based on this range of evidence, climate scientists estimate that the total amount of vegetation on land was reduced by roughly 25% (from 2,160 to 1,630 billion tons) during the last glacial maximum 20,000 years ago compared to interglacial times (see Figure 11-7). Because of various uncertainties, the actual

reduction could have been anywhere between 15% and 30%, but the decrease was substantial. Clearly, the carbon removed from the atmosphere did not go into land vegetation, and now we face the added problem of explaining the carbon missing from not just the atmosphere but also the land.

The other place where the missing carbon could have been stored is the ocean, either the small surface reservoir, or the much larger deep-ocean reservoir, or both. But the surface ocean is not the answer. Ocean surface waters exchange all of their carbon with the atmosphere in just a few years. Because of this rapid exchange of CO_2 gas, most areas of the surface ocean today have CO_2 values within 30 ppm of the value in the overlying atmosphere. Such rapid exchanges mean that if CO_2 values were 30% lower in the glacial atmosphere, they must have been lower by nearly the same average amount in the glacial surface ocean. This estimate of reduced carbon in the surface ocean adds another 300 billion tons to the growing list of carbon that was "missing" during glacial times (see Figure 11-7).

The only remaining near-surface carbon reservoir left is the deep ocean. The missing glacial carbon removed from the atmosphere, vegetation, and surface waters must have been stored or sequestered there. The total amount of carbon missing from these other reservoirs adds up to about 1,000 billion tons (see Figure 11-7). This amount must have ended up in the deep ocean during the last glaciation. But the deep ocean is such an enormous carbon reservoir (~38,000 billion tons) that this additional carbon would only have increased the amount present during interglacial times by about 2.7%.

11-5 δ¹³C Evidence of Carbon Transfer

The amount of carbon transferred from the land to the deep ocean can be estimated independently from $\delta^{13}C$ measurements of foraminifera that lived in the ocean. Organic carbon in terrestrial vegetation is tagged with an average $\delta^{13}C$ value of −25‰, whereas the large amount of inorganic carbon in the ocean has an average value near 0‰. Although a small fraction of the terrestrial organic carbon delivered to the ocean remains as organic matter, most of it is converted to inorganic carbon that retains the very negative (−25‰) $\delta^{13}C$ composition.

During glacial times (Figure 11-8A), ^{12}C-rich organic carbon that was transferred from the land to the deep ocean and converted to inorganic carbon should have made the $\delta^{13}C$ value of the inorganic carbon in the ocean more negative. These lower $\delta^{13}C$ values should correlate with times of more positive $\delta^{18}O$ values due to more glacial ice and colder deep-ocean temperatures. The opposite pattern should have occurred during interglacial times (Figure 11-8B), with more positive oceanic $\delta^{13}C$ values because of the return of the negative carbon to the land, and more negative $\delta^{18}O$ values because of the melting of ^{16}O-rich ice.

We can use a mass balance calculation to estimate the effect of adding very negative (^{12}C-enriched) carbon to the deep sea during maximum glaciations:

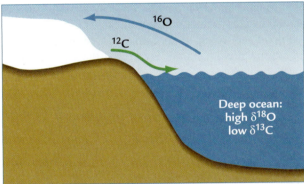

A Glacial climate

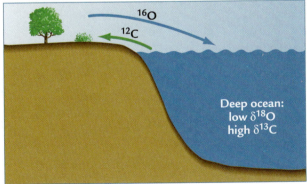

B Interglacial climate

FIGURE 11-8
Glacial transfers of ^{12}C and ^{16}O

During glaciations, ^{12}C-enriched organic matter is transferred from the land to the ocean at the same time that ^{16}O-enriched water vapor is extracted from the ocean and stored in ice sheets (A). During interglaciations, ^{12}C-rich carbon returns to the land as ^{16}O-rich water flows back into the ocean (B).

$$(38{,}000) \quad (0‰) \quad + \quad (530) \quad (−25‰) \quad = \quad (38{,}530) \quad (x)$$

Inorganic C in ocean	Mean $\delta^{13}C$	C added from land	Mean $\delta^{13}C$	Glacial inorganic carbon total	Mean $\delta^{13}C$

where x is the average $\delta^{13}C$ value of glacial inorganic carbon in the ocean, and the sizes of the various carbon reservoirs shown are in billions of tons of carbon. Solving for x, we find that the mean $\delta^{13}C$ value of inorganic carbon in the glacial ocean should have shifted from ~0‰ to −0.34‰ because of the addition of ^{12}C-rich carbon transferred from the land.

This estimated −0.34‰ shift has been tested by comparing it to $\delta^{13}C$ values in the $CaCO_3$ shells of glacial-age benthic foraminifera in available cores distributed across the world ocean over a range of water depths. The measurements document the $\delta^{13}C$ composition of the deep waters on a region by region basis. With all of these analyses combined into a global average, the $\delta^{13}C$ change estimated for the entire ocean is −0.35‰ to −0.4‰, close to the value calculated above. In view of the uncertainties involved, the good agreement between the two methods indicates that carbon isotopes are useful tracers of past carbon shifts among Earth's reservoirs.

This evidence suggests that we should be able to trace carbon transfers during earlier glacial cycles using $\delta^{13}C$ changes measured in the shells of benthic foraminifera. Because the Pacific Ocean contains by far the largest volume of water in Earth's oceans, scientists have used changes in its deep-water $\delta^{13}C$ values through time as the best single "quick" index of average $\delta^{13}C$ changes in the global ocean. The longer-term $\delta^{13}C$ record in Figure 11-9 confirms the basic relationship expected from the transfers shown in Figure 11-8: large amounts of terrestrial carbon with negative (−25‰) $\delta^{13}C$ values were transferred into the ocean every time ice sheets grew, and then were returned to the land when the ice melted.

The most negative $\delta^{13}C$ values are spaced at intervals close to 100,000 years apart during the last 0.9 million years, and at cycles averaging 41,000 years in length prior to 0.9 million years ago. Although the correlation between the two signals is far from perfect, time series analysis confirms that the $\delta^{13}C$ (carbon transfer) and $\delta^{18}O$ (ice volume) signals have varied at the same periods and with the same approximate timing over millions of years.

On closer inspection, many of the $\delta^{13}C$ variations shown in Figure 11-9 exceed the −0.35‰ to −0.4‰ shift that can easily be attributed to transfers of carbon from land to sea. For example, unusually light $\delta^{13}C$ values occur just after one million years ago. Unexpectedly large $\delta^{13}C$ responses like these tell us that additional transfers of ^{12}C-rich terrestrial carbon must have affected oceanic $\delta^{13}C$ values through time, although the reason for these transfers is not clear.

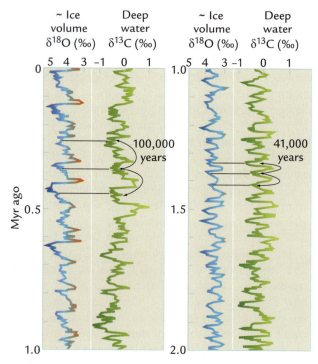

FIGURE 11-9
Long-term δ¹³C signals of carbon transfer

A sediment core from the deep Pacific Ocean shows more negative $\delta^{13}C$ values during glaciations ($\delta^{18}O$ maxima), mainly because ^{12}C-enriched carbon is transferred from the land into the ocean. (ADAPTED FROM D. W. OPPO ET AL., "A $\delta^{13}C$ RECORD OF UPPER NORTH ATLANTIC DEEP WATER DURING THE PAST 2.6 MILLION YEARS," *PALEOCEANOGRAPHY* 10 [1995]: 373–94.)

▶ How Did the Carbon Get into the Deep Ocean?

The answer to this seemingly simple question is complicated and not yet resolved. Scientists are actively investigating several factors that seem likely to have played a role in this transfer.

11-6 Increased CO₂ Solubility in Seawater

Changes in the average temperature of the ocean during glaciations would have altered the chemical solubility of CO_2 in seawater and thereby affected the amount of CO_2 left in the atmosphere. Because CO_2 dissolves more readily in colder seawater, atmospheric CO_2 levels fall by ~10 ppm for each 1°C of ocean cooling. As we will see in Chapter 12, surface-ocean temperatures cooled by an estimated average of 2° to 4°C in the tropics during the last glacial maximum 20,000 years ago, and by larger amounts in high-latitude regions. The much larger volume of water in the deep ocean cooled by an average of 2° to 3°C. As a result of this cooling, the atmospheric

CO_2 concentration should have fallen by ~20–30 ppm because of increased CO_2 solubility in seawater.

The altered salinity of the glacial ocean would also have affected atmospheric CO_2, but in the opposite direction. CO_2 dissolves more easily in seawater with a lower salinity, but the average glacial ocean was saltier than it is today because of the amount of freshwater taken from the ocean and stored in ice sheets. Although some high-latitude ocean surfaces (such as the North Atlantic) became less salty during glaciations, the average salinity of the entire ocean increased by about 1.1‰, enough to cause an estimated glacial CO_2 increase of 11 ppm.

The 11-ppm CO_2 increase caused by higher salinity would have offset just under half of the 20–30 ppm decrease caused by ocean cooling, for a net CO_2 drop of ~14 ppm. Most of the observed CO_2 decrease of 90 ppm must have occurred by means of other mechanisms.

11-7 Biological Transfer from Surface Waters

One way to move carbon from surface to deep waters is through biological activity. Photosynthesis occurs in sunlit surface waters where sufficient nutrients (phosphorous and nitrogen) are available to stimulate growth of marine phytoplankton (Figure 11-10). CO_2 is extracted

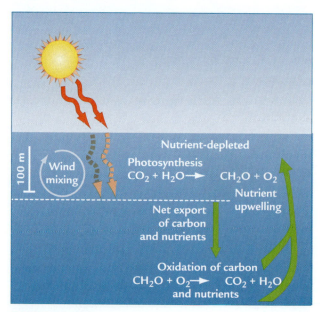

FIGURE 11-10
Photosynthesis in the ocean

Sunlight that penetrates the surface layers of the ocean causes photosynthesis in microscopic marine plankton. After the plankton die, some of their organic tissue sinks to the seafloor and is oxidized to nutrient form. Upwelling returns the nutrients (including organic carbon) to the ocean surface.

from surface waters and incorporated in organic tissue (CH_2O), some of which sinks to the deep ocean:

$$CO_2 + H_2O \rightarrow CH_2O + O_2$$

If the rate of photosynthesis increases, CO_2 concentrations in surface waters will decrease because of greater downward export of carbon held in dead organic matter. To increase this transfer, sometimes referred to as the **carbon pump**, more nutrients must become available to the plant plankton in surface waters to stimulate greater productivity. If greater productivity occurred during glacial times, the increased downward transfer of organic carbon would have reduced CO_2 values in surface waters and in the overlying atmosphere. One way to track past changes in this carbon pump is to use paired $\delta^{13}C$ measurements of foraminifera that lived in surface and deep waters (Box 11-1).

Most ocean surface waters have very small amounts of nutrients, which are quickly consumed by planktic organisms. As a result, annual productivity in many mid-ocean regions is very low in the current interglacial climate (Figure 11-11). Because a range of evidence indicates that many of these low-productivity regions remained nutrient-depleted during glaciations, increased carbon pumping did not occur in these regions.

In contrast, the deep ocean is loaded with nutrients because the nitrogen and phosphorus and carbon contained in organic matter falling from surface waters are oxidized back to mineral form (Figure 11-10). But these nutrients remain unused in the darkness of the deep sea unless they are delivered back to the surface waters by upwelling or other processes that stimulate surface productivity.

High productivity occurs today in several kinds of regions: coastal areas, where rivers and upwelling deliver nutrients; narrow regions of upwelling near the equator, especially in the Pacific Ocean; and high latitudes of the North Pacific Ocean and the Southern Ocean around Antarctica (see Figure 11-11). Large-scale changes in biological pumping of carbon to the deep sea can occur in these productive regions.

The high-latitude oceans, especially those near Antarctica, are particularly promising locations for increased carbon pumping because they contain large amounts of unused nutrients in the current interglacial climate. Solar radiation in these areas is relatively weak, and active photosynthesis is limited to a relatively brief summer season. As a result, photosynthesis does not utilize most of the available nutrients, and high concentrations persist even during the productive season. This modern excess of nutrients means that increased productivity during glacial intervals could have transferred additional carbon to deep waters compared to today, leaving surface waters with reduced CO_2 levels.

Scientists are investigating several mechanisms that might have increased glacial productivity and downward transfer of carbon. One intriguing explanation is linked to an increase in delivery of nutrients from the continents by stronger glacial winds

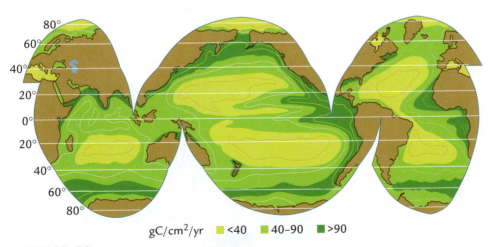

gC/cm²/yr   ■ <40   ■ 40–90   ■ >90

FIGURE 11-11

Annual carbon production in the modern surface ocean

Primary production of carbon (in grams/m²/yr) is highest in shallow coastal regions, in high-latitude oceans (especially the Southern Ocean), and across equatorial upwelling belts, but lower in central ocean gyres. (ADAPTED FROM W. H. BERGER ET AL., "OCEAN CARBON FLUX: GLOBAL MAPS OF PRIMARY PRODUCTION AND EXPORT PRODUCTION," IN *BIOGEOCHEMICAL CYCLING AND FLUXES BETWEEN THE DEEP EUPHOTIC ZONE AND OTHER OCEANIC REALMS*, NATIONAL UNDERSEA RESEARCH PROGRAM REPORT 88-1 [ASHEVILLE, NC: NOAA, 1987].)

Looking Deeper into Climate Science

Box 11-1

Using $\delta^{13}C$ to Measure Carbon Pumping

As photosynthesis of planktic organisms proceeds in sunlit ocean waters, fractionation preferentially removes the ^{12}C isotope and stores it in organic matter, which sinks toward the seafloor once the plankton die. Extraction of ^{12}C-rich carbon from the surface waters leaves the inorganic carbon enriched in ^{13}C, leaving the inorganic $\delta^{13}C$ values more positive. In contrast, as more organic carbon is oxidized back to inorganic form in the deep ocean, deep waters are enriched in ^{12}C and $\delta^{13}C$ values become more negative.

These opposing $\delta^{13}C$ trends provide a way to measure the strength of the ocean carbon pump: by analyzing changes in carbon isotopic values in foraminifera. The shells of planktic foraminifera record changes in surface-water $\delta^{13}C$ values through time, while the shells of benthic foraminifera record ongoing (and opposite) changes in deep-water $\delta^{13}C$ values. The changing difference between these two trends through time is a monitor of the changing strength of the carbon pump. If nutrient delivery to surface waters increases during glacial times, productivity and carbon pumping should also increase. The result should be an increase in the *difference* between the more positive $\delta^{13}C$ values in surface waters and the more negative values in deep waters.

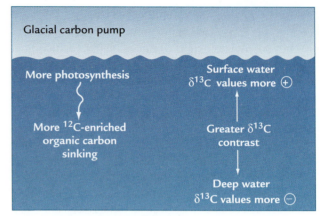

Measuring changes in the ocean carbon pump
Greater photosynthesis in surface waters during glaciations should pump more organic carbon to the deep sea and reduce atmospheric CO_2 levels. Past changes in this process can be measured by the increased difference between the ^{13}C-enriched carbon (higher $\delta^{13}C$ values) in the shells of planktic foraminifera living in surface waters and the ^{12}C-enriched carbon (lower $\delta^{13}C$ values) in the shells of benthic foraminifera living on the deep-ocean floor.

(Figure 11-12). The marine scientist John Martin proposed that iron, a trace element, is critical to marine life, as it is to humans. Because erosion of the land is the main source of iron for the oceans, he suggested that ocean regions should receive an iron "boost" if extra dust is blown in by winds. This concept is called the **iron fertilization hypothesis**.

Stronger glacial winds blowing from semiarid and arid continental areas should have carried greater amounts of iron to both coastal and mid-ocean areas than today, stimulating greater productivity over broad areas. Iron fertilization might have stimulated productivity in two regions that today have leftover nutrient excesses: the high latitudes of the North Pacific Ocean, which receive large dust influxes from central Asia; and the Atlantic sector of the Southern Ocean, which receives smaller influxes from the Patagonian tip of South America. Extra iron arriving in these two areas during glaciations could have

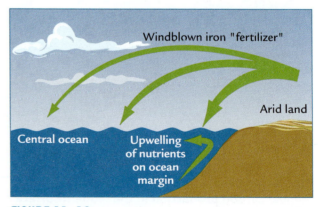

FIGURE 11-12
Iron fertilization of ocean surface waters

Dust rich in trace elements like iron is blown from continental interiors to the ocean during glaciations. The addition of this and other nutritional supplements stimulates productivity across broad regions.

stimulated greater productivity and carbon pumping to the deep ocean.

Martin's hypothesis is still being debated. Oceanic field tests have shown that adding iron to surface waters can stimulate greater short-term productivity of ocean phytoplankton. Estimates of the effect of iron fertilization on atmospheric CO_2 concentrations range all the way from negligible (a few ppm) to sizeable (several tens of ppm). Whether or not a persistent long-term increase in iron fertilization would have stimulated a large enough increase in carbon pumping and glacial productivity to alter atmospheric CO_2 concentrations remains unclear. Some scientists have also noted that dust from the land also delivers other key elements that might have stimulated ocean productivity.

11-8 Changes in Deep-Water Circulation

Another mechanism for transferring more carbon to the deep ocean is by changing the deep-ocean circulation. Today, most of the water that fills the deep oceans forms in the Southern Ocean and the subpolar North Atlantic Ocean (see Chapter 2). North Atlantic deep water flows southward through much of the mid-level Atlantic, while colder and denser water formed near Antarctica fills the Pacific and Indian ocean basins.

Evidence from $\delta^{13}C$ measurements in bottom-dwelling foraminifera indicates a different circulation pattern during glacial times. We saw earlier that the average $\delta^{13}C$ composition of the entire ocean became more negative during glacial times, but in the case considered here, our focus is on *regional $\delta^{13}C$ variations in the ocean*. Varying degrees of photosynthesis (see Box 11-1) in different polar areas of the ocean give inorganic carbon distinctively different $\delta^{13}C$ values in north polar versus south polar areas. As a result, the deep waters that form from surface waters in these regions begin their downward trip with distinctively different $\delta^{13}C$ values (Figure 11-13).

At one extreme are the relatively positive $\delta^{13}C$ values (at or above 0.8‰) in the North Atlantic Ocean near 2,000–4,000 meters depth. These values are positive because North Atlantic deep water is formed from surface waters that have been greatly enriched in ^{13}C by photosynthetic removal of ^{12}C (see Box 11-1). A plume of water with positive $\delta^{13}C$ values at depths of 2,000 to 4,000 meters defines the core of this southward flow (Figure 11-14A).

In contrast, the waters that form in the Antarctic region and flow to a wide range of depths have $\delta^{13}C$ values lower than 0.5‰. Because photosynthetic fractionation does not go to completion in the Southern Ocean, extremely positive $\delta^{13}C$ values do not develop in the surface waters that feed the deeper flow. The

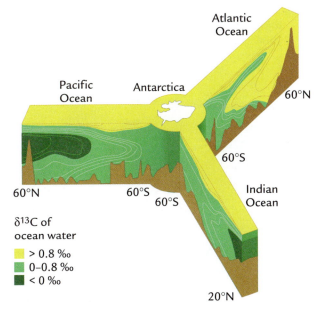

FIGURE 11-13
Modern deep-ocean $\delta^{13}C$ patterns

In today's ocean, photosynthesis and carbon isotope fractionation drive $\delta^{13}C$ values higher in surface waters compared to deep waters. (ADAPTED FROM C. D. CHARLES AND R. G. FAIRBANKS, "EVIDENCE FROM SOUTHERN OCEAN SEDIMENTS FOR THE EFFECT OF NORTH ATLANTIC DEEPWATER FLUX ON CLIMATE," *NATURE* 355 [1992]: 416–19.)

contrast between the low-$\delta^{13}C$ water coming from the Antarctic and the high-$\delta^{13}C$ water from the North Atlantic makes them relatively easy to trace.

A second process that affects the regional $\delta^{13}C$ pattern is called **$\delta^{13}C$ aging**. As deep waters flow from their source regions to other parts of the world ocean, their $\delta^{13}C$ values gradually become more negative. This gradual shift results from the continual downward rain of ^{12}C-rich carbon ($\delta^{13}C = -22‰$) from surface waters along the paths of deep-water flow (see Box 11-1).

This aging effect is evident mainly in the Pacific Ocean, where the flow of deep water is slow enough to give the water time to age significantly (see Figure 11-13). The most negative $\delta^{13}C$ values are found in the North Pacific, far from the Atlantic and Antarctic sources. In contrast, deep water circulates quickly enough through much of the Atlantic Ocean that it has little time to age. Regional variations in $\delta^{13}C$ values in the deep Atlantic are determined mainly by physical mixing of water from the North Atlantic and Antarctic sources.

During the maximum glaciation 20,000 years ago, the $\delta^{13}C$ pattern in the deep Atlantic Ocean changed considerably (see Figure 11-14B). The tongue of high-$\delta^{13}C$ (northern) source water that today lies at 2,000–4,000 meters shifted above 1,500 meters depth.

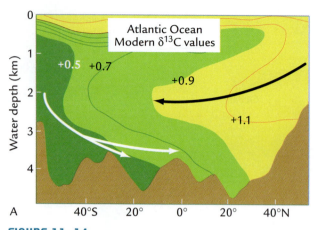

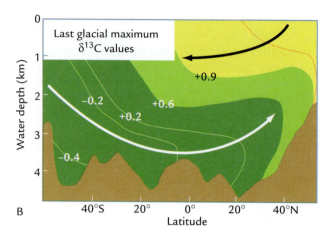

FIGURE 11-14

Change in deep Atlantic circulation during glaciation

In contrast to the modern distribution of δ^{13}C in the Atlantic Ocean (A), the axis of high-δ^{13}C water formed in the north flowed south at shallower levels during the last glacial maximum (B). (MODIFIED FROM J.-C. DUPLESSY AND E. MAIER-REIMER, "GLOBAL OCEAN CIRCULATION CHANGES," IN *GLOBAL CHANGES IN THE PERSPECTIVE OF THE PAST*, ED. J. A. EDDY AND H. OESCHGER [NEW YORK: JOHN WILEY, 1993].)

This pattern suggests that water chilled in northern latitudes during glacial times did not become dense enough to sink as deep as it does today, reaching only intermediate depths. The lower δ^{13}C values below 1,500 meters indicate the presence of a larger water mass that originated near Antarctica and flowed north.

These changes in deep circulation patterns can be traced through time by examining δ^{13}C records from benthic foraminifera in tropical Atlantic sediment cores. The equatorial Atlantic seafloor near 3,000 meters depth is an ideal place for this kind of study because it lies in the region and at the depths that were strongly affected by δ^{13}C changes between the glacial and interglacial patterns (see Figure 11-14A and B).

Measurements of δ^{13}C values from this region can be used to calculate the relative contributions of high-δ^{13}C water formed in the North Atlantic and low-δ^{13}C water formed in the Southern Ocean. Changes in δ^{13}C values in these two source areas through time can be determined by analyzing benthic foraminifera in deep-sea sediments from those regions. The δ^{13}C values measured in foraminifera from the tropical sediments can then be expressed as percentage contributions from these two sources. This analysis removes the whole-ocean δ^{13}C change caused by carbon transfers from land to sea and isolates the relative contributions from the northern and southern sources through time.

A long record from the tropical Atlantic shows oscillations in the percentages of northern and southern source water (Figure 11-15). Northern source waters generally dominated during δ^{18}O minima (interglaciations), while Antarctic source waters dominated during δ^{18}O maxima (glaciations). The

measured fluctuations in deep-water sources occurred mainly at a period near 100,000 years after 0.9 million

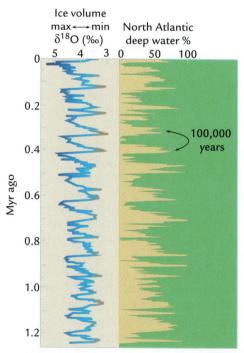

FIGURE 11-15

Changing sources of Atlantic deep water

The percentage of deep water originating in the North Atlantic and flowing to the equator during the last 1.25 Myr has been consistently lower during glaciations than during interglaciations. (ADAPTED FROM M. E. RAYMO ET AL., "THE MID-PLEISTOCENE CLIMATE TRANSITION: A DEEP-SEA CARBON ISOTOPIC PERSPECTIVE," *PALEOCEANOGRAPHY* 12 [1997]: 546–59.)

years ago. This $\delta^{13}C$ trend indicates that a link exists between the size of northern hemisphere ice sheets and the pattern of deep-water flow in the North Atlantic Ocean.

These changes in the pattern of deep-water circulation could have affected CO_2 concentrations in the atmosphere in several ways. One possibility is that stronger overturn of Antarctic source waters at intermediate depths could have delivered more nutrients to surface waters and stimulated more biological productivity and downward pumping of carbon. The carbon sent to the deep ocean could then reside there "hidden" from the atmosphere.

A second potential link to atmospheric CO_2 is through the carbonate chemistry of the ocean. Concentrations of dissolved CO_2 in the surface waters of the world ocean are linked to the amount of the carbonate ion CO_3^{-2} present. These ions are produced when corrosive deep water dissolves $CaCO_3$ on the seafloor. When the CO_3^{-2} returns to the surface in the large-scale ocean circulation, it combines chemically with dissolved CO_2 in the surface waters to produce the bicarbonate ion HCO_3^{-1}. This process removes CO_2 from surface waters. As a result, if increased amounts of $CaCO_3$ are dissolved on the seafloor, the CO_2 concentrations will be reduced in surface waters and in the overlying atmosphere.

On a global average basis, this mechanism acts slowly (over thousands of years) because of the time required to dissolve deep-sea carbonates and because of the slow overturn of the deep ocean. In addition, deep-ocean regions experienced different and often opposing changes in $CaCO_3$ dissolution during glaciations (increases in most of the Atlantic, decreases in most of the Pacific). Because these changes more or less cancelled each other out at a global scale, little if any change in the average CO_3^{-2} content of deep waters occurred.

Nevertheless, regional changes in CO_3^{-2} could still have altered atmospheric CO_2 concentrations. Today, deep water from North Atlantic sources comes to the surface in the Antarctic region (see Figure 11-14A). Because modern North Atlantic deep water is not very corrosive, it dissolves relatively little $CaCO_3$ on the seafloor and delivers relatively small concentrations of CO_3^{-2} to the surface waters of the Southern Ocean. As a result, dissolved CO_2 concentrations can remain at relatively high levels in the Southern Ocean and the overlying atmosphere.

During glaciations, however, the water that formed in the North Atlantic sank to much shallower depths and did not intersect as much of the seafloor (see Figure 11-14B). Instead, an expanded area was bathed by southern source water that was more corrosive, dissolved more $CaCO_3$, and eventually returned more CO_3^{-2} to Antarctic surface waters. The geochemists

Wallace Broecker and Tsung-Hung Peng proposed that this change in circulation would have reduced the concentration of dissolved CO_2 in the Southern Ocean, along with the amount of CO_2 in the overlying atmosphere. They estimated that this mechanism, called the **polar alkalinity hypothesis**, might explain as much as 40 ppm of the observed 90-ppm decrease in atmospheric CO_2 during glacial times. Because the deep flow in the Atlantic Ocean is relatively rapid, this mechanism could influence atmospheric CO_2 concentrations within a few hundred years.

> **In Summary**, several factors contributed to the reduction in atmospheric CO_2 concentrations during glacial intervals: reduced CO_2 solubility in colder waters, greater biological pumping of carbon from surface to deep waters, and changes in deep-ocean circulation. The size of the latter two contributions is highly uncertain, and this problem is currently an area of intense investigation.

Orbital-Scale Changes in CH₄

In contrast to the oxidized carbon common in most of Earth's environments, the carbon in methane (CH_4) occurs in reduced form in environments where oxygen is absent. Most natural methane originates in wetlands, where decaying plant matter uses up the available oxygen and creates the necessary reducing conditions. Most natural wetlands are located in the tropics and in circumarctic regions.

In the tropics, monsoon fluctuations determine the extent of wetland sources of methane (see Chapter 9, Figure 9-12). Heavy rainfall saturates the ground, reduces its ability to absorb water, and increases the amount of standing water in swampy areas that lack natural drainage outlets. Vegetation that grows and decays each summer in these wetlands uses up the oxygen dissolved in the water and creates the reducing conditions needed to generate methane (Figure 11-16). The extent of these boggy areas expands during wet monsoon maxima and shrinks during monsoon minima at the 23,000-year precession cycle.

Wetlands in circumarctic regions also emit substantial amounts of methane (Figure 11-17). In these far-northern bogs, summer warmth is the main control on methane releases, rather than the precipitation control prevailing in tropical monsoon regions. For much of the year, temperatures in the far north are too cold for plants to grow, but the warmth of the short summer season allows plant growth and

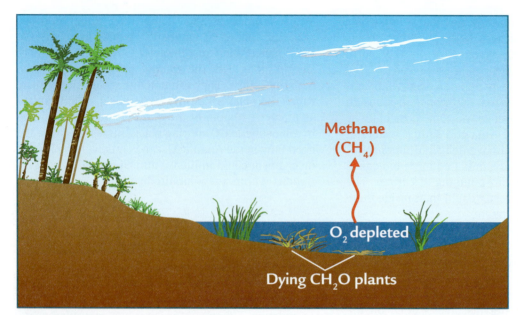

FIGURE 11-16
Tropical methane sources
Carbon-rich vegetation growing in tropical wetlands rots in oxygen-depleted water and releases CH_4 to the atmosphere.

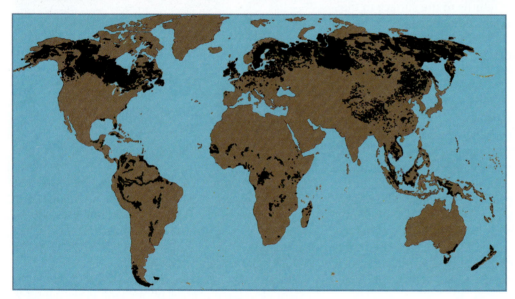

FIGURE 11-17
Arctic wetland methane sources
Large wetlands cover cold circumarctic regions. (ADAPTED FROM Z. YU ET AL., "GLOBAL PEATLAND DYNAMICS SINCE THE LAST GLACIAL MAXIMUM," *GEOPHYSICAL RESEARCH LETTERS* 37 [2010], L13402, DOI:10.1029/2010GL043584.)

methane release. Circumarctic wetlands can respond to the 23,000-year insolation heating of central Asia and to the 41,000-year chilling of the Arctic region.

Methane concentrations in ice cores show a series of cyclic variations between maxima of 650 to 700 ppb and minima of 350 to 450 ppb (Figure 11-18). The CH_4 signal shows some relationship to changes in ice volume, but not as close a link as the CO_2 record (see Figure 11-6). The strongest methane maxima generally fall at or near times of peak interglacial warmth, but large CH_4 variations also occur during times when ice volume is not large.

A strong 23,000-year CH_4 signal is evident during the last 350,000 years (see Figure 11-18). Age models based on ice flow show that the peaks in methane concentration during this interval fall very close to times of maxima in northern hemisphere summer

insolation at the 23,000-year cycle of orbital precession. Small adjustments in these age models bring the CH_4 peaks into alignment with July insolation maxima. This midsummer 23,000-year timing for the methane peaks is further supported by the 11,000- to 10,500-year age of the most recent CH_4 maximum in annually layered Greenland ice, coincident with the time of the most recent July insolation maximum.

The presence of strong 23,000-year CH_4 variations with a midsummer (July) timing is consistent with two explanations linked to orbital precession: (1) control of the size of north-tropical wetlands by summer monsoon rainfall; and (2) control of emissions from circumarctic bogs by midsummer warmth driven by insolation heating of northern continents. These explanations are further supported by the more detailed match between the size of most

CH₄ (ppb)

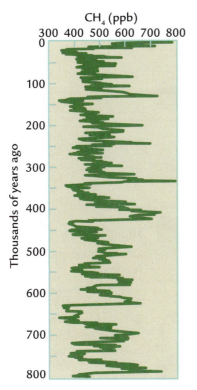

FIGURE 11-18

An 800,000-year record of atmospheric methane from Dome C, Antarctica

(ADAPTED FROM L. LOULERGUE ET AL., "ORBITAL AND MILLENNIA-SCALE FEATURES OF ATMOSPHERIC CH₄ OVER THE PAST 800,000 YEARS," *NATURE* 453 [2008]: 383–6, DOI:10.1038/NATURE06950.)

individual insolation maxima and methane peaks (both indicating eccentricity modulation of the precession signal).

Although peaks in the 23,000-year CH₄ signal are generally well matched in amplitude to the strength of summer insolation maxima, methane minima are not. Many of the minima reach more or less similar values despite the fact that the insolation minima vary widely in amplitude. These lower values in the methane trend have a truncated or "clipped" look, apparently because some tropical wetlands survive even the strongest insolation forcing toward extreme drying. Likely candidates are wetlands very near the equator, which remain in the tropical wet zone for any climate. Growth of new wetlands north of Australia in regions exposed by sea level lowering may also counter large CH₄ minima.

Prior to 350,000 years ago, the 23,000-year precession signal is much less obvious in the ice core record, and a 41,000-year signal more apparent (see Figure 11-18). Although the reason for this change is not clear, one possibility is that circumarctic wetland temperatures were more influenced by 41,000-year climatic changes in Arctic regions during this somewhat different climatic regime.

In Summary, prominent variations in methane are linked primarily to changes in strength of the summer monsoon in tropical and subtropical regions and to variable heating of Asia and its northern wetlands.

▶ Orbital-Scale Climatic Roles: Forcing or Feedback?

Because the basic timing of CO₂ and CH₄ variations during the last several hundred thousand years is known, we can attempt to assess the role these gases played in orbital-scale climatic variations, specifically those of northern hemisphere ice sheets. Milankovitch proposed that summer insolation drives ice sheets at the 41,000-year and 23,000-year cycles with a lag of approximately 5,000 years (Figure 11-19A), so the question at hand is how the greenhouse-gas changes fit this framework.

We consider here two relationships that span the likely range of possibilities. In one case, the gas concentrations respond immediately to insolation changes and acquire the same early phase as insolation. Changes in the gas concentrations then join summer insolation as an additional "early" forcing of ice sheets, and the ice sheets respond with their characteristic lag of ~5,000 years to this joint forcing (Figure 11-19B). Note that in this scenario, the gas concentrations clearly lead the ice response.

The other possibility is that the greenhouse gases do not respond directly to summer insolation changes but instead react through a positive feedback that is tightly linked to the ice sheets (Figure 11-19C). As summer insolation changes drive the ice-sheet response, the ice sheets then create conditions that control the concentrations of the greenhouse gases. As ice sheets grow, they reduce gas concentrations in the atmosphere, and the lower gas concentrations cool the climate and help the ice grow even faster. When the ice melts, the gas concentrations rise, climate warms, and ice melting accelerates. During both ice growth and melting, the gases play a positive feedback role. In this scenario, the gas concentrations lag well behind the insolation forcing but are in phase with the ice response.

Which of these possible scenarios best agrees with the available evidence? The answer to this question appears to be different at the orbital cycles of precession and tilt, and different as well for the main response of the two greenhouse gases.

The clearest example of an early greenhouse-gas forcing role is the 23,000-year methane signal, which clearly leads minima in the ice volume record

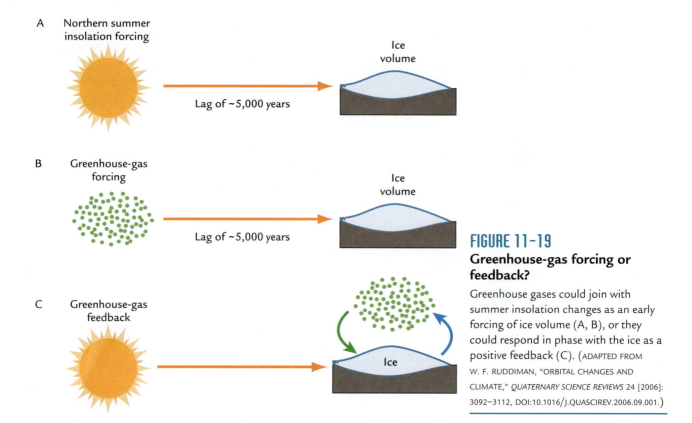

A   Northern summer insolation forcing

Ice volume

Lag of ~5,000 years

B   Greenhouse-gas forcing

Ice volume

Lag of ~5,000 years

C   Greenhouse-gas feedback

Ice

FIGURE 11-19

Greenhouse-gas forcing or feedback?

Greenhouse gases could join with summer insolation changes as an early forcing of ice volume (A, B), or they could respond in phase with the ice as a positive feedback (C). (ADAPTED FROM W. F. RUDDIMAN, "ORBITAL CHANGES AND CLIMATE," *QUATERNARY SCIENCE REVIEWS* 24 [2006]: 3092–3112, DOI:10.1016/J.QUASCIREV.2006.09.001.)

during the well-dated interval of the last 150,000 years (Figure 11-20). This early timing of methane changes is consistent with insolation forcing of tropical monsoons (and to some extent of north Asian wetlands). However, this 23,000-year signal is less evident prior to 350,000 years ago, for reasons not yet understood.

The clearest example of a late greenhouse-gas feedback role is the 41,000-year CO_2 signal, which is in phase with ice volume. For example, the strong glacial maxima that occurred near 21,000 and 63,000 years ago are both manifestations of the 41,000-year ice volume cycle, and each of these ice volume maxima corresponds to a CO_2 minimum (see Figure 11-20). This relationship suggests that the ice sheets were the ultimate control on the in-phase CO_2 changes, and that the gases provided positive feedback to the ice sheets at the 41,000-year tilt cycle.

For the prominent oscillations at a period near 100,000 years, the greenhouse-gas role is not obvious. The amount of insolation forcing at the 100,000-year eccentricity period is so small that it has a negligible effect on the ice sheets (see Chapter 10). Without any significant insolation forcing, the only timing comparison we can make for the 100,000-year signal is between changes in the ice sheets (based on the $\delta^{18}O$ record) and the gas concentrations, with the focus here on CO_2.

As is obvious in Figure 11-6, the overall shape of the $\delta^{18}O$ and CO_2 signals correlate closely, with both signals dominated by changes near a period of 100,000 years. In addition, both signals have phases at that period that fall very close to changes in eccentricity: greater eccentricity correlates with higher CO_2 and smaller ice sheets. Overall, CO_2 has a very small lead relative to changes in eccentricity, while the ice sheets have a very small lag. As a result, the gas responses lead the ice sheet changes by 1,000 to 3,000 years. Most of this relative timing is determined by the rapid CO_2 rises that occur early in the deglacial ice sheet melting (during terminations). This comparison is complicated by the fact that the $\delta^{18}O$ signal used as a proxy for ice volume also contains a sizeable temperature overprint.

This small difference in timing does not fit cleanly into the interpretation of greenhouse gases as either pure "forcing" or pure "feedback." The CH_4 lead of ~5,000 years at the 23,000-year cycle is appropriate to a forcing-and-slow-response relationship because the delay amounts to nearly one-quarter (25%) of the length of the cycle. In contrast, the CO_2 lead of ~2,000 years is only 2% of the much longer 100,000-year period. The only way that a 2,000-year lead could be consistent with a forcing-and-response relationship is if the ice sheet response at the 100,000-year period were very fast, in fact more than twice as fast as the response at the 23,000-year cycle. But it is hard to imagine why this would be the case (or how). On the other hand, the opposite possibility—that the ice sheets fully control

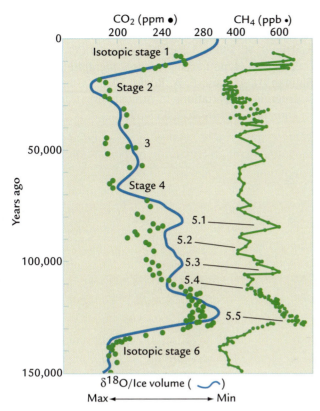

FIGURE 11-20
The last 150,000 years

The phasing of the greenhouse gases relative to ice volume at the orbital periods is evident in the changes of the last 150,000 years. (ADAPTED FROM W. F. RUDDIMAN, "ICE-DRIVEN CO_2 FEEDBACK ON ICE VOLUME," *CLIMATE OF THE PAST* 2 [2006]: 43–66.)

the gas responses at ∼100,000 years—also seems to be contradicted by the fact that the gases do lead the ice (even if just by 1,000–3,000 years).

One possible resolution of this problem is that the small lead of the greenhouse gases relative to the ice sheets at the ∼100,000-year period reflects a combination of both roles, partly early gas forcing of the slow ice response, but mainly ice control of the gas concentration with no lead or lag. Because the gases and the ice sheets are so nearly in phase at the ∼100,000-year period, the in-phase feedback role for the gases appears to be larger than the forcing role.

In Summary, the greenhouse gases act as a forcing of the ice sheets at the 23,000-year period, but as an ice-driven feedback at the 41,000-year period. The reason why the gases have such different roles at the two periods is not understood. The role of the greenhouse gases in the major oscillations near ∼100,000 years appears to be a combination of a large feedback role and a smaller forcing role.

Key Terms

ice flow model (p. 217)
sintering (p. 217)
carbon isotopes (p. 218)
phytoplankton (p. 218)
carbon pump (p. 223)

iron fertilization hypothesis (p. 224)
$\delta^{13}C$ aging (p. 225)
polar alkalinity hypothesis (p. 227)

Review Questions

1. How are ice cores dated?

2. Why are air bubbles in ice cores younger than the ice in which they are sealed?

3. What features of the ice core CH_4 signal suggest a link to tropical monsoons?

4. To what extent does a cooler glacial ocean explain lower CO_2 levels in the glacial atmosphere?

5. Where did the carbon (CO_2) removed from the atmosphere go during glaciations?

6. What effect does fractionation have on carbon isotope values?

7. Explain how carbon isotope ($\delta^{13}C$) changes trace shifts of carbon from the land to the ocean.

8. How does the oceanic carbon pump reduce CO_2 levels in the atmosphere?

9. What evidence indicates a different glacial deep-water flow pattern in the North Atlantic?

10. How could a change in deep-water circulation in the Atlantic Ocean alter atmospheric CO_2?

11. What characteristics of $\delta^{18}O$ and greenhouse-gas records tell us whether the gases act as ice sheet forcing or feedback?

Additional Resources

Basic Reading

EPICA Community Members. 2004. "Eight Glacial Cycles from an Antarctic Ice Core." *Nature* 429: 623–28.

Advanced Reading

Broecker, W. S., and T.-H. Peng. 1989. "The Cause of the Glacial to Interglacial Atmospheric CO_2 Change: A Polar Alkalinity Hypothesis." *Global Biogeochemical Cycles* 3: 215–39.

Kohfeld, K., C. Le Quéré, S. P. Harrison, and R. F. Anderson. 2005. "Role of Marine Biology in Glacial-Interglacial Cycles." *Science* 308: 74–78.

Loulergue, L., et al. 2008. "Orbital and Millennial-Scale Features of Atmospheric CH_4 over the Past 800,000 Years." *Nature* 453: 383–86. doi:10.1038/nature06950.

Luthi, D., et al. 2008. "High-Resolution Carbon Dioxide Concentration Record 650,000–800,000 Years Before Present." *Nature* 453: 379–82. doi:10.1038/nature06949.

Martin, J. H. 1990. "Glacial-Interglacial CO_2 Change: The Iron Hypothesis." *Paleoceanography* 5:1–13.

Sigman, D. M., and E. A. Boyle. 2000. "Glacial/Interglacial Variations in Atmospheric Carbon Dioxide." *Nature* 407: 859–69.

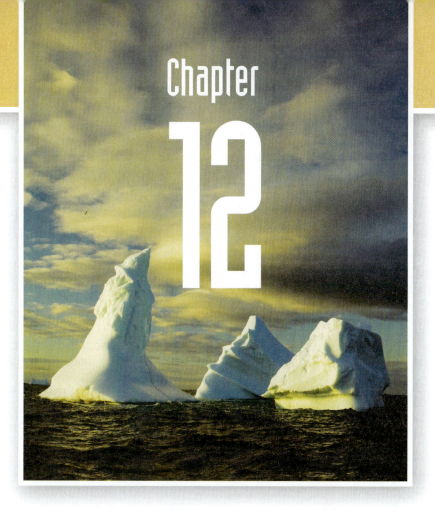

Chapter

12

Orbital-Scale Interactions, Feedbacks, and Unsolved Mysteries

The previous chapters in this section described insolation forcing at orbital time scales and the ways in which it drives responses by Earth's monsoons, ice sheets, and greenhouse gases. This chapter assesses the larger picture of their regional interactions. Did the responses remain regional in extent or did they send their climatic signatures far afield, and if so, what were the key mechanisms? As noted in Chapter 10, the major unsolved mystery of the last 2.75 million years is the behavior of the northern ice sheets. Although a wide range of evidence supports the general validity of the Milankovitch theory, a major question has not been fully answered: Why did the ice sheets vary at rhythms that do not appear to have matched the tempos of the insolation forcing?

Orbital-Scale Climatic Interactions

Chapters 9 and 10 identified two major parts of the climate system that play major roles at orbital time scales: monsoons and northern hemisphere ice sheets.

Monsoon variations (see Chapter 9) are largest in the Northern Hemisphere because the large northern landmasses respond strongly to insolation forcing, and because of the high topography in southern Asia. Lake levels in Africa and accurately dated $\delta^{18}O$ signals in $CaCO_3$ deposits in caves from China and India show that the monsoonal response at low northern latitudes has been mainly driven by 23,000-year insolation forcing. The deposition of Mediterranean sapropels at the 23,000-year period over at least the last 15 million years points to the long-term persistence of low-latitude insolation forcing of North African monsoons, as do lake-level variations revealed by sediment sequences lying between well-dated basalt deposits in East Africa (see Chapter 9, Box 9-1).

The geographic reach of northern monsoons is limited to regions of strong summer heating at tropical to subtropical latitudes. Similarly, monsoon changes in the Southern Hemisphere (South America) react to the oppositely phased summer insolation heating of lower latitudes in that hemisphere. The northern monsoon signal at orbital time scales does not reach across the equator, nor does the southern one reach north.

One monsoon-related response does have a global reach: the 23,000-year component of the CH_4 trend, which has a phase that indicates a strong influence of the northern summer monsoons. But the effect of these 23,000-year CH_4 changes on global climate is not very large because CH_4 changes represent a minor component of total greenhouse forcing.

Ice sheet changes (see Chapter 10) have drawn far more attention from scientists than monsoons. Milutin Milankovitch's theory correctly predicted two characteristics of ice responses to orbital variations that have been confirmed in recent decades: (1) the presence of ice volume changes at the major periods of orbital variations; and (2) ice volume changes with a timing consistent with northern hemisphere summer insolation as the critical pace setter. Despite this evidence in favor of Milankovitch's hypothesis, however, the tempo of the ice sheet variations does not match the rhythms of the insolation changes (Figure 12-1).

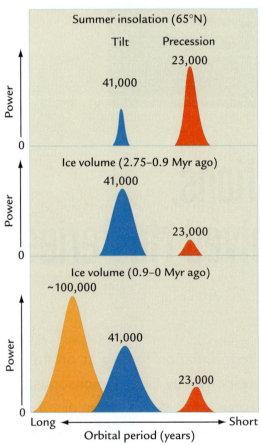

FIGURE 12-1
Spectral analysis: insolation and ice volume

Summer insolation changes at 65°N are strongest at the 23,000-year period, weaker at 41,000 years, and negligible at 100,000 years (top). The $\delta^{18}O$ signal of northern hemisphere ice volume (and deep-ocean temperature) between 2.75 and 0.9 Myr ago varied mainly at 41,000 years with little response at 23,000 years (middle). Since 0.9 Myr ago, a $\delta^{18}O$ signal at or near 100,000 years has been dominant (bottom).

12-1 Climatic Responses Driven by the Ice Sheets

As ice sheets respond to summer insolation forcing, changes in their size produce a variety of **ice-driven responses** that propagate widely through the climate system. Several of these processes work at the regional scale of high northern latitudes (Figure 12-2). Albedo-temperature feedback from ice sheets is a particularly important mechanism (see Chapter 2). Because bright ice surfaces reflect much more incoming sunlight than darker ice-free land, the loss of solar heating in regions covered by ice sheets cools the air not just above the ice sheets but also in nearby areas. A related factor is the height of the ice sheets. They protrude thousands of meters into the air and form massive obstacles to the free circulation of winds in the lower atmosphere, thereby rearranging the flow of air at middle and high latitudes.

Climate scientists have run sensitivity test experiments to isolate these effects by inserting ice sheets

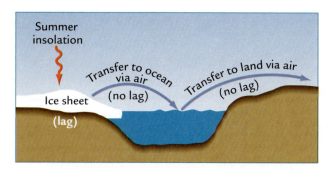

FIGURE 12-2
Ice-driven responses
Orbital-scale ice sheet rhythms may be quickly transferred to other parts of the climate system via the atmosphere and ocean. (ADAPTED FROM W. F. RUDDIMAN, "NORTHERN OCEANS," IN *NORTH AMERICA AND ADJACENT OCEANS DURING THE LAST DEGLACIATION*, ED. W. F. RUDDIMAN AND H. E. WRIGHT, GEOLOGICAL SOCIETY OF AMERICA DNAG VOL. K-3 [BOULDER, CO: GEOLOGICAL SOCIETY OF AMERICA, 1987].)

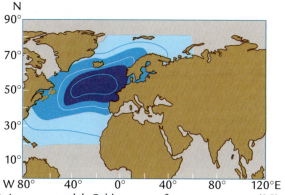

A Input to model: Colder sea-surface temperatures (°C)

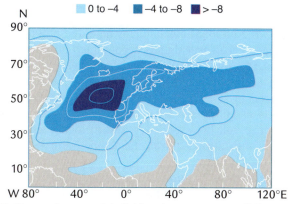

B Output from model: Colder air temperatures (°C)

FIGURE 12-3
Surface-ocean sensitivity test
Sensitivity tests with a GCM show that inserting cold ocean temperatures into an interglacial world (A) produces colder air temperatures over Europe and Asia (B). (ADAPTED FROM D. RIND ET AL., "THE IMPACT OF COLD NORTH ATLANTIC SEA SURFACE TEMPERATURES ON CLIMATE: IMPLICATIONS FOR THE YOUNGER DRYAS," *CLIMATE DYNAMICS* 1 [1986]: 3-33.)

into general circulation models (GCMs) that otherwise have modern boundary conditions. These simulations show that the ice sheet over North America has a large influence on nearby ocean temperatures. The colder temperatures over the ice sheet, combined with a clockwise flow of winds around the central ice dome, sends very cold winds blowing southeastward over the western North Atlantic Ocean.

As the ice sheets cool the North Atlantic Ocean, the climatic effects from the chilled sea surface project farther eastward into Europe. Additional GCM sensitivity tests have been run to evaluate how changes in North Atlantic temperature alter Europe's climate. In these experiments, the only change in boundary-condition input was a reduction of the temperature of the North Atlantic Ocean to near-glacial values north of 20°N (Figure 12-3A). The rest of Earth's surface was left in its current interglacial state, with no ice sheets present except the small one now on Greenland.

The model simulation shows that the cold North Atlantic sea surface sends very cold air into the western maritime parts of Europe, with some cooling reaching as far east as the interior of Eurasia (Figure 12-3B). Precipitation in these chilled regions also falls significantly because the colder ocean gives off less water vapor to the atmosphere. This experiment indicates that an ice sheet signal initially transferred to the North Atlantic will then be transferred into Europe and even parts of Asia.

Albedo-temperature feedback can add to these downwind climatic effects from the ice sheets. If temperatures fall enough to allow sea ice to cover the Norwegian Sea and the southeastern margin of the Arctic Ocean, northern Eurasia will experience very harsh winters because the Siberian high-pressure cell

strengthens. Even in the current interglacial warmth, the Siberian high is the winter season center of cooling and a powerful source of cold dry winds across northern Asia (see Chapter 2). In a world with a completely ice-covered Arctic and Norwegian Sea, the loss of the moderating influence of ocean moisture and temperature would greatly strengthen the bitter cold in this high-pressure cell.

These model experiments and other evidence indicate that ice sheets have a large downstream climatic impact across a broad area of the Northern Hemisphere at high and middle latitudes (Figure 12-4). The effects reach farthest south during winter because of very strong wind flow during that season. In summer, wind strength drops and local heating of the land by the Sun becomes more important. These ice-driven circulation changes are also propagated to lower latitudes. The intertropical convergence zone (ITCZ) can be displaced

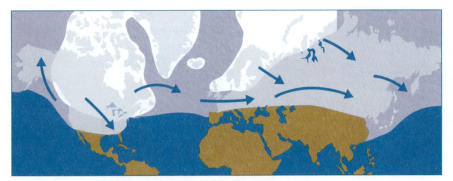

FIGURE 12-4
Regions of ice-driven responses
High and middle latitudes of the Northern Hemisphere show evidence of climate responses controlled by changes in the sizes of ice sheets.

southward, especially when the North Atlantic Ocean is covered by winter sea ice south to unusually low latitudes.

Two additional mechanisms have the potential to propagate ice sheet signals beyond high northern latitudes. Storing ocean water in the ice sheets causes sea level to drop and exposes more land in regions now covered by a few tens of meters of water. Regional effects of falling sea level will vary, but the general response will be a tendency toward locally drier climates with stronger seasonal temperature extremes (see Chapter 6).

A far more important mechanism, and one with a fully global reach, is the changes in CO_2 concentrations that accompany ice sheet growth and melting. The ice and CO_2 signals have varied nearly in tandem for the entire 800,000 years of available ice core CO_2 measurements (see Chapter 11, Figure 11-6), and this nearly in-phase timing indicates the existence of a positive feedback relationship. Because CO_2 is a major greenhouse gas, it provides a means to transfer the ice sheet signal across the entire planet.

Because the rhythms of insolation forcing do not match the tempo of ice sheet responses (see Figure 12-1), most scientists infer that processes internal to the climate system must account for the mismatches. These processes are thought to amplify relatively small amounts of insolation forcing at particular orbital periods by means of positive feedbacks. The ice sheet interactions and feedbacks listed above are good candidates for internal mechanisms to explain the mismatches.

The Mystery of the 41,000-Year Glacial World

The evidence for 41,000-year variations in ice sheets during the interval between 2.75 and 0.9 million years ago is based on variations in marine $\delta^{18}O$ values in benthic foraminifera (see Chapter 10, Figure 10-13). Temperature overprints on the $\delta^{18}O$ signal are thought to account for perhaps half of the variations during

that time, yet ice volume still varied mainly at the 41,000-year cycle. But this tempo does not match the relatively strong insolation forcing at 23,000 years (see Figure 12-1). Three explanations have been proposed for this mismatch.

12-2 Explanation 1: Insolation Varied Mainly at 41,000 Years

For the last few decades, many climate scientists have been using an index of insolation changes based on values during summer months (usually June) or even days (the June 21 solstice) for comparison against the ice sheet responses. Because these monthly or daily indices show insolation changing almost entirely at the 23,000-year period, they suggest a nearly total mismatch between the 41,000-year ice volume signal and the 23,000-year insolation signal.

This mismatch is not as large as it would have been if climate scientists had followed Milankovitch's preference. He chose a summer half-year index that integrates the total insolation over those 182 days of the year for which insolation levels are higher than the other 182 days of the year. The specific 182 calendar days in each caloric season shift slowly over time because of the extra insolation "boosts" introduced at different times of the year by Earth's precession (see Chapter 8, Figure 8-18). Milankovitch's index tacitly assumed that the 182 days of highest ("summer") insolation would cover the part of the year when most ice melting was occurring in a regime of strong ablation. As shown in the upper panel of Figure 12-1, Milankovitch's insolation index has significant 41,000-year power, although still less than at 23,000 years. By switching to the more convenient (more easily calculated) 23,000-year monthly insolation index, later scientists inadvertently overemphasized the insolation/ice mismatch.

The atmospheric scientist Peter Huybers proposed a different solution for the mismatch. He acknowledged that large variations in the amplitude of insolation changes occur at the 23,000-year period during summer, but he claimed that these changes are cancelled by reductions in the length of the summer season.

This explanation originates with work centuries ago by the astronomer Johannes Kepler. Kepler's second "law" states that planetary bodies moving in an elliptical orbit vary in angular speed with their distance from the Sun. When Earth is close to the Sun (at perihelion), it moves faster than it does at other times in the orbit. When the eccentricity of Earth's orbit is unusually high, Earth's precessional motion brings it even closer to the Sun at perihelion during the summer season, and it moves even faster. At such times, the length of the summer season is reduced by Earth's greater speed.

Huybers proposed that the net effect of Koepler's law is that higher-than-normal levels of insolation caused by Earth being unusually close to the Sun at perihelion are offset by the shorter-than-normal length of the summer season. The result is that the extra insolation at the 23,000-year cycle is cancelled by the shortening of the season (Figure 12-5). If this explanation is correct, no mismatch may actually exist between the key insolation forcing and the ice sheet responses, with both varying mainly at 41,000 years.

All choices of an insolation index require simplifying assumptions. Huyber's index assumes that only insolation changes above an arbitrary threshold value drive ice ablation. The other indices—both the monthly mean values and Milankovitch's integrated summer index—also require arbitrary choices of the season over which summer insolation is summed.

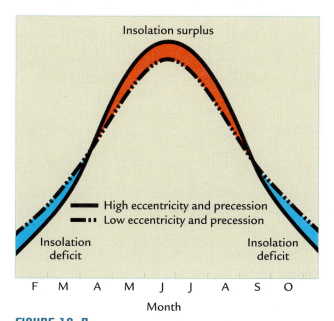

FIGURE 12-5

Intensity versus length of season

Variations in the intensity of summer insolation at the 23,000-year precession cycle are balanced by changes in the duration of the summer season.

One complicating issue is the fact that a 23,000-year ice response is clearly evident in the $\delta^{18}O$ (and coral reef) signals during the last several hundred thousand years (see Figure 12-1; also see Chapter 10). This evidence poses a challenge to Huyber's proposed explanation: Why would the 23,000-year signal be almost completely absent from 2.75 to 0.9 million years ago, yet clearly present since then? In any case, Huyber's findings suggest that at least part of the apparent mismatch between insolation and ice volume prior to 0.9 million years ago may well be an artifact of an incorrect choice of insolation index.

12-3 Explanation 2: Antarctic Ice Changes at 23,000 Years Cancel Northern Ones

The marine geologist Maureen Raymo proposed another explanation for the insolation/ice mismatch between 2.75 and 0.9 million years ago. She claimed that northern hemisphere ice sheets actually have responded to insolation forcing with a moderately strong 23,000-year signal, but that this response was cancelled out in the globally averaged signal by a 23,000-year response of Antarctic ice with opposite timing.

From a seasonal point of view, changes in precession occur at opposite times in the northern and southern hemispheres (see Chapter 8). For example, the Northern Hemisphere is currently very near a summer insolation minimum because Earth is in its aphelion (distant pass) position at that time of year. But, six months later in early January, which is summer in the Southern Hemisphere, Earth has moved into its perihelion (close pass) position. As a result, the Southern Hemisphere is currently very near a 23,000-year summer insolation maximum, exactly opposite the configuration in the Northern Hemisphere. Through time, the two hemispheres maintain these exactly opposite seasonal insolation trajectories at the 23,000-year cycle.

Raymo proposed that these oppositely phased insolation changes at the 23,000-year precession cycle drove similarly opposed ice volume responses in the two hemispheres between 2.75 and 0.9 million years ago (Figure 12-6). When northern ice sheets were growing, Antarctic ice was shrinking. As a result, the effects of the northern and southern ice sheets on the global $\delta^{18}O$ signal were directly opposed. In addition, the average $\delta^{18}O$ composition of the Antarctic ice sheet was considerably more negative (−50 to −55‰) than that of the northern ice sheets (−30 to −35‰, based on present Greenland ice). As a result, changes in the volume of Antarctic ice at 23,000 years only had to be about 60% as large as those in the Northern Hemisphere to cancel them out in the global average value recorded in marine $\delta^{18}O$ records (32‰ divided by 52‰).

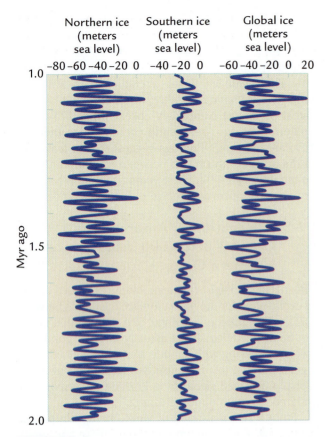

FIGURE 12-6
Interhemispheric cancellation

The response of northern hemisphere ice sheets to summer insolation forcing at the 23,000-year cycle may have been offset by a response of Antarctic ice to summer insolation with opposite timing.

A key issue for this hypothesis is whether or not the large East Antarctic ice sheet would have responded to local orbital forcing during the interval between 2.75 and 0.9 million years ago. Even during the interglacial climate of the last several thousand years, Antarctica has remained deeply refrigerated, with temperatures barely reaching the freezing point around the margins of the continent even in midsummer. In this deep-frozen world, most of the weak incoming insolation is reflected by the high-albedo ice surface. But according to Raymo's hypothesis, the East Antarctic ice sheet did not extend all the way to the coast between 2.75 and 0.9 million years ago, leaving ice-free land exposed to summer insolation heating at both the 41,000-year and 23,000-year cycles. The issue of East Antarctic ice sheet vulnerability to 23,000-year insolation forcing during that interval is unresolved.

One problem with this explanation comes from sediment cores in the nearby North Atlantic Ocean. During the interval from 1.5 to 1.2 million years ago, sea-surface temperature changes derived from assemblages of planktic foraminifera correlate peak for peak with the $\delta^{18}O$ signal (Figure 12-7). Because sea-surface temperatures in this region are one of the key ice-driven responses (see Figures 12-2 and 12-4), the dominance of the 41,000-year signal in the temperature record suggests that the northern ice sheets did vary mainly at the 41,000-year tempo, as indicated by the $\delta^{18}O$ trend.

12-4 Explanation 3: Positive CO₂ Feedback at 41,000 Years

As noted earlier, the $\delta^{18}O$ and CO_2 signals correlate closely for the 800,000 years for which both are available (see Chapter 11, Figure 11-6). This close correlation suggests that a positive feedback relationship exists between ice volume and CO_2 levels.

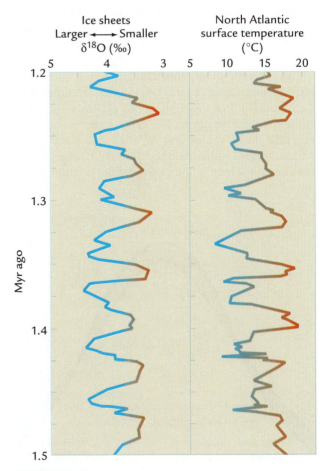

FIGURE 12-7
North Atlantic surface response to ice

Climate signals in a North Atlantic sediment core show that sea-surface temperatures between 1.5 and 1.2 Myr ago closely tracked $\delta^{18}O$ (~ice volume) fluctuations. (ADAPTED FROM W. F. RUDDIMAN ET AL., "PLEISTOCENE EVOLUTION: NORTHERN HEMISPHERE ICE SHEETS AND NORTH ATLANTIC OCEAN CLIMATE," *PALEOCEANOGRAPHY* 4 [1989]: 353–412.)

Although no ice cores have yet reached back to the 41,000-year glacial regime, this same relationship seems likely to have prevailed during the 2.75–0.9-million-year interval. A conceptual model of this relationship shown in Figure 12-8 uses Milankovitch's summer half-year insolation index, in which insolation variations at the 23,000-year cycle are larger than those at the 41,000-year cycle. Because the ice sheets have almost twice as long to grow at the 41,000-year cycle as they do at the 23,000-year cycle (41,000/23,000 = 1.8), the 41,000-year component of the ice volume response to insolation forcing gains in relative strength. If positive CO_2 feedback is arbitrarily assumed to cause a doubling in strength of the insolation-driven ice response at 41,000 years, the 41,000-year component would then become larger than the ice response at 23,000 years (which receives no positive feedback from CO_2). As in the "small glaciation" phase (see Chapter 10), ice sheets would appear during summer insolation minima (mainly at the 41,000-year cycle), but melt away during the next insolation maximum.

One problem with this explanation is that climate scientists do not know how exactly to weigh the relative effects on ice sheets of changes in insolation compared to changes in greenhouse gases. Expressed in units of W/m^2, the relative heating effects of the insolation changes are much larger than those of the greenhouse gases. On the other hand, the greenhouse-gas changes persist throughout the year, whereas the insolation changes trend in opposite directions during summer and winter. A second problem is that the reason CO_2 acts as an ice-driven feedback at the 41,000-year cycle, but not at the 23,000-year signal, is not known.

In Summary, the reason for the dominance of the 41,000-year signal in $\delta^{18}O$ (and presumably ice volume) variations prior to 0.9 million years ago is not fully resolved. As summarized above, three explanations are now under consideration, and each may have contributed part of the answer.

The Mystery of the ~100,000-Year Glacial World

After 0.9 million years ago, the second great mystery of the ice ages emerged: fluctuations centered near a period of 100,000 years. Both $\delta^{18}O$ trends and coral reef (sea level) positions confirm that ice sheets fluctuated at this tempo (see Chapter 10). The central problem in this mystery is that insolation forcing provides almost no direct forcing at this period. Orbital eccentricity does vary at several periods that average roughly 100,000 years, but these act mainly to modulate variations at the precession period rather than as direct forcing (see Chapter 8).

Climatic changes in regions of proximal ice-driven responses (see Figures 12-2 through 12-4) show strong 100,000-year power, along with the characteristic sawtoothed shape of slow cooling and then rapid warming (terminations). Changes in sea-surface temperature in the North Atlantic reconstructed from assemblages of planktic foraminifera (Figure 12-9) track the 100,000-year $\delta^{18}O$ (ice volume) signal, just as they did during the prior interval of 41,000-year glacial changes. Climatic oscillations at or near a period of 100,000 years can also be traced into nearby Europe. Pollen from European lakes vary between assemblages of diverse trees adapted to warmth and abundant moisture, such as those in present European forests, and assemblages of grasses and herbs that indicate much drier and colder conditions during glacial intervals. Both the sea-surface temperature and pollen records show the characteristic deglacial terminations that bring the slowly intensifying glacial climates to an abrupt end.

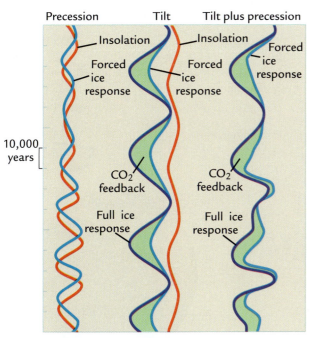

Precession   Tilt   Tilt plus precession

Insolation   Insolation
Forced ice response   Forced ice response   Forced ice response

10,000 years

CO_2 feedback CO_2 feedback

Full ice response   Full ice response

Minimum ← Insolation → Maximum
Maximum ← Ice volume → Minimum

FIGURE 12-8
CO_2 feedback
The ice volume response to insolation forcing at the 41,000-year cycle may have been amplified by CO_2 feedback. (ADAPTED FROM W. F. RUDDIMAN, "ICE-DRIVEN CO_2 FEEDBACK ON ICE VOLUME," *CLIMATE OF THE PAST* 2 [2006]: 43–78.)

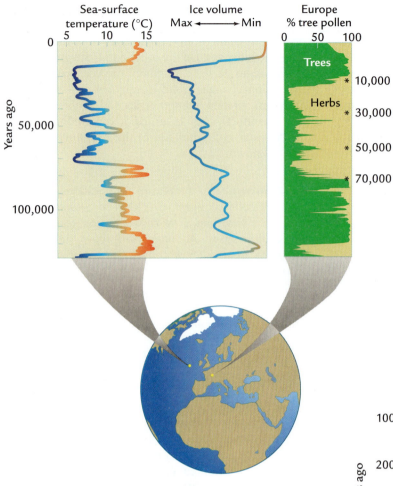

FIGURE 12-9
North Atlantic sea-surface temperature and European vegetation

Variations in $\delta^{18}O$ (~ice volume) are similar to changes in North Atlantic sea-surface temperature (estimated from planktic foraminifera) and European vegetation (based on pollen assemblages). (MIDDLE: ADAPTED FROM D. MARTINSON ET AL., "AGE DATING AND THE ORBITAL THEORY OF THE ICE AGES: DEVELOPMENT OF A HIGH-RESOLUTION 0 TO 300,000-YEAR CHRONOSTRATIGRAPHY," *QUATERNARY RESEARCH* 27 [1987]: 1–29. LEFT AND RIGHT: ADAPTED FROM C. SANCETTA ET AL., "CLIMATIC RECORD OF THE PAST 130,000 YEARS IN THE NORTH ATLANTIC DEEP-SEA CORE V23-82: CORRELATION WITH THE TERRESTRIAL RECORD," *QUATERNARY RESEARCH* 3 [1973]: 110–16.)

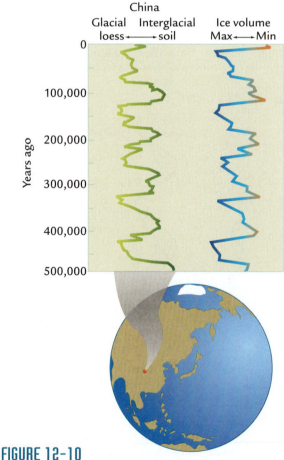

Signals that look similar can be found far to the east in the loess plateau of southern China. During the last several hundred thousand years, interglacial soils that developed in relatively warm moist conditions and windblown glacial loess that were deposited in cold dry windy climates have alternated at a period near 100,000 years (Figure 12-10). The correlation to the $\delta^{18}O$ (ice volume) record suggests that these loess/soil variations were ice-driven. This ice-driven imprint from Asia can even be tracked to the Greenland ice sheet, where dust particles in glacial-age layers come mainly from north-central Asia, where they were lifted by strong winds and blown eastward at jet stream elevations.

In Summary, large oscillations in ice volume at a period near 100,000 years transferred an ice sheet signal across high and middle latitudes of the Northern Hemisphere. These ice-driven responses resulted mainly from the chilling effect of the high-albedo ice sheet surfaces and from altered wind patterns caused by rerouting of atmospheric flow around the topographic obstacle of the ice sheets.

FIGURE 12-10
Responses of windblown debris in Southeast Asia to ice volume

Alternating layers of windblown loess and soils in Southeast Asia match variations in $\delta^{18}O$ (~ice volume). (ADAPTED FROM G. KUKLA ET AL., "PLEISTOCENE CLIMATES IN CHINA DATED BY MAGNETIC SUSCEPTIBILITY," *GEOLOGY* 16 [1988]: 811–14.)

12-5 How is the Northern Ice Signal Transferred South?

General circulation model experiments show that most of the tropics and all of the Southern Hemisphere lie beyond the reach of the ice sheet effects on temperature and rearrangements of wind circulation. Yet many climatic records from far-distant regions show climatic responses that are surprisingly similar to those near the ice sheets.

A long sediment core from a lake in the eastern Colombian Andes of South America shows cyclic alternations between pollen produced by trees and pollen from high mountain grasslands (Figure 12-11). Dating of this record by radiocarbon analysis and the presence of volcanic ash layers indicate that the major pollen fluctuations occurred at a period near 100,000 years. Tree pollen increased during interglaciations, while grass pollen increased during glaciations, the same pattern shown by records in Europe. Similarly, a marine sediment core from east of New Zealand contains a history of variations between tree pollen and mountain grassland types (Figure 12-12). In this case, because both the marine $\delta^{18}O$ signal and the terrestrial pollen signal are recorded in the same core, correlating them is not a problem. Both of these southern pollen responses resemble the northern ice sheet signal.

This persistence of a sawtoothed oscillation at a ~100,000-year period extending well into the Southern Hemisphere is a surprise. These responses cannot be "ice-driven," at least not in the sense of being created by changes in northern sea ice or atmospheric winds. Nor can they be caused by the effects of northern hemisphere changes in displacing the intertropical convergence zone. Pushing the ITCZ south would produce a southern signal reversed in sense from that in the north, with the warmth of the southern ITCZ margin pushed south into cooler middle latitudes. Instead, the southern pollen signals in Figures 12-11 and 12-12 have the same sense of timing as those in the north (for example, the rapid warmings occur at the same time in both regions).

The sawtoothed oscillations are also unlikely to have been produced by fluctuations in the Antarctic ice sheet, which covers 97% of the continent even during the present warm interglacial interval and cannot have covered a much larger area during glacial climates. During glaciations, sea level fell because of ocean water stored in the northern ice sheets, and Antarctic ice expanded out to the newly exposed continental margins, but the resulting difference in extent of the ice sheet was not large.

Two mechanisms remain for transferring a signal of northern origin to the south. Changes in sea

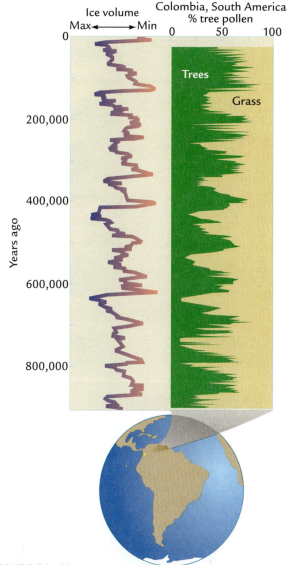

FIGURE 12-11
Vegetation response in South America

A long lake core from the eastern Andes Mountains in Colombia shows major shifts between forest and grassland pollen that match 100,000-year glacial-interglacial ice volume changes in the Northern Hemisphere. (ADAPTED FROM H. HOOGHIEMSTRA ET AL., "FREQUENCY SPECTRA AND PALEOCLIMATIC VARIABILITY OF THE HIGH-RESOLUTION 30–1450 KYR FUNZA I POLLEN RECORD," *QUATERNARY SCIENCE REVIEWS* 12 [1993]: 141–56.)

level tied to storage and release of ocean water in the northern ice sheets can affect climate in coastal regions that are alternately flooded and exposed by vertical movements of shallow seas (see Chapter 6). Climates in and near such areas may vary between relatively temperate maritime conditions as the ocean rises and harsher, more continental environments as it falls. The problem with this explanation is that many regions that show sawtoothed oscillations are located at high altitudes well into continental

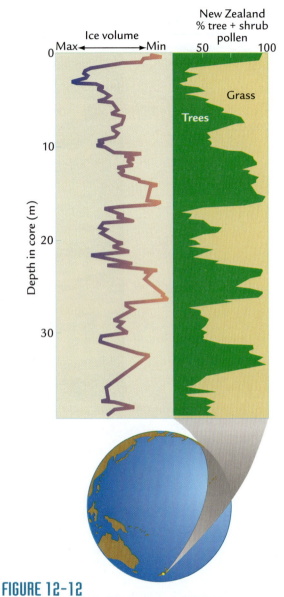

FIGURE 12-12
Vegetation response in New Zealand

A marine sediment core from the east coast of New Zealand shows major 100,000-year shifts between forest and grassland pollen that match glacial-interglacial ice volume ($\delta^{18}O$) signals. (ADAPTED FROM L. E. HEUSSER AND G. VAN DER GEER, "DIRECT CORRELATION OF TERRESTRIAL AND MARINE PALEOCLIMATIC RECORDS FROM FOUR GLACIAL-INTERGLACIAL CYCLES—DSDP SITE 594, SOUTHWEST PACIFIC," *QUATERNARY SCIENCE REVIEWS* 13 [1994]: 275-82.)

reduction that occurred during the last glacial maximum (combined with the 50% reduction in methane) would have cooled the Southern Hemisphere, especially in south polar regions where changes in sea ice amplify climatic responses (Figure 12-13). Changes in greenhouse-gas concentrations would also have affected interior regions and mountainous areas of continents at lower and middle latitudes.

The relative timing of the changes in CO_2 and ice volume at the ~100,000-year period is consistent with a positive feedback link between the two (see Chapter 11). The phasing is so nearly the same that it argues against CO_2 forcing of slow-responding ice sheets (Box 12-1). The small (~2,000-year) CO_2 lead indicates that the situation is not quite this simple, and that some minor CO_2 forcing of ice sheets probably occurred. But the overall similarity in timing of the gas and ice signals suggests that the main relationship is one of positive feedback.

Ice sheets could have altered CO_2 concentrations by any or all of several mechanisms (see Chapter 11). If strong winds picked up dust formed along the ice margins or in regions of cold, dry, ice-driven climates such as central Asia and blew it out to the ocean, fertilization by iron and other trace elements could have strengthened the carbon pump mechanism and reduced atmospheric CO_2 concentrations. If ice sheets caused a decrease in the depth of sinking of North Atlantic deep water, the resulting drop in the CO_3^{-2} ion concentration in south polar waters could have caused a CO_2 decrease. And if ice sheet growth helped to cool deep-water temperatures, the CO_2 solubility mechanism would have reduced atmospheric CO_2 concentrations. Because the evidence examined to this point suggests that ice sheets did drive dust

interiors (for example, the record from the Andes in Figure 12-11). Climate in such regions would not be highly sensitive to sea level changes in distant coastal regions.

A far more plausible way to link the two hemispheres is through variations in greenhouse gases (primarily CO_2). Sensitivity test experiments with general circulation models suggest that the 30% CO_2

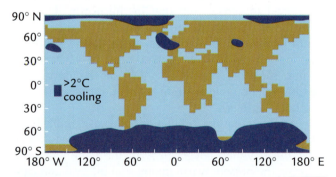

FIGURE 12-13
Southern hemisphere response to CO_2 changes

GCM experiments show that the Southern Hemisphere responds to lower CO_2 levels during glacial times, especially at higher latitudes. (ADAPTED FROM A. J. BROCCOLI AND S. MANABE, "THE INFLUENCE OF CONTINENTAL ICE, ATMOSPHERIC CO_2, AND LAND ALBEDO ON THE CLIMATE OF THE LAST GLACIAL MAXIMUM," *CLIMATE DYNAMICS* 1 [1987]: 87-100.)

Looking Deeper into Climate Science

Box 12-1

The Link Between Forcing and the Time Constants of Ice Response

The relationship between the forcing of ice sheets and their response can be quantified by the following equation, provided that the variations under investigation are sine waves.

$$\varphi = \arctan 2\pi f T$$

where arctan means "that angle whose tangent is . . .," φ is the lag of the ice response behind the forcing in degrees (out of a full 360° circle), f is the frequency under investigation, in 1/years (f being the inverse of the period), and T is the response time (the time constant) of the ice sheets, in years.

If φ is known, the equation can be used to solve for T, and conversely. For example, consider the case of summer insolation forcing of ice sheets at the 41,000-year tilt cycle. Ice sheets are thought to have a time constant T of response somewhere in the range of 5,000 to 15,000 years. If we assume a mid-range value of 10,000 years for T, and we use $f = 1/41,000$ years ($= 0.000024$), then

$$\varphi = \arctan 2 \times 3.14 \times 0.000024 \times 10,000$$

So: $\varphi = \arctan 1.51$

Geometry tables show that the angle whose tangent is 1.51 is 56.5°. This angle, expressed as a portion of a full 360° circle, needs to be converted to the number of years as a portion of a full 41,000-year cycle:

$$56.5°/360° \times 41,000 = 6,400 \text{ years}$$

The estimated lag of ice sheets with a 10,000-year time constant behind summer insolation forcing at the 41,000-year cycle is thus 6,400 years. This value is close to the observed lag of the 41,000-year component of the $\delta^{18}O$ (~ice volume) signal behind the summer insolation forcing at 41,000 years.

---

concentrations in the Northern Hemisphere (see Figure 12-10) and also affected the flow of North Atlantic deep water (see Figure 11-15), both of these mechanisms are plausible.

To this point, we have established that ice sheets varied at or near a 100,000-year tempo, that they sent that signal across high latitudes of the Northern Hemisphere by ice-albedo feedback and circulation changes, and that they further transmitted the signal into the Southern Hemisphere by positive CO_2 feedback. But a central question remains unresolved. Why did the ice sheets fluctuate at a period near 100,000 years in the first place, especially given the lack of direct insolation forcing at that tempo? And in view of the sawtoothed shape of these cycles, what explains the slow buildup of ice over periods averaging 100,000 years and the abrupt deglacial terminations that ended those intervals of gradual ice buildup?

▶ Proposed Mechanisms for ~100,000-Year Ice Buildup

12-6 Ice Sheets Interaction with Bedrock

One proposed mechanism for building larger ice sheets focuses on the character of the substrate over which the ice moved and its friction with the basal ice layers. Glacial geologists have found a few ice-deposited moraines in Iowa and Nebraska that lie beyond the geographic limits of the very large North American ice sheet that existed at the most recent glacial maximum (~20,000 years ago). Layers of volcanic ash date these older moraines to about 2 million years ago, early in the long interval of 41,000-year ice sheet cycles.

Evidence from $\delta^{18}O$ trends shows that the combined effect of ice volume and deep-ocean temperatures at that time was smaller (see Chapter 10, Figure 10-13). Based on this evidence, glacial maximum ice sheets were almost certainly smaller in volume 2 million years ago than they have been during the 100,000-year world of the last 0.9 million years. But those ice sheet deposits from 2 million years ago show that those smaller-volume 41,000-year ice sheets were capable of reaching maximum extents comparable to or even larger than those of the later, larger-volume 100,000-year ice sheets. If the ice sheets 2 million years ago were similar in extent to the more recent ones but smaller in volume, they must have been thinner.

The glacial geologist Peter Clark proposed a mechanism to explain those thinner ice sheets. Under normal circumstances, the weight of overlying ice melts the bottom layers of ice sheets, and the meltwater trickles down into the underlying sediments. Sediments saturated with water are more easily deformed by the overlying ice and can cause the base of the ice sheet to slip and slide.

Clark proposed that the ice sheets formed earlier in the northern hemisphere glaciation sequence built up on top of thick layers of soil and sedimentary rock that had accumulated on a North American continent that had not been glaciated for the preceding hundreds of millions of years. Water saturating these thick underlying layers would have reduced the friction with the bottom layers of the ice sheets and caused frequent sliding that carried large amounts of ice away from central ice sheet areas and outward to ice margins at lower latitudes (Figure 12-14A). High ablation in these warmer regions would have melted the ice lobes that kept sliding south. This sliding and melting could have kept the North American ice sheet low, thin, and small in volume. These thinner ice sheets would also have been easier to melt during relatively weak insolation maxima (as in the "small glaciation" phase of Chapter 10).

Today, after almost 3 million years of repeated glacial erosion, very little of the original soil and sedimentary rock cover across north-central Canada is left, even though thick sequences still exist in unglaciated regions of the east-central United States south of the ice limits. The surface in Canada is mostly

bare crustal bedrock, with scattered areas of coarse ice-eroded debris. Without the soft substrate, later ice sheets have not been as susceptible to slipping, and the reduction in sliding could have allowed them to grow much thicker (Figure 12-14B). Reconstructions of ice sheets at the last glacial maximum 20,000 years ago indicate a broad interior region where the ice was thick and frozen to its base, making sliding unlikely. Thicker ice sheets also stand a better chance of surviving through relatively weak insolation maxima and growing to larger size (the "large glaciation" phase of Chapter 10).

Eroded material preserved both in old moraines in the United States and in sediments pushed into the nearby ocean show that ancient soils and sedimentary rocks were the main type of debris eroded prior to the last 1.5 to 1 million years, while freshly pulverized debris from hard bedrock has predominated since that time. This evidence supports Clark's hypothesis that the ancient substrate was gradually eroded by the early ice sheets between 2.75 and 0.9 million years ago. The bedrock sliding mechanism does not explicitly predict that the long-term ice sheet response will change from 41,000 to 100,000 years, but it does provide a plausible explanation for a change from smaller (thinner) to larger (thicker) ice sheets.

12-7 Long-Term Cooling and CO_2/ Ice-Albedo Feedback

Another explanation for the shift from the 41,000-year glacial world to a regime of oscillations centered near 100,000 years begins with the slow cooling that had been underway for millions of years, as shown by long $\delta^{18}O$ records from the deep Pacific Ocean (Figure 12-15). This gradual cooling would have slowly reduced ice ablation in high northern latitudes. Between 2.75 and 0.9 million years ago, the amplitude of the 41,000-year $\delta^{18}O$ changes gradually increased because of larger ice growth and greater cooling.

In this earlier 41,000-year regime with higher ablation, ice sheets formed mainly at the 41,000-year period, but all of the ice melted during the next summer insolation maximum (Figure 12-16). In contrast, in the glacial world that developed after 0.9 million years ago, ice sheet growth still occurred mainly during 41,000-year insolation minima, but some ice remained unmelted during relatively weak summer insolation maxima. The unmelted ice that survived would then have grown thicker during the subsequent intervals of low insolation, producing glacial cycles of longer duration (as in the proposed "large glaciation" phase in Chapter 10).

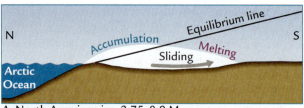

A North American ice 2.75–0.9 Myr

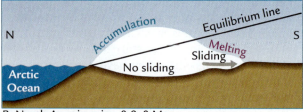

B North American ice 0.9–0 Myr

FIGURE 12-14
Ice slipping may control ice sheet volume
During earlier glaciations of North America, ice sheets may have been thin because they slid on water-saturated soils and sedimentary rocks toward lower elevations and warmer temperatures (A). Later, after ice sheets stripped off most of the softer underlying substrate, their central regions could grow higher because they no longer slid (B).

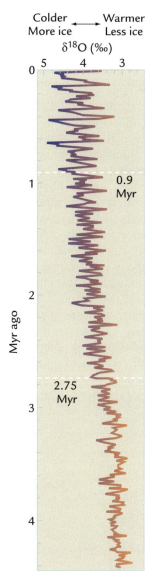

FIGURE 12-15
Changes in δ¹⁸O in the last 4.5 Myr

A δ¹⁸O record from a sediment core in the eastern Pacific Ocean shows a slow increase in δ¹⁸O values over the last 4.5 Myr. (ADAPTED FROM A. C. MIX ET AL., "BENTHIC FORAMINIFER STABLE ISOTOPE RECORD FROM SITE 849 [0–5MA]: LOCAL AND GLOBAL CLIMATE CHANGES," *OCEAN DRILLING PROGRAM, SCIENTIFIC RESULTS* 138 [1995]: 371–412.)

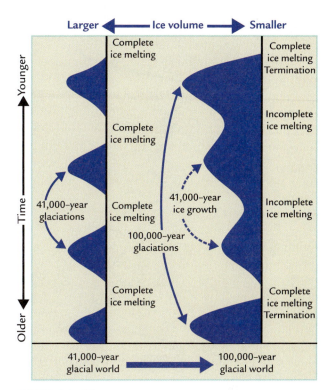

FIGURE 12-16
Transforming 41,000-year cycles to ~100,000-year oscillations

With decreasing global temperature and reduced ablation, glacial maxima that disappeared in the 41,000-year glacial world could have persisted for longer intervals and created glacial cycles averaging ~100,000 years.

The most recent ice growth cycle provides one example of how these longer cycles developed (Figure 12-17). Subsequent to the last interglaciation 120,000 years ago, the δ¹⁸O record shows five ice volume maxima near 115,000, 95,000, 72,000, 45,000 and 30,000 years ago. Two of these intervals—the ones near 95,000 and 45,000 years ago—were a response to summer insolation changes at the 23,000-year precession cycle, and the additional ice that formed at those times was removed during the next phase of higher summer insolation. As a result, no net ice growth occurred during those intervals.

Instead, all of the ice growth occurred during the three other oscillations—those near 115,000, 72,000, and 30,000 years ago. These ice growth episodes were separated by approximately 41,000 years, they occurred during insolation minima at the tilt cycle, and they were accompanied by large decreases in atmospheric CO_2 concentrations.

Based on this evidence, net ice growth during this most recent sawtoothed oscillation occurred only during 41,000-year episodes and was aided by positive CO_2 feedback (along with ice sheet albedo), much like the ice growth episodes in the earlier 41,000-year glacial world. (see Figure 12-16). The major difference from the 41,000-year glacial world is that only about one-third of the new ice that accumulated during the ice growth episodes 115,000 and 72,000 years ago melted during the next summer insolation maximum. The unmelted ice remained in place as new ice was

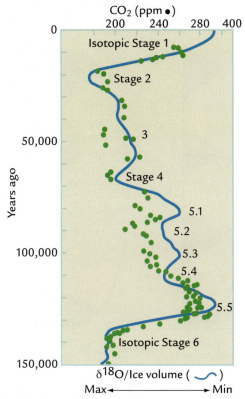

FIGURE 12-17

Changes in ice volume and CO₂ during the last 150,000 years

During the last full interglacial-glacial oscillation, sawtooth-shaped changes in CO_2 and ice volume ($\delta^{18}O$) occurred at similar times. (ADAPTED FROM W. F. RUDDIMAN, "ICE-DRIVEN CO₂ FEEDBACK ON ICE VOLUME," *CLIMATE OF THE PAST* 2 [2006]: 43–78.)

added during the next 41,000-year insolation cycle. In effect, the persistence of unmelted ice transferred part of the ice sheet response from a 41,000-year cycle to a longer-period (~100,000-year) response.

In Summary, part of the explanation for the growth of larger ice sheets during the last 0.9 million years is a reduction in ablation that transformed what had been 41,000-year ice-sheet cycles between 2.75 and 0.9 million years into a longer-wavelength response.

▶ Proposed Mechanisms for ~100,000-Year Ice Melting

The most distinctive feature of the longer-term ice sheet cycles at ~100,000 years is the abrupt deglacial terminations that end the intervals of gradual ice growth. These terminations define the basic timing

of the large oscillations averaging 100,000 years in length. But what governs their timing?

12-8 Timing of Deglacial Terminations

One reason why large glaciations would have been spaced at approximately 100,000 years is evident in the summer insolation trends in Figure 12-18. Although these insolation changes are dominated by changes at the 23,000-year precession cycle, modulation of the precession signal by eccentricity at a period near 100,000 years is obvious. This modulation produces clusters of unusually high insolation maxima at intervals of approximately 100,000 years. As a result, ice sheets that are able to grow to a relatively large size during intervals when the insolation maxima are relatively small would then be vulnerable to rapid melting when insolation maxima grew larger, at intervals of ~100,000 years.

This 100,000-year response is possible because ice melting is sensitive only to the "upper side" of the envelope of modulation by eccentricity. Insolation minima are equally prominent on the opposite side of this envelope, but they are irrelevant to deglacial terminations because ice melting occurs only during insolation maxima. In this way, major deglaciations could be

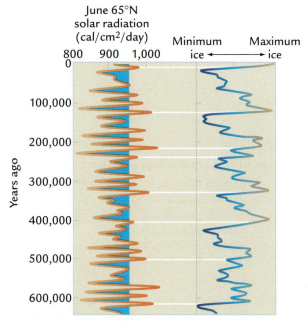

FIGURE 12-18

Strong summer insolation peaks pace rapid deglaciations

Strong summer insolation peaks at the precession cycle resulting from eccentricity modulation (left) match rapid deglacial terminations indicated by $\delta^{18}O$ (~ice volume) signals (right). (ADAPTED FROM W. S. BROECKER, "TERMINATIONS," IN *MILANKOVITCH AND CLIMATE*, ED. A. L. BERGER ET AL. [DORDRECHT: D. REIDEL, 1984].)

produced at intervals of ~100,000 years from the upper envelope of a highly modulated 23,000-year cycle.

Although major deglaciations (terminations) occur at an average spacing of ~100,000 years, none of the intervals between successive terminations is actually close to 100,000 years. The spacing between the last five terminations falls in the range of ~84,000 to ~116,000 years, but in clusters that are either longer than or shorter than the 100,000-year value. This timing makes sense if deglaciations are constrained to occur on or near one of the major insolation maxima at the precession cycle and only after the ice sheets had grown to a large size. Given these constraints, terminations would naturally tend to occur after either four or five 23,000-year insolation maxima, or at intervals of either ~92,000 (4 × 23,000) or ~115,000 (5 × 23,000) years.

A second factor that affects the timing of deglaciations is the net growth of ice sheets at 41,000-year intervals (see Figures 12-16 and 12-17). Because a relatively large volume of ice obviously has to accumulate in order for a major deglaciation to be possible, this constraint tends to position terminations at multiples of the 41,000-year ice growth cycle, at either 82,000 or 123,000 years (equivalent to two or three 41,000-year intervals).

Both precession and obliquity are likely to have determined the timing of terminations, which should tend to occur in one of two time clusters: either every 82,000–92,000 years or every 115,000–123,000 years. The intervals separating the last five terminations all fall in one of these two time clusters. The fastest and largest terminations also tend to occur when insolation peaks are particularly strong because the effects from precession and tilt are closely aligned.

In Summary, even though ice sheets grew larger during 41,000-year episodes, they would have melted at intervals of 82,000–92,000 or 115,000–123,000 years, averaging near 100,000 years.

12-9 Proposed Local Causes of Abrupt Deglacial Terminations

One of two groups of explanations for accelerated ice melting during terminations looks to local interactions between ice sheets and their environment. The hypothesis of basal sliding of ice sheets doesn't specifically predict an average spacing of terminations, but it does provide an internal source of stored energy that can be released during deglaciations. Because of the great height of ice sheets during the last 0.9 million years, the force of gravity acting on the ice is available to accelerate melting by sliding along the margins south of the central ice sheet during times of increased insolation.

A closely related idea calls on the fact that large ~100,000-year ice sheets developed extensive margins that fronted on the ocean and rested on bedrock lying below sea level (similar to the modern West Antarctic ice sheet). As rising insolation levels initiated melting, these marine margins would have become vulnerable to sea level rises that "lifted" them off their bedrock bases. As a result, they could have responded more quickly to climate changes than the more sluggish land-based ice sheets deep in the continental interiors. Evidence from marine sediments suggests that ice sheets first became vulnerable to rising sea level along the Arctic margin north of Norway near 0.9 million years ago. One of these ice sheets—the Barents ice sheet just north of Scandinavia (see Chapter 10, Figure 10-4)—was also among the earliest northern ice to melt during the most recent deglaciation. The early disappearance of this ice sheet might have helped warm nearby regions and hastened the melting of the slower-responding parts of the northern ice sheets.

Another proposed idea is that the cooling produced by large northern ice sheets had caused large amounts of sea ice to expand across nearby high-latitude oceans. This increased sea-ice cover would have reduced the extraction of moisture that could be delivered to the ice sheets, and the ice sheets could have melted faster because they were starved for moisture. But glacial geologists note that ice accumulation is a relatively weak term in the mass balance of ice sheets compared to ablation (see Chapter 10, Figure 10-1). As a result, moisture starvation should have had a relatively small effect on changes in ice volume during terminations.

Another proposal is that the windy, dusty world of early melting just after glacial maxima would have produced a thin coating of dust on the lower southern margins of the ice sheets. Because dust has a lower albedo than ice, it would have absorbed more incoming solar radiation and warmed the surface of the ice. The absorbed heat would have promoted greater melting and more rapid deglaciation. One problem with this idea is that just a slightly thicker coating of dust can have the opposite effect of insulating the ice surface from solar heating and reducing the rate of ablation. As ice melted, the layer of residual dust seems likely to have grown thicker and opposed rapid melting.

12-10 CO₂ and Ice Sheet Albedo Feedback During Terminations

Greenhouse gases (primarily CO_2) and ice sheet albedo are widely viewed as the two most important feedback processes that make glacial climates colder than interglacial ones. But these two feedbacks can also be thought of in the opposite sense: as a stored source of positive feedback that can accelerate warming when ice sheets start melting. As ice sheets melt, their highly

Looking Deeper into Climate Science

Box 12-2

An Antarctic Role in Long-Term CO_2 and $\delta^{18}O$ Trends?

Hints of an independent Antarctic role in long-term CO_2 and $\delta^{18}O$ trends have emerged in recent years. Detecting such a role is difficult because insolation changes over the Antarctic continent and the higher latitudes of the Southern Ocean at the 41,000-year cycle have the same phasing as those over north polar regions (see Chapter 8).

One clue to a possible Antarctic role lies in the trends of CO_2 and $\delta^{18}O$ over the last 800,000 years (see Chapter 11, Figure 11-6). Prior to 400,000 years ago, interglaciations had weaker $\delta^{18}O$ and CO_2 minima than those during the five subsequent interglaciations. One possible interpretation is that the earlier interglaciations were not as warm and northern hemisphere ice loss was not as complete as during the more recent interglaciations. But evidence from Eurasia indicates that the earlier interglaciations were very warm, suggesting that ice probably disappeared from North America and Eurasia at those times.

Another possible explanation lies in the Antarctic region, which is also a source of CO_2 and $\delta^{18}O$ changes. Bottom waters that form in the Southern Ocean dominate the world ocean, and changes in their temperature alter the $\delta^{18}O$ values measured in sediments on the seafloor.

Because temperature changes account for about 40% of marine $\delta^{18}O$ signals, colder bottom water during the interglaciations before 400,000 years ago could have made the $\delta^{18}O$ values more positive than in subsequent interglaciations, even though northern ice volume was reaching similar minima.

Bottom water formed in the Southern Ocean also alters atmospheric CO_2 concentrations, both by changes in deep circulation and through the effect of CO_2 solubility (see Chapter 11). Larger volumes of colder bottom water formed in the Southern Ocean could have driven down atmospheric CO_2 values during interglaciations prior to 400,000 years ago. If so, both the $\delta^{18}O$ and CO_2 signals would register "weak" interglaciations, even though interglaciations in the Northern Hemisphere were too warm for ice sheets to survive.

This explanation, while plausible, would lead to another question: Why would the Southern Ocean have formed larger amounts of colder bottom water prior to 400,000 years ago, when the long-term climatic trend over the last few million years has been just the opposite: toward colder climates in more recent times?

---

reflective surfaces retreat to the north, exposing darker landmasses to solar heating and warming high northern latitudes. At the same time, CO_2 levels rise and further warm the entire surface of the Earth. In effect, feedback from ice sheet albedo and CO_2 are powerful mechanisms for amplifying rates of deglaciation. On most terminations, this amplification is so effective that ice sheets melt entirely from Eurasia and North America, along with some of the ice on Greenland.

The most likely initial trigger of these two important feedback processes lies in the insolation maxima spaced at intervals of 82,000–92,000 or 115,000–123,000 years (see Section 12-8). The insolation levels at these times do not have to be much higher than during the previous tens of thousands of years to start the melting process. Once underway, ice sheet melting activates the powerful CO_2 and ice-albedo feedbacks.

In Summary, the large ice sheet cycles near a period of 100,000 years may have a deceptively simple explanation. Long-term global cooling allows ice that had melted entirely during the 41,000-year glacial regime to instead survive and accumulate to larger ice volumes in the 100,000-year glacial world. The larger ice sheets that grow in a series of 41,000-year steps eventually produce internal responses (either in underlying bedrock or in the climate system) that hasten their own destruction during intervals of increased summer insolation. CO_2 and ice-albedo feedback are likely to have played important positive feedback roles both in ice growth and melting. Yet another possibility is that the Antarctic Ocean and its connection to bottom-water formation plays an independent role in these variations (Box 12-2).

A final issue is the "normal" state of climate during the last several hundred thousand years. Some climate scientists see the sawtooth-like behavior of the ice sheets as characteristic of a **relaxation oscillator**. Relaxation oscillators vary between a slow destabilization phase (in this analogy, the gradual buildup of ice sheets) and a fast relaxation phase (rapid ice melting). This concept is analogous to the behavior of a spring, with a slowly imposed stretching and a rapid return to a relaxed state. Implicit in this concept is the idea that the source of the fast relaxation rebound lies in the system itself, as in the tension imposed by stretching the spring.

The relaxation oscillator concept views deglaciations as returning the system to its stable interglacial state, but this claim is questionable. Because ice sheets of some size have been present in the Northern Hemisphere for 90% of the last 0.9 million years, it seems more likely that the true stable state of the climate system has been one with at least moderate-sized ice sheets present. From this point of view, the full interglaciations may be the most *unstable* intervals of the last 0.9 million years, with the system generally tending to exist in a partly glaciated state.

Key Terms

ice-driven response (p. 234)

relaxation oscillator (p. 249)

Review Questions

1. In what sense are ice sheets both a climatic response and a source of climatic forcing?

2. Name an ice-driven response and explain its origin.

3. Summarize three possible explanations for the unexpected strength of the 41,000-year response of ice sheets between 2.75 and 0.9 million years ago.

4. How could the northern hemisphere ice sheets affect climate in the Southern Hemisphere?

5. What evidence suggests that orbital-scale changes in northern hemisphere ice volume drive changes in atmospheric CO_2, rather than the opposite?

6. How could 100,000-year cycles in the size of northern hemisphere ice sheets be paced by summer insolation changes that occur only at cycles of 41,000 and 23,000 years?

7. If a sine wave CO_2 signal with a period of 100,000 years forces an ice sheet response at the same period, and if the ice has a time constant of 10,000 years, what should be the size of the ice sheet lag behind the forcing?

Additional Resources

Basic Reading

Imbrie, J., and K. P. Imbrie. 1979. *Ice Ages: Solving the Mystery.* Short Hills, NJ: Enslow.

Advanced Reading

Broecker, W. S. 1984. "Terminations." In *Milankovitch and Climate,* ed. A. L. Berger et al., Dordrecht: Reidel, 687–98.

Broccoli, A. J., and S. Manabe. 1987. "The Influence of Continental Ice, Atmospheric CO_2, and Land Albedo on the Climate of the Last Glacial Maximum." *Climate Dynamics* 1: 87–100.

Clark, P. U., and D. Pollard. 1998. "Origin of the Middle Pleistocene Transition by Ice Sheet Erosion of Regolith." *Paleoceanography* 13:1–9.

Huybers, P. 2006. "Early Pleistocene Glacial Cycles and the Integrated Summer Insolation Forcing." *Science* 313: 508–11.

Manabe, S., and A. J. Broccoli. 1985. "The Influence of Continental Ice Sheets on the Climate of an Ice Age." *Journal of Geophysical Research* 90: 2167–90.

Raymo, M. E., L. E. Lisiecki, and K. H. Nisancogliu. 2006. "Plio-Pleistocene Ice Volume, Antarctic Climate, and the Global $\delta^{18}O$ Signal." *Science* 313; 492–95.

Rind, D., D. Peteet, W. S. Broecker, A. McIntyre, and W. F. Ruddiman. 1986. "The Impact of Cold North Atlantic Sea Surface Temperatures on Climate: Implications for the Younger Dryas Cooling (11–10K)." *Climate Dynamics* 1: 3–33.

Ruddiman, W. F. 2006. "Ice-Driven CO_2 Feedback on Ice Volume." *Climate of the Past* 2, 43–78.

Glacial/Deglacial Climate Change

The last several tens of thousands of years have seen enormous changes in climate, some of them in the time span of recorded human civilization. The largest change was the transition from the last glaciation near 20,000 years ago (Chapter 13) to the warmth of the present interglaciation.

Changes during this interval are recent enough to be dated by the radiocarbon method, and thousands of ^{14}C-dated records from the land can be added to hundreds of records from the ocean to provide nearly global coverage. Most terrestrial ^{14}C dates come from lakes or bogs, where watery environments preserve organic carbon. Radiocarbon dates must be converted to calendar years, which for most intervals are older by a few thousand years. Throughout the chapters in this Part, all ages are in calendar ("true") years unless ^{14}C ages are specified.

The ability to reconstruct regional climatic responses across most of Earth's surface also gives scientists the opportunity to look at climate change from a full geographic perspective. Mapping climate change at specific times ("time slices") gives researchers insights into specific regional processes and allows comparisons of proxy data with model simulations. Boundary conditions based on observations from the geologic record can be used to run model simulations of past climates, and the results (output) from these simulations can be compared with other (independent) geologic data assembled into map form. The major boundary conditions that have driven climate change during the last 21,000 years have been changes in the size of ice sheets, in seasonal insolation, and in the levels of greenhouse gases in the atmosphere (Chapter 14).

In climate archives with sufficiently high resolution, climatic oscillations that lasted a few thousand years are apparent during the last glacial interval, especially in and around the North Atlantic Ocean (Chapter 15). These "millennial oscillations" cannot be explained by orbital forcing and appear to result from processes internal to the climate system, although their origin remains unclear.

In this part, we address the following important questions:

part IV

Deglacial climatic forcing For the past 21,000 years, model experiments based on known climatic forcing (insolation, ice sheets, and CO_2) can be compared with observations from the geologic record. (ADAPTED FROM J. E. KUTZBACH ET AL., "CLIMATE AND BIOME SIMULATIONS FOR THE PAST 21,000 YEARS," *QUATERNARY SCIENCE REVIEWS* 17 [1998]: 473–506.)

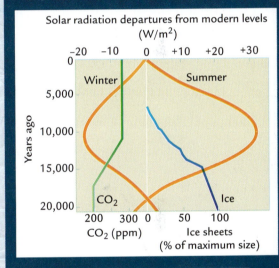

- To what extent does the timing of ice melting in the last 21,000 years support the Milankovitch theory that ice sheets are forced by orbital insolation?

- What does the cooling of the tropics at the last glacial maximum 21,000 years ago tell us about Earth's sensitivity to the concentration of atmospheric CO_2?

- To what extent do changes in tropical moisture during the last 21,000 years support the Kutzbach theory that orbital insolation controls the strength of summer monsoons?

- How has Earth's climate system responded to insolation changes during the several thousand years since the North American and Eurasian ice sheets melted?

- What is the origin of the brief oscillations of climate that occur at intervals much shorter than changes in Earth's orbital configurations?

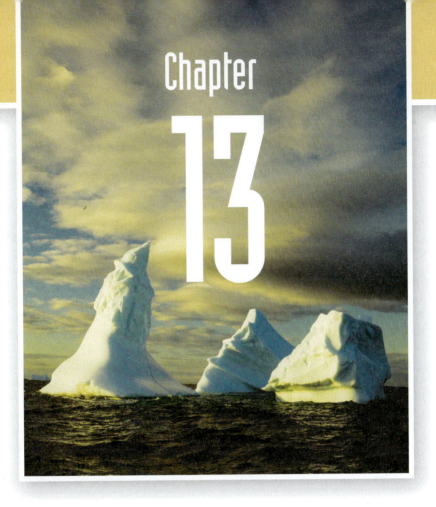

Chapter 13

The Last Glacial Maximum

When the most recent of the ~100,000-year glacial oscillations culminated 21,000 years ago, much of Earth's surface looked very different from its present appearance. Ice sheets 2 km high or more covered Canada, the northern tier of the United States, northernmost Europe, and parts of Eurasia. Global sea level was roughly 120 m lower, joining together modern islands between Asia and Australia and connecting Britain to mainland Europe. South of the ice sheets, conditions were cold and windy, with dust blowing across many areas. Many regions of North America, Europe, and Asia south of the ice sheets that are forested today were then tundra or grasslands, and the lower concentrations of atmospheric CO_2 and CH_4 caused cooling and drying across the tropics and the Southern Hemisphere.

In this chapter, we focus on aspects of this glacial world, such as the extent and thickness of the ice sheets and the debris they produced. Then we explore how changes in the distribution of life-forms, especially land vegetation and ocean plankton, allow us to run climate simulations on general circulation models. Finally, we examine a controversy that has implications for future changes in climate: How cold were the tropics at the last glacial maximum?

Glacial World: More Ice, Less Gas

Even though the glacial world was icy, cold, dry, windy, and sparsely vegetated, it was, from a long tectonic perspective, nearly identical to today. The continents had moved to their modern positions, and plateaus and mountains were at very nearly the same elevations as today. In a sense, this glacial maximum world represents an alternative version of our present Earth, the product of a giant experiment run by the climate system in response to changes in several forcing factors. As we saw in Part III, the three factors with the greatest potential to account for departures from modern climate conditions over this interval of time were larger ice sheets, lower CO_2 levels, and changes in seasonal insolation.

Surprisingly, summer and winter insolation levels 21,000 years ago were close to those today. This seemingly counterintuitive configuration results from the fact that intervals of lower summer insolation had helped to build the ice sheets thousands of years earlier, and the ice sheets had responded with their usual lags by slowly growing to maximum size by 21,000 years ago. By the time the ice sheets reached their maximum size, however, summer insolation had reversed direction and had already risen close to today's level, heading toward the higher levels that would soon begin to melt the ice (Figure 13-1).

Because the seasonal insolation levels 21,000 years ago were close to those today, insolation is not a major explanation of the differences in climate between the two intervals. Two other factors are likely explanations of the colder and drier glacial maximum climates: the larger size of the ice sheets and the lower concentrations of greenhouse gases. This chapter focuses on the ability of these two factors to account for the very different glacial maximum world.

13-1 Project CLIMAP: Reconstructing the Last Glacial Maximum

Cooperative efforts to reconstruct past climates began in the 1970s, when a large interdisciplinary effort called the **CLIMAP (Climate Mapping and Prediction) Project** reconstructed the surface of Earth at the last glacial maximum. Led by the marine geologists John Imbrie and Jim Hays and the geochemist Nick Shackleton, CLIMAP drew on the expertise of dozens of scientists with specialized knowledge of ice sheets, windblown deposits, marine sediments, vegetation, and climate modeling.

In the years before CLIMAP, relatively few scientists worked on past climates, usually on individual research projects using widely differing techniques. CLIMAP succeeded in bringing many of these scientists together and combining their skills into a single interdisciplinary effort. As a result, a generation of climate scientists in these many disciplines learned to see Earth's climate as an integrated whole. Today, interdisciplinary alliances are common in the study of climate science.

The CLIMAP group published a first map of the ice-age Earth in 1976 and then a revised version in 1981 (Figure 13-2). Their maps show conditions at Earth's surface during typical seasons (summer in this case) at the last glacial maximum. Because individual years cannot be resolved in glacial-age records, the maps portray an *average* northern summer during the millennia centered on the glacial maximum 21,000 years ago.

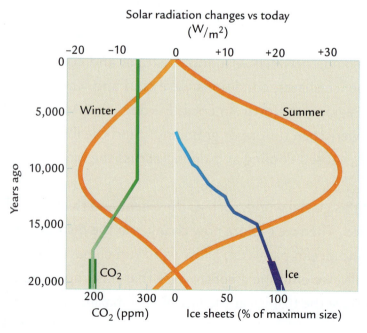

FIGURE 13-1

Boundary conditions for glacial maximum climate

Climate models simulate glacial maximum climate by using larger ice sheets (thick blue line) and lower levels of greenhouse gas (thick green line) as boundary-condition inputs. (ADAPTED FROM J. E. KUTZBACH ET AL., "CLIMATE AND BIOME SIMULATIONS FOR THE PAST 21,000 YEARS," *QUATERNARY SCIENCE REVIEWS* 17 [1998]: 473–506.)

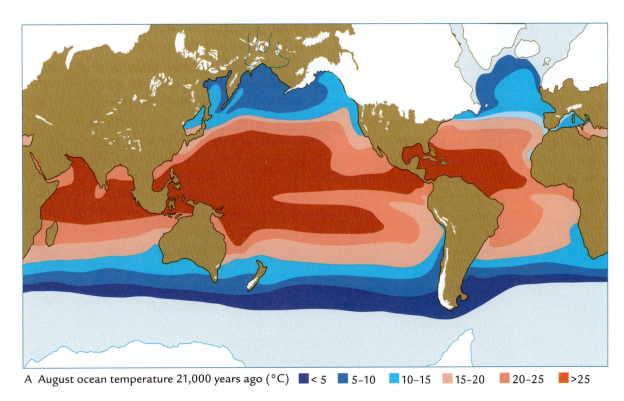

A August ocean temperature 21,000 years ago (°C)   ■ < 5   ■ 5–10   ■ 10–15   ■ 15–20   ■ 20–25   ■ >25

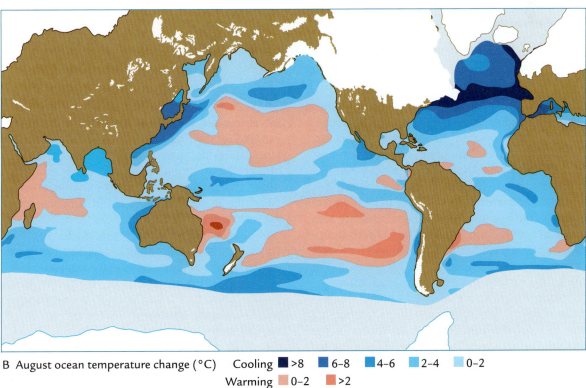

B August ocean temperature change (°C)   Cooling ■ >8   ■ 6–8   ■ 4–6   ■ 2–4   ■ 0–2
                                            Warming ■ 0–2   ■ >2

FIGURE 13-2
The CLIMAP reconstruction of glacial maximum ocean temperatures

CLIMAP produced the first global-scale reconstruction of Earth's surface at the most recent glacial maximum, including ice sheet size and ocean surface temperature. These maps show conditions during an average summer during the glacial maximum (A) and the changes in temperature between then and today (B). (ADAPTED FROM CLIMAP PROJECT MEMBERS, *SEASONAL RECONSTRUCTION OF THE EARTH'S SURFACE AT THE LAST GLACIAL MAXIMUM*, MAP AND CHART SERIES MC-36 [BOULDER, CO: GEOLOGICAL SOCIETY OF AMERICA, 1981].)

Table 13-1   Approximate Volumes of Ice and Amounts of Water Stored in Glacial Ice Sheets by Lowering Sea Level beneath Today's Position

| | | | SEA LEVEL | |
Ice sheet	Location	Excess ice volume (million km³)	Amount (m)	Change (m)[a]
Laurentide	East-central Canada	25–34[b]	72–100	50–70
Cordilleran	Western North America	1.8	5	3.5
Greenland	Greenland	2.6[c]	7	5
Britain	England, Scotland, Ireland	0.8	2	1.5
Scandinavian	Northern Europe	7.3	21	15
Barents/Kara	Shelf north of Eurasia	6.9	20	14
East Antarctic	Eastern Antarctica	13.3[d]	9	6
West Antarctic	Western Antarctica	16.5[d]	18	13
Others	Various	1.2	3	2
All ice sheets		55–64	155–183	109–129

[a]Net sea level changes are 30% smaller than the volumes of seawater removed from the ocean because ocean bedrock rises when the weight of water is removed.
[b]The higher estimate shown is for a thick ice sheet like that in the CLIMAP maximum reconstruction; the lower estimate is for a thin ice sheet.
[c]Present volume of ice on Greenland is 3 million km³.
[d]Present volume of ice on Antarctica is 29 million km³.

Source: Adapted from G. H. Denton and T. J. Hughes, *The Last Great Ice Sheets* (New York: John Wiley, 1981)

The most striking features of the CLIMAP reconstruction are the continent-sized ice sheets covering North America as far south as 37°N latitude, Scandinavia down to 48°N, and the Arctic margins of Eurasia. Today, ice sheets on Antarctica and Greenland cover a combined area of about 14.2 million square km, equivalent to just under 3% of Earth's surface area and about 10% of its land surface. The CLIMAP reconstruction for the last glacial maximum shows ice sheets covering an area of 35 million square km, equal to 7% of Earth's total surface and 25% of its land surface.

The North American ice sheet was by far the largest of the northern hemisphere ice sheets, accounting for over 55% of the volume of ice in excess of the amount on Earth today (Table 13-1). This great ice sheet was roughly equivalent in volume to the amount of ice on Antarctica today. Most of the ice in North America was stored in the **Laurentide ice sheet**, centered on east-central Canada, with the rest in the much smaller **Cordilleran ice sheet**, over the Rockies in the American West. The **Scandinavian ice sheet**, in northern Europe, and the **Barents ice sheet**, on the northern Eurasian continental shelf, together represented about 22% of the extra ice compared to today. The remainder was accounted for by seaward expansion of the

ice sheets on Antarctica and Greenland across land exposed by the drop in sea level. Expansion of mountain glaciers dramatically transformed the appearance of high terrain around the world but contributed little to the change in ice volume. In some regions, such as Patagonia in southernmost South America, mountain glaciers merged into small ice caps.

The glacial world as reconstructed by CLIMAP was almost 4°C colder than today. The North Atlantic cooled by 8°C or more, and sea ice was more extensive (Figure 13-2B). Farther from the ice sheets, the North Pacific cooled by a lesser amount, with expanded winter sea ice in the northwest. Sea ice also advanced well beyond its present limits in the Southern Ocean, with a band of cooler ocean-surface temperatures north of the expanded limit of sea ice around Antarctica. The CLIMAP reconstruction indicates low-latitude ocean temperatures only slightly cooler than today's, and in some regions even slightly warmer.

In the decades since it was published, other groups have challenged and refined the CLIMAP reconstruction. Some of its features now appear to be incorrect, but this reconstruction of the ice-age Earth still remains the reference point against which these later attempts are weighed.

13-2 How Large Were the Ice Sheets?

The glacial geologists George Denton and Mikhail Grosswald and the glaciologist Terry Hughes headed the CLIMAP ice sheet reconstructions. Of the two alternative reconstructions they published, one broke with conventional thinking by portraying the ice sheets at their maximum plausible size (the "maximum reconstruction"). Three aspects of this reconstruction proved highly controversial.

One debate arose over the lateral extent of the ice sheets, which in most places were shown extending to the maximum limits that ice had reached at any time in the history of northern hemisphere glaciation (Figure 13-2). The southern limits of most continental ice sheets at the last glacial maximum had not been in much doubt for many years prior to the CLIMAP reconstruction, but larger uncertainties remained about whether or not ice had actually reached the ocean in some higher-latitude regions. Some glacial geologists argued that the glacial maximum ice sheets in the north were less extensive than they had been during earlier parts of the glacial cycles because they had been starved for moisture. The main reason for this uncertainty about the ice limits was the scarcity of organic carbon in the cold, dry Arctic, which made it difficult to find suitable deposits for ^{14}C dating.

Estimates of the northern ice margins generally disagreed by a few hundred kilometers, amounts that were important to arguments about the physical state of the ice sheets in specific regions but were actually rather small in comparison with the full lateral dimensions of ice sheets (thousands of kilometers) and with the size of grid boxes then used in climate modeling (a few hundreds of kilometers on a side). As it turned out, careful ^{14}C dating along the northern margins of the ice sheets has subsequently confirmed that most of them were indeed at or near their maximum limits 21,000 years ago.

A second major disagreement about the extent of glacial maximum ice sheets centered on marine ice sheets, those ice sheets that formed on shallow continental shelves with their bases lying below sea level (see Chapter 2). The CLIMAP reconstruction placed two marine ice sheets along the northern margin of Eurasia, a large one over the Barents Sea directly north of Scandinavia, and a smaller one over the Kara Sea, located farther east and north of western Russia. This aspect of the reconstruction caused years of controversy. Subsequent surveys of these shallow seas collected evidence (glacial debris in sediment cores and depth soundings showing the extent of submerged moraine ridges) proving that a large marine ice sheet did exist in the Barents Sea and in the westernmost Kara Sea at the glacial maximum, but not in the eastern Kara Sea. The ice sheets in land areas south and east of the Kara Sea proposed in the CLIMAP reconstruction are not supported by these more recent reconstructions of the most recent glacial maximum.

Another controversial aspect of the CLIMAP reconstruction was the thickness (and height) of the ice sheets. Height and thickness are usually related in a simple way: as an ice sheet grows, it eventually weighs down the underlying bedrock by an amount equal to about 30% of its thickness (see Chapter 10). As a result, the height of the ice above the surrounding landscape usually represents about 70% of its actual thickness.

The CLIMAP maximum reconstruction showed relatively high (thick) ice sheets, especially over North America. This reconstruction was based on the assumption that the ice sheets had existed at or near their maximum extent long enough to build up slowly to full equilibrium thickness and that they consisted of stiff ice that was frozen to the underlying bedrock. These two assumptions produced ice sheets with profiles that rose relatively steeply from their outer margins to high elevations and great thicknesses in their central regions (Figure 13-3A and B).

This aspect of the CLIMAP reconstruction is still somewhat controversial. One line of evidence has come from measurements of the size of the fall in global sea level caused by water stored in the ice sheets. Coral reefs can be used as dipsticks to measure past changes in sea level (see Chapter 10), and reefs in different areas indicate a sea level drop of somewhere between 110 and 125 m during the last glacial maximum. The higher estimate agrees with the large-volume ice sheets in the CLIMAP reconstruction, but the lower estimate falls close to the smaller estimate of ice volume (see Table 13-1).

If the ice sheets were as extensive as those CLIMAP proposed but held a smaller volume of water taken from the ocean, one or more of the ice sheets must have been thinner than the CLIMAP reconstruction indicated. Scientists naturally focused on the North American ice sheet because of its huge size: if this ice sheet were thinner by 30%, it would fit the lower estimate of sea level change.

Independent evidence supports the possibility of thinner North American ice. Ice sheets slip and slide across soft, saturated sediments and unconsolidated sedimentary rock (see Chapters 10 and 12). Although the central part of the North American ice sheet rested on hard bedrock, the southern margins sat on easily deformed sediments at lower, warmer elevations. As a result, the southern margins of the ice sheet were probably thinned by frequent sliding (see Figure 13-3C). This sliding could also have thinned some of the inner portion of the ice sheet by drawing ice out toward the margins.

A second line of evidence comes from the amount of bedrock rebound that has occurred since the ice sheets melted. Bedrock weighed down by ice sheets

A Ice sheet extent

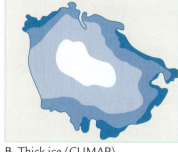

B Thick ice (CLIMAP)

Ice elevation (km)

■ <1   ■ 1–2   ■ 2–3   □ >3

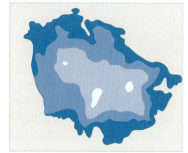

C Thin ice

FIGURE 13-3

How thick were the ice sheets?

The limits of the North American ice sheet are well known (A), but one CLIMAP reconstruction showed a thick high-elevation ice sheet (white) during the glacial maximum (B), while some recent reconstructions have favored a thinner ice sheet (C). (ADAPTED FROM P. CLARK ET AL., "NUMERICAL RECONSTRUCTION OF A SOFT-BEDDED LAURENTIDE ICE SHEET DURING THE LAST GLACIAL MAXIMUM," *GEOLOGY* 24 [1996]: 679–82.)

has a slow viscous response, and some of its rebound is delayed for thousands of years after ice melting (see Chapter 10). As a result, the bedrock retains a "memory" of how much ice was once on the land and when it began to melt. Scientists can exploit this behavior by examining regions in which the land is slowly rising out of the sea even now, leaving a trail of fossil beach shorelines up the side of the rising land. By radiocarbon dating the shells that formed when these beaches were at sea level and then measuring the present elevations of the old beaches above sea level, scientists can attempt to reconstruct the history of bedrock rebound. Based on this method, geophysicists have estimated how thick the ice sheets originally were. This technique has produced thickness estimates for the North American ice sheet intermediate between the two CLIMAP reconstructions, although the bedrock "memory" of ice sheet thickness 21,000 years earlier has by now grown a bit dim.

13-3 Glacial Dirt and Winds

The ice sheets were prolific producers of debris in sizes ranging from large boulders to fine clay. Ice sheets grind across the landscape, scraping and dislodging soils and relatively unconsolidated sedimentary rock. The weight of the ice sheets provides a pressure force that uses debris carried in the bottom layer of the ice to grind and gouge out small pieces of even the hardest bedrock. In areas where basal layers of ice alternately freeze and thaw, water trickles down into cracks in the rock when

the ice melts and then expands when it freezes again, breaking off large chunks. This freeze-thaw process quarries large slabs of bedrock and incorporates them in the ice for further grinding and fragmentation.

These and other processes erode large volumes of debris of all sizes. The ice sheets carry this material and deposit it along their margins when the ice melts. Much of the unsorted debris is piled into moraines (see Chapter 3). Running water from melting ice or local precipitation reworks the debris, extracting finer sediments and producing **glacial outwash**. Ice margins usually have little vegetation because the constant supply of new debris buries new growth and because meltwater inundates much of the landscape in summer. The lack of vegetation exposes this debris to further erosion.

Winds then rework these deposits, creating a gradation of grain sizes away from the ice margins. The coarsest debris (boulders, cobbles, and pebbles) remains in place, but strong winds can transport medium to fine sand over short distances. Winds also lift and carry finer silt-sized sediment farther from source regions, leaving loess deposits that become thinner and finer away from the glacial outwash (Figure 13-4). The loess patterns suggest that winds carried this debris mainly from the west-northwest to the east-southeast in both North America and Europe south of the ice.

Winds can carry even finer (clay-sized) dust completely around the world. Glacial-age layers in the Greenland ice sheet contain ten times as much fine dust as interglacial layers. Chemical analysis of this dust indicates that its main source region was Asia

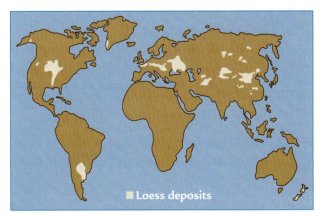

FIGURE 13-4
Glacial maximum loess

Ice sheets and mountain glaciers eroded large amounts of debris of all sizes and carried it to their margins. Winds picked up silt-sized material and deposited it as loess downwind of these sources. (ADAPTED FROM K. PYE, "LOESS," *PROGRESS IN PHYSICAL GEOGRAPHY* 8 [1984]: 176-217.)

rather than nearby North America. Glacial ice also contains more Na^{+1} and Cl^{-1} ions, an indication that far more salt was lifted from stormy glacial sea surfaces and deposited in the ice than today.

Dust transport was also greater at lower latitudes during the last glacial maximum. Today, the North African and Arabian deserts and their semiarid margins produce some of the largest dust storms on Earth. By comparison, sediment cores from the Indian Ocean east of the Arabian Peninsula show that dust accumulated five times faster during the last glacial maximum and during several previous glaciations than it does today. Cores from the equatorial Atlantic Ocean reveal that dust was also deposited in that region at higher rates during glaciations.

The extremely arid cores of the deserts were also affected (Figure 13-5). Regions of Arabia and North Africa identified as deserts on modern maps have actually varied in moisture level through time: mobile sand dunes and loose soil are prevalent during extremely arid intervals, but the dunes are stabilized by sparse desert vegetation during moist monsoonal intervals (see Chapter 9). In both of these areas and in Australia as well, moving sand dunes were much more extensive during the last glaciation than they are today because the winds were stronger and the climate was drier.

Even the South Pole was dustier. Glacial-age layers in ice cores from Antarctica contain more than ten times as much dust as interglacial layers. Geochemical fingerprinting of likely source areas suggests that much of this dust came from the southernmost tip of South America (Patagonia), where a small ice sheet complex generated ground-up debris. The increased flux of dust to Antarctic ice probably resulted both

from greater production of debris at the sources and from a more turbulent atmosphere that lifted dust and carried it farther south than today.

Almost everywhere we look, the glacial world shows evidence that more debris was blowing across Earth's surface. One way of testing climate model simulations of the last glacial maximum is to examine the distribution of various kinds of debris carried by winds, ranging from desert sands to windblown silts (loess) to fine (clay-sized) dust. Because climate models can simulate the strength and direction of winds from the surface up to jet stream altitudes, the potential exists to compare model simulations with the observed patterns of windblown glacial debris.

Unfortunately, even though the current generation of climate models does a fairly good job of simulating the large-scale circulation of the atmosphere, this success does not yet extend to the smaller spatial scales needed to simulate dust transport. The models are less successful at simulating the processes that actually lift and transport silt and dust from Earth's surface, such as local wind gusts along frontal systems or small-scale eddies of wind.

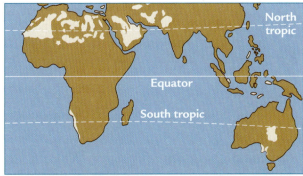

A Sand dunes active today

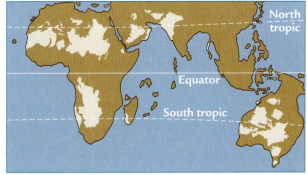

B Sand dunes active at glacial maximum

FIGURE 13-5
Glacial maximum sand dunes

Moving sand dunes occur today only in Africa, Arabia, and Australia (A). At the last glacial maximum, drier climates and stronger winds created more extensive sand dunes (B). (ADAPTED FROM M. SARNTHEIN, "SAND DESERTS DURING GLACIAL MAXIMUM AND CLIMATIC OPTIMUM," *NATURE* 272 [1978]: 43-46.)

Testing Model Simulations Against Biotic Data

So far, we have examined only the physical aspects (ice and dirt) of the glacial maximum world, but living organisms also have a story to tell. They also allow us to test the performance of climate models on a world quite different from our present one.

13-4 Project COHMAP: Data-Model Comparisons

During the 1980s, an interdisciplinary project called **COHMAP** (*Cooperative Holocene Mapping Project*) used a combined data-model approach to examine the last glacial maximum and the subsequent change to interglacial conditions. Led by the meteorologists John Kutzbach and Tom Webb, the paleoecologist Herb Wright, and the geographer Alayne Street-Perrott, COHMAP brought together scientists from countries around the world to pool information from hundreds of individual ^{14}C-dated records of lake levels and pollen in lake sediments in order to examine regional patterns.

The first step in the COHMAP approach was to assemble records of the changing boundary conditions that have driven climate over the last 21,000 years (Figure 13-6). As noted earlier, the largest departures of boundary conditions compared to those today were

the larger ice sheets and the lower greenhouse-gas concentrations (see Figure 13-1).

The COHMAP researchers then ran model simulations of climate at intervals of several thousand years between the glacial maximum and the present to determine how changes in the major boundary conditions drove regional patterns of climate change. The COHMAP team focused on the role of orbital-scale changes in climate over intervals of thousands of years, rather than on shorter-term fluctuations superimposed on this gradual trend.

The climate data produced as output from these model simulations were then tested against climate reconstructions based on ^{14}C-dated records of pollen from lake cores and plankton shells from ocean sediment cores. Modern relationships between the abundances of species and climatic variables can be measured, quantified, and used to reconstruct past climates from fossil organisms. By comparing these fossil-based estimates of past climate with the changes simulated by the models, scientists can test the reliability of both approaches (see Figure 13-6).

13-5 Pollen: An Indicator of Climate on the Continents

Precipitation and temperature determine the larger-scale vegetation units such as forests, grasslands, and deserts, and also the distribution of particular species within those units. Pollen is carried mainly by winds, and to a lesser extent by water and insects. Some pollen comes to rest in lakes and settles into the mud, where its resistant outer layer aids preservation. The preserved pollen reflects the average composition of vegetation over a region extending tens of kilometers from the lakes, with some bias toward tree types that are abundant close to the lakes. The pollen percentages are generally similar to those of the actual vegetation, although "overproducers" such as pine trees leave disproportionately large amounts of pollen compared to "underproducers" such as maples. Climate scientists adjust for this disproportionate representation.

The northern Midwest of the United States is a useful region for demonstrating a clear climatic control on vegetation (Figure 13-7). The percentage of pollen from prairie grasses and herbs is higher in modern lake sediments from more arid regions west of the Mississippi River than in the wetter, tree-dominated area to the east. In the eastern forests, cold-tolerant spruce pollen is more abundant in the north, while oak pollen is more abundant in warmer southern latitudes. These climatic controls can also be demonstrated by plotting pollen percentages against different combinations of seasonal and annual temperature and precipitation (Figure 13-8).

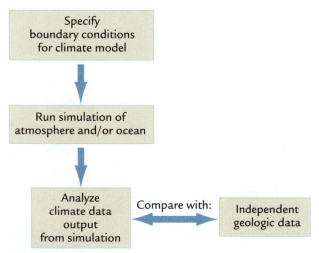

FIGURE 13-6
Data-model comparisons

Past climates can be estimated by running climate model simulations with boundary conditions different from those of today and comparing the model output against estimates derived from pollen in lake sediments or other climatic data.

(ADAPTED FROM J. KUTZBACH ET AL., "CLIMATE AND BIOME SIMULATIONS FOR THE PAST 21,000 YEARS," *QUATERNARY SCIENCE REVIEWS* 17 [1998]: 473–506.)

A Annual precipitation (cm)

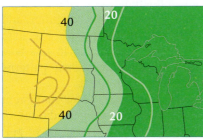

B Annual temperature (°C)

C Prairie pollen (%)

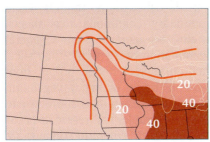

D Spruce pollen (%)

E Oak pollen (%)

FIGURE 13-7

Pollen distributions and climate

Today, the distribution of pollen in the northern midwest of the United States reflects control by precipitation (A) and temperature (B). Prairie grasses and herbs are most abundant where rainfall is low (C), and tree pollen is more common in wetter eastern regions. Spruce trees proliferate in the colder north (D), and oak in the warmer south (E). (ADAPTED FROM T. WEBB III, "HOLOCENE PALYNOLOGY AND CLIMATE," IN *PALEOCLIMATE ANALYSIS AND MODELING*, ED. A. HECHT [NEW YORK: JOHN WILEY, 1985].)

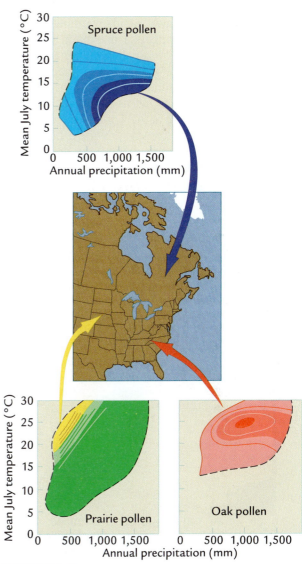

FIGURE 13-8

Pollen percentages and climate

The abundances of spruce, oak, and prairie pollen follow distinct temperature and precipitation patterns. Colors indicate the same pollen abundances as in Figure 13-7.

(ADAPTED FROM T. WEBB III ET AL., "CLIMATIC CHANGE IN EASTERN NORTH AMERICA DURING THE PAST 18,000 YEARS: COMPARISONS OF POLLEN DATA WITH MODEL RESULTS," IN *NORTH AMERICA AND ADJACENT OCEANS DURING THE LAST DEGLACIATION*, ED. W. F. RUDDIMAN AND H. E. WRIGHT [BOULDER, CO: GEOLOGICAL SOCIETY OF AMERICA, 1987].)

These modern relationships are a useful basis for understanding the past. The bottom layers of sediment in a ^{14}C-dated core from Minnesota are late-glacial in age, dating from the time just after the North American ice sheet melted back from this region, whereas the upper layers of mud record the postglacial climate of the present interglaciation (Figure 13-9). Most of the pollen in the older layers is from spruce trees, indicating conditions colder than today. An abrupt switch from spruce pollen to warm-adapted oak near 10,000 years ago (~12,000 years ago in calendar years) indicates rapid warming in this region. Subsequent changes to maximum values of dry-adapted herb and grass pollen culminating near 6,000 years ago indicate a climate drier than today's.

Many hundreds of similar cores have been examined in North America, as well as additional hundreds in Europe and elsewhere in the world (see Figure 3-2). Viewed together, these records provide a broad geographic perspective on the pattern of pollen (and vegetation) distribution at the last glacial maximum and during the deglaciation. This larger map perspective can be compared with map patterns produced by model simulations.

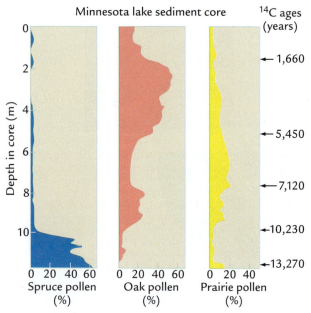

Minnesota lake sediment core

FIGURE 13-9
Pollen in a lake core

A ^{14}C-dated sediment core from a Minnesota lake shows a transition in climate near 10,000 years ago, from colder conditions (abundant spruce) to a warmer climate (abundant oak). High percentages of prairie grasses near 6,000 years ago indicate a drier climate. (ADAPTED FROM H. E. WRIGHT ET AL., "TWO POLLEN DIAGRAMS FROM SOUTHEASTERN MINNESOTA: PROBLEMS IN THE LATE- AND POSTGLACIAL VEGETATION HISTORY," *GEOLOGICAL SOCIETY OF AMERICA BULLETIN* 74 [1963]: 1371–96.)

13-6 Using Pollen for Data-Model Comparisons

Data-model comparisons focus on the distribution of pollen at specific intervals in the past across geographic regions. Counts of pollen percentages in lake sediments in these regions produce mapped patterns of "observed" pollen abundance. These observed patterns are then compared with pollen distributions simulated by climate models for the same interval in the past.

These model-simulated pollen distributions result from several steps. First, boundary conditions are chosen and used in model simulations that yield estimates of past temperature and precipitation. Then the model-derived estimates of temperature and precipitation are used to generate estimates of the percentage abundance of each type of pollen based on the modern relationship between climate and pollen (for example, see Figure 13-8). Each estimate of annual precipitation and mean July temperature simulated for a specific grid box in the initial simulation yields a specific estimate of the percentage of oak, spruce, and prairie pollen for that particular location. The map patterns of pollen abundance estimated in this way can then be compared directly with the map patterns based on pollen counts from lake cores.

For example, observations today show maximum amounts of spruce pollen in the cold conditions of northeastern Canada (Figure 13-10A). Counts of pollen in lake sediments during the last glaciation show spruce concentrated in the east-central United States just south of the ice sheet, implying a cold environment there at that time (Figure 13-10B). These analyses agree fairly well with climate model simulations of the cold regions where spruce should have occurred at the glacial maximum (Figure 13-10C). As we will see shortly, the agreement between models and observations is not always this good.

Comparisons of observed and modeled vegetation for times in the past with different boundary conditions can also be made using **biome models**. Once again, different boundary conditions (ice sheets, greenhouse gases, etc.) are used to drive a general circulation model simulation of a specific time slice. Based on the temperature-precipitation output from the model, the possible range of major vegetation types that can occur in each region is narrowed to larger biome units (for example, no trees can occur in model grid boxes for which hyperarid climates are simulated, but grass and desert scrub vegetation can).

In the second modeling step, the surviving vegetation units in each grid box compete for the resources necessary for growth and reproduction, such as water, nutrients, and light. Both steps are based on observed modern relationships between vegetation and the environment. Because the first step in the

Distribution of spruce pollen

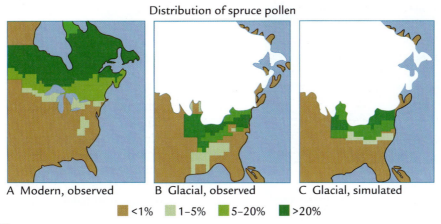

A Modern, observed   B Glacial, observed   C Glacial, simulated

■ <1%   ■ 1–5%   ■ 5–20%   ■ >20%

FIGURE 13-10
Modern and glacial maximum spruce

Today, spruce pollen is most abundant in the cold climate of northeastern Canada (A). At the glacial maximum, spruce pollen is found mainly in lake sediments from the northern United States (B). Model simulations confirm that the large North American ice sheet produced temperatures cold enough for spruce to flourish in the northeastern United States (C). (ADAPTED FROM T. WEBB III ET AL., "LATE QUATERNARY CLIMATE CHANGE IN EASTERN NORTH AMERICA: A COMPARISON OF POLLEN-DERIVED ESTIMATES WITH CLIMATE MODEL RESULTS," *QUATERNARY SCIENCE REVIEWS* 17 [1998]: 587–606.)

biome method encompasses all of the major vegetation groupings on Earth, this approach can simulate changing patterns of vegetation on any continent.

▶ Data-Model Comparisons of Glacial Maximum Climates

In this section, we examine matches and mismatches between model simulations of the climate of the last glacial maximum and observations based on the actual climate record.

13-7 Model Simulations of Glacial Maximum Climates

Glacial ice sheets are a critical boundary condition for simulations of glacial climate (see Chapters 10 and 12). Their central domes protruded upward as massive icy plateaus, redirecting the flow of air above the ice. Climate model simulations suggest that a high-domed ice sheet over North America could have split the winter jet stream into two branches at the glacial maximum.

In modern winters, a single jet stream enters North America near the border between Canada and the United States. Storms associated with this jet bring wet, rainy winters to the coasts of Oregon, Washington State, and British Columbia and snow to the higher terrain (Figure 13-11A). In contrast, during glacial times, the jet stream splits into a northern branch located along the northern flank of the ice sheet and a southern branch over the American Southwest (Figure 13-11B).

Ice sheets did not literally poke high enough into the atmosphere to block the flow of the jet stream. The ice domes reached elevations of 2 to 3 km, whereas winter jet streams flow at altitudes of 10 to 15 km. But the ice blocked the lower-level atmospheric flow, and the disruption was propagated higher into the atmosphere. This disruption, along with the tendency of jet streams to align their axis of flow along regions of strong temperature gradients at Earth's surface, caused the split jet. Model simulations using a high ice sheet (like that of the CLIMAP "maximum" reconstruction) split the jet to a much greater degree than do simulations based on lower-elevation ice.

Climate model simulations indicate that ice sheets caused other major changes in atmospheric circulation at Earth's surface (Figure 13-11B). The models simulate a clockwise downward spiral of cold air around the ice sheets in winter. Cold air flowing eastward along the northern flank of the North American ice sheet as part of this circulation blew southeastward over the western North Atlantic, chilling the ocean surface (one of the ice-driven responses mentioned in Chapter 12). In addition, a narrow layer of cold winds blew from east to west across the northernmost United States, reversing the west-to-east flow that dominates that region today. In Alaska, the clockwise pattern around the ice sheet produced a south-to-north wind flow during the glacial maximum that probably prevented climate in the ice-free Alaskan interior from becoming even harsher than it is today.

A similar clockwise spiral of winds over the Scandinavian ice sheet brought cold, dry air southward

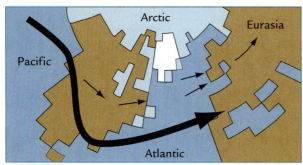

A Modern winters

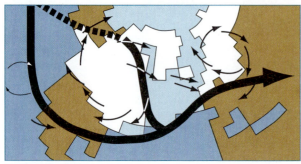

B Glacial winters

| ☐ Sea ice | ↗ Surface winds |
| ☐ Ice sheets | ➤ Jet stream |

FIGURE 13-11
GCM simulation of climate near the northern ice sheets

Simulations run on climate models reproduce the modern path of the winter jet stream over North America (A). For the last glacial maximum, a high-elevation ice sheet over North America splits the jet stream into two branches, one south and one north of the ice (B). At the surface, cold winds flow down off the North American and Scandinavian ice sheets and spiral in a clockwise pattern. (ADAPTED FROM COHMAP PROJECT MEMBERS, "CLIMATIC CHANGES OF THE LAST 18,000 YEARS: OBSERVATIONS AND MODEL SIMULATIONS," *SCIENCE* 241 [1988]: 1043–52.)

into Europe (Figure 13-11B). In addition, a strong upper-level jet stream crossed the Atlantic Ocean along latitudes between 45° and 50°N and entered Europe south of the ice sheet.

13-8 Climate Changes Near the Northern Ice Sheets

The most dramatic changes in climate at the glacial maximum were those in regions closest to and most directly influenced by the ice sheets. Most of the climate changes simulated by the models are consistent with independent geologic evidence.

The CLIMAP reconstruction of sea-surface temperatures based on the shells of planktic organisms shows the largest cooling occurring in the North Atlantic

Ocean (see Figure 13-2B). Frigid water and sea ice reached much farther south than today. The warm waters of the Gulf Stream and North Atlantic Drift flowed eastward toward Portugal instead of penetrating northeastward toward Scandinavia (Figure 13-12A). The flow of cold winds off the North American ice sheet was an important cause of this glacial cooling of the North Atlantic Ocean. Simulations using climate models that allow the ocean surface to react to the cold winds predict changes in sea-surface temperature similar to those estimated by CLIMAP. In summer, the sea ice retreated to the north and the water warmed somewhat, but remained well below modern temperatures.

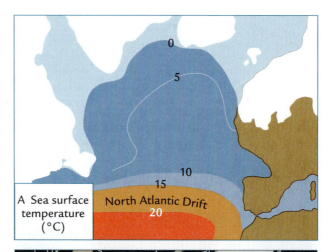

A Sea surface temperature (°C)   North Atlantic Drift

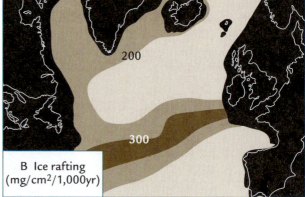

B Ice rafting (mg/cm²/1,000yr)

FIGURE 13-12
A cooler glacial North Atlantic Ocean

The region of largest ocean cooling in the CLIMAP reconstruction is the northern North Atlantic, which is surrounded by ice sheets (A). Highest rates of deposition of ice-rafted debris occurred near 50°N, where southward-floating icebergs first encountered warm waters and melted (B). (A: ADAPTED FROM A. MCINTYRE ET AL., "GLACIAL NORTH ATLANTIC 18,000 YEARS AGO: A CLIMAP RECONSTRUCTION," *GEOLOGICAL SOCIETY OF AMERICA MEMOIR* 145 [1976]: 43–76; B: ADAPTED FROM W. F. RUDDIMAN, "NORTH ATLANTIC ICE RAFTING: A MAJOR CHANGE AT 75,000 YEARS B.P.," *SCIENCE* 196 [1977]: 1208–11.)

Later studies have indicated a larger summer retreat of sea ice and warmer temperatures than those in the CLIMAP reconstruction.

Other large changes accompanied the North Atlantic cooling. A broad band of ice-rafted debris deposited in deep-ocean sediments near 50°N latitude shows that icebergs broke off from continental ice sheets and drifted southward until encountering warm water and melting (see Figure 13-12B).

CHANGES IN NORTH AMERICA An impressive example of agreement between observations and model simulations for the last glacial maximum occurs in the southwestern United States. Today, this area is arid semidesert, except for deep winter snow pack on the mountains, and small lakes maintained by meltwater runoff into the basins. Most runoff is trapped in the basins and never reaches the ocean (Figure 13-13A).

At the last glacial maximum, this region was strikingly transformed, with hundreds of large new lakes where none exist today (see Figure 13-13B). The most prominent of these, glacial Lake Bonneville, was ten times larger than today's Great Salt Lake. Dissolved salt that precipitated out of the brackish water into the lake muds created the Bonneville salt flats.

Climate model simulations of the last glacial maximum provide an explanation for this dramatically wetter regional climate. The southern branch of the split jet stream entered North America over south-central California (see Figure 13-13B). This configuration produced two responses favorable to a moister climate and expansion of lakes: more precipitation caused by winter storms following the path of the jet stream, and reduced evaporation caused by greater cloud cover and cooler temperatures.

In contrast to the wetter southwest, the climate of the Pacific Northwest was drier during the glacial maximum. In that region today, frequent winter storms from the Pacific Ocean bring moisture that sustains lush forests, including rain forests on the western slopes of the Olympic Mountains in coastal Washington. At the last glacial maximum, this region was covered by grass and herb vegetation indicative of much drier conditions and reduced Pacific moisture.

The climate model simulations (see Figure 13-11B) suggest two reasons for this change. First, the shift of the winter jet stream to the southwestern United States displaced the main storm track and associated precipitation away from this region. In addition, the clockwise flow of cold, dry winds around the North American ice sheet produced more frequent low-level winds blowing westward from the dry mid-continent and replacing the flow of moist west-to-east winds from the Pacific.

A

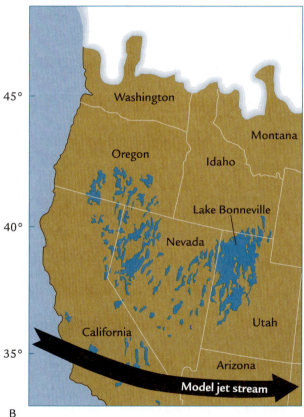

B

FIGURE 13-13
The glacial Southwest was wetter

Today, most basins in the southwestern United States, such as Death Valley, are dry or are occupied only occasionally by temporary lakes (A). At the last glacial maximum, lakes filled hundreds of basins because the southward displacement of the jet stream from Canada brought increased rain and cloud cover (B). (A: PETER KRESAN; B: ADAPTED FROM G. I. SMITH AND F. A. STREET-PERROTT, "PLUVIAL LAKES OF THE WESTERN UNITED STATES," IN *LATE QUATERNARY ENVIRONMENTS OF THE UNITED STATES*, ED. S. C. PORTER [MINNEAPOLIS: UNIVERSITY OF MINNESOTA PRESS, 1983].)

The region with the most extensive coverage of lake cores and pollen data for testing climate models is eastern North America, today an area of temperate deciduous forests. In this region, scientists can test the performance of climate models by checking for data-model agreement or disagreement about the *magnitude* of climate changes, not just the direction.

Pollen from east-central North America south of the ice sheet at the glacial maximum comes from a number of sources: spruce trees, several kinds of deciduous trees, and grasses and herbs. This mixture indicates a region of discontinuous tree cover interrupted by grassy openings. The more continuous forest cover south of 35°N was mainly pine and various deciduous trees.

Although the model-simulated pattern and the observed pattern of spruce in the northern United States at the glacial maximum match reasonably well (see Figure 13-10), this match does not hold up for several pollen types farther to the south. Pollen produced by warm-adapted deciduous trees such as oak and elm is scarce in lake sediments from this region (Figure 13-14A) but far more abundant in the climate model simulations (Figure 13-14B).

This mismatch indicates that the model-simulated cooling for the southeastern United States underrepresents the cooling that actually occurred. One possible cause of this mismatch may have been an unusual geographic configuration in which large volumes of cold meltwater from the southern margin of the great Laurentide ice sheet flowed down the Mississippi River and emptied directly into the Gulf of Mexico at subtropical and tropical latitudes. If the sea-surface boundary conditions used in the model had incorporated this cold inflow, the model might have simulated cooler temperatures across a broad region of the southeastern United States influenced by air masses from the nearby Gulf. Disagreements like these in initial data-model comparisons point the way toward future improvements in model simulations and data interpretation.

CHANGES IN EURASIA Europe was completely transformed at the glacial maximum. The conifer and deciduous forests typical of today's interglacial climate (Figure 13-15A) were eliminated from most of Europe south of the Scandinavian ice sheet. In their place, grass-covered steppes and herb-covered tundra vegetation covered much of the continent, with forest remnants scattered in the south (Figure 13-15B). The moderate maritime climate of today was then a far harsher continental climate, more like that of modern northern Asia. These changes in vegetation agree with the dry, windy conditions indicated by the greater prevalence of windblown loess (see Figure 13-4).

Biome models simulate glacial vegetation in Europe similar to that observed, with Arctic tundra replacing forest in the north near the ice sheets and grassy steppe vegetation prevailing farther to the south and east. One reason for this harsh glacial climate was the large chilling of the North Atlantic Ocean, which suppressed its moderating influence on winters in Europe (see Figure 12-3). A second reason was the clockwise outflow of cold winds from the Scandinavian ice sheet (Figure 13-11B).

One of the most striking features of the last glacial maximum was the vast extent of steppe and tundra that covered much of northern Asia (Figure 13-16). A region covered today by forests of larch, birch, and alder trees was at that time a treeless expanse of grasses and herbs. Forests were completely absent from the entire northern part of Asia, an indication that this region had an even harsher continental climate than it does today.

The harsh winter cold and sparse snow cover caused the ground to freeze tens of meters deep, forming **permafrost**, but in most regions the surface layer thawed in summer and allowed grasses and herbs to proliferate. South of the year-round and seasonal permafrost were extensive grassy steppes (see Figure 13-16A). The greater area covered by tundra and steppe (rather than today's forests) made the ground surface much more reflective in the snowy season and further cooled this region.

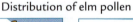

Distribution of elm pollen

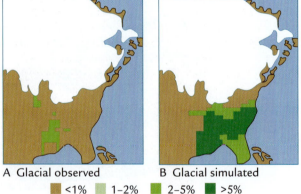

A Glacial observed   B Glacial simulated

<1%   1–2%   2–5%   >5%

FIGURE 13-14

Data-model mismatch in the southeastern United States

Observed abundances of warm-adapted deciduous pollen such as elm in the southeastern United States during the glacial maximum (A) are much smaller than the amounts simulated by climate models (B). (ADAPTED FROM T. WEBB III ET AL., "LATE QUATERNARY CLIMATE CHANGE IN EASTERN NORTH AMERICA: A COMPARISON OF POLLEN-DERIVED ESTIMATES WITH CLIMATE MODEL RESULTS," *QUATERNARY SCIENCE REVIEWS* 17 [1998]: 587–606.)

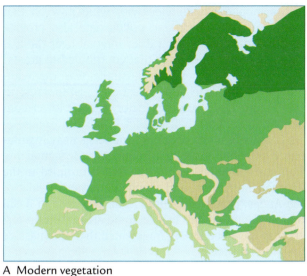

A Modern vegetation

☐ Ice   ■ Arctic forest   Mediterranean scrub
  Tundra and   ■ Deciduous   Grassland and steppe
  mountain   and conifer forest

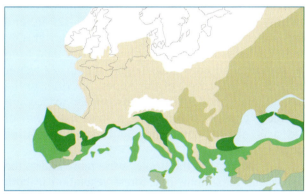

B Glacial vegetation

FIGURE 13-15
Glacial north-central Europe was treeless

Natural vegetation in modern Europe is dominated by forest, with conifers in the north and deciduous trees to the south (A). At the glacial maximum, Arctic tundra covered a large area south of the ice sheet, with grassy steppe farther south and east, and forests reduced to patches near the Mediterranean coasts (B). (ADAPTED FROM R. F. FLINT, *GLACIAL AND QUATERNARY GEOLOGY* [NEW YORK: JOHN WILEY, 1971].)

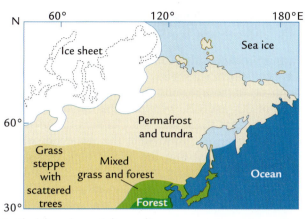

A Glacial maximum (observed)

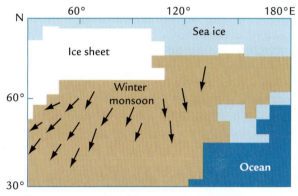

B Glacial maximum (model)

FIGURE 13-16
Glacial northern Asia was treeless

At the last glacial maximum, Asia was covered by permafrost and tundra in the north and steppe in the south, with little forest left (A). Climate models indicate that this distribution of vegetation resulted from a much stronger and colder winter high-pressure cell in northern Asia, with stronger cold winds blowing to the south (B). (A: ADAPTED FROM V. P. GRICHUK, "LATE PLEISTOCENE VEGETATION HISTORY," IN *LATE QUATERNARY ENVIRONMENTS OF THE SOVIET UNION*, ED. A. A. VELICHKO ET AL. [MINNEAPOLIS: UNIVERSITY OF MINNESOTA PRESS, 1984]; B: ADAPTED FROM J. E. KUTZBACH ET AL., "SIMULATED CLIMATIC CHANGES: RESULTS OF THE COHMAP CLIMATE MODEL EXPERIMENTS," IN *GLOBAL CLIMATES SINCE THE LAST GLACIAL MAXIMUM*, ED. H. E. WRIGHT ET AL. [MINNEAPOLIS: UNIVERSITY OF MINNESOTA PRESS, 1993].)

Climate model simulations suggest that the main reason for this colder, drier glacial climate in Asia was a much stronger winter high-pressure cell in Siberia that greatly increased the outflow of cold, dry air southward (see Figure 13-16B). In addition, icy conditions in the North Atlantic cut off much of that other moisture source to the Asian interior.

The effects of these harsh winters in northern Asia extended into southeastern Asia. Today, the influence of the warm, moist summer monsoon allows forests to grow along the Pacific coasts of China and Japan. At the glacial maximum, the stronger and more persistent winter monsoon outflow sent cold air southward from Siberia and pushed the northern forest limit to the south (see Figure 13-16). The North Pacific was also colder than it is today, with much more extensive winter sea ice along the coast of Asia and in the Bering Sea because of cold Siberian air blowing out over the ocean.

13-9 Climate Changes Far from the Northern Ice Sheets

Farther from the direct influence of the great northern hemisphere ice sheets, climate changes were less dramatic. In these regions, the lower glacial concentrations of CO_2 and CH_4 were probably the major cause of the changes in climate (see Chapters 11 and 12).

Large changes occurred in the Antarctic, where CLIMAP estimated that the winter limit of sea ice expanded northward by several degrees of latitude in the far-southern Atlantic and Indian oceans (Figure 13-17). Later reconstructions have reduced these limits, but only slightly. Associated with this shift in sea ice was a northward displacement of the region of strongest upwelling and highest surface-water productivity, but productivity decreased in regions nearer Antarctica where the cover of sea ice persisted longer during summer.

The record of climate change on arid southern hemisphere continents remains sparse. Expanded desert dunes in Australia (see Figure 13-5) suggest an even more arid climate and an intensification

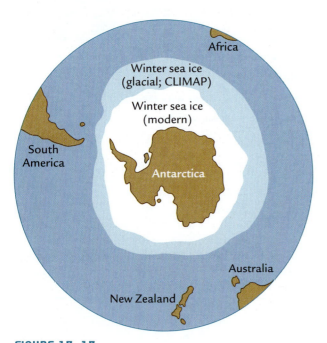

FIGURE 13-17
Glacial Antarctica was surrounded by more sea ice

The CLIMAP glacial maximum reconstruction indicated that the seasonal maximum limit of sea ice in late winter and early spring expanded northward around Antarctica. (ADAPTED FROM J. D. HAYS, "A REVIEW OF THE LATE QUATERNARY HISTORY OF ANTARCTIC SEAS," IN *ANTARCTIC GLACIAL HISTORY AND WORLD PALEOENVIRONMENTS*, ED. E. M. VAN ZINDEREN BAKKER [ROTTERDAM: A. A. BALKEMA, 1978]; AND FROM L. H. BURCKLE ET AL., "DIATOMS IN ANTARCTIC ICE CORES: SOME IMPLICATIONS FOR THE GLACIAL HISTORY OF ANTARCTICA," *GEOLOGY* 16 [1988]: 326–29.)

of the modern counterclockwise wind flow on that continent. Lake levels and pollen data also provide supporting evidence for a drier interval near the glacial maximum. One factor that contributed to greater glacial aridity in northern Australia was the withdrawal of the ocean from a vast area just to its north (see Figure 13-2). Another factor was the lower concentrations of greenhouse gases in the atmosphere.

Climate in much of South America is heavily influenced by winds from nearby oceans. Most moisture from the Atlantic Ocean to the east is dropped in the Amazon rain forest and along the eastern flanks of the Andes. Scattered records from the Amazon Basin hint at the possibility that in some areas the rain forest may have been more fragmented than the massive area forested today.

Along the Andes, where most lake-sediment records from South America exist, pollen data generally indicate drier conditions at the glacial maximum. This drying is probably the combined result of reduced extraction of water vapor from the cooler oceans, the lowering of sea level by 110–125 m, and the cooler temperatures resulting from lower CO_2 and CH_4 levels in the atmosphere. Pollen data from far southern latitudes indicate glacial climates wetter than those today west of the Andes, but drier to the east. Climate model simulations appear to account for these changes by showing a southward shift of the axis of westerly winds and moisture-bearing storms.

Because most of the tropics were more arid at the last glacial maximum, rain forest vegetation in both South America and Africa was probably less extensive than it is today. Yet despite this drier climate, total tropical biomass was not necessarily lower. The large drop in global sea level exposed vast expanses of new land across continental shelves of Southeast Asia (see Figure 13-2). Because this region lay in the moist intertropical convergence zone, it would likely have supported tropical rain forest vegetation. The increase in rain forest in this region may have offset the loss of biomass elsewhere in the tropics, but it could not offset the enormous decrease in forest biomass at high northern latitudes. As a result, the total glacial biomass on Earth's continents was reduced by about 25%.

How Cold Were the Glacial Tropics?

For almost two decades, climate scientists have argued about the size of the temperature changes in the tropics and subtropics. Tropical sea-surface temperatures reconstructed by CLIMAP on the basis of fossil shells of ocean plankton averaged just 1°–2°C cooler than they are today, and in some regions, such as in the subtropical Pacific, the ocean was estimated to have been more than 1°C warmer.

In contrast, other evidence suggests that temperatures over tropical landmasses and parts of the tropical ocean may have been 4°–6°C cooler than at present, far larger than the difference in the CLIMAP reconstruction. This discrepancy in estimates has large ramifications concerning Earth's basic sensitivity to changes in atmospheric CO_2 and other greenhouse gases.

The tropics lie too far from the immediate thermal impact of the ice sheets to have been cooled by changes in atmospheric circulation or by ice-albedo feedback (see Chapter 12). In addition, solar insolation values at the last glacial maximum were close enough to those today that they could not have been a major factor in the glacial cooling of the tropics. What does explain the cooling in the tropics?

By process of elimination, the main cause must have been the 30% lower (190 ppm) level of carbon dioxide, along with the 50% drop in methane (Figure 13-18). When greenhouse-gas concentrations are lower, less of the outgoing back radiation from Earth's surface is trapped in the atmosphere and the temperature falls. As a result, the amount of glacial cooling in the tropics should be largely a measure of the sensitivity of this part of the climate system to changes in carbon dioxide and methane. With half of Earth's surface area lying between 30°N and 30°S, this

cooling should give us a measure of the fundamental sensitivity of the climate system.

In Chapter 6, we examined Earth's response to higher levels of CO_2 in a greenhouse world. The last glacial maximum now provides a complementary perspective on the same relationship: Earth's response to lower levels of CO_2 in an "icehouse" world. This analysis is directly relevant to future climate change because the warming we face in the future will be caused by human-induced increases in atmospheric CO_2 and CH_4, and we need to know how large this warming will be.

13-10 Evidence for a Small Tropical Cooling

The evidence for a small tropical cooling in the CLIMAP reconstruction was based on the small changes in planktic fauna and flora in low-latitude oceans. CLIMAP's technique for reconstructing sea-surface temperatures used the assumption that the distribution of species and assemblages of plankton is mainly determined by the temperature of the water in which they live. At higher northern latitudes during the glacial maximum, cold-adapted species moved into areas where warm-adapted species prevail today, indicating a large cooling in these regions. Across most low-latitude regions, however, the species that existed at the glacial maximum were not much different from the warm-adapted forms found there today (Figure 13-19). This lack of change in tropical plankton was the main reason CLIMAP concluded that ocean temperatures in the tropics cooled by an average of only 1.5°C at the glacial maximum.

13-11 Evidence for a Large Tropical Cooling

A different view emerges from other indicators, most of which come from continental records. The most compelling evidence is the descent of the lower limit of mountain glaciers by 600 to 1,000 m throughout the tropics and middle latitudes (Figure 13-20). This drop in the elevation of the ice line has been interpreted as requiring a cooling of 4°–6°C over tropical mountains.

The lower limit of mountain glaciers today is determined mainly by temperature and secondarily by factors such as the amount of precipitation and the degree to which local mountain topography shelters the glaciers from direct sunlight. Glaciers exist today on tropical mountains higher than 5 km because the atmosphere cools by 6.5°C or more per kilometer of elevation, resulting in subfreezing temperatures at higher elevations (see Chapter 2). Based on this relationship, the observed lowering of tropical mountain

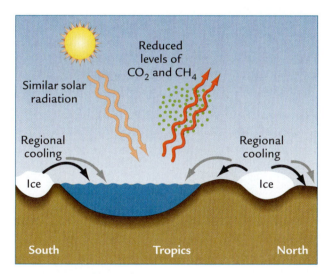

FIGURE 13-18
Lower CO_2 and CH_4 levels cooled the glacial tropics

The tropics were too distant from the glacial ice sheets to feel their direct influence, and insolation values in summer and winter were close to those today. Lower levels of atmospheric CO_2 and CH_4 were the main cause of tropical cooling.

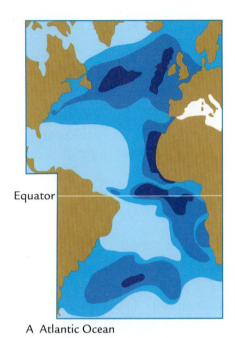

Equator

A Atlantic Ocean

B Indian Ocean

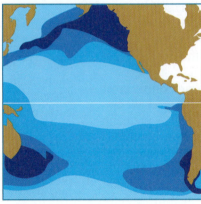

C Pacific Ocean

Glacial plankton vs. plankton today
(Percent difference)

■ > 50   ■ 25–50   ■ 10–25   ■ < 10

FIGURE 13-19

Planktic fauna of the glacial maximum versus that of today

The CLIMAP method of reconstructing glacial maximum ocean temperatures was based on temperature-sensitive plankton assemblages. Planktic assemblages in most low-latitude regions of the Atlantic (A), Indian (B), and Pacific (C) oceans differed only slightly from those of today, indicating little glacial cooling. (ADAPTED FROM T. C. MOORE ET AL., "THE BIOLOGICAL RECORD OF THE ICE-AGE OCEAN," *PALAEOGEOGRAPHY, PALAEOCLIMATOLOGY, PALAEOECOLOGY* 35 [1981]: 357–70.)

glaciers by 600 to 1,000 m during the glacial maximum would require a cooling of 4°–6°C. Additional evidence for larger glacial cooling comes from the descent of the upper tree limit and other specific kinds of vegetation high on tropical mountains. In the harsh conditions on the upper flanks of mountains, temperature limits the growth of many kinds of vegetation, and the vertical drop in high-altitude vegetation limits during the last glaciation equals or even exceeds that of the mountain glaciers (see Figure 13-20).

13-12 The Actual Cooling was Medium-Small

After years of disagreement, a resolution of this problem seems to be emerging—the tropical cooling was neither as small as CLIMAP claimed, nor as large as the critics initially thought, but about midway between the two estimates. One reason the CLIMAP estimates were too small is that plankton are less sensitive to changes in temperature at low latitudes than to changes in the availability of food. The low-latitude surface ocean is depleted in nutrients, and plankton are forced to adopt strategies for surviving where food is scarce. In some regions, this requirement overwhelms the relatively small dependence on temperature.

In addition, the CLIMAP reconstruction for the Pacific Ocean was based on samples in which $CaCO_3$ had been extensively dissolved on the seafloor, thereby altering the assemblages of foraminifera and coccoliths. The remnants of siliceous organisms (radiolaria and diatoms) left in the sediments are also different from those that originally lived in the surface waters during the peak glaciation because of chemical

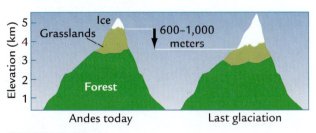

FIGURE 13-20

Descent of tropical mountain glaciers and forests

The limits of mountain glaciers in the Andes were 600 to 1,000 m lower during the last glaciation than they are today, and the upper limits of forests were similarly lower. These major shifts indicate a tropical cooling of at least 4°C, much larger than the 1°–2°C suggested by CLIMAP. (ADAPTED FROM T. VAN DER HAMMEN, "THE PLEISTOCENE CHANGES OF VEGETATION AND CLIMATE IN TROPICAL SOUTH AMERICA," *JOURNAL OF BIOGEOGRAPHY* 1 [1974]: 3–26.)

attack. The estimates of surface temperatures in the Pacific derived from these different types of plankton often disagree, an indication that at least some estimates are unreliable in those areas, if not all.

Evidence obtained from the biochemical composition of plankton shells has been used to test the CLIMAP estimates. One technique is based on the relative abundance of complex organic molecules called **alkenones** that constitute small fractions of tiny plant plankton (coccolithophores). The past abundances of these molecules can be measured in small $CaCO_3$ plates (coccoliths) deposited in ocean sediments (see Figure 3-14). The relative amounts of two types of alkenone molecules are sensitive to temperature in the modern ocean and can be used to reconstruct past temperatures.

In a north-south transect of cores across the western Indian Ocean, the cooling indicated by the two methods (the CLIMAP census counts and the alkenone analyses) is generally less than 2°C, confirming the small CLIMAP cooling for most of this region (Figure 13-21). But at latitudes north of 15°N, the alkenone method indicates a cooling of 2°C or more, while many CLIMAP estimates indicate a warming. Temperatures estimates based on Mg/Ca ratios measured in surface-water foraminifera (see Chapter 7) have also been made for several regions, with a similar result: some areas are cooler than the CLIMAP estimates, and other areas are in agreement.

The large-cooling view also has its own problems. The drier glacial climate in most of the tropics would have steepened the lapse rate from its present 6.5°C/km toward the 9.8°C/km rate typical of very dry air. A steeper lapse rate in the drier glacial

tropics could account for part of the discrepancy between the ocean and land evidence. In addition, the evidence from mountain glaciers is poorly dated. Only a handful of regions have ^{14}C dates that closely constrain the glacial lowering of 600–1,000 m to the exact time of the maximum glaciation. Some glacial moraines initially thought to date from the glacial maximum have turned out to be 30,000 years old or earlier, a time of cooler but also wetter climates. Lower glacier limits at such times could have been caused at least in part by greater snowfall.

Other factors contribute to the apparent discrepancy between land and ocean temperature changes. One factor is sea level lowering. A sea level drop of 110–125 m increases the "height" of the mountains (relative to the lowered sea level) by that amount. For a lapse rate of 6.5°C per 1,000 m, this change would make high mountain elevations cooler by ~0.75°C relative to the sea surface, without any actual change in climate. Another factor is the greater responsiveness of the land than the ocean to climatic forcing. Climate model simulation of land surface reactions to seasonal changes tend to exceed those of the ocean, which integrates year-round forcing. Still another factor is the fact that the very low glacial CO_2 values may fail to provide trees with enough CO_2 for photosynthesis and thereby cause the upper tree line to drop even without changes in temperature.

In Summary, the most likely resolution to the controversy over the glacial tropical cooling is that the cooling in some ocean regions (especially the Pacific Ocean) was larger than the CLIMAP estimate, and that the cooling in many tropical land areas was not as large as the evidence from the mountains suggests.

What does all this evidence imply about Earth's sensitivity to CO_2 and other greenhouse gases? We know for certain from measurements of air bubbles in ice cores that the glacial atmospheric CO_2 concentration was 90 ppm lower than the typical interglacial value of ~280 ppm, while the glacial CH_4 concentration was about half of typical interglacial values of ~700 ppb. And we can estimate from the various climate indicators that the actual tropical cooling was about 3°C, roughly midway between the CLIMAP estimate (1.5°C) and the initial land-based estimate (5°C).

A 3°C drop in tropical temperatures is close to the cooling that general circulation models simulate for this amount of lowering of greenhouse-gas concentrations. The evidence from the last glacial maximum indicates that the models capture the effect of greenhouse gases on climate reasonably well. This match gives climate scientists confidence that these same models are useful in forecasting future climate change caused by increases in CO_2 and other greenhouse gases.

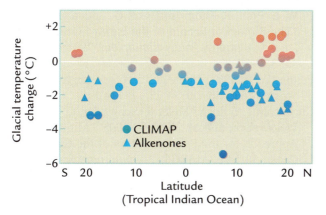

FIGURE 13-21

Glacial cooling in the Indian Ocean

A biochemical ("alkenone") method of estimating past sea-surface temperatures indicates a small cooling of much of the tropical Indian Ocean, similar to the pattern found by CLIMAP, but a larger cooling north of 15°N. (ADAPTED FROM E. BARD ET AL., "INTERHEMISPHERIC SYNCHRONY OF THE LAST DEGLACIATION INFERRED FROM ALKENONE PALEOTHERMOMETRY," *NATURE* 385 [1997]: 707-10.)

Key Terms

CLIMAP (*Climatic Mapping* and *Prediction*) Project (p. 254)

Laurentide ice sheet (p. 256)

Cordilleran ice sheet (p. 256)

Scandinavian ice sheet (p. 256)

Barents ice sheet (p. 256)

glacial outwash (p. 258)

COHMAP (*Cooperative Holocene Mapping Project*) (p. 260)

biome models (p. 262)

permafrost (p. 266)

alkenones (p. 271)

Review Questions

1. What is the major uncertainty about the total volume of ice sheets at the glacial maximum?

2. Earth's radius (r) is 6,371 km, and its surface area ($4/3\pi r^3$) is 70% water. If sea level fell by 120 m during the most recent glacial maximum, and if ice sheets expanded from a modern surface area of 15 million km^2 to 45 million km^2, what was the average thickness of the ice sheets?

3. In what ways did ice sheets make the glacial world a "dirtier" place?

4. How does the composition of pollen in lake sediments tell us about climate change?

5. How and why did the glacial climate of the southwestern United States differ from the climate there today? Did the changes there generally agree with those in most other regions?

6. How and why did glacial climates of Europe and northern Asia differ from the climate there today?

7. What caused the cooling of the tropics during the last glacial period?

8. Explain why a large versus small tropical cooling during the last glacial maximum is important for understanding our future.

Additional Resources

Basic Reading

CLIMAP Members. 1981. *Seasonal Reconstruction of the Earth's Surface at the Last Glacial Maximum.* Map and Chart Series MC-36. Boulder, CO: Geological Society of America.

COHMAP Members. 1988. "Climatic Changes of the Last 18,000 Years: Observations and Model Simulations." *Science* 241: 1043–62.

Imbrie, J., and K. P. Imbrie. 1979. *Ice Ages: Solving the Mystery.* Short Hills, NJ: Enslow.

Rind, D., and D. Peteet. 1985. "Terrestrial Conditions at the Last Glacial Maximum and CLIMAP Sea-Surface Temperature Estimates: Are They Consistent?" *Quaternary Research* 24: 1–22.

Advanced Reading

Clark, P. U., J. M. Licciardi, D. R. MacAyeal, and J. W. Jenson. 1996. "Numerical Reconstruction of a Soft-Bedded Laurentide Ice Sheet During the Last Glacial Maximum." *Geology* 24: 679–82.

Fairbanks, R. G., and P. H. Wiebe. 1980. "Foraminifera and Chlorophyll Maximum: Vertical Distribution, Seasonal Succession, and Paleoceanographic Significance." *Science* 209: 1524–26.

Mix, A. C., E. Bard, and R. Schneider. 2001. "Environmental Processes of the Ice Age: Land, Oceans, Glaciers (EPILOG)." *Quaternary Science Reviews* 20: 627–57.

Moore, T. C., Jr., W. H. Hutson, N. Kipp, J. D. Hays, W. L. Prell, P. Thompson, and G. Boden. 1981. "The Biological Record of the Ice-Age Ocean." *Palaeogeography, Palaeoclimatology, Palaeoecology* 35: 357–70.

Peltier, W. R. 1994. "Ice Age Paleotopography." *Science* 265: 195–201.

Kucera, M., A. Rosell-Melé, R. Schneider, C. Walebroeck, and M. Weinelt. 2005. "Multiproxy Approach for the Reconstruction of the Glacial Ocean Surface (MARGO)." *Quaternary Science Reviews* 24: 813–19.

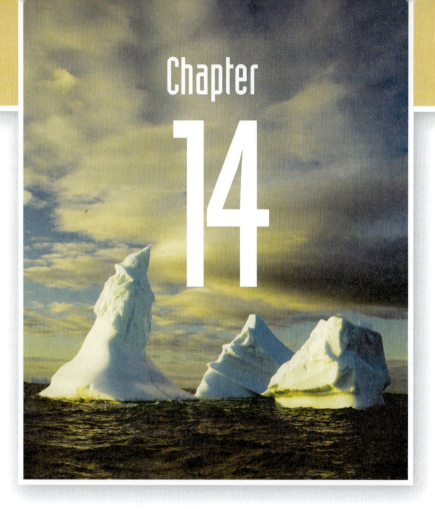

Chapter 14

Climate During and Since the Last Deglaciation

Earth was transformed following the last glacial maximum. The melting ice sheets sent enough water to the ocean to raise global sea level by 110–125 meters. The rising ocean submerged glacial maximum links between continents and islands, and it flooded basins that had earlier been cut off from the sea. Meltwater lakes formed in bedrock depressions left by the retreating ice. Ice lobes dammed these lakes but were periodically breached, sending catastrophic floods across some regions. Forests and tundra gradually moved north to occupy broad regions abandoned by the ice, in some regions penetrating beyond their present limits before retreating in recent millennia. Tropical monsoons strengthened until 10,000 years ago and then weakened.

The abundance of well-dated records permits testing of two theories proposed as explanations of major climatic changes: the Milankovitch theory that insolation controls the size of ice sheets and the Kutzbach theory that low-latitude insolation controls the strength of tropical monsoons. In general, the data confirm both theories: rising summer insolation in the Northern Hemisphere initiated melting of high-latitude ice sheets and strengthened tropical monsoons. Subsequent weakening of monsoons and cooling of high northern latitudes during the last 7,000 years are consistent with decreasing summer insolation in the Northern Hemisphere.

Fire and Ice: A Shift in the Balance of Power

Two factors largely explain why climate 21,000 years ago was different from today: the presence of northern hemisphere ice sheets and lower atmospheric greenhouse-gas concentrations (primarily CO_2). During the subsequent deglaciation, an important shift occurred in the balance of power among the factors that had been controlling global climate (Figure 14-1). Summer and winter insolation values that had been near modern levels at the last glacial

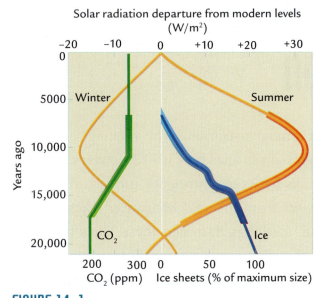

FIGURE 14-1

Causes of climate changes during deglaciation

During the deglacial interval between 17,000 and 6,000 years ago, climate changes were driven by rising summer insolation and by increased concentrations of CO_2 in the atmosphere; as the ice sheets shrank, their ability to influence climate diminished. (ADAPTED FROM J. E. KUTZBACH ET AL., "CLIMATE AND BIOME SIMULATIONS FOR THE PAST 21,000 YEARS," *QUATERNARY SCIENCE REVIEWS* 17 [1998]: 473–506.)

maximum began to change (the former increasing, the latter decreasing). By 10,000 years ago, the angle of tilt of Earth's axis had reached a maximum at the same time that Earth's precessional motion moved it closest to the Sun on June 21. These orbital changes combined to produce a summer insolation maximum at all latitudes of the Northern Hemisphere, with important climatic consequences.

The rising summer insolation at higher northern latitudes triggered melting of the northern ice sheets. As the ice sheets melted, their influence on climate began to diminish, and the insolation anomalies became more important. The major features of the most recent deglaciation are mainly related to this shift in the balance of power from ice (sheets) to fire (solar insolation). A second important change during this deglaciation was the increase in atmospheric CO_2 concentrations from 190 to 280 ppm (see Figure 14-1), along with a doubling of methane levels. As noted in previous chapter, the increases in greenhouse gases coincided closely with ice melting.

14-1 When Did the Ice Sheets Melt?

Abundant evidence available from the recent deglaciation gives scientists an unusual opportunity to test explanations about how and why deglaciations occur. The Milankovitch theory (see Chapter 10) predicts that the orbital maximum in summer insolation near 10,000 years ago in the Northern Hemisphere should have caused high rates of ice melting.

At first, it might seem that the way to quantify the rate of deglacial ice melting would be to measure the gradual retreat of the ice sheet margins. Radiocarbon dating of organic material found in, under, or atop hundreds of ice-deposited moraines shows that the retreat of the large ice sheet in North America began near 15,000 ^{14}C years ago, reached a midpoint near 10,000 ^{14}C years ago, and ended by 6,000 ^{14}C years ago (Figure 14-2). The smaller Scandinavian ice sheet disappeared a few thousand years earlier. The timing of these retreats agrees with the Milankovitch theory.

Unfortunately, this method is not sufficient. Knowing the area covered by the retreating ice is a good start, but a complete analysis requires that measurements be converted to ice *volume*. To make this conversion, we need to know the thickness of the ice as it retreated (thickness × area = volume). To complicate this calculation, the thickness of an ice sheet can be affected by the conditions in its basal layer. Portions of ice sheets that repeatedly slide on their basal layers are thin and relatively low in volume; areas frozen to their beds are thicker and larger in volume. Because of this uncertainty about thickness, records of changing ice area through time do not guarantee valid records of ice volume.

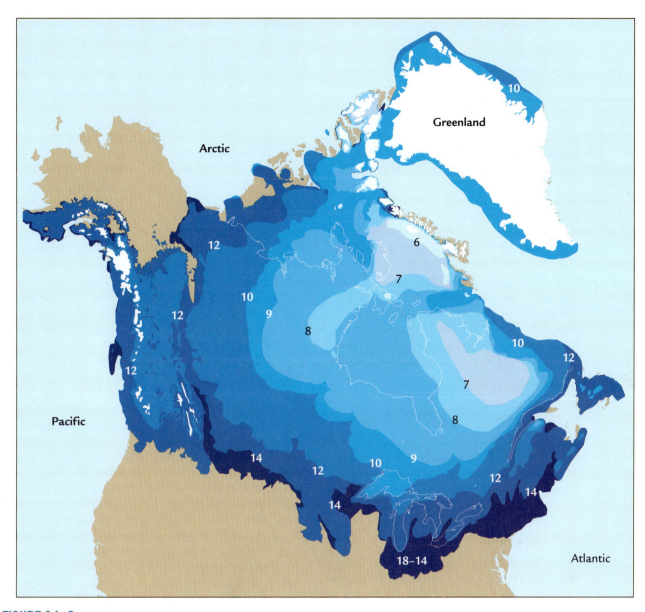

FIGURE 14-2
Retreat of the North American ice sheets
Radiocarbon dating of organic remains shows that the margins of ice sheets in North America began to retreat near 14,000 ^{14}C years ago, and the ice disappeared completely shortly after 6,000 years ago. The numbers indicate ^{14}C-dated ice limits in thousands of years. True "calendar-year" ages for these positions would be a few thousand years older. (COURTESY OF ARTHUR DYKE, GEOLOGICAL SURVEY OF CANADA, OTTAWA.)

14-2 Coral Reefs and Rising Sea Level

The best record of ice sheet melting comes from tropical coral reefs far from the polar ice sheets (see Chapter 10). Because several coral species grow just below sea level, the current elevation of older reefs built by corals can be used as a measure of past sea level, once the effect of tectonic movement of bedrock since the corals died is removed.

Changes in sea level are related to changes in global ice volume for the simple reason that continental ice sheets are made of water taken from the sea. Coral reef measurements of lower sea level during the last glacial maximum and the subsequent deglaciation can be converted to a record of global ice volume, with each 1-meter rise of sea level equivalent to about 0.4 million km^3 of ice.

In the late 1980s, the marine geochemist Richard Fairbanks drilled and ^{14}C-dated a series of now-submerged coral reefs off Barbados, a Caribbean island. These reefs yielded a history of sea level rise from its low extreme at the last glacial maximum to its higher position during the modern interglaciation (Figure 14-3). Barbados is a region of slow tectonic uplift, and the modern depth of each dated coral reef had to be adjusted downward by a few meters to remove this effect. The ^{14}C-dated deglacial sea level curve at Barbados supports the Milankovitch theory in a general way: the middle of the deglaciation occurred near the insolation maximum 10,000 years ago, as expected. But the story is not that simple because the ^{14}C dates on the corals do not represent their true ages.

The evidence presented in Box 14-1 indicates that the dates of the rise in sea level from the Th/U method are more accurate than those from the ^{14}C method. The Th/U chronology shifts the timing of the middle part of the deglaciation back in time by about 2,000 years, and the earlier parts of the deglaciation by as much as 3,500 years (see Figure 14-3). In contrast, the timing of the summer insolation signal remains fixed by the independent (and highly accurate) astronomical time scale. As a result, the major rise in sea level shifts back somewhat earlier in time compared to the insolation curve.

Does this earlier timing for the deglaciation invalidate the Milankovitch theory? In a larger sense, it does not. Milankovitch chose summer as the critical season of insolation control of ice sheets, and the last deglaciation still occurred during a time when summer insolation was higher than it is now, although somewhat earlier in that interval than the Milankovitch theory predicts.

14-3 Rapid Early Deglaciation

One important feature of the coral reef record is the fact that sea level did not rise smoothly throughout the deglaciation (see Figure 14-3). After a moderately rapid rate of rise between 19,000 and 14,500 years ago (an interval constrained by very few data points), sea level rose quickly between 14,500 and 13,000 years ago, after which the rate was considerably slower. The earlier and more rapid sea level rise accounts for the somewhat earlier sea level response compared to the insolation increase, but where and why did this early ice melting occur?

One way to monitor the melting of individual ice sheets is to look for local pulses of meltwater delivery to the oceans. Because the $\delta^{18}O$ values of the northern ice sheets were −30‰ to −35‰, whereas those in the surface ocean were near 0‰, major influxes of meltwater should be registered as distinct pulses of lower $\delta^{18}O$ values in the shells of planktic organisms living in the ocean.

Planktic foraminifera in the northeastern Norwegian Sea record a pulse of unusually negative $\delta^{18}O$ values early in the deglaciation (Figure 14-4). Because the species composition of surface-dwelling microfossils preserved in the same cores rules out the possibility that major temperature fluctuations caused this $\delta^{18}O$ oscillation, the only alternative left is an episode of early melting and low-$\delta^{18}O$ meltwater from the nearby Barents ice sheet, north of Scandinavia. This evidence from marine sediments suggests that the moderately large marine ice sheet, with its base lying below sea level in the Barents Sea, was vulnerable to early collapse when summer insolation began to rise. A similar low-$\delta^{18}O$ pulse in cores from the Gulf of Mexico indicates a short-term increase in meltwater flow down the Mississippi River from the southern margin of the North American ice sheet (Figure 14-4).

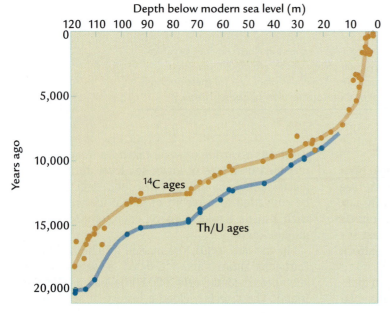

FIGURE 14-3
Deglacial rise in sea level

Submerged corals off Barbados, in the Caribbean, show the deglacial history of the rise in sea level caused by the return of meltwater from the ice sheets to the ocean. (ADAPTED FROM R. G. FAIRBANKS, "A 17,000-YEAR GLACIO-EUSTATIC SEA LEVEL RECORD: INFLUENCE OF GLACIAL MELTING RATES ON THE YOUNGER DRYAS EVENT AND DEEP-OCEAN CIRCULATION," *NATURE* 349 [1989]: 637–42; AND FROM E. BARD ET AL., "CALIBRATION OF THE ^{14}C TIME SCALE OVER THE PAST 30,000 YEARS USING MASS SPECTROMETRIC U-TH AGES FROM BARBADOS CORALS," *NATURE* 345 [1990]: 405–10.)

Tools of Climate Science

Box 14-1

Deglacial ^{14}C Dates Are Too Young

When the same Barbados corals that had been dated by the ^{14}C method were dated by the thorium (Th)/uranium (U) technique, the Th/U ages turned out to be older than the ^{14}C ages by amounts that increased back in time. For samples with a ^{14}C age of 8,000 years, the Th/U age was older by 1,000 years, while for samples with a ^{14}C age of 18,000 years, the age difference increased to 3,500 years.

With both sets of ages giving consistent-looking trends, scientists faced the problem of deciding which (if either) was correct. Fortunately, independent evidence was available from earlier work on the annual rings in long-lived trees. Individual rings in these trees had been dated both by the ^{14}C method and by counting backward year by year from the modern ring during the year the tree was cored.

The ages derived from counting rings turned out to be older than those from the ^{14}C analyses. Near 8,000 to 9,000 years ago, the ages from the tree ring counts were older than those from the ^{14}C analyses by 1,000 years, the same as the offset of the Th/U coral ages from the ^{14}C ages. This agreement of two independent methods suggests that the ^{14}C ages are in error (younger than the actual ages).

The main reason the ^{14}C ages are too young is that the rate of production of ^{14}C atoms in Earth's atmosphere has varied in the past. Atoms of ^{14}C are produced when cosmic particles entering our atmosphere from other galaxies transform ^{14}N atoms into radioactive ^{14}C atoms, which then slowly decay away. The ^{14}C dating method is based on the assumption that ^{14}C has been produced in the atmosphere at a constant rate through time.

Because the rate of cosmic bombardment was higher during glacial times than now, more ^{14}C atoms were produced. Although many of those ^{14}C atoms have undergone radioactive decay, some still remain in the carbon-bearing material used for ^{14}C dating. This surviving excess of ^{14}C atoms caused by higher production makes it look as if less ^{14}C has decayed (and less time has elapsed) than is actually the case.

Why would cosmic bombardment have been higher in the past? Earth is partly shielded from cosmic rays by its magnetic field. If this shield has at times been weaker than it is today, it would have provided less protection against cosmic bombardment. Two observations confirm this explanation. First, other elements known to be the

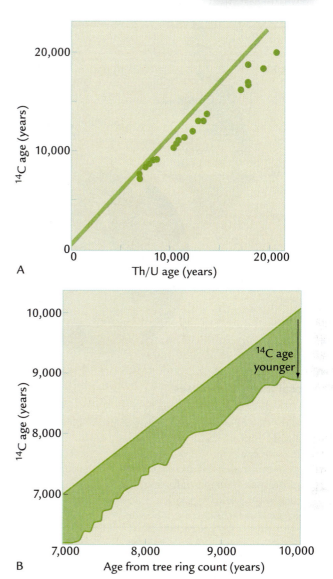

A

B

Offset of ^{14}C ages Th/U dates from Barbados corals are older than ^{14}C ages by as much as 3,500 years (A). Ages derived by counting tree rings backward in time are offset from ^{14}C dates on the same layers by a similar amount (B). (A: ADAPTED FROM E. BARD ET AL., "CALIBRATION OF THE ^{14}C TIME SCALE OVER THE LAST 30,000 YEARS USING MASS SPECTROMETRIC U-TH AGES FROM BARBADOS CORALS," *NATURE* 345 [1990]; 405–10; B: ADAPTED FROM M. STUIVER ET AL., "CLIMATIC, SOLAR, OCEANIC, AND GEOMAGNETIC INFLUENCES ON LATE-GLACIAL AND HOLOCENE ATMOSPHERIC ^{14}C/^{12}C CHANGE," *QUATERNARY RESEARCH* 35 [1991]: 1–24.)

product of cosmic bombardment were also more abundant before 7,000 years ago, confirming weaker magnetic shielding. Second, direct measurements of the past magnetic field in Earth's rocks and sediments show that it was weaker at that time.

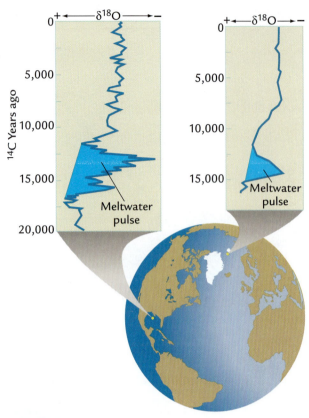

FIGURE 14-4
Local meltwater pulses

CaCO₃ shells of ocean plankton from the Norwegian Sea and the Gulf of Mexico record pulses of low-$\delta^{18}O$ meltwater delivered from nearby ice sheets. (TOP LEFT: ADAPTED FROM A. LEVENTER ET AL., "DYNAMICS OF THE LAURENTIDE ICE SHEET DURING THE LAST DEGLACIATION: EVIDENCE FROM THE GULF OF MEXICO," *EARTH PLANETARY SCIENCE LETTERS* 59 [1982]: 11–17; TOP RIGHT: ADAPTED FROM G. JONES AND L. D. KEIGWIN, "EVIDENCE FROM FRAM STRAIT (78°N) FOR EARLY DEGLACIATION," *NATURE* 336 [1988]: 56–59.)

In addition, North Atlantic sediment cores contain a distinctive layer of sediment deposited 17,000 to 14,500 years ago rich in ice-rafted sand grains but nearly barren of the planktic foraminifera and coccoliths normally found in that region. This layer is evidence of a large influx of icebergs to the North Atlantic Ocean early in the deglaciation. This evidence suggests that the major continental ice sheets lost a significant amount of their mass early in the deglaciation by calving icebergs to the ocean.

14-4 Mid-Deglacial Cooling: The Younger Dryas

The slowing in the rate of deglacial ice melting after 13,000 years ago (see Figure 14-3) coincided with the onset of a cold climatic oscillation that is especially evident in records near the subpolar North Atlantic Ocean (Figure 14-5 top). Temperatures in this region had warmed part of the way toward interglacial levels, but this reversal brought back almost full glacial cold. The first evidence for this event came from pollen records in Europe, where a distinctive flowering plant *Dryas* moved south from the Arctic into Europe. This cold episode is called the **Younger Dryas**.

Later work on sediments in the North Atlantic Ocean also detected a clear Younger Dryas imprint: a rapid oscillation in the regional extent of icy polar water (Figure 14-5A). At the glacial maximum, polar water reached southward across the North Atlantic to 45°N. The southern margin of this cold water was defined by the **polar front**, a zone of rapid transition to the more temperate waters to the south (see Chapter 13). Early in the deglaciation, near 15,000 years ago, the polar front had rotated back to the northwest around a hinge point near Newfoundland in eastern Canada. During this change, warm water had begun to flow northward along the European coast and to moderate climate enough that temperate species of trees had begun to advance northward from their full-glacial positions in far-southern Europe.

Near 12,900 years ago the polar front abruptly advanced back to the south, almost reaching its full-glacial position. At the same time, with the North Atlantic Ocean again cooler, Arctic vegetation (including *Dryas*) returned to northern Europe (Figure 14-5B). Later, near 11,700 years ago, the polar front abruptly retreated to the north, and forests began their final advance into north-central Europe.

The Younger Dryas readvance of the polar front represents a major reversal in Atlantic circulation. The estimated sea-surface cooling west of Ireland was at least 7°C, close to the difference between fully glacial and typically interglacial extremes. A similar cooling has been estimated from changes in the fossil remains of temperature-sensitive insect populations in England (Figure 14-5C).

Ice cores from Greenland contain a remarkably detailed record of the Younger Dryas event (Figure 14-6). During fully glacial climates, snow (ice) had been accumulating slowly, but the rates increased abruptly near 15,000 years ago, when the North Atlantic Ocean warmed. Accumulation rates then slowed during the transition into the Younger Dryas event but again increased when it ended 11,700 years ago. Some of these transitions occurred in less than a century, with much of the change concentrated in a single decade. Similar changes occurred in ice core concentrations of windblown dust, which peaked during the cold, dry, windy Younger Dryas climate and then abruptly decreased.

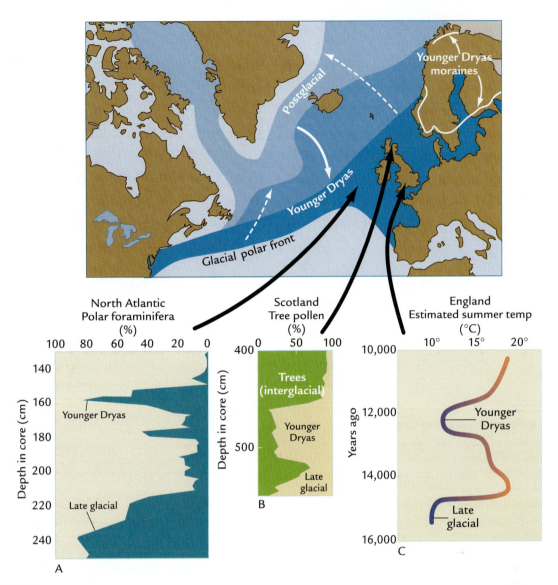

FIGURE 14-5
The Younger Dryas cold interval

Evidence of a cold episode that interrupted the general deglacial warming comes from: a southward readvance of polar water in the North Atlantic (A), a reversal toward Arctic vegetation in Europe (B), and a return to cooler temperatures indicated by fossil insect populations in Britain (C). (TOP: ADAPTED FROM W. F. RUDDIMAN AND A. MCINTYRE, "THE NORTH ATLANTIC OCEAN DURING THE LAST DEGLACIATION," *PALAEOGEOGRAPHY, PALAEOCLIMATOLOGY, PALAEOECOLOGY* 35 [1981]: 145–214; A: ADAPTED FROM W. F. RUDDIMAN, C. D. SANCETTA, AND A. MCINTYRE, "GLACIAL/ INTERGLACIAL RESPONSE RATE OF SUBPOLAR NORTH ATLANTIC WATERS TO CLIMATIC CHANGE: THE RECORD IN OCEANIC SEDIMENTS," *PHILOSOPHICAL TRANSACTIONS OF THE ROYAL SOCIETY OF LONDON B* 280 [1977]: 119–42; B: ADAPTED FROM G. R. COOPE AND G. LEMDAHL, "REGIONAL DIFFERENCES IN THE LATE GLACIAL CLIMATE OF NORTHERN EUROPE," *JOURNAL OF QUATERNARY SCIENCE* 10 [1995]: 391–95; C: ADAPTED FROM T. C. ATKINSON ET AL., "SEASONAL TEMPERATURES IN BRITAIN DURING THE PAST 20,000 YEARS, RECONSTRUCTED USING BEETLE REMAINS," *NATURE* 325 [1987]: 587–92.)

During the Younger Dryas event, the ice sheets in Scandinavia stopped retreating and in some regions readvanced a few hundred kilometers (see Figure 14-5 top). These pauses or small advances are thought to have been a response to the large regional cooling. The deglacial sea level curve derived from coral reefs (see Figure 14-3) shows that global ice volume continued to shrink during the Younger Dryas, although at a much slower rate than before.

What caused the Younger Dryas oscillation? The geochemist Wally Broecker proposed changes in the path of meltwater flow from the North American ice sheet as the cause (Figure 14-7). He suggested that the major meltwater route southward into the Gulf of Mexico was abruptly diverted eastward into the North Atlantic Ocean during the Younger Dryas. This diversion would have delivered a pulse of low-salinity water that altered the circulation of the North

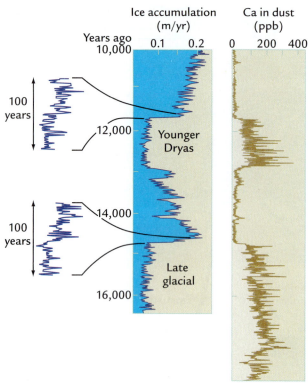

FIGURE 14-6
Deglacial ice accumulation in Greenland

Rates of accumulation of ice in the Greenland ice sheet abruptly decreased during the Younger Dryas cold event and then increased when it ended, with the major changes occurring within 100 years. Concentrations of windblown dust increased during the Younger Dryas and decreased afterward. (MODIFIED FROM R. B. ALLEY ET AL., "ABRUPT INCREASE IN GREENLAND SNOW ACCUMULATION AT THE END OF THE YOUNGER DRYAS EVENT," *NATURE* 362 [1993]: 527–29).

FIGURE 14-7
Routes of meltwater flow

During deglaciation, the main direction of drainage of the North American ice sheet changed: southward to the Gulf of Mexico early in the deglaciation, and then north into the Arctic Ocean late in the deglaciation. (ADAPTED FROM J. T. TELLER, "MELTWATER AND PRECIPITATION RUNOFF TO THE NORTH ATLANTIC, ARCTIC, AND GULF OF MEXICO FROM THE LAURENTIDE ICE SHEET AND ADJACENT REGIONS DURING THE YOUNGER DRYAS," *PALEOCEANOGRAPHY* 5 [1990]: 897–905.)

Atlantic Ocean by lowering the density of its surface waters enough to prevent them from sinking and forming deep water. Because ocean surface waters give off heat when deep water forms in the North Atlantic Ocean, cutting off this process could have cooled climate in the North Atlantic and in Europe.

At first, this explanation looked promising. Before the Younger Dryas event, the major pathway of meltwater flow had been down the Mississippi River and into the Gulf of Mexico, as recorded by a negative $\delta^{18}O$ signal in sediments from the Gulf of Mexico (see Figure 14-4). Then, during the Younger Dryas event (shown in ^{14}C ages as 11,000 to 10,000 years ago, but actually 13,000 to 11,500 years ago in calendar years), the negative $\delta^{18}O$ pulse in the Gulf of Mexico largely disappeared because the meltwater flow weakened.

Subsequent investigations have raised questions about this hypothesis. Examination of drainage patterns from large meltwater lakes near the southern ice sheet margin have failed to find convincing evidence of an unusual outflow of freshwater eastward to the Atlantic during the onset of the Younger Dryas event. More recent studies, however, have found evidence for a strong outflow of meltwater from the northwestern Canadian ice sheet margin down the McKenzie River and into the Arctic Ocean at about 13,000 years ago. As a result, the possibility of a meltwater-outflow explanation of the Younger Dryas remains alive, although via a different pathway.

Earlier claims that the Younger Dryas cooling was global in extent have proven incorrect. A small cooling evident in Antarctic ice cores that was once interpreted as correlative with the Younger Dryas event is not. Instead, at least part of the Antarctic was undergoing a slow warming throughout the Younger Dryas interval. The origin of the Younger Dryas cooling remains an enigma.

14-5 Positive Feedbacks to Deglacial Melting

Most climate scientists agree that rising summer insolation values driven by changes in Earth's orbital tilt and precession initially set in motion the melting of the great northern hemisphere ice sheets. But here we need to revisit an important question explored in Chapter 12 and assess how well it applies to this most recent deglaciation, about which we know so much more than the previous ones: How did so small an insolation maximum melt so much ice so quickly? The first (and simplest) answer to this question is that positive feedbacks accelerated the loss of ice. But which feedbacks, and when, and where?

Two widely acknowledged feedbacks would have been in action throughout the deglaciation, one of which was the rising concentration of CO_2 in the atmosphere. As the ice sheets melted, the CO_2 level rose in near lockstep, warming the planet through the greenhouse effect and adding to the initial effect of increasing summer insolation (Figure 14-8). The CO_2 increase led the sea level (ice volume) rise by an average of 1,000 years, too small an amount for CO_2 to have been an independent driver of the ice sheets, but consistent with the idea that CO_2 acted as a positive feedback to ice melting (see Chapter 11).

The change in surface albedo as the ice sheets melted would have been a second major source of feedback. Large ice sheets with very high albedos reflected much of the incoming solar radiation back out to space, but the darker surfaces exposed as the ice melted would have absorbed most of the incoming radiation, except when snow-covered in winter. This decrease in surface albedo as the ice retreated would have warmed the region and helped to melt more ice.

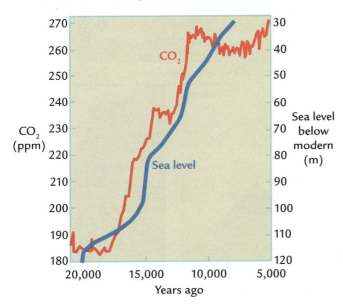

FIGURE 14-8
Relative timing of CO_2 rise and ice melting

The deglacial increase of CO_2 in the atmosphere slightly preceded the melting of ice sheets. (CO₂ CONCENTRATIONS ADAPTED FROM EPICA COMMUNITY MEMBERS, "EIGHT GLACIAL CYCLES FROM AN ANTARCTIC ICE CORE," *NATURE* 429 [2004]: 623–628; SEA LEVEL RISE ADAPTED FROM E. BARD ET AL., "CALIBRATION OF THE ¹⁴C TIME SCALE OVER THE PAST 30,000 YEARS USING MASS SPECTROMETRIC U-TH AGES FROM BARBADOS CORALS," *NATURE* 345 [1990]: 405–10.)

Retreating sea ice over nearby oceans would have further enhanced the albedo feedback.

As noted earlier, one noteworthy feature of the most recent deglaciation was the rapid early rate of sea level rise (see Figure 14-3) compared to the gradual increase in the summer insolation driver (see Figure 14-1). Part of this offset in timing was presumably due to the fact that a much larger amount of ice was available to melt earlier in the deglaciation than later. When summer insolation rose to values high enough to melt ice *before* the insolation maximum near 10,000 years ago, the ice sheets were still very large and vulnerable to melting. In contrast, when declining summer insolation fell back to the same values *after* the insolation maximum, much less ice was left to melt, no matter how warm the climate had become. This bias would naturally tend to produce fast rates of ice melting before the summer insolation maximum.

The rapid early rise in sea level between 19,000 and 14,000 also suggests that important positive feedbacks may have been in action during this time of rapid ice melting. The negative $\delta^{18}O$ pulse in Norwegian Sea cores (see Figure 14-4) indicates that a substantial part of the marine ice sheet over the Barents Sea melted early in the deglaciation. Because its base lay below sea level, it was probably especially vulnerable to early destabilization.

Other observations summarized in section 14-3 point to thinning of the Laurentide ice sheet on North America early in the deglaciation. The major influx of icebergs to the North Atlantic Ocean early in the deglacial sequence arrived at a time when Laurentide ice had not melted back far from its glacial position, but when global ice volume (as defined by the sea level curve) was rapidly decreasing. These observations suggest that the marine margin of the ice sheet was losing volume through ice streams, but with little loss of areal extent.

In addition, the elevations of moraine deposits along the southern ice lobes of the Laurentide ice sheet indicate that the ice profiles were very thin during the early-middle parts of the deglaciation, consistent with the idea that ice was sliding on a water-lubricated sediment base in that region. Ice that flowed across the land to the southern margins of the ice sheets would have melted relatively rapidly in the relative warmth of the bedrock depression left behind by the once-thicker ice. Again, some of the ice sheet interior could have been thinned without any major retreat of the margins.

14-6 Deglacial Lakes, Floods, and Sea Level Rise

As the ice sheets melted back, the land in front of them remained depressed for some time, rather than immediately rebounding to its former (ice-free) level (see Chapter 10). Into these depressions poured meltwater from the retreating ice sheets, forming **proglacial lakes**. Because of the large volumes of meltwater arriving each summer, the lakes frequently cut new channels and overflowed into other lakes and then into rivers that carried water to the ocean.

The proglacial lakes existed in a highly dynamic landscape (Figure 14-9). As deglaciation proceeded, the ice margins fluctuated, with lobes of ice retreating and sliding forward for brief intervals but gradually shrinking farther to the north over time. Each time the ice lobes retreated, they left behind new regions of depressed bedrock that formed the deepest parts of new proglacial lakes. All the while, the slow rebound of bedrock caused the parts of the lakes farther south of the ice margins to shallow and eventually disappear. Through time, the locations of the proglacial lakes moved north across the landscape, following the wave of depressed bedrock left south of the retreating ice.

The largest of the proglacial lakes in North America was Lake Agassiz, in western Canada. Over its entire existence, it flooded an area greater than 500,000 km² (Figure 14-10A), but much smaller areas at specific intervals during the ice retreat sequence (Figure 14-10B). At its maximum size, it covered more than

200,000 km² to a depth of 100 m or more, forming a reservoir of 20,000 km³. Even this amount represented only a tiny fraction of the tens of millions of cubic kilometers of water stored in the ice sheets at the glacial maximum. Still, the water impounded in proglacial lakes and then released transformed parts of the landscape (Box 14-2).

The deglacial rise in sea level altered Earth's surface on a very large scale. Many regions of the world's continental shelves had been exposed during the low sea level at the glacial maximum, and many continents or ocean islands had been linked by land connections (Figure 14-11). One particularly large

A Total area covered by deglacial lakes

B Lakes during deglaciation

FIGURE 14-10
Glacial Lake Aggasiz

During the several thousand years of its existence, glacial Lake Agassiz flooded a total of more than 500,000 km² in western Canada, although a much smaller area was flooded at any one time. (A: ADAPTED FROM J. T. TELLER, "LAKE AGASSIZ AND ITS CONTRIBUTION TO FLOW THROUGH THE OTTAWA-ST. LAWRENCE SYSTEM," GEOLOGICAL ASSOCIATION OF CANADA SPECIAL PAPER 35 [1987]: 281–89; B: ADAPTED FROM J. T. TELLER, "GLACIAL LAKE AGASSIZ AND ITS INFLUENCE ON THE GREAT LAKES," GEOLOGICAL ASSOCIATION OF CANADA SPECIAL PAPER 30 [1985]: 1–16.)

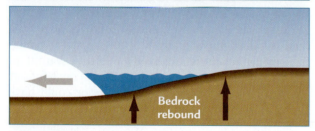

FIGURE 14-9
Proglacial lakes moving north

Proglacial lakes develop in bedrock depressions left by melting ice sheets. Over time, the lakes move north behind the ice sheets, while the land farther south rebounds toward its undepressed elevation.

Land exposed by drop in sea level
Land created under ice by drop in sea level

FIGURE 14-11
Deglacial flooding of coastlines

The regions shown in dark brown were exposed at the last glacial maximum but flooded by the rise in sea level when the ice melted. (ADAPTED FROM CLIMAP PROJECT MEMBERS, *SEASONAL RECONSTRUCTION OF THE EARTH'S SURFACE AT THE LAST GLACIAL MAXIMUM*, MAP AND CHART SERIES MC-36 [BOULDER, CO: GEOLOGICAL SOCIETY OF AMERICA, 1981].)

area was the expanse of dry land that joined Australia with New Guinea to the north. Land connections also linked the modern southeast Asian mainland with islands as far south as Borneo and joined northeastern Asia (Siberia) and westernmost Alaska across the modern Bering Strait. England and Scotland were linked to the European mainland during the glacial maximum just south of the ice sheet. The return of some 44 million km³ of meltwater to the oceans during deglaciation submerged all of these land corridors.

The lower level of the glacial ocean had also transformed smaller seas around the margins of the oceans, especially in the western Pacific. The present Yellow Sea was dry land, and other seas in the western Pacific were more isolated from the open ocean because sea level was lower. Rising sea level flooded these seas and rejoined them to the open ocean.

▶ Other Climate Changes During and After Deglaciation

Scientists have investigated two important changes during the late-deglacial and postglacial interval—the strength of north tropical monsoons and the warmth of summers in north polar latitudes. Monsoons grew stronger and summers warmer as summer insolation rose toward maximum values 10,000 years ago and as the effects of the ice sheets and the reduced greenhouse-gas levels diminished. By 7,000 years ago, with the northern hemisphere ice sheets melted and the greenhouse gases close to full interglacial levels, the major factor left to influence northern hemisphere climate was the ongoing drop in summer insolation toward modern values, along with the accompanying increase in winter insolation.

14-7 Stronger, Then Weaker Monsoons

Monsoons were strong near 10,000 years ago because of Earth's orbital configuration (see Chapter 8). Summer insolation values over tropical and subtropical landmasses of the Northern Hemisphere at that time were 8% higher than they are today (Figure 14-12). According to Kutzbach's orbital

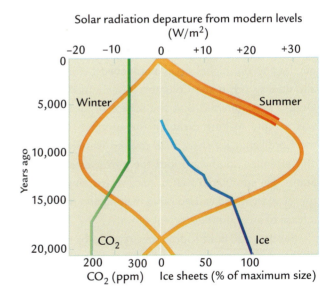

Solar radiation departure from modern levels (W/m²)

FIGURE 14-12
Causes of climate change since deglaciation

During the last 6,000 years, with the ice sheets melted and CO₂ levels near interglacial levels, the main orbital-scale factor affecting climate was the gradual change in solar insolation toward present values. (ADAPTED FROM J. E. KUTZBACH ET AL., "CLIMATE AND BIOME SIMULATIONS FOR THE PAST 21,000 YEARS," *QUATERNARY SCIENCE REVIEWS* 17 [1998]: 473-506.)

Climate Interactions and Feedbacks

Giant Deglacial Floods

In an unusual landscape called the **channeled scablands** in Idaho and east-central Washington State, the bedrock consists of thick sequences of basalt deposited by lava flows during a time of heightened volcanic activity some 15 million years ago. As the North American ice sheets were melting during the most recent deglaciation, the surfaces of these ancient lava flows were eroded into shapes suggesting the violent action of water on an immense scale. In the scabland region, deep canyons with nearly vertical walls were gouged into bedrock. At some locations, these now-dry channels end in steep cliffs that plunge into larger channels with depressions like those found at the base of modern waterfalls (but much larger). Huge boulders and displaced gravel and sand lie in the channels, but upland areas nearby have a thin cover of windblown loess typical of much of the rest of the Pacific Northwest outside the scabland region.

The geologist Harlen J. Bretz, working in the 1920s and 1930s, came to the conclusion that these erosional features must have resulted from a flood of immense proportions, one that within a few days carried a volume of water equivalent to all of Earth's rivers today. He suggested that the water in this flood ran wildly across the landscape, gouging and eroding the lower terrain but leaving the higher areas untouched before eventually flowing down the Columbia River into the Pacific Ocean. Bretz inferred that the source of all this water was a rapid melting event on the southern margin of the Cordilleran ice sheet that covered western Canada and extended southward into the northwestern United States. He speculated that a volcano erupting beneath the ice margin had caused rapid melting and the sudden release of an enormous torrent of water.

For decades, Bretz's ideas were rejected by geologists. At that time, most geologists took too literally the principle of uniformitarianism, the concept that slow geologic processes working today have, over immense spans of time, shaped and molded all of the features we see on Earth. Over-enthusiastic application of this otherwise useful concept left no room for infrequent catastrophic phenomena; these events were rejected because no examples were available from the historical record. The problem with this view is that the human life span, and indeed all of recorded human history, is extremely short in relation to the age of the Earth, and so our perspective on "normal" processes is very narrow. With much longer life spans, we would naturally take a much broader view of what is normal.

In the 1950s, aerial photography revealed that the scablands were covered by giant ripple marks and gravel bars up to 50 feet high and spaced at intervals of 500 feet. Working on foot, Bretz had not recognized these enormous features because they were masked by scrubby vegetation. This new evidence convinced most geologists that the scablands had indeed been flooded, and further studies suggested that a discharge of some 25,000 to 30,000 m^3 (750,000 ft^3) per second flowing at 80 km (50 m) per hour was required to carve such a landscape. All the features Bretz had noted were indeed the result of rushing water on an unimaginably large scale.

A likely source of the water was glacial Lake Missoula, a proglacial lake ponded against the side of the ice sheet in Idaho. Although Bretz proposed only a single flood, the multiple layers of sediment left by the waters now imply dozens of floods. One possibility is that each time a lobe of Cordilleran ice advanced far enough south to act as a dam, Lake Missoula filled up with water, later releasing it in catastrophic bursts when the blocking ice lobe melted. Geologists have also found evidence that large lakes hundreds of kilometers wide and tens of meters deep existed underneath the thin western lobes of the ice sheet in the Canadian provinces of Alberta and Saskatchewan, and that these lakes could have periodically released large volumes of water to the north.

Box 14-2

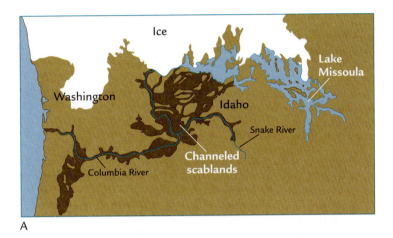

A

B

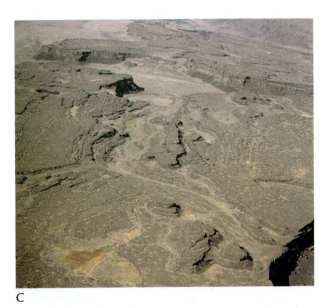

C

The channeled scablands Massive erosion by giant deglacial floods affected a large region in eastern Washington State and Idaho. One source of meltwater for the floods may have been glacial Lake Missoula, ponded against a lobe of the ice sheet in Idaho and southwest Canada (A). Evidence of erosion in the channeled scablands by floods of meltwater includes huge boulders, gravel, and sandbars on an immense scale and channels that lead to the edges of cliffs (B, C). (A: ADAPTED FROM G. A. SMITH, "MISSOULA FLOOD DYNAMICS AND MAGNITUDES INFERRED FROM SEDIMENTOLOGY OF SLACKWATER DEPOSITS ON THE COLUMBIA PLATEAU, WASHINGTON," *GEOLOGICAL SOCIETY OF AMERICA BULLETIN* 105 [1993]: 77–100; B AND C: PHOTOGRAPHS BY V.R. BAKER.)

monsoon theory, increased insolation should have driven a stronger summer monsoon circulation and produced higher tropical lake levels near 10,000 years ago compared to those today.

The COHMAP project (see Chapter 13) compared climate model simulations against ground-truth geological data during this interval. Their simulations used summer insolation values from 9,000 years ago, when northern hemisphere ice sheets were greatly reduced in size. The model simulated stronger monsoons across the entire north-tropical region of North Africa, southern Arabia, and southern Asia (Figure 14-13A).

Geologic evidence matches this simulation (Figure 14-13B and C). Lake levels determined by ^{14}C dating of old shorelines were substantially higher than they are today between 9,000 and 7,000 ^{14}C years ago (roughly 10,000 and 8,000 calendar years ago) across a wide area: most of North Africa between 15° and 30°N, the southern half of Arabia, and southeastern Asia. Lake Chad, in northern Africa, expanded to 300,000 km^2, an area comparable to the modern Caspian Sea.

Evidence in sediment cores from the Arabian Sea indicates that stronger monsoonal winds blew along the coast of Somalia in eastern Africa, and along the southeast coast of Arabia 9,000 years ago (Figure 14-14). These intensified winds drove water offshore and caused upwelling, which was recorded by key species of planktic foraminifera.

Scientists have also found that ancient rivers flowed across regions that today are hyperarid desert. In central Arabia, a river flowed more than 500 km northeastward through the present Arabian desert (see Chapter 9, Figure 9-7 for a similar example in the Sahara desert). Many parts of North Africa and Arabia that are now extremely dry were once grassy river valleys dotted with freshwater lakes and occupied by hippopotamuses, crocodiles, turtles, rhinoceroses, giraffes, and buffaloes.

Although climate models and evidence from the geologic record confirm Kutzbach's theory of stronger north-tropical monsoons 10,000 years ago, a closer look reveals a mismatch in amplitude. The increase in rainfall in the models was small compared to the amounts indicated by geologic evidence from lake levels and pollen assemblages. Subsequent modeling efforts have explored positive feedback processes that could have amplified the small response simulated by the early models.

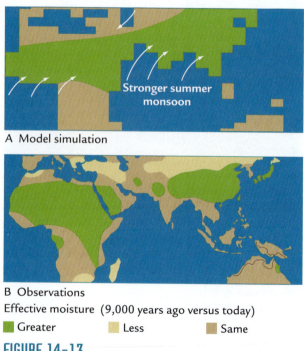

A Model simulation

B Observations

Effective moisture (9,000 years ago versus today)

■ Greater   ■ Less   ■ Same

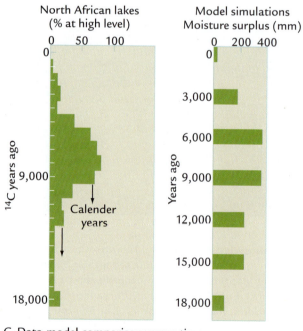

C Data-model comparison versus time

FIGURE 14-13

Tropical monsoon maximum

Climate model simulations of stronger summer monsoons in the north tropics near 9,000 years ago agree with evidence in the geologic record, such as higher lake levels. (A, B: ADAPTED FROM COHMAP PROJECT MEMBERS, "CLIMATIC CHANGES OF THE LAST 18,000 YEARS: OBSERVATIONS AND MODEL SIMULATION," *SCIENCE* 241 [1988]: 1043–52; C: ADAPTED FROM J. E. KUTZBACH AND F. A. STREET-PERROTT, "MILANKOVITCH FORCING OF FLUCTUATIONS IN THE LEVEL OF TROPICAL LAKES," *NATURE* 317 [1985]: 130–34.)

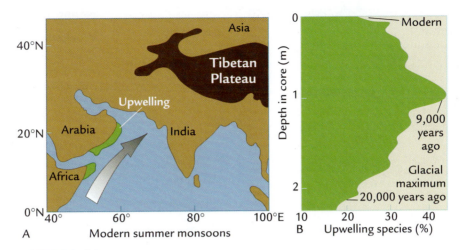

FIGURE 14-14
Upwelling in the Arabian Sea

Climate model simulations of stronger summer monsoons over India 9,000 years ago are supported by evidence from ocean cores indicating greater upwelling along the coasts of East Africa and Arabia in response to strong winds pushing surface waters offshore. (A: ADAPTED FROM W. L. PRELL, "MONSOONAL CLIMATE OF THE ARABIAN SEA DURING THE LATE QUATERNARY: A RESPONSE TO CHANGING SOLAR RADIATION," IN *MILANKOVITCH AND CLIMATE*, ED. A. BERGER ET AL. [DORDRECHT: D. REIDEL, 1984]; B: W. L. PRELL, "VARIATION OF MONSOONAL UPWELLING: A RESPONSE TO CHANGING SOLAR RADIATION," IN *CLIMATE PROCESSES AND CLIMATE SENSITIVITY*, ED. J. E. HANSEN AND T. TAKAHASHI [WASHINGTON, DC: AMERICAN GEOPHYSICAL UNION, 1984].)

One important feedback is the increase in recycling of water vapor provided by evapotranspiration from vegetation (see Chapter 2). The initial increase in monsoonal rains in the early models was sufficient to allow trees to advance northward into grasslands, and grasses to move northward into deserts. In more recent models, however, the advancing vegetation was allowed to draw moisture out of the soil and transfer it to the atmosphere through evapotranspiration. With more water vapor in the atmosphere, more rain fell, especially farther to the north (Figure 14-15). The vegetation-precipitation feedback that resulted produced model simulations with wetter landscapes and higher lakes, in closer (but still not full) agreement with the geologic evidence. One feedback not yet incorporated in models is increased recycling of moisture from small lakes and low swampy regions near rivers that are too small to be represented in modern climate model grid boxes.

After reaching a peak near 10,000 years ago, summer insolation values at lower latitudes of the Northern Hemisphere have fallen continuously (Figure 14-16A). This decrease has occurred because Earth's precessional motion has continued to carry it from a June 21 position close to the Sun 10,000 years ago to a June 21 position far from the Sun today (see Chapter 8).

By 6,000 years ago, summer insolation values in the northern tropics were still about 5% higher than the modern levels but were falling toward modern values. This slow decrease should have produced a corresponding decline in the strength of the tropical monsoons. Direct observations and ^{14}C dates of lakes across North Africa confirm a major drop in water

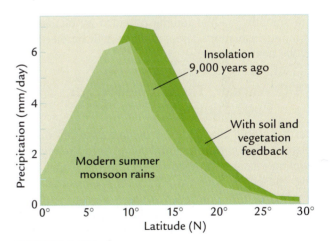

FIGURE 14-15
Vegetation-precipitation feedback in North Africa

Climate model simulations indicate that higher summer insolation 9,000 years ago caused stronger summer monsoons and greater northward penetration of moisture into Africa. Additional model experiments include positive moisture feedback from wetter soils and increased vegetation showing greater penetration of moisture into the continental interior. (MODIFIED FROM J. E. KUTZBACH ET AL., "VEGETATION AND SOIL FEEDBACKS ON THE RESPONSE OF THE AFRICAN MONSOON TO ORBITAL FORCING IN THE EARLY TO MIDDLE HOLOCENE," *NATURE* 384 [1986]: 623–26.)

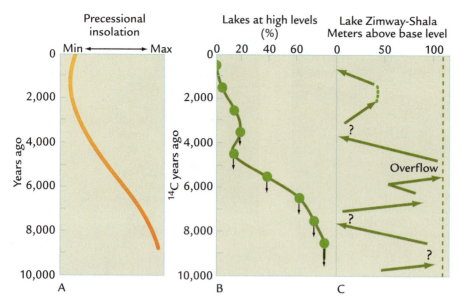

FIGURE 14-16

Weakening monsoons

Low-latitude summer insolation has slowly decreased since reaching a maximum value 10,000 years ago (A). The decrease in summer insolation has weakened the summer monsoons and caused lake levels in North Africa to fall (B, C). (A, B: ADAPTED FROM J. E. KUTZBACH AND F. A. STREET-PERROTT, "MILANKOVITCH FORCING OF FLUCTUATIONS IN THE LEVEL OF TROPICAL LAKES," *NATURE* 317 [1985]: 130–34; C: ADAPTED FROM R. GILLESPIE ET AL., "POSTGLACIAL ARID EPISODES IN ETHIOPIA HAVE IMPLICATIONS FOR CLIMATE PREDICTION," *NATURE* 306 [1983]: 680–83.)

levels during the last 9,000 years (Figure 14-16B). Today, lakes are lower than they were 9,000 years ago, and many have completely dried out.

Examined individually, most lake-level histories in North Africa and India show large and abrupt changes during the overall transition to lower levels (Figure 14-16C). As was the case for the irregular rates of melting of northern ice sheets, these short-term changes in lake levels represent a type of climate response that cannot be directly attributed to the smooth, gradual forcing provided by changes in summer insolation. These changes will be discussed in the next chapter.

14-8 Warmer, Then Cooler North Polar Summers

At the glacial maximum, the main controls on climate at high northern latitudes had been the regional cooling effects of the ice sheets and the global cooling caused by low CO_2 (and CH_4) concentrations. As deglaciation proceeded, rising summer insolation values increasingly warmed land areas located far from the ice sheets and in time overcame the cooling effects of the shrinking ice sheets. Summer insolation values in the Northern Hemisphere reached a peak 10,000 years ago, with much less ice still present.

NORTHWARD SHIFTS IN VEGETATION The last deglaciation dramatically transformed the vegetation on the northern hemisphere continents. In North America, cold-tolerant spruce trees retreated from their glacial position in the central United States to their modern position in northeastern Canada (Figure 14-17A). Warm-tolerant trees such as oak moved a smaller distance from their glacial location in the far

southeastern United States to their modern concentrations in the mid-Atlantic states (Figure 14-17B).

The climate changes that occurred midway through the deglaciation produced unusual mixtures of plants called **no-analog vegetation** because no similar combination exists today. For example, spruce trees grew with hardwood deciduous trees (such as ash) in the northern Midwest of the United States early in the deglaciation, even though ash and other deciduous trees are rare today in regions where spruce trees grow.

No-analog mixtures developed because each vegetation type responded to a different combination of environmental variables. These individualistic responses make it impossible to analyze past vegetation changes by simply lumping pollen together into larger communities or assemblages; each vegetation type has to be analyzed on its own.

The distributions of spruce and oak pollen simulated by climate models are also shown in Figure 14-17 for comparison with the observed distributions. Both sets of maps show the same large-scale northward relocation of spruce, and they agree on the existence of a mid-deglacial interval when spruce became rare throughout eastern North America. Both sets of maps also show a similar northward expansion of oak, but the model simulates more oak in the southeast during deglaciation than the pollen data show, a mismatch similar to that noted for the glacial maximum (see Chapter 13).

PEAK WARMTH Once CO_2 values had risen to full interglacial levels and only remnants of the ice sheets were left, summer insolation values became the main variable controlling climate responses, particularly for vegetation. The high summer insolation values that

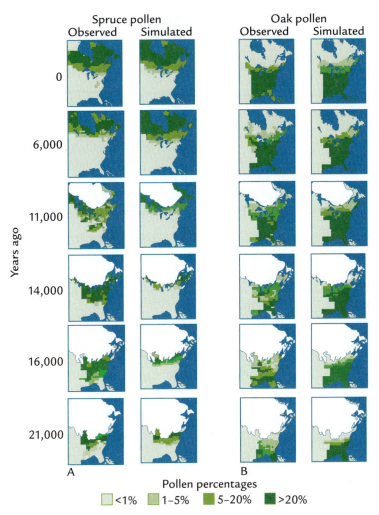

| Spruce pollen | | Oak pollen | |
| Observed | Simulated | Observed | Simulated |

Years ago: 0, 6,000, 11,000, 14,000, 16,000, 21,000

A   B

Pollen percentages
<1%   1–5%   5–20%   >20%

FIGURE 14-17
Data-model vegetation comparisons

Pollen in lake sediments indicates large-scale changes in the distribution of spruce and oak pollen during the last deglaciation. Model simulations of climate and vegetation reproduce many but not all aspects of these observed patterns. (ADAPTED FROM T. WEBB III ET AL., "LATE QUATERNARY CLIMATE CHANGE IN EASTERN NORTH AMERICA: A COMPARISON OF POLLEN-DERIVED ESTIMATES WITH CLIMATE MODEL RESULTS," *QUATERNARY SCIENCE REVIEWS* 17 [1998]: 587–606.)

had triggered ice melting remained moderately high, but had begun to decrease toward modern levels (see Figure 14-12). As a result, the warmest temperatures of the last several thousand years were registered soon after the regional chilling effect of the ice sheets was removed, but before the renewed cooling effect caused by falling insolation levels had gained much strength. Some climate scientists refer to this warmer-than-modern interval as the *hypsithermal,* but the time of greatest warmth actually varies widely from region to region, depending on when the nearby ice melted and its regional cooling effect was removed.

With summer insolation values still 5% higher than those today 6,000 years ago, the northern limit of Arctic (spruce and larch) forest in Siberia and west-central Canada had moved as much as 300 km north of its modern position, reducing the fringe of tundra bordering the Arctic Ocean. This expansion of forest beyond its modern limits confirms that summer temperatures on the northern continents were warmer than they have been in recent millennia. Winter insolation values lower than those today may

have produced cooler winter temperatures, but the northern limit of Arctic forest is mainly sensitive to temperature during the summer growing season.

Climate models have been used to assess how much warmer the high-latitude summer temperatures were 6,000 years ago. With summer insolation values 5% above those today, the models simulated summers that were warmer in parts of northern Canada by as much as 2° to 3°C, but by smaller amounts in central Asia (Figure 14-18A). Additional simulations were then run to incorporate the positive feedback effect of vegetation on temperature (see Chapter 2). In regions where Arctic forest advanced northward beyond its modern limits, the low albedo (25%) typical of these dark-green tree canopies replaced the high albedo (60%) typical of scrubby tundra vegetation under seasonal snow cover. The low-albedo trees absorbed more sunlight than the underlying snow and further warmed the climate. Because far-northern regions are snow-covered for much of the year, this additional warming extended through most of the year and affected broad areas of northern Canada and

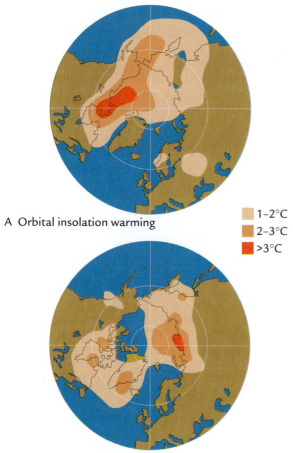

A Orbital insolation warming

■ 1–2°C
■ 2–3°C
■ >3°C

B Vegetation feedback warming

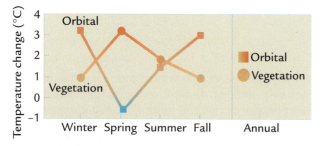

C Seasonal and annual averages

FIGURE 14-18
Peak deglacial warmth

Climate model simulations indicate that ==higher insolation 6,000 years ago warmed high latitudes, especially in Canada in the summer== (A). Additional simulations show that positive feedback caused by northward expansion of low-albedo spruce forest into high-albedo tundra may have almost doubled this regional warming, with the largest changes in Asia during the spring (B, C). (ADAPTED FROM TEMPO (J. E. KUTZBACH ET AL.), "POTENTIAL ROLE OF VEGETATION IN THE CLIMATIC SENSITIVITY OF HIGH-LATITUDE REGIONS: A CASE STUDY AT 6000 YEARS B.P.," *GLOBAL BIOGEOCHEMICAL CYCLES* 6 [1996]: 727–36.)

Asia (Figure 14-18B). The net effect of this vegetation-albedo feedback almost doubled the initial insolation warming of high northern latitudes (Figure 14-18C).

Sea ice also contributed to these far-northern climate changes. High summer insolation caused the sea ice margin in the model to thin and retreat northward, and this change propagated into the rest of the yearly cycle, with delayed refreezing of seasonal sea ice in autumn, thinner and less extensive sea ice in winter, and earlier melting of sea ice in spring. As a result, despite the fact that the winter insolation values were considerably lower than those today, the reduced sea ice cover and larger areas of open water near the coasts moderated the winter cooling of the continents. As a result, the simulated annual average change in this region showed a considerable warming.

RENEWED COOLING IN THE LAST SEVERAL THOUSAND YEARS   During the last 6,000 years, Earth's tilt has continued to slowly decrease and its precessional motion has moved the northern hemisphere summer solstice toward the aphelion (distant pass) position. These combined orbital changes have produced a 5% decrease in summer insolation and a 5% increase in winter insolation at high latitudes since 6,000 years ago (see Figure 14-12). As a result, summer temperatures have fallen significantly during the last several thousand years in several regions at high northern latitudes (Figure 14-19).

Evidence of cooler summers comes from ice cores taken from small ice caps in several parts of the Arctic. Ice from the tiny Agassiz Ice Cap on Ellesmere Island, in far northern Canada, shows that summer melting episodes were far more frequent before 5,000 years ago than they have been since that time (Figure 14-19A). This evidence supports a trend toward cooler summers.

A second region where cooling is evident over the last several thousand years is the high-latitude Atlantic Ocean off the coast of Greenland, a region that today has a sea ice cover in winter. Ocean sediment cores from this area contain diatom shells that record a history of the species that once lived in these waters. The diatom record before about 5,000 years ago indicates that sea ice was absent or scarce in this region but has since increased in extent (see Figure 14-19B).

A third piece of evidence that indicates cooling in recent millennia is the increase in size of small glaciers on Arctic islands. Glacier margins on Arctic islands were located well back from their modern positions between 8,000 and 3,500 years ago (see Figure 14-19C), and some of the ice caps melted entirely. These glaciers have reappeared or grown since 3,500 years ago, oscillating toward progressively larger sizes, consistent with cooler summer temperatures and less melting.

A fourth indication of cooling is changes in the abundance of temperature-sensitive diatom species off the southwest coast of Norway. The estimated

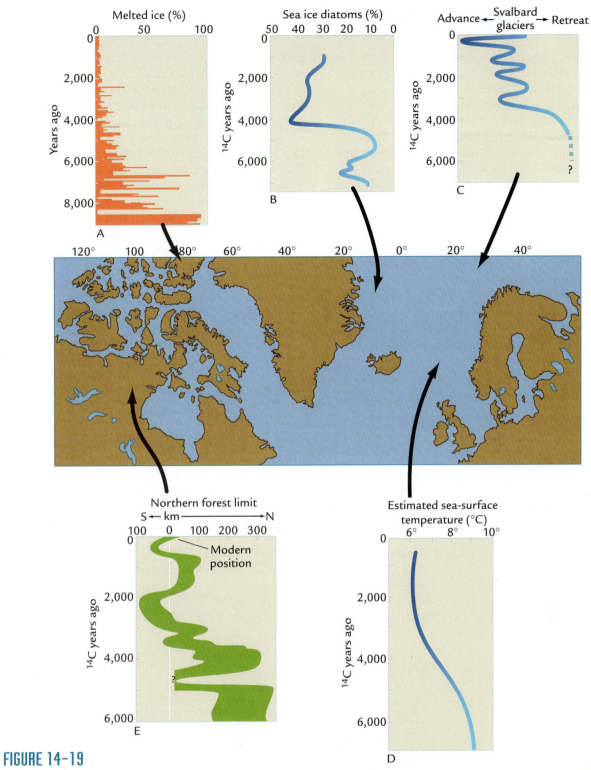

FIGURE 14-19
Cooling toward the present

Evidence that high northern latitudes have cooled during the last several thousand years includes: less frequent summer melting episodes in ice caps on Arctic islands (A); more frequent sea ice off Greenland (B); advances of ice caps on Arctic islands north of Europe (C); lower temperatures in the Atlantic Ocean west of southern Norway (D); and a southward shift of the boundary between tundra and spruce forest in northern Canada (E). (A: ADAPTED FROM R. M. KOERNER AND D. A. FISHER, "A RECORD OF HOLOCENE SUMMER CLIMATE FROM A CANADIAN HIGH-ARCTIC ICE CORE," *NATURE* 343 [1990]: 630–31; B AND D: ADAPTED FROM N. KOC ET AL., "PALEOCEANOGRAPHIC RECONSTRUCTIONS OF SURFACE OCEAN CONDITIONS IN THE GREENLAND, ICELAND, AND NORWEGIAN SEAS THROUGH THE LAST 14 KA BASED ON DIATOMS," *QUATERNARY SCIENCE REVIEWS* 12 [1992]: 115–40; C: ADAPTED FROM J. LUBINSKI, S. L. FORMAN, AND G. H. MILLER, "HOLOCENE GLACIER AND CLIMATE FLUCTUATIONS ON FRANZ JOSEPH LAND, ARCTIC RUSSIA," *QUATERNARY SCIENCE REVIEWS* 18 [1999]: 87–109; E: ADAPTED FROM H. NICHOLS, "PALYNOLOGICAL AND PALEOCLIMATIC STUDY OF THE LATE QUATERNARY DISPLACEMENT OF THE BOREAL FOREST-TUNDRA ECOTONE IN KEEWATIN AND MACKENZIE, N.W.T.," INSTITUTE OF ARCTIC AND ALPINE RESEARCH OCCASIONAL PAPER NO. 15 [BOULDER, CO, 1975].)

changes in sea-surface temperature reconstructed from these assemblages show a gradual cooling since 6,000 years ago (see Figure 14-19D).

The boundary between tundra to the north and Arctic forest to the south is yet another climatic indicator in Asia and North America. This boundary in northern Canada was well north of its present limit 6,000 years ago but has since advanced southward by up to 300 km (see Figure 14-19E). This shift from forest to tundra vegetation suggests cooler summers and a shorter growing season.

Still another indication of summer cooling in the last several thousand years (not shown) comes from mountain glaciers. Like the much larger ice sheets, mountain glaciers at high latitudes will melt if summer insolation increases, but mountain glaciers can respond to climate changes in just a few decades. Before 5,000 years ago, mountain glaciers were small, but after that time their size increased in an oscillating way in most regions, consistent with the evidence of progressive cooling driven by a long-term decrease in summer insolation.

▶ Current and Future Orbital-Scale Climatic Change

Astronomy tells us not only the changes in Earth's orbit that have already occurred but also those that will occur in the future. This knowledge of the future gives us a good basis for predicting the course climate would follow if it responded only to orbital forcing (changes in precession and tilt). [Factors related to human interference in the natural operation of the climate system will be examined in Part V].

Today, June 21 occurs near the July 4 aphelion (distant pass) position in Earth's eccentric orbit around the Sun. In another 10,000 years, Earth will have returned to the opposite configuration. June 21 will occur at perihelion, when Earth is closest to the Sun, just as it did 10,000 years ago (Figure 14-20 left). Because the 23,000-year cycle of orbital precession controls changes in the tropical monsoons, this shift in Earth's orbit will slowly increase the amount of summer insolation across the northern tropics and drive a progressively stronger monsoon over North Africa and southern Asia.

Predicting the effect of orbital changes on higher latitudes is more difficult. In the next 10,000 years, the tilt of Earth's axis will have fallen to the next minimum (Figure 14-20 center), and the resulting decrease in summer insolation will tend to cool climate. But the insolation decrease caused by the change in tilt will be opposite in direction to the increase caused by the change in precession (Figure 14-20 left). One estimate of the combined signal (Figure 14-20 right) shows a larger effect of precession than of tilt, although other estimates predict a larger relative influence of tilt (see Chapter 12).

Climate scientists would like to be able to predict when the next glaciation will begin (under natural

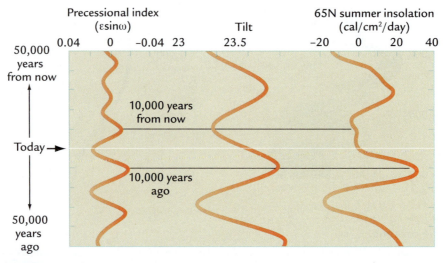

FIGURE 14-20
Future summer insolation trends
During the next 10,000 years, precession-dominated insolation at low latitudes of the Northern Hemisphere will return to another maximum (left), while tilt-dominated insolation at very high northern latitudes will continue to fall toward the next minimum (center). The combined effects of tilt and precession will increase summer insolation (right), which will not fall well below modern levels for another 50,000 years. (ADAPTED FROM A. BERGER AND M.-F. LOUTRE, "MODELING THE CLIMATIC RESPONSE TO ASTRONOMICAL AND CO₂ FORCINGS," *C. R. ACAD. SCI. PARIS* 323 [1996]: 1–16.)

conditions, without human interference), with ice caps starting to grow on North America, Eurasia, or both. Over the last million years, ice sheets have been present on northern hemisphere continents for about 90% of the time (see Chapter 10, Figure 10-13), and intervals without any ice on both North America and Europe have lasted for 10,000 years or less. This evidence suggests that the current interglaciation must now be near its natural end.

One way to estimate when new ice sheets will grow at high northern latitudes is to examine what has happened at similar times in the past. Records of $\delta^{18}O$ changes covering the last 900,000 years are an approximate index of ice volume, but they have a sizeable temperature overprint. The current interval of low $\delta^{18}O$ values (minimal ice volume and warm temperatures) has lasted as long as previous interglaciations, and some models that simulate longer-term ice sheet responses predict that we are at or near the point when ice sheets should start growing again because of low summer insolation.

Other models predict that glaciation is not imminent and indeed may not occur for the next 50,000 years because summer insolation is now near a minimum and will not be significantly lower for the next 50,000 years (Figure 14-20 right). In this view, if new ice sheets have not begun to form during the present insolation minimum, why would they do so at higher insolation levels during the next 50,000 years? If this view is correct, the ice-age cycles seem to have "skipped a beat."

But an ice-free interval 50,000 years in length would pose a major problem for scientists. No ice-free interval that long has occurred in the entire 2.75 million years of northern hemisphere glaciation, so why would one occur now? Has the natural operation of the climate system begun to shift in recent millennia toward unprecedented warmth and an ice-free state? Such an interpretation would seem to completely contradict the evidence for gradual cooling during the last few million years (see Chapters 7 and 10).

Another option will be considered in Chapter 16. The climatic trend of the last several thousand years may not have been natural because a new factor has arisen that may have prevented a new glaciation that would otherwise be underway by now. That new factor is agricultural activities of humans.

Key terms

Younger Dryas (p. 278)
polar front (p. 278)
proglacial lakes (p. 282)

channeled scablands (p. 284)
no-analog vegetation (p. 288)

Review Questions

1. What is the best method of measuring the melting of ice sheets over the last 17,000 years?

2. To what degree does the timing of ice sheet melting support the Milankovitch theory that orbital insolation controls the size of ice sheets?

3. To what degree do changes in intensity of summer monsoons in the last 17,000 years support the Kutzbach theory that orbital insolation controls the intensity of monsoons?

4. What evidence suggests that variations in orbital insolation were not the only cause of climate changes during the last 17,000 years?

5. Describe how and why proglacial lakes travel slowly across the landscape behind melting ice sheets.

6. Why were summer temperatures at high northern latitudes warmer 6,000 years ago than they are today? Would they also have been warmer at high southern latitudes?

7. What do orbital trends imply about future changes in monsoons and northern ice sheets?

Additional Resources

Basic Reading

COHMAP Members. 1988. "Climatic Changes of the Last 18,000 Years: Observations and Model Simulations." *Science* 241:1043–62.

Kutzbach, J. E., and F. A. Street-Perrott. 1985. "Milankovitch Forcing of Fluctuations in the Level of Tropical Lakes from 18 to 0 Kyr B.P." *Nature* 317:130–34.

Roberts, N. 1998. *The Holocene.* Oxford: Blackwell.

Advanced Reading

Bard, E., B. Hamelin, R. G. Fairbanks, and A. Zindler. 1990. "Calibration of the ^{14}C Time Scale over the Last 30,000 Years Using Mass Spectrometric U-Th Ages from Barbados Corals." *Nature* 345: 405–10.

Dyke, A. S., and V. K. Prest. 1987. "Late Wisconsinan and Holocene History of the Laurentide Ice Sheet." *Géographie Physique et Quaternaire* 41: 237–63.

Fairbanks, R. G. 1989. "A 17,000-Year Glacio-eustatic Sea Level Record: Influence of Glacial Melting on the Younger Dryas Event and Deep-Ocean Circulation. "*Nature* 342: 637–42.

Teller, J. T. 1987. "Proglacial Lakes and the Southern Margin of the Laurentide Ice Sheet." In *North America and Adjacent Oceans During the Last Deglaciation*, ed. W. F. Ruddiman and H. E. Wright. Geology of North America, K-3. Boulder, CO: Geological Society of America.

Webb, T., III. 1998. "Late Quaternary Climates: Data Synthesis and Model Experiments." *Quaternary Science Reviews* 17: 587–606.

Wright, H. E. Jr., et. al. 1993. *Global Climates Since the Last Glacial Maximum*. Minneapolis: University of Minnesota Press.

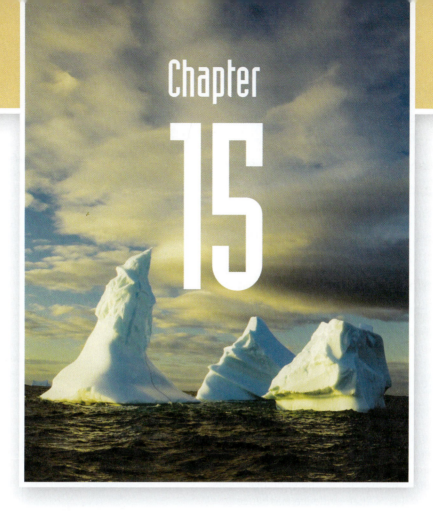

Chapter 15

Millennial Oscillations of Climate

Large climatic oscillations have occurred over intervals considerably shorter than the orbital cycles. Because they last for a few thousand years, they are called **millennial oscillations**. Some of these fluctuations ended within decades, fast enough to be of possible relevance to human concerns about future climate, but their cause remains uncertain. This chapter describes the accumulating evidence about their distribution in time and space, and their possible origin.

The oscillations were largest when glacial ice sheets existed in the Northern Hemisphere and much smaller or absent during interglacial climates like the current one. Especially large oscillations have been found in Greenland ice cores and in North Atlantic sediments, and similar fluctuations have also appeared in numerous other records from the Northern Hemisphere. Much smaller fluctuations occur in south polar regions, with a different (opposite) timing compared to the fluctuations in the north. The oscillations are largely random, rather than cyclic. Scientists have narrowed the range of likely explanations to two possibilities. They could be driven by internally generated fluctuations in the margins of northern hemisphere ice sheets that alter conditions in and around the North Atlantic Ocean, with the signal then propagated southward through the atmosphere, or they could result from more complex interactions between the Northern and Southern Hemispheres linked to the redistribution of ocean heat.

Millennial Oscillations During Glaciations

The first critical clue that the climate system is capable of large changes over short intervals came from studies of the deglacial Younger Dryas event, which lasted less than 1,500 years and began and ended very abruptly (see Chapter 14). Other evidence has since emerged that an ongoing series of short-term oscillations is superimposed on the orbital-scale climatic cycles, mainly during glacial intervals.

15-1 Oscillations Recorded in Greenland Ice Cores

Long ice cores taken on Greenland in the 1970s recovered records spanning much of the last interglacial-glacial cycle (Figure 15-1). The upper portions of these records were dated by counting annual layers, while the age of the lower section was estimated from models of the flow of layers deeper within the ice.

Two signals from this record attracted particular interest—the $\delta^{18}O$ composition of the ice and the concentration of dust. Signals of $\delta^{18}O$ in ice cores record changes in the composition of the ^{16}O-rich water vapor that falls as snow and consolidates into ice (see Appendix A). Of the several processes that can affect the $\delta^{18}O$ composition of ice (Table 15-1), local air temperature is the primary control for the signals preserved in the ice sheets. Chemical analysis of the dust in Greenland ice shows that the main source region was northern Asia. The transport path may have followed the northern branch of the split jet stream that moved across the Canadian margin of the North American ice sheet (see Chapter 13, Figure 13-11B).

Both records show two distinct patterns. One trend is the slow underlying change from low dust

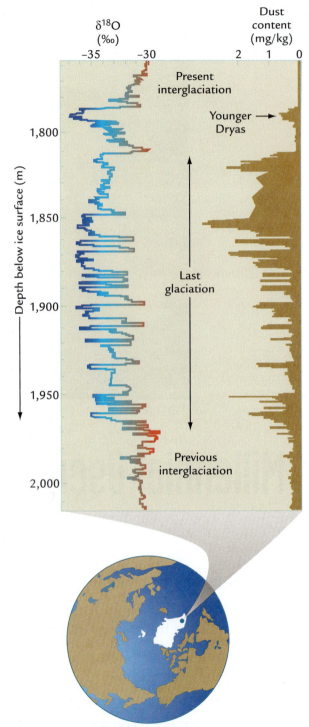

FIGURE 15-1
Millennial oscillations in ice cores

An ice core drilled through the Greenland ice sheet in the 1970s contained records of $\delta^{18}O$ and dust concentrations. Large oscillations occur in the glacial portion of both signals, but not in the present interglaciation, or in the previous one. (ADAPTED FROM W. DANSGAARD ET AL., "NORTH ATLANTIC CLIMATIC OSCILLATIONS RECORDED BY DEEP GREENLAND ICE CORES," IN *CLIMATE PROCESSES AND CLIMATE SENSITIVITY*, ED. J. E. HANSEN AND T. TAKAHASHI [WASHINGTON, DC: AMERICAN GEOPHYSICAL UNION, 1984].)

Table 15-1 Causes of δ¹⁸O Changes Recorded in Ice Cores

Negative	Change in $\delta^{18}O$ values	Positive
Colder	Air temperature over ice	Warmer
Distant	Proximity of source region	Close
Low $\delta^{18}O$	$\delta^{18}O$ composition of source	High $\delta^{18}O$
High	Elevation of ice	Low
Winter	Primary season of precipitation	Summer

concentrations and relatively positive (low-^{16}O) $\delta^{18}O$ values at the top, to higher dust concentrations and more negative (high-^{16}O) $\delta^{18}O$ values in the middle part, and a return to positive $\delta^{18}O$ values and little dust in the bottom part of the record. The imprecise dating available at the time these features were discovered indicated that the upper section spans the current interglaciation, the middle part represents the last glacial interval, and the bottom section is part of the previous interglaciation. The current and previous interglacial intervals were warmer (with more positive $\delta^{18}O$ values) and relatively free of dust compared to the cold, dusty ^{16}O-enriched glacial interval in the middle.

But these slower orbital-scale changes are to some extent masked by a far more prominent pattern—the rapid oscillations over much shorter intervals between high and low dust concentrations and between negative and positive $\delta^{18}O$ values. The $\delta^{18}O$ fluctuations of 4‰ to 6‰ represent a large fraction of the difference between the full glacial and full interglacial values. Temperatures over Greenland during glacial intervals oscillated rapidly between extremely cold intervals called "stadials" and relatively mild intervals called "interstadials." The amplitude of the temperature variation ranged between 8° and 16°C, with a tendency toward a sawtoothed shape: fast warming and slower cooling. These oscillations are often referred to as **Dansgaard-Oeschger oscillations** in honor of the ice-core scientists Willi Dansgaard and Hans Oeschger, who first identified them. Their work suggested that the oscillations were spaced at intervals that ranged from as little as 1,000 years to almost 9,000 years in length.

Each fluctuation toward more negative (glacial) $\delta^{18}O$ values is matched by an abrupt increase in dust concentrations in the ice. As in the case of the $\delta^{18}O$

changes, the range of variation in dust concentrations is a large fraction of the total difference between the glacial and interglacial values. Geochemical analysis of the dust shows that most of it comes from distant source regions in Asia, not from nearby North America. The size of the dust particles is larger in the cold intervals than in the milder ones, indicating that strong winds lifted and transported the dust when climate was very cold. The colder intervals also contain larger amounts of sea salt (Na^{+1} and Cl^{-1} ions) plucked from salty sea spray above the turbulent ocean during cold and windy intervals and carried to the ice.

In the late 1980s, two long sequences were drilled on the summit of the Greenland ice sheet at sites named GISP and GRIP. These sites were carefully positioned over areas of smooth underlying bedrock to minimize the impact of changes in ice flow that can disturb deeper ice layers, and they were drilled less than 30 kilometers apart to see whether or not they would reveal similar climate histories. The annual layering in these new cores was used to date the deeper ice layers, although stretching and thinning introduced growing uncertainties deeper in the ice. The combined use of several annually deposited signals (dust, $\delta^{18}O$, and others) reduced layer miscountings. The estimated counting errors increase from a few decades for ice 10,000 years old to several thousand years for ice 50,000 years old, and considerably more for older layers.

Both the GISP and GRIP ice core sequences yielded nearly identical long-term climate records to within 200 meters of the underlying bedrock. Most of one of these records is shown in Figure 15-2 (right). Because the two cores recorded nearly identical climatic signals, scientists had no doubt that both were reliable records of climate.

15-2 Oscillations Recorded in North Atlantic Sediments

Millennial oscillations have also been detected in North Atlantic sediments. Deep-ocean sediments are normally not a promising archive for monitoring short-term climatic fluctuations because deposition rates are usually no greater than 1 or 2 centimeters per 1,000 years, and small burrowing animals stir and mix the sediments to depths of 5 to 10 centimeters (see Chapter 3). As a result, mixing usually obliterates climate oscillations that are shorter than 2,500 to 5,000 years.

Fortunately, places exist in the North Atlantic where deposition rates can be as high as 10 to 20 centimeters per 1,000 years. Bottom currents carry fine sediments away from locations subjected to swift flow and deposit them as large lens-shaped piles (called **sediment drifts**) in regions where the currents slow. The process is the same as the one that creates snowdrifts by scouring snow from exposed regions where

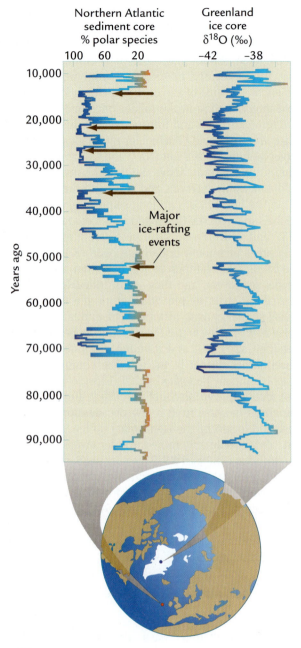

FIGURE 15-2

Millennial oscillations in the North Atlantic Ocean

Millennial-scale fluctuations in the composition of North Atlantic foraminifera and in ice-rafting influxes (left) match $\delta^{18}O$ changes in Greenland ice cores (right). (MODIFIED FROM S. STANLEY, *EARTH SYSTEM HISTORY*, © 1999 BY W. H. FREEMAN AND COMPANY, AFTER G. BOND ET AL., "CORRELATIONS BETWEEN CLIMATIC RECORDS FROM NORTH ATLANTIC SEDIMENTS AND GREENLAND ICE," *NATURE* 365 [1993]: 143–47.)

that rapidly bury and preserve the climatic signals in the coarser material.

In the mid-1980s, studies of rapidly deposited sediments in the North Atlantic Ocean first detected shorter climate oscillations. The marine geologist Hartmut Heinrich found episodes of unusually abundant ice rafting separated by as little as 7,000 years to as much as 12,000 years. These episodes are called **Heinrich events**. Later, the geologist Gerard Bond discovered even shorter-term (2,000–3,000 year) variations in two climatic indices: (1) the abundance of the one polar species of foraminifera as a fraction the total population, and (2) the relative amounts of the shells of foraminifera compared to the sand-sized ice-rafted grains. As was the case for orbital-scale studies (see Chapter 12), higher percentages of the polar species of foraminifera and larger concentrations of ice-rafted debris are an indication of the presence of colder North Atlantic waters carrying larger numbers of icebergs.

Radiocarbon dating of the North Atlantic sediments back to 30,000 years ago (adjusted to calendar years) indicates a correlation in time with the oscillations dated in Greenland ice by counting annual ice layers. Times of cold air (more negative $\delta^{18}O$ values in Greenland ice) correlate with times of cold ocean temperatures (larger percentages of the polar species) in the nearby ocean (see Figure 15-2).

Both records show a tendency toward a similar sawtoothed pattern: repeated slow drifts toward colder, more glacial conditions, followed by an influx of ice-rafted debris, and then an abrupt shift back to warmer conditions. Although the major Heinrich events show the largest ice-rafted influxes, the smaller Dansgaard-Oeschger oscillations also have lesser ice-rafted pulses. Prior to 30,000 years ago, both the marine and ice-core dating methods become too uncertain to allow firm time correlations, but the very strong resemblance of the two signals suggests that the same close relationship persists.

Initially, one question about the North Atlantic signal was whether the *relative* increases in the amount of ice-rafted debris compared to the foraminifera were caused by greater delivery of ice-rafted debris, by reduced rates of deposition of foraminifera, or both. Radiocarbon dating (again adjusted to calendar years) of the $CaCO_3$ shells of foraminifera in several of the younger ice-rafting layers indicated tenfold or larger increases in the rate of deposition of ice-rafted debris, compared to much smaller decreases in the rate of deposition of foraminifera (generally by less than half). The major factor in these variations turned out to be the increased influxes of ice-rafted debris.

Other questions concerned the source or sources of the ice-rafted debris. Although most of the ice sheets surrounding the North Atlantic contributed mineral debris to the ice-rafting influxes, a large fraction of the

the winds are strongest and piling it in regions where the winds weaken. In the ocean, the coarser sand-sized sediments such as foraminifera and ice-rafted debris are not easily moved by bottom currents and tend to stay in place as a reliable record of climate changes, while bottom currents deliver fine sediments

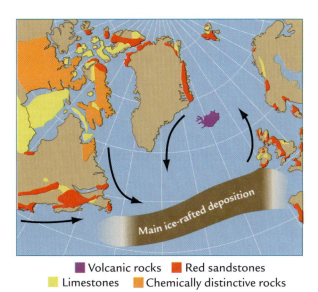

■ Volcanic rocks ■ Red sandstones
■ Limestones ■ Chemically distinctive rocks

FIGURE 15-3
Sources and deposition of ice-rafted debris

Highest rates of deposition of ice-rafted debris occur in the North Atlantic Ocean between 45° and 50°N. During smaller ice-rafting episodes, sources of debris include volcanic rocks on Iceland and red sandstone rocks on several coastal margins. During large ice-rafting events, massive amounts of material come from eastern North America, including limestone from Hudson Bay and fragments from other regions with distinctive chemical signatures.

(ADAPTED FROM G. BOND ET AL., "EVIDENCE OF MASSIVE DISCHARGES OF ICEBERGS INTO THE NORTH ATLANTIC DURING THE LAST GLACIAL PERIOD," *NATURE* 360 [1992]: 245–49.)

grains deposited in the primary ice-rafting zone at 45° to 50°N latitude during most of the Heinrich events came from the northeastern margin of the Laurentide ice sheet covering North America (Figure 15-3).

Initial investigations of limestone fragments during the largest ice-rafting (Heinrich) events showed that rocks in and north of Hudson Bay in Canada were the major source of this debris. Further evidence supporting this conclusion came from geochemical (isotopic) analysis of ice-rafted mineral grains that pointed to bedrock sources north and east of Hudson Bay. Detailed sampling of these large ice-rafting events showed, however, that the first debris deposited often came from smaller ice sheets around the North Atlantic Ocean. Only later did the distinctive limestone debris from North America arrive.

During the fluctuations, ice-rafted debris came from a range of source regions, two of which left particularly distinctive evidence (Figures 15-3 and 15-4). Fragments of volcanic glass originated mainly from eruptions on Iceland. In addition, iron-stained quartz grains came from several sources around the North Atlantic where outcrops of Pangaean-age sandstone contain quartz grains stained red by iron oxidation during ancient monsoonal climates (see Chapter 5). Other regions around the Atlantic margins also delivered debris during these oscillations.

Long sequences of ocean sediments recovered by the Ocean Drilling Program (see Chapter 3, Figure 3-4)

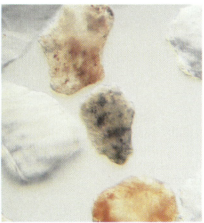

A B C

FIGURE 15-4
Ice-rafted grains

Sand-sized grains ice-rafted into the North Atlantic: volcanic debris from Iceland (A) and iron-stained quartz grains from sandstone rocks around the Atlantic margins (B). Sources of iron-stained quartz grains include red sandstones from the Orkney Islands, off northern Scotland (C). (A AND B: COURTESY OF G. BOND, LAMONT-DOHERTY EARTH OBSERVATORY OF COLUMBIA UNIVERSITY; C: DEREK CROUCHER/GETTY IMAGES.)

show that millennial fluctuations also occurred during previous glaciations, both during the 41,000-year cycles prior to 0.9 million years ago, and the subsequent ice sheet fluctuations at ~100,000 years. The largest millennial oscillations occurred during times when ice sheets were of medium to large size, with negligible fluctuations during full interglacial climates.

Because the millennial oscillations are so well developed in North Atlantic surface waters, scientists searched for evidence that this signal had penetrated into deep water formed in this region. The method exploited was the same one used to measure similar changes at orbital scales (see Chapter 11): more negative $\delta^{13}C$ values in the $CaCO_3$ shells of bottom-dwelling (benthic) foraminifera mark times when deep water from North Atlantic sources was replaced by bottom water formed in the Southern Ocean. Because these deep-water $\delta^{13}C$ signals are measured in the same cores containing the planktic foraminifera used to monitor changes in the surface waters, the *relative* timing of the two kinds of changes can be determined, even without accurate knowledge of absolute ages.

The $\delta^{13}C$ evidence showed that deep-water formation in the North Atlantic was rapid during warm interglacial climates, slower and shallower during glacial climates, and especially slow during these ice-rafting episodes. Apparently, the changes at the surface were felt in the deeper North Atlantic Ocean.

> **In Summary**, apparently synchronous millennial oscillations of very large amplitude are recorded in Greenland ice and North Atlantic sediments. These oscillations indicate coupled changes in several of the most important components of Earth's climate system: air and surface-ocean temperature, ice sheet margins, and ice rafting.

15-3 Detecting and Dating Other Millennial Oscillations

The verification of similar millennial oscillations in both Greenland ice and North Atlantic sediments set off an energetic search for similar fluctuations in other regions. Scientists who took part in this search faced two major questions at the outset: (1) Is the climatic archive that is being examined capable of recording such brief oscillations? (2) How accurately can the oscillations be dated?

Resolution of climate signals varies from archive to archive and from region to region. The ideal archive is one that allows resolution of annual changes and provides a record stretching well back into the last glaciation or preferably beyond. Unfortunately, few archives combine these characteristics, but many archives that reach far back in time can resolve

climate changes at a resolution of tens to hundreds of years. This resolution is sufficient for detecting the presence of millennial oscillations, even if inadequate to define their full amplitude.

The second (and more difficult) problem is determining how the oscillations detected in other archives correlate in time with those found in Greenland ice and North Atlantic sediments. The ^{14}C method used to date most continental records (after correcting to calendar years) has analytical uncertainties of several thousand years for most glacial-age material. Because these dating errors grow comparable in size to the length of the oscillations prior to 30,000 years ago, it is more often than not impossible to determine how the observed oscillations actually correlate.

Consider the example of an oscillation that lasts for 2,000 years and is perfectly dated in one region but has a dating uncertainty of $+/-1,000$ years in another (Figure 15-5). Within this uncertainty, the two signals could be synchronous, changing at totally opposite tempos, or varying with subtle leads and lags that are neither synchronous nor opposite. At orbital time scales, leads and lags between climate signals provide critical clues to cause-and-effect relationships (see Chapter 12). In the same sense, scientists would like to have accurate information about leads and lags among millennial-scale records, but dating uncertainties make this goal difficult and often impossible.

It is much easier to show that millennial-scale oscillations are present or absent in a given climate record, whatever their exact ages and correlations to other records. In a few such cases, the millennial oscillations match the pattern of the changes in Greenland and the North Atlantic so closely that little doubt exists that we are looking at the same oscillations, even though the dating is not very accurate. But even in these cases, small leads or lags between the signals could exist.

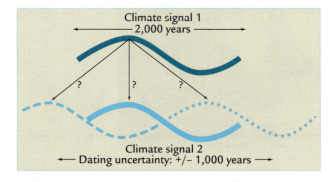

FIGURE 15-5
Uncertainties in dating millennial oscillations
Typical dating uncertainties of 2,000 years make it difficult to determine whether millennial oscillations in two separate climate records are synchronous, exactly opposite in timing, or offset slightly in timing.

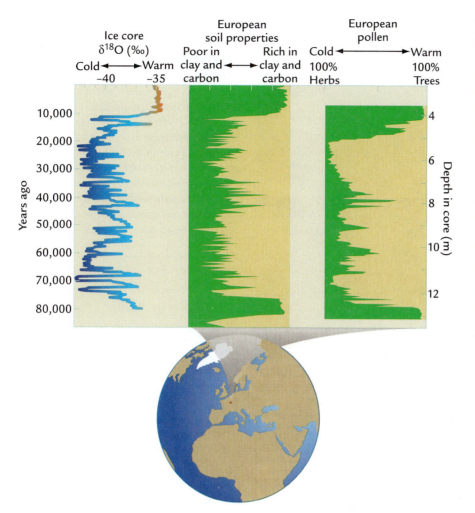

FIGURE 15-6

Millennial-scale climate changes in Europe

Similar to δ18O changes in Greenland ice (left), millennial-scale fluctuations occur in European soils (middle) and pollen (right). (LEFT: ADAPTED FROM P. GROOTES ET AL., "COMPARISON OF OXYGEN ISOTOPE RECORDS FROM THE GISP AND GRIP GREENLAND ICE CORES," *NATURE* 366 [1993]: 552–54; MIDDLE: ADAPTED FROM N. THOUVENY ET AL., "CLIMATE VARIATIONS IN EUROPE OVER THE PAST 140 KYR DEDUCED FROM ROCK MAGNETISM," *NATURE* 371 [1994]: 503–6; RIGHT: ADAPTED FROM G. M. WOILLARD AND W. G. MOOK, "CARBON-14 DATES AT GRANDE PILE: CORRELATION OF LAND AND SEA CHRONOLOGIES," *SCIENCE* 215 [1982]: 159–61.)

15-4 Oscillations Elsewhere in the Northern Hemisphere

Other regions of the Northern Hemisphere also show millennial-scale oscillations, including western Europe (Figure 15-6). European soils were richer in organic material during warmer episodes but almost free of organic carbon during colder oscillations. In addition, European pollen records that show orbital-scale changes from interglacial forests to glacial tundra also reveal short-term fluctuations within glacial intervals from full tundra to mixed grass steppe and forest vegetation. Although the correlations of the European records with the Greenland δ18O fluctuations are not particularly obvious, 14C dating of the younger fluctuations indicates that cold-adapted vegetation occurred in Europe during times of colder air over Greenland and colder temperatures in the North Atlantic Ocean.

Short-term fluctuations have also been discovered in the glacial sections of windblown loess deposits in China. Coarser loess-rich layers that indicate physical weathering during very cold intervals alternated with finer, clay-rich soils that indicate greater chemical weathering during warmer episodes. Although the Asian soil/loess sequences are not well dated, changes in that region are generally thought to match oscillations in and around the North Atlantic.

Climate scientists have also found millennial oscillations in regions far from the North Atlantic Ocean. In the Santa Barbara Basin along the Pacific coast of North America at 35°N, several 14C-dated climatic signals match the δ18O fluctuations in the Greenland ice sheet (Figure 15-7). Short-term oscillations in δ18O values measured in the shells of planktic foraminifera indicate large (4°C or more) temperature changes in near-surface waters. In addition, the type of sediment deposited in the Santa Barbara Basin fluctuates between layers mixed by burrowing animals and intervals with varve-like layering still intact. Oxygen must have been absent from the deep basin during the warmer intervals to prevent small creatures from burrowing and churning the delicate layering. Instead, sediment mixing occurred during cold-climate oscillations, and during these times burrowing activity was vigorous. The obvious match in pattern between the records in the Santa Barbara Basin records and those in Greenland ice makes it clear that very similar millennial oscillations affected both regions.

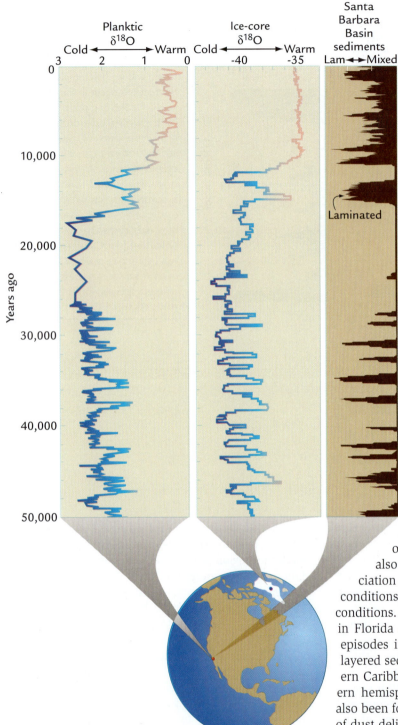

FIGURE 15-7

Millennial fluctuations in the Santa Barbara Basin

The pattern of δ¹⁸O changes in planktic foraminifera from surface waters of California's Santa Barbara Basin (left) closely matches δ¹⁸O changes in Greenland ice (middle). Fluctuations between intervals of varved sediments in the Santa Barbara Basin and intervals disturbed by burrowing animals show the same pattern (right). (ADAPTED FROM I. L. HENDY AND J. P. KENNETT, "LATEST QUATERNARY NORTH PACIFIC SURFACE-WATER RESPONSES IMPLY ATMOSPHERE-DRIVEN CLIMATIC INSTABILITY," *GEOLOGY* 27 [1999]: 291–94.)

and the Colorado Rocky Mountains. In the Midwestern United States, far from the Atlantic margins of the Laurentide ice sheet, several ice lobes appear to have fluctuated in rough synchrony with the larger ice-rafting episodes, hinting at a possible link between the land and ocean margins. A pollen record from Florida also shows fluctuations during the last glaciation between pine pollen, indicative of wetter conditions, and grass and oak pollen that indicate dry conditions. The fluctuations toward wetter conditions in Florida appear to correlate with major ice-rafting episodes in the North Atlantic. Well-dated annually layered sediments in the Cariaco Trench in the southern Caribbean also show oscillations with the northern hemisphere timing. Millennial fluctuations have also been found in the northern tropics in: the amount of dust delivered to the Arabian Sea, the deposition of organic carbon off the coast of Pakistan, the level of North African lakes, and the abundance of plankton in the equatorial Atlantic.

Cave calcite deposits from the northern and southern tropics also show clear millennial-scale oscillations that are very accurately dated by the Th/U method. At orbital time scales, the δ¹⁸O variations in these deposits show out-of-phase responses between the hemispheres at the 23,000-year period of precession (see Chapter 9, Figure 9-13). This behavior

Other indications of millennial-scale oscillations in western North America come from fluctuations of glacial Lake Bonneville in Utah, where the younger ¹⁴C-dated lake level maxima appear to correlate with major ice-rafting events in the North Atlantic. Also, millennial-scale advances of mountain glaciers have been found in the Sierra Nevadas of California, the Cascades of Oregon and Washington,

reflects out-of-phase summer insolation forcing of low-latitude monsoons in the two hemispheres, resulting in migrations of the intertropical convergence zone (ITCZ) across tropical regions.

At the millennial scale, southern hemisphere $\delta^{18}O$ oscillations recorded in Brazilian speleothems are somewhat reduced in amplitude compared to those in the northern tropics and are out of phase with those in the north. Brazil is warm and wet when the north is cold and dry. This pattern appears to be the result of short-term ITCZ migrations from more northerly positions to more southerly ones. Southward movement of the ITCZ leaves the northern tropics and subtropics drier and cooler, but makes the southern tropics and subtropics warmer and wetter.

15-5 Oscillations in Antarctica

The presence of millennial oscillations with similar timing across much of the Northern Hemisphere raised the obvious question of whether or not they were global in extent, and ice cores from Antarctica are the obvious place to look for an answer. Antarctic ice does show short-term $\delta^{18}O$ (and other) oscillations, but they are much smaller in size than those in Greenland ice and more symmetrical in shape compared to the

sawtoothed oscillations in the north. Unfortunately, Antarctic ice lacks the annual layering needed for sufficiently accurate dating to determine the relative phasing with the changes in the Northern Hemisphere, but both Antarctic and Greenland ice contain a common signal that can be used to correlate the two records very closely in a relative sense: millennial-scale changes in atmospheric methane concentrations (Figure 15-8).

With the two trends synchronized based on the methane variations, the temperature oscillations over Antarctica turn out to be roughly opposite in timing to those in the north. In general, slow warming trends in Antarctica were underway when Greenland ice recorded the coldest northern temperatures, and the fastest rates of warming in Greenland occurred when Antarctica had already reached maximum warmth and was beginning to cool. The two regions seem to be nearly, but not precisely, out of phase.

This offset in timing is confirmed by an ocean core from near Portugal, where the $\delta^{18}O$ signal recorded in planktic (surface-dwelling) foraminifera follows the pattern of North Atlantic oscillations, but the $\delta^{18}O$ changes in benthic (bottom-dwelling) foraminifera more closely resemble the oscillations found in Antarctic ice. The fact that both signals are present in the same core allows scientists to compare their

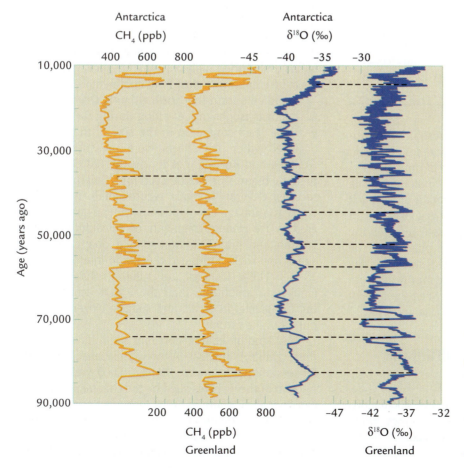

FIGURE 15-8
Opposed millennial oscillations in Antarctic and Greenland ice

Records from Antarctic and Greenland ice that are correlated based on common variations in methane trends show that Antarctica tends to be warm (more positive $\delta^{18}O$ values) when Greenland is cold (more negative $\delta^{18}O$ values). (ADAPTED FROM T. BLUNIER AND E. J. BROOK, "TIMING OF MILLENNIAL-SCALE CLIMATE CHANGE IN ANTARCTICA AND GREENLAND DURING THE LAST GLACIAL PERIOD," *SCIENCE* 291 [2001]: 109–12.)

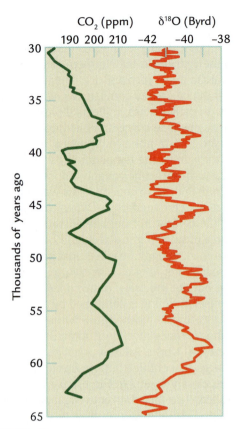

FIGURE 15-9
Millennial CO₂ changes

CO₂ changes at millennial time scales in Antarctic ice lag slightly behind changes in air temperature shown by δ¹⁸O trends. (ADAPTED FROM J. AHN AND E. J. BROOK, "ATMOSPHERIC CO₂ AND CLIMATE ON MILLENNIAL TIME SCALES DURING THE LAST GLACIAL PERIOD," *SCIENCE* 322 [2008]: 83–85.)

relative timing, even without high-resolution radiocarbon dating. Again, the northern and southern signals are offset (and almost opposite) in timing.

Antarctic ice cores also show millennial-scale changes in atmospheric CO₂ concentration of as much as 20–25 ppm (Figure 15-9). The CO₂ signal is out of phase with the Dansgaard-Oeschger oscillations in the Northern Hemisphere, and roughly in phase with the temperature changes in Antarctica. The small CO₂ lag relative to that of Antarctic temperature may result from errors in correcting for differences between the ages of the gases (CO₂) and the solid ice (see Chapter 11).

Millennial Oscillations During the Present Interglaciation

In contrast to the large millennial-scale fluctuations found in many climate records during times when northern hemisphere ice sheets were relatively large, fluctuations were muted or even absent from the inter-

glacial portions of the same records. The last two oscillations of significant size during the present interglaciation occurred during the late stages of melting of the northern ice sheets, not during full interglacial times.

The large-amplitude Younger Dryas episode was not associated with any obvious release of meltwater eastward through the Saint Lawrence Seaway, although a large release northward to the Arctic Ocean may have occurred (see Chapter 14). A second oscillation occurred near 8,200 years ago, by which time the Scandinavia ice sheet had completely melted and the Laurentide ice sheet in North America had shrunk to a small area (see Figure 14-2). This last oscillation occurred at the same time as a large release of meltwater from glacial lakes in the Hudson Bay region eastward through Hudson Strait. This cold interval lasted only a few hundred years, but temperatures dropped markedly over Greenland and Europe.

Since the 8,200-year cold oscillation, fluctuations at millennial scales have been much smaller in amplitude. They have also differed in pattern both from region to region and among different climatic indices within the same region (and even within the same sediment or ice archives).

Millennial-scale δ¹⁸O oscillations are not obvious in Greenland ice cores during the last 8,000 years (see Figure 15-2), although small fluctuations do occur in the amount of sea salt (Na^{+1} and Cl^{-1} ions) from the ocean (Figure 15-10). These oscillations are interpreted as an indication of changes in wind strength. Salty spray from the sea surface is lifted by strong winter and spring winds and carried high onto the ice sheet, where it is deposited along with dust from continental sources.

Sediments from the North Atlantic Ocean indicate no major episodes of ice rafting during the last 8,000 years, at least compared to the size of the ones that occurred when ice sheets were present on North America and Eurasia. Carbon in the $CaCO_3$ shells of planktic organisms can be used for ^{14}C dating of these records. Several intervals record small increases in concentration of two kinds of mineral grains delivered by ice rafting: fragments of volcanic glass from Iceland and grains of iron-stained quartz from red sandstone rocks around the Atlantic margins (Figure 15-11). These influxes are 1,000 to 100,000 times smaller in amplitude than those that occurred during full glacial intervals and suggest extremely small pulses of ice rafting.

The kind of ice transport that could have delivered these grains is unclear. Although icebergs have been far less abundant during the last 8,000 years, a small fraction of the icebergs shed by the Greenland ice sheet can float as far south as 40°N. Icebergs from Greenland could have carried iron-stained quartz grains.

Sea ice is also capable of picking up and carrying small amounts of debris, either because sediment freezes

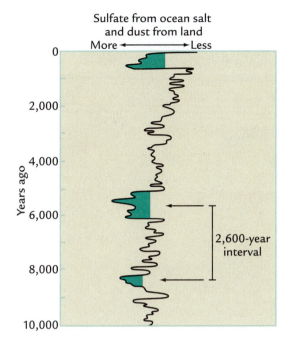

Sulfate from ocean salt
and dust from land
More ←——————→ Less

FIGURE 15-10
Changes in sea salt

Over the last 10,000 years, ice cores from Greenland show nearly identical short-term fluctuations in the amount of sea salt (Na^{+1} and Cl^{-1} ions) and the amount of dust from the continents, with a faint suggestion of a cycle near 2,600 years. (ADAPTED FROM S. R. O'BRIEN ET AL., "COMPLEXITY OF HOLOCENE CLIMATE AS RECONSTRUCTED FROM A GREENLAND ICE CORE," *SCIENCE* 270 [1995]: 1962–64.)

onto the bottom ice layers along coastlines or because material is deposited on top of the ice, either by volcanoes that erupt glass fragments into the atmosphere, or by spring floods that wash debris onto coastal sea ice around the Arctic margins. Because sea ice is at most only a few meters thick, it cannot carry debris far into a warm ocean before melting. Yet sea ice is common today along the east coast of Greenland, the north coast

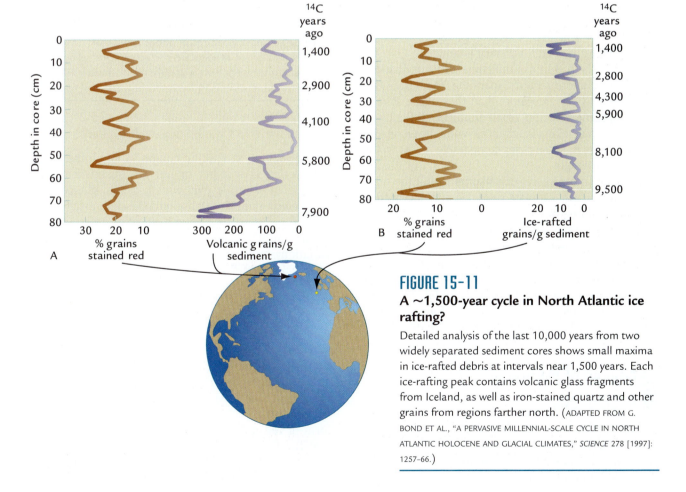

FIGURE 15-11
A ~1,500-year cycle in North Atlantic ice rafting?

Detailed analysis of the last 10,000 years from two widely separated sediment cores shows small maxima in ice-rafted debris at intervals near 1,500 years. Each ice-rafting peak contains volcanic glass fragments from Iceland, as well as iron-stained quartz and other grains from regions farther north. (ADAPTED FROM G. BOND ET AL., "A PERVASIVE MILLENNIAL-SCALE CYCLE IN NORTH ATLANTIC HOLOCENE AND GLACIAL CLIMATES," *SCIENCE* 278 [1997]: 1257–66.)

of Iceland, and around the Arctic island of Spitsbergen (Svalbard), and sea ice could have transported many of the sand-sized grains measured in these cores.

Curiously, the signal shown in Figure 15-11 is not registered in other records spanning the last 8,000 years from the high-latitude North Atlantic Ocean. Changes in $CaCO_3$ concentrations from cores near Iceland do not show the changes that would be expected if surface water $CaCO_3$ productivity varied at the millennial scale. Influxes of noncarbonate silts and clays around the Atlantic margins also fail to register major millennial fluctuations. One index of bottom-current strength near Iceland does show oscillations at or near intervals of 1,500 years.

A similar problem appears in compilations of advances and retreats of mountain glaciers, which have been superimposed on a slow drift toward slightly colder conditions during the last 8,000 years. One compilation shows numerous short glacial advances (Figure 15-12A), while another reconstruction shows fewer and more widely spaced advances (Figure 15-12B). These differences may reflect different choices of glaciers used in the compilations as well as the difficulty in obtaining reliable ^{14}C (or other) dates to constrain the upper and lower age limits of each ice advance. Many other regions show what appear to be millennial-scale fluctuations during the last 8,000 years. For example, North African lake levels have fluctuated markedly during that time (see Chapter 14, Figure 14-16C), but dating these changes with high accuracy is difficult.

Overall, the patterns of millennial-scale oscillations in high-resolution records spanning the last 8,000 years seem to disagree more than they agree. Part of this disagreement could result from the difficulty in detecting small climatic changes with measurement errors that are in some cases comparable in size to the signal being sought.

Another (more likely) explanation is that the separate parts of Earth's climate system simply "went their own way" during the last 8,000 years. In the absence of a strong central driving force such as ice sheets, smaller-scale components of the climate system may have acted relatively independently of each other during this interval. If so, the weaker millennial-scale oscillations found in various regions during the current interglaciation may not represent the same phenomenon as the large and widespread fluctuations that occurred during times when ice sheets were present.

▶ Causes of Millennial Oscillations

Several explanations for millennial-scale oscillations have been proposed, and these can be evaluated using the evidence summarized above. Some can be eliminated, but others remain viable.

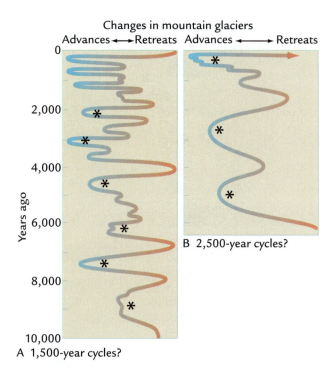

FIGURE 15-12

Millennial-scale oscillations of mountain glaciers

Two attempts to synthesize advances and retreats of mountain glaciers over the last several millennia have produced different interpretations. One (left) hints that advances may have occurred at intervals of 1,500 years, but the other (right) indicates advances separated by about 2,500 years. (A: ADAPTED FROM F. ROTHLISBERGER, *10,000 JAHRE GLETSCHERGESCHICHTE DER ERDE* [FRANKFURT: SAUERLÄNDER, 1997]; B: ADAPTED FROM G. H. DENTON AND S. C. PORTER, "NEOGLACIATION," *SCIENTIFIC AMERICAN* 222 [1970]: 100-10.)

15-6 Are the Oscillations Periodic?

Before looking at specific causes, a key issue worth considering is whether or not the observed millennial variations are cyclic, which, at first glance, they may seem to be. The Dansgaard-Oeschger oscillations shown in Figure 15-2 have been often referred to as "Dansgaard-Oeschger *cycles*," and the longer-term variations in polar foraminifera and ice-rafted debris (which include the Heinrich events) are sometimes called "Bond *cycles*," after the geologist Gerard Bond who first detected them.

In fact, however, the case for cyclic behavior is very weak. As noted earlier, the difference in time between successive peaks or valleys in these oscillations ranges from as little as 1,000 years to as much as 9,000 years. No strong evidence of a preferred timing within that range is evident.

Instead, these oscillations have the characteristic form of **red noise**. The word *noise* conveys the idea that the fluctuations are random and unpredictable, rather than cyclic and predictable. The term *red* refers to a characteristic behavior in which the longer-duration oscillations are generally larger in size than the shorter-term oscillations.

Red noise is common in nature. As an analogy, consider a late spring afternoon with distinct clouds of many sizes randomly drifting across the sky. Small clouds that block the Sun for a few minutes may cool temperatures at Earth's surface in a barely noticeable way. Larger clouds that block the sunlight for an hour or more will cool a warm afternoon in a more obvious way. A band of clouds that persists for most of the afternoon in one area (rather than in another area nearby) will reduce the peak temperature for that day even more. In this example of red noise, the passage of clouds in front of the Sun is random, but the clouds alter local surface temperature in direct proportion to how long they block the Sun. A similar tendency is present in the northern hemisphere millennial oscillations.

Some scientists claim to have found a ~1,500-year cycle in ice sheet and ocean records, but in general the evidence is sparse. Time-series analysis of the Greenland ice core $\delta^{18}O$ record indicates that only a very small fraction of the observed variations falls within a band centered near 1,500 years. Most of the power at or near 1,500 years appears to come from three $\delta^{18}O$ variations that occurred from 30,000 to 35,000 years ago, but the 1,500-year period is absent or weak in the rest of the record (Figure 15-13).

The glacial geologist Richard Alley has proposed that millennial oscillations could fall somewhere between the extremes of cyclic behavior and red noise. His term for this intermediate behavior, **stochastic resonance,** is a composite of two concepts that seem in conflict with each other because the word *stochastic* means random, whereas *resonance* implies cyclicity.

In his view, resonance is evident in the fact that oscillations at a cycle near 1,500 years do appear at some times in some records (Figure 15-14). At other times, however, the climate system skips past (fails to register) individual oscillations at 1,500 years because of interference from the effects of random noise. When this happens, northern hemisphere climate continues to drift slowly toward colder conditions until it eventually responds to one of the subsequent multiples of the 1,500-year cycle (at 3,000; 4,500; 6,000; 7,500; or 9,000 years) and shifts abruptly back to warmer climates. Because these abrupt transitions occur over a wide range of intervals from 1,500 to 9,000 years, they appear to be random (stochastic) but could actually be an expression of an irregular cyclic behavior. In any case, noise plays an important role in this concept.

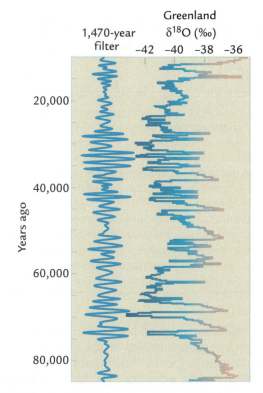

FIGURE 15-13
A millennial cycle in Greenland ice?

Filtering of the full $\delta^{18}O$ record from Greenland ice GISP 2 ice covering the interval 85,000 to 10,000 years ago (right) reveals that a 1,470-year cycle (left) is present at times but weak or absent during others. (ADAPTED FROM M. STUIVER, T. F. BRAZUNIAS, P. M. GROOTES, AND G. A. ZIELINSKI, "IS THERE EVIDENCE FOR SOLAR FORCING OF CLIMATE IN THE GISP2 OXYGEN ISOTOPE RECORD?" *QUATERNARY RESEARCH* 48 [1997]: 259–66.)

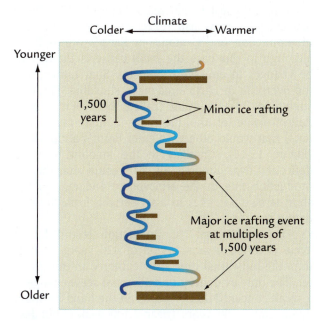

FIGURE 15-14
Millennial-scale North Atlantic cycles?

One view of millennial-scale changes in the North Atlantic is that short cooling cycles 1,500 years in length gradually drift toward colder conditions and occasionally culminate in major ice-rafting episodes, followed by an abrupt return to warmer conditions.

In Summary, the evidence that millennial oscillations occur as regular cycles is weak. The oscillations appear to be largely the product of random behavior (red noise) in the climate system.

15-7 Are the Oscillations Forced by the Sun?

Most of the energy that Earth receives from the Sun arrives at visible or near-visible (infrared or ultraviolet) wavelengths. Unfortunately, direct records of past changes in this incoming solar radiation do not extend back into the last glaciation.

Instead, several proxies provide scientists with other potential measures of solar fluctuations. As noted in Chapter 14, the difference between the ages derived by counting (matching) tree rings and those derived by ^{14}C dating of the same rings are one such index. The gradually increasing discrepancy between the ages derived from these two methods (see Chapter 14, Box 14-1) indicates that Earth's magnetic field was weaker near 15,000 to 20,000 years ago, and that the weaker solar shielding permitted more bombardment by charged cosmic particles (protons) and faster production of ^{14}C atoms.

Shorter-term changes in age offsets are also apparent within the last 10,000 years (Figure 15-15). These discrepancies may also reflect changes in the rate of production of ^{14}C atoms in Earth's atmosphere, although in this case the main cause is thought to be changes in emissions from the Sun, rather than the overprint of Earth's magnetic shielding. Particles streaming from the Sun (called the "solar wind") deflect some of the incoming cosmic rays (protons) that would otherwise enter Earth's atmosphere (Figure 15-16). Changes in the amount of solar deflection over hundreds of years could alter the ^{14}C production rate in the atmosphere and explain the short-term differences in ages derived from the two dating methods.

This 10,000-year ^{14}C production record has a cycle centered near a period of 420 years, shorter than the millennial time scale explored in this chapter. The only evidence of a millennial-scale change is a weak concentration of power in the band between 2,000 and 2,500 years (near 2,100 years), but few millennial oscillations in Earth's climate show major responses at this period. One reconstruction of the timing of mountain glacier advances over the last 7,000 years does show a hint of a response near 2,100 years (see Figure 15-12B), but the other reconstruction does not (see Figure 15-12A). A hint of changes at intervals near 2,600 years occurs in the

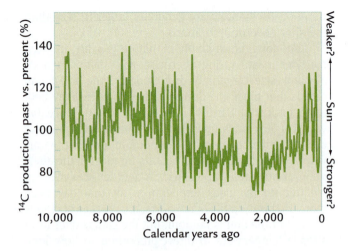

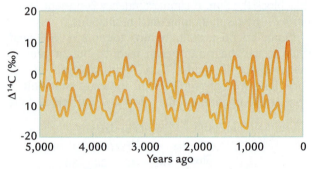

FIGURE 15-15

Millennial-scale changes in Sun strength?

Ages determined by counting individual tree rings can be compared against ages determined from ^{14}C analyses (A). The age differences reflect changing ^{14}C production in the atmosphere. Comparison of ^{14}C production and ^{10}Be production during the last 5,000 years after smoothing to remove long-term magnetic changes and the effects of processes in the carbon system on the ^{14}C signal (B). (A: ADAPTED FROM M. STUIVER ET AL., "CLIMATIC, SOLAR, OCEANIC, AND GEOMAGNETIC INFLUENCES ON LATE-GLACIAL AND HOLOCENE ATMOSPHERIC $^{14}C/^{12}C$ CHANGE," *QUATERNARY RESEARCH* 35 [1991]: 1–24. B: ADAPTED FROM J. BEER ET AL., "INFORMATION ON PAST SOLAR ACTIVITY AND GEOMAGNETISM FROM ^{10}Be IN THE CAMP CENTURY ICE CORE," *NATURE* 331 [1988]: 675–79.)

sea salt signal in Greenland ice (see Figure 15-10), but no such response occurs in the ice-rafting signals in the nearby ocean (see Figure 15-13). This evidence refutes the existence of a *cyclic* millennial-scale link between the Sun and Earth's climate, although it does not eliminate the possibility of a noncyclic link.

Another proposed solar proxy is the ^{10}Be (beryllium) isotope, which is produced by collisions with cosmic particles. As in the case of ^{14}C, solar emissions can deflect the cosmic particles and modulate production of ^{10}Be in Earth's atmosphere. Ice-core records show large millennial-scale changes in ^{10}Be concentration

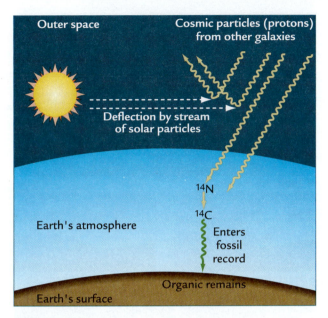

FIGURE 15-16
Changes in strength of solar emissions?
Changes in rates of ^{14}C production can be caused by changes in solar shielding of Earth's atmosphere from cosmic-ray bombardment.

during the interval from 40,000 to 25,000 years ago, with an obvious correlation between ^{10}Be maxima and $\delta^{18}O$ minima that are indicative of colder climates. At first, this correlation would seem to point to a clear solar-climate link, but the variations in ^{10}Be concentration are actually an artifact of climatic changes. During cold oscillations, accumulation rates on ice sheets fall. With slower ice accumulation, the concentration of ^{10}Be in the ice increases even if its rate of production in the atmosphere remains constant. Similarly, intervals of lower ^{10}Be concentrations during warm climates ($\delta^{18}O$ maxima) reflect dilution of the ^{10}Be signal by more rapidly accumulating ice. As a result, these older ^{10}Be fluctuations cannot be used to infer a connection to solar deflection of cosmic rays.

Changes in ^{10}Be concentration in ice cores have also occurred over the last 5,000 years, and they have been compared against changes in ^{14}C production determined from the difference between tree ring counts and ^{14}C analyses (see Figure 15-15B). In making this comparison, allowance has to be made for differences in how the two signals are registered by the climate system. In effect, the ^{10}Be signal must be smoothed in such a way as to mimic the more complicated processes that alter the ^{14}C record. The very good match of the two signals suggests that both the ^{10}Be and the ^{14}C signals may have varied in response to changes in solar deflection of the incoming cosmic rays. But another possible explanation is that both oscillations are a response to

changes in the internal operation of the climate system such as the circulation of the ocean. Because these changes occurred mainly over intervals of centuries and decades, we will return to this issue in Chapter 17.

In Summary, the evidence argues against a strong cyclical effect of solar variability on climate. A link between random (noncyclic) variations in Sun strength and climate at shorter (decadal to century) time scales remains a possibility.

15-8 Are the Oscillations Caused by Natural Ice Sheet Instabilities?

Because millennial fluctuations only occur when ice sheets are present in the Northern Hemisphere, some kind of causal link to the ice sheets must exist. This conclusion is further supported by the fact that the largest oscillations occur in the vicinity of the northern ice sheets. But are the ice sheets the "initiator" of a process that produces millennial oscillations, or are they simply a feedback amplification within a more complex process?

Considered as a whole, the great masses of ice lying on the continents have very slow response times measured in thousands of years (see Chapter 10). But the ocean margins of the ice sheets are capable of much faster changes because they slide on soft sediments, respond to sea level changes, and interact with ocean waters that contain large amounts of heat. Two mechanisms that favor an independent initiator role for ice sheets focus on the massive episodic releases of icebergs to the North Atlantic Ocean.

Along the marine margins of ice sheets, ice flows over bedrock with irregular bumps and depressions (Figure 15-17A). The bottom layers of ice scrape against higher-standing areas called **bedrock pinning points**, and the resulting friction slows the flow of ice. The bottom layers of ice can also freeze to the bedrock and slow the flow even more. Ocean water can produce the opposite effect: because the ice margins float in seawater, changes in sea level can lift the ice off its pinning points.

One idea is that the very slow natural release of small amounts of heat from Earth's interior can melt the lower ice layers along ice margins (see Figure 15-17B). The melting produces meltwater that trickles into the soft underlying sediments and makes them unstable, causing the ice margins to surge forward into the ocean. The surges release icebergs, which float away elsewhere to melt, and the thinned ice sheet margins retreat well inland from their previous positions. The new margins then slowly thicken and advance until the buildup of heat from below again destabilizes them.

A second idea, proposed by the glacial geologist Doug MacAyeal, focuses on a different kind of interaction between ice margins and bedrock (see Figure 15-17C). Over time, as the ice margins thicken, their weight depresses the underlying bedrock, which gradually sinks. After thousands of years, depression of the bedrock with respect to sea level reaches the point that the ocean can lift the ice off its pinning points. At that point, because the ice margin is no longer anchored to the bedrock, it begins to flow faster, and the ice streams release icebergs to the ocean. Once this outward flow of ice is exhausted, the ice stream retreats to another bedrock pinning point farther upstream (see Figure 15-17D).

Both of these hypotheses are consistent with the timing of the observed ice-rafting episodes in the North Atlantic Ocean compared to the prevailing regional climate. The largest episodes occurred when the air and ocean were very cold, rather than during the warmer intervals that would seem more likely to cause fast melting and ice margin collapse. This evidence argues against local warming as the driver of the iceberg pulses but allows for mechanisms based on internal ice sheet instabilities.

The composition of the debris deposited by icebergs in the North Atlantic shows that it came from many distinct source regions (North America, Europe, and Iceland). But why would so many ice sheet margins be simultaneously releasing icebergs? One possible link is sea level (see Figure 15-17C). If one ice margin surges and sends icebergs into the ocean, the melting icebergs raise sea level. A rise in sea level could then destabilize ice shelves on other coastal margins off their bedrock pinning points and cause them to discharge some of their ice into the ocean.

This explanation raises the issue of whether or not the millennial-scale rises in sea level were large enough to link the ice sheets. Variations in sea level based on coral reefs along the slowly uplifting coast of New Guinea suggest that sea level changes could have been as large as 10–15 meters during major millennial oscillations (the Heinrich events). In addition, marine $\delta^{18}O$ signals recorded in the shells of benthic foraminifera from the deep tropical Pacific Ocean show millennial-scale variations of 0.1‰ or slightly larger during these intervals. The coral reef and $\delta^{18}O$ evidence both point to sea level changes of 10 meters or more during the largest millennial oscillations.

The massive North American ice sheet is the best candidate for causing sea level rises large enough to trigger reactions in the other ice sheets during major ice-rafting events. Yet the evidence in North Atlantic sediments suggests that the first sand grains deposited

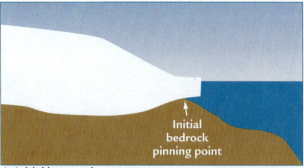

A Initial ice margin

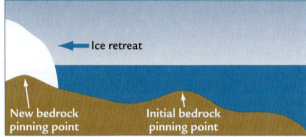

B Heat from below   C Depression of bedrock

D New ice margin

FIGURE 15-17
Natural oscillations of ice margins

Marine margins of ice sheets end in thin ice shelves flowing across upward-protruding bedrock knobs (A). This ice can be dislodged from these pinning points either by Earth's heat escaping from below and melting ice (B) or by the gradual weighing down of bedrock under the heavy load of growing ice (C). Either way, the ice margin retreats inland and stabilizes over another bedrock pinning point (D).

during major ice-rafting pulses often came from the smaller ice sheets on Iceland and coastal regions in Europe, with icebergs from North America only arriving later. This sequence brings into question the idea of calling on North American ice as the initial sea level trigger during at least some large ice-rafting events.

Changes in size of the northern ice sheets during the smaller and shorter millennial oscillations probably produced sea level fluctuations of no more than a few meters. It is more difficult to see how sea level changes this small could have provided a link among the northern ice sheet margins. Yet several ice margins contributed ice-rafted debris during the smaller oscillations as if they were linked.

Another possibility is that the ice streams delivering icebergs to the Atlantic Ocean pulled enough ice out of the interior of the North American ice sheet to alter atmospheric circulation. Over orbital time scales, climate responds to splitting of the jet stream because of changes in the elevation of the North American ice sheet (see Chapter 13, Figure 13-11B). Some climate scientists have suggested that this explanation might also apply to shorter millennial-scale circulation shifts in response to ice volume changes of 1% to 10%. For this explanation to be viable, atmospheric circulation would have to have been *extremely* sensitive to small changes in the elevation of the North American ice sheet, probably requiring the existence of a critical threshold height.

If northern ice sheets are the initiator of millennial oscillations, then the question is: How were millennial-scale oscillations transmitted to other regions? The evidence summarized to date suggests an answer. Discharges of icebergs and meltwater into the North Atlantic would lower its salinity and enhance formation of sea ice, which would have acted as a very powerful positive climate feedback and cooled high northern latitudes (Figure 15-18). From this northern "center of action," the transmission of the oscillations through the rest of the climate system may then have occurred by means of rapid-acting atmospheric processes. In southern China and in India, well-dated $\delta^{18}O$ trends in cave deposits show millennial intervals of weakened north-tropical monsoons and cooler/drier climate that correlate closely with changes in the North Atlantic. This close timing and rapid response must have happened through an atmospheric link. As part of this response, the northern margin of the intertropical convergence zone (ITCZ) was pushed south.

Southward ITCZ displacements were also recorded in the Southern Hemisphere, but in this case with climatic effects that were the reverse of those in the north. Well-dated speleothem $\delta^{18}O$ trends in Brazil show rapid millennial responses, but toward warmer and wetter climates. This opposing response makes sense if we look at the large-scale pattern of dislocation of the ITCZ. The same southward ITCZ shifts that weakened the north-tropical monsoons and produced

drier and cooler climates also displaced the southern ITCZ margin to the south and produced warmer and wetter climates along its southern margin in the southern hemisphere tropics and subtropics.

An important remaining question is whether the millennial-scale warming episodes registered in the southern ITCZ margin in Brazil are responsible for the warm intervals in the Antarctic. The Antarctic warm intervals are fewer in number and smoother in form than the oscillations in the north, but this may result from the fact that the ocean-dominated climate system around Antarctica smoothes out the signals arriving from the north and mutes the weaker oscillations. At this point, the evidence at hand seems consistent with a scenario in which signals originating in the North Atlantic region reached all the way south to Antarctica, thereby producing a global response (see Figure 15-18).

> **In Summary**, natural internal oscillations within northern ice sheets could have initiated changes that spread all the way to Antarctica via atmospheric linkages, with a reversal in sign across the equator.

15-9 Are the Oscillations Caused by Interhemispheric Climate Instabilities?

An alternative possibility is that the millennial oscillations were the product of slower-acting interhemispheric exchanges of heat. In this view, changes in the amount of northward redistribution of heat by the Atlantic Ocean produce a response pattern called the **bipolar seesaw**.

In most oceans today, the typical pattern of heat transport removes excess heat from the warm tropics and carries it toward the cold poles (see Chapter 2). The Indian and Pacific Oceans both follow this pattern, but the Atlantic Ocean does not. Instead, heat from the South Atlantic Ocean crosses the equator and moves into the high latitudes of the North Atlantic Ocean. The marine geologist Tom Crowley first proposed that changes in the northward transport of heat in the surface Atlantic Ocean could explain the bipolar seesaw pattern (Figure 15-19). Greater cross-equatorial transport of heat would warm the North Atlantic Ocean but leave the Southern Ocean cold. Weaker transport would cool the North Atlantic Ocean but leave more heat in the Southern Ocean.

The geochemist Wally Broecker proposed that changes in the rate of freshwater and/or iceberg fluxes from northern ice sheets could control the amount of deep water formed in the North Atlantic Ocean. During

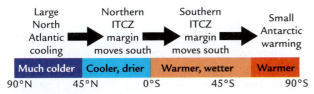

FIGURE 15-18

A northern center of action for millennial oscillations?

One explanation for millennial-scale oscillations is that they originate in ice margin fluctuations that cool northern latitudes and displace the ITCZ into the Southern Hemisphere, where the sign of the change reverses from that in the Northern Hemisphere.

times of large influxes, the low salinity produced in the surface Atlantic layers could have reduced the formation of deep water and the amount of ocean heat "pulled" northward into the Atlantic, leaving the Southern Ocean warmer (and conversely). Although many experiments with simplified ocean models have supported the idea that freshwater discharges can suppress deep-water formation and alter surface heat patterns, the models that incorporate the most complete array of processes tend to indicate weaker (if any) effects. Physical oceanographers have also criticized this explanation by noting that the large-scale circulation of the surface ocean is largely driven by winds, not by rates of deep-water formation.

Another problem with the hemispheric seesaw hypothesis is the lack of an explanation for the sudden onset of ice-rafted events in the North Atlantic Ocean. Shallow waters cannot be responsible for melting the ice margins because they were already very cold before and during these episodes. As noted in Section 15-5, the deep-water circulation in the Atlantic does have a southern (Antarctic) timing, but these changes occur at depths too far removed from the surface to have played any role in melting the ice sheet margins. The possibility that heat delivered northward in shallow subsurface Antarctic waters melted the ice has been explored but not convincingly demonstrated.

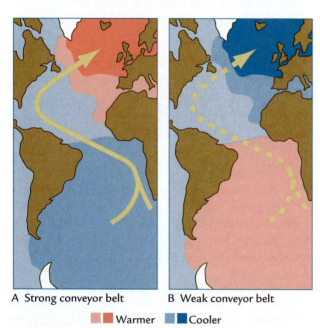

A Strong conveyor belt   B Weak conveyor belt

■■ Warmer   ■■ Cooler

FIGURE 15-19

Opposite hemispheric responses caused by ocean heat transport?

When cross-equatorial heat flow in the Atlantic is strong, it warms the North Atlantic but cools south polar regions (A). When cross-equatorial flow weakens, the temperature responses are reversed (B).

In Summary, the hypothesis of interhemispheric linkages is in accord with some of the evidence, but a key link in the proposed bipolar seesaw appears to be missing: a mechanism for initiating northern iceberg discharges. The available evidence can be explained equally well by a signal initiated by the ice sheets in the north and rapidly transmitted to the south, but with its sign reversing across the equator (see Figure 15-18). This explanation would account for the opposite-phased temperature changes around Antarctica which could then propagate north in the bottom-water flow but not affect changes at the surface in the North Atlantic.

15-10 What Role Did the Greenhouse Gases Play?

Because greenhouse gases play a major role in climatic changes at tectonic and orbital time scales, their behavior during millennial oscillations is worth considering. How large were the millennial-scale oscillations in carbon dioxide and methane, and what kind of role did they play in the millennial oscillations?

Methane concentrations in ice cores show clear millennial oscillations at amplitudes as large as 200 ppb (see Figure 15-8). Because these changes lag slightly behind the $\delta^{18}O$ (temperature) fluctuations in the north, they appear to have been the result of the northern oscillations and consequently a source of positive feedback, rather than a driver of those changes.

Carbon dioxide varied by as much as 20–25 ppm over millennial intervals (see Figure 15-9). These CO_2 changes are roughly in phase with the Antarctic temperature responses, which probably indicates a causal link to deep-water formation and sea-ice extent in the Southern Ocean, both of which can alter CO_2 concentrations in the atmosphere. Because the timing of these CO_2 variations roughly matches the temperature changes in the Southern Hemisphere, they provide positive feedback, but because they are opposite in sense to the temperature oscillations in the Northern Hemisphere, they act as negative feedback.

In Summary, both CH_4 and CO_2 vary at millennial time scales, but with opposite timing and climatic roles (CH_4 amplifying the changes initiated in the north, but CO_2 opposing them). To a first approximation, a 20-ppm change in CO_2 has about twice the effect on global temperature of a 200-ppb change in CH_4, so the combined global effect of these two greenhouse gases has the timing of the CO_2 changes, warming the Northern Hemisphere during its cold oscillations and warming the Southern Hemisphere during its warm oscillations.

15-11 Implications for Future Climate

Because millennial oscillations occur much faster than orbital-scale changes, they have the potential to have a more immediate impact on our climatic future. For obvious reasons, scientists and policy planners would like to know whether natural oscillations could cause climate to warm or to cool in future decades. Some scientists have speculated that a natural millennial-scale warming is underway now. If this view is correct, it means that the observed warming of the last century or so could, in part, be the result of natural processes rather than human activities.

This claim is unjustified for several reasons. Because millennial oscillations are either completely random (red noise) or at best quasi-periodic, their present and future course cannot be predicted with any confidence. More critically, the largest oscillations have occurred only during glacial climates, whereas the changes during the past 8,000 years of warm interglacial climate have been small and local in scale. This observation argues against natural oscillations playing a major role in recent or future climate changes.

Of course, the Greenland and Antarctic ice sheets are still in place, and they remain susceptible to some degree of melting in the warmer climate of the future. Because ice sheets played an important role in the glacial-age millennial oscillations, partial melting of the present ice sheets because of human activities could conceivably trigger changes in the climate system in the future, although likely at a much smaller scale. In this case, however, the ultimate cause of these future changes will still be human activities, not natural variations within the climate system.

Key Terms

millennial oscillations (p. 295)

Dansgaard-Oeschger oscillations (p. 297)

sediment drifts (p. 297)

Heinrich events (p. 298)

red noise (p. 306)

stochastic resonance (p. 307)

bedrock pinning points (p. 309)

bipolar seesaw (p. 311)

Review Questions

1. How do the processes that control $\delta^{18}O$ changes in ice sheets differ from those that control $\delta^{18}O$ fluctuations in ocean cores?

2. Why is it difficult to correlate millennial climatic oscillations in records from different regions?

3. What other regions show millennial oscillations like those in the North Atlantic and Greenland?

4. Are millennial oscillations true cycles?

5. How strong is the evidence that solar changes drive millennial oscillations?

6. What is the evidence for and against internal ice sheets processes causing millennial oscillations?

7. Could changes in the North Atlantic region account for opposite changes in the Southern Hemisphere?

8. Could changes in ocean heat transport explain opposing millennial oscillations north and south of the equator?

Additional Resources

Basic Reading

Alley, R. B. 2000. *The Two-Mile Time Machine*. NJ, Princeton University Press.

Advanced Reading

Alley, R. B., et al. 2006. "Outburst Flooding and the Initiation of Ice-Stream Surges in Response to Climatic Cooling: A Hypothesis." *Geomorphology* 75: 76–89.

Beer, J., et al. 1988. "Information on Past Solar Activity and Geomagnetism from [10]Be in the Camp Century Ice Core." *Nature* 331: 675–79.

Bond, G., et al. 1993. "Correlations Between Climatic Records from North Atlantic Sediments and Greenland Ice." *Nature* 365: 143–47.

Boyle, E. A. 2000. "Is Ocean Thermohaline Circulation Linked to Abrupt Stadial-Interstadial Transitions?" *Quaternary Science Reviews* 19: 255–72.

Brook, E. J., et al. 2005. "Timing of Millennial-Scale Climate Change at Siple Dome, West Antarctica, During the Last Glacial Period." *Quaternary Science Reviews* 24: 1333–43.

Crowley, T. J. 1992. "North Atlantic Deep Water Cools the Southern Hemisphere." *Paleoceanography* 7: 489–97.

Hendy, I. L., and J. P. Kennett. 1999. "Latest Quaternary North Pacific Surface-Water Responses Imply Atmospheric-Driven Climatic Instability." *Geology* 27: 291–94.

MacAyeal, D. R. 1993. "Binge/Purge Oscillations of the Laurentide Ice Sheet as a Cause of North Atlantic's Heinrich Events." *Paleoceanography* 8: 775–84.

Stuiver, M., and T. F. Brazunias. 1993. "Sun, Ocean, Climate, and Atmospheric $^{14}CO_2$: An Evaluation of Causal and Spectral Relationships." *Holocene* 3: 289–305.

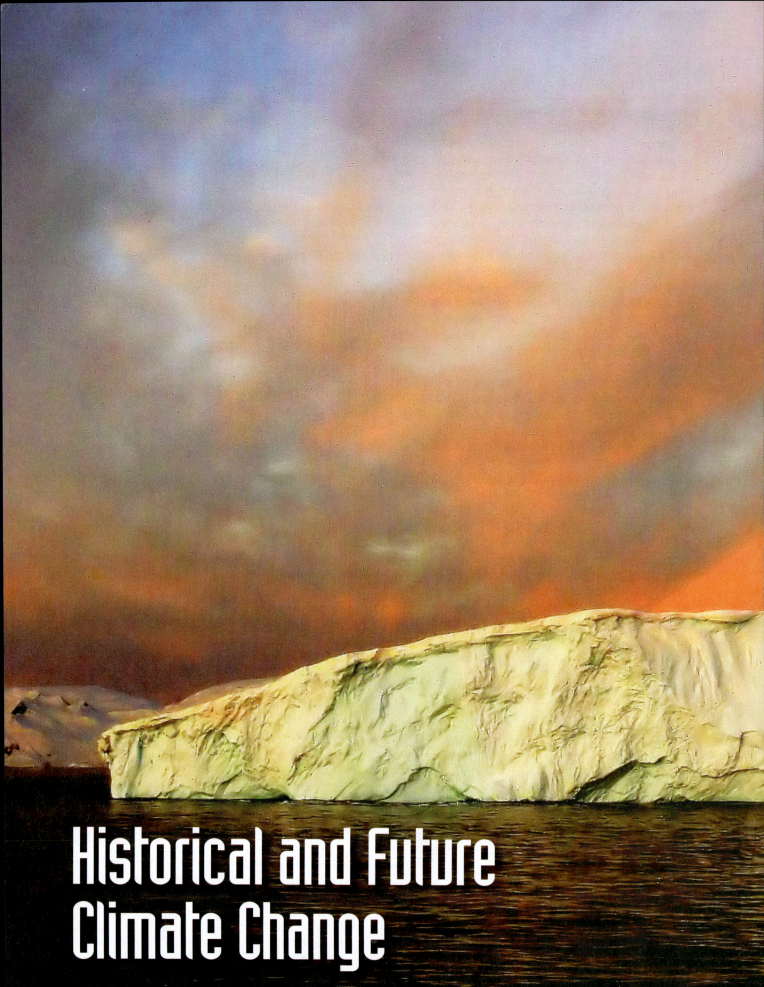

Historical and Future Climate Change

During the last 12,000 years, a new era in Earth's history emerged. Prior to that time, scattered bands of Stone Age people lived a migratory hunter-gatherer-fisher existence. Then agriculture was discovered, and humans began to live in permanent dwellings. A new hypothesis suggests that early farmers began to alter climate thousands of years ago by emitting greenhouse gases. By 2,000 years ago, something like modern life had emerged: iron tools, towns and cities, writing and religions. By 150 years ago, early in the industrial era, humans were already moving more rock and soil per year than all of the water, ice, and wind in the climate system.

Climate changes over this interval are reconstructed from a wide array of archives and proxy indicators. The best archives for intervals spanning a few thousand years are annually layered ice cores and lake sediments. For recent centuries, annually layered tree rings and corals provide additional coverage, and historical observations are useful in a few regions. For the last 100 years, and especially in the satellite era of the last few decades, measurements made by instruments have become the major source of climatic histories.

Natural climate changes over the last 1,000 years were much smaller than those over tectonic, orbital, and millennial time scales, amounting to less than 0.5°C on a global basis. Within the last 125 years, industrialization has had a rapidly accelerating effect on global climate, warming it by 0.8°C. The last decade has seen accelerations of sea-ice retreat and thinning in the Arctic Ocean, melting of the Greenland ice sheet, and sea level rise. Our growing overprint on climate seems destined to continue for centuries to come, and many parts of Earth's surface (particularly the Arctic) will be totally transformed by future warming.

To evaluate the extent of both natural and human causes of climate change during the last century and in the future, we explore these issues:

A major component of Earth's climate system, Arctic sea ice had thinned and retreated significantly by the first decade of the twenty-first century, an early response to a global climate warming that will intensify in the future. (Darrell Gulin/ Photodisc/Getty Images.)

▷ When did humans begin to play a role in climatic change?

▷ What caused climate change during the last 1,000 years?

▷ Was the warming since the late 1800s caused by humans or by natural factors?

▷ What is the sensitivity of Earth's climate system to the by-products of the Industrial Revolution, including greenhouse gases (CO_2 and CH_4) and sulfur dioxide (SO_2)?

▷ What kinds of climate changes lie in Earth's future?

Part V

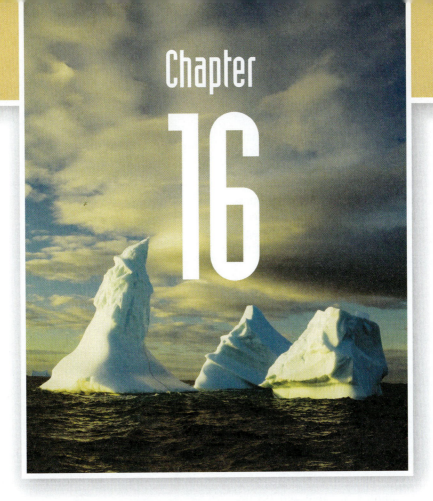

Chapter 16

Humans and Preindustrial Climate

Our species arrived very late in Earth's long history. The first somewhat humanlike creatures that walked on two legs and used stone tools appeared only within the last 4 million years, equivalent to less than 1/10,000 of Earth's 4.55-billion-year age. Subsequent milestones in human history were marked off at intervals that grew shorter by factors of approximately ten: (1) the initial appearance of our species, *Homo sapiens* within the last 200,000 years; (2) the development and spread of agriculture within the last 12,000 years; and (3) the origin and growth of the industrial era within the last 200 years.

Over most of our time on Earth, we humans and our immediate predecessors were affected by climate but did not have any measurable impact on the climate system. With the spread of agriculture during the last several millennia of the current interglacial period, however, we began to alter climate by adding greenhouse gases to the atmosphere in amounts sufficient to offset part of a natural cooling. In this chapter, we trace the progression of humans from passive participants in climate change to active contributors to the operation of the climate system.

Climate and Human Evolution

Scientists in several disciplines agree that humans evolved in Africa. Most of the earliest evidence for our ancient ancestors comes from plateaus along the eastern side of the continent (Figure 16-1). Volcanic activity in this region over millions of years deposited basalt layers that can be dated by radiometric (K/Ar) methods. These dated basalts bracket the ages of intervening sediment layers that hold much of the record of human evolution.

16-1 Evidence of Human Evolution

Anthropologists focus either on distinctive events that break the continuous process of evolution into separate stages or on quantitative traits that can be measured as they gradually change. Human evolution is marked by five distinctive developments (Figure 16-2): (1) the initial branching off from primitive apes between 6 and 4 million years ago; (2) the onset of bipedalism (a preference for moving upright on two legs) near 4 million years ago; (3) the use of stone tools beginning near 2.5 million years ago; (4) the branching of the prehuman line into the genus *Homo* and other forms by 2 million years ago; and (5) the

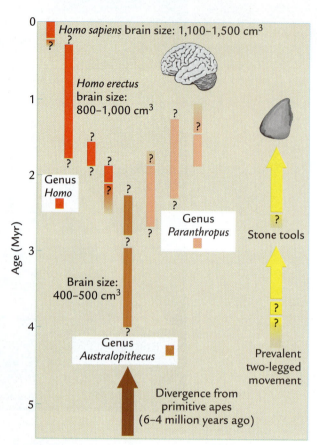

FIGURE 16-2

Human evolution

The last 5 million years span the evolutionary line from primitive apes to the australopithecines ("southern apes") to our own genus, *Homo*, and finally our own species, *Homo sapiens* ("intelligent man"). The onset of two-legged walking appeared early in this evolutionary progression, followed by the first use of stone tools more than a million years later. (ADAPTED FROM P. B. DEMENOCAL, "PLIO-PLEISTOCENE AFRICAN CLIMATE," *SCIENCE* 270 [1995]: 53–59.)

development of large brains in the *Homo* genus since 2 million years ago.

APPEARANCE OF HUMAN ANCESTORS Human evolution can be traced back to small shrewlike mammals that evolved during the millions of years after the massive extinctions caused by the asteroid impact 65 million years ago (see Chapter 6). These primitive mammals eventually evolved features we now associate with monkeys, such as grasping front paws and long prehensile tails, which led to the lemur family (a monkeylike tree-climbing animal). Modern lemurs are shown in Figure 16-3.

By 10 million years ago, one such line had further evolved to primitive apes. Subsequently, a group of apes that included both our human ancestors and chimpanzees branched off from the primitive apes,

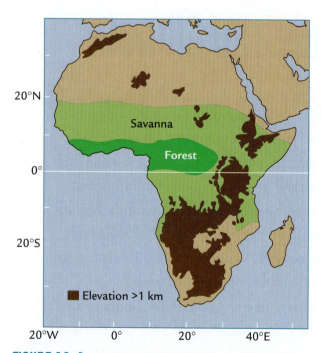

FIGURE 16-1

African topography and rain forests

Eastern Africa is a region of broad plateaus at elevations not far above 1 km. Most rain forest vegetation occurs today in the wet intertropical convergence zone near the equator, encircled by a broad band of semi-arid savanna.

the line that led to the modern great apes between 6 and 4 million years ago (Figure 16-2).

WALKING UPRIGHT   The evolutionary line that led to modern humans, called hominins (from the family *Hominidae*, meaning humanlike), appeared by 4 million years ago. Fossil remains of ankle bones with a distinctive structure suggest that walking had become the primary means of movement by 4.3 million years ago. These creatures were considerably smaller than modern humans and had chimpanzee-like faces with large, strong jaws. They probably spent most of their time in trees, gathering fruits and nuts and avoiding predators, but also moved on the ground when necessary.

A remarkable deposit dated to 3.6 million years ago in Tanzania holds footprints of individuals walking across a freshly fallen layer of volcanic ash that had cooled (Figure 16-4). The tracks suggest that one of the creatures turned, perhaps to look back at something, and then walked on. These humanlike apes of the genus *Australopithecus* (known as australopithecines) walked upright to travel. Scientists argue whether these creatures developed the ability to walk in order to exploit food resources on the grassy savanna lying between stands of trees, or whether they developed upright postures in order to stand on the lower tree limbs (or on the ground) and reach up for fruits and nuts.

FIGURE 16-3
Early mammals
The line of early mammals from which humans evolved included creatures resembling modern lemurs. (TONY CAMACHO/ SCIENCE SOURCE.)

with modern apes evolving along the other branch. Our prehuman ancestors are thought to have foraged for food in and near woodlands, moving at times on two legs. Radiometric dating of volcanic rocks in East Africa initially suggested that this branching occurred sometime between 10 and 5 million years ago, but for a long time the timing was difficult to constrain more precisely.

A new source of evidence—molecular biology—has now reduced this uncertainty. Molecular biologists measure the composition of DNA molecules in the protein of living organisms. They make use of DNA as an evolutionary clock: the longer the time that has elapsed since two organisms branched off from a common ancestor, the more dissimilar their DNA will have become. Assuming that this DNA dissimilarity increases through time at a constant rate, the amount of dissimilarity can be used as a clock to measure elapsed time. Molecular biologists have concluded that the line that led to humans diverged from

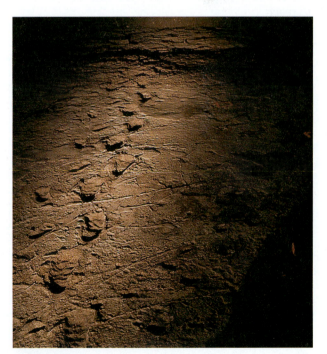

FIGURE 16-4
Footprints from 3.6 Myr ago
Hominin (humanlike) creatures that walked across fresh volcanic ash 3.6 Myr ago in East Africa left their footprints, now fossilized. (KENNETH GARRETT, NATIONAL GEOGRAPHIC SOCIETY IMAGE COLLECTION.)

USE OF STONE TOOLS The first firm evidence that hominins used stone tools dates to about 2.5 million years ago. These early tools, used to butcher dead animals, produced marks on the animal bones that indicate crushing and scraping but are sometimes difficult to distinguish from similar marks made by the teeth of carnivores (lions, leopards, and cheetahs).

One early hypothesis suggested that humans evolved mainly as "killer apes" because of the aggressive use of tools to kill their prey. But many anthropologists today believe that hominins simply made opportunistic use of the remains of animals previously killed by lions and leopards, and that they were constantly forced to contend with hyenas and other scavengers for this food. Brandishing tools as weapons may have helped them drive off competitors.

Use of tools for butchering implies an important change in diet, which previously must have consisted mainly of items collected from the environment: fruits, nuts, leaves, and small insects (mainly grasshoppers and termites). The use of crude cutting tools would have allowed these early hominins to extract not just meat but also bone marrow and other internal animal parts. These protein-rich food sources would have more readily satisfied their energy needs. Stone tools could also have helped hominins dig out buried roots and tubers that were otherwise difficult to reach.

The available evidence suggests that toolmaking followed more than 1 million years after the ability to walk upright. Some scientists infer that toolmaking was a natural evolutionary development for creatures whose hands were freed for other uses when they began to walk on two legs.

APPEARANCE OF *HOMO* Sometime after 2.5 million years ago, the ancestral australopithecines evolved into several new forms (see Figure 16-2). One line led to the genus *Paranthropus*, stout creatures with large teeth and strong jaws used to crush protein-rich palm nuts and other hard food in their vegetarian diet. This group became extinct by 1 million years ago.

The other major group carries the name of our own genus, *Homo*. These were more graceful (lean-bodied) creatures with larger heads and braincases. The earliest of these creatures is dated (with some uncertainty) to 2.4 to 2.3 million years ago, and our human ancestor *Homo erectus* ("upright man") was definitely present by just after 2 million years ago (see Figure 16-2). The patterns of wear on their teeth indicate a broader-based diet of meat, fruits, and vegetables.

The first appearance of these humanlike creatures appears to follow closely after the earliest use of stone tools, but the precise timing is difficult to determine because the fossil record is so fragmentary. The use

Table 16-1	Growth in Size (Volume) of Hominin Braincases	
Type of hominin	**Age (Myr ago)**	**Braincase (cm³)**
Homo sapiens	0.2–0	1,100–1,500
Homo erectus	2.4–1.8	800–1,000
Australopithecus	4.1–3.1	400–500

of tools is likely to have favored those prehumans who were able to move easily across the landscape in pursuit of a large variety of seasonally changing food sources. Frequent movement may also have called on a greater use of intellect and imagination.

BRAIN SIZE Over time, our human ancestors developed larger brains, shown by the increasing size of preserved skulls that encased and protected the brain. In broad outline, the volume of the braincase tripled over the last 3 or 4 million years (Table 16-1).

The bipedal australopithecines had braincase volumes of 400 to 500 cm^3. The *Homo erectus* ancestors who first used stone tools had braincases twice as large, roughly 800 to 1,000 cm^3. Fully modern humans (*Homo sapiens*, or intelligent man) first appeared between 200,000 and 100,000 years ago, with braincases ranging in size between 1,100 and 1,500 cm^3. A tripling of brain size in 4 million years is unusually rapid compared to most evolutionary changes.

16-2 Did Climate Change Drive Human Evolution?

Several hypotheses have been put forward to explain the rapid evolution of our ancestors during the last 3 or 4 million years. Some ideas focus on the impact of new technologies, such as tools to facilitate gathering of food and to serve as weapons for defense and hunting. Others emphasize changes in social factors, such as the development of communication skills and the need for early humans to gather food for infants who spent several years in a defenseless state.

Still other hypotheses focus on climate change as the key control, and several components of the climate system are potentially relevant to environmental changes in Africa that could have had an impact on humans. For example, over the last 4 million years, orbital-scale variations in monsoonal summer rainfall have occurred in the Nile River headwaters, leaving a long sequence of organic sapropels in eastern

Mediterranean sediments (see Chapter 9). Evidence of ancient summer monsoon fluctuations is also recorded by lake-level fluctuations in East African sediment sequences dated by interlayered basalts.

Northern hemisphere ice sheets are another potential factor. As summarized in Chapter 10, a long-term cooling shown by increasing $\delta^{18}O$ values in marine foraminifera culminated in the first appearance of moderate-sized northern ice sheets by 2.75 million years ago and larger ones by 0.9 million years ago (Figure 16-5 left). Pollen counts in a lake deposits in the mountains of East Africa suggest that mountain vegetation descended to lower elevations during some glaciations.

Marine sediment cores also show that south equatorial water masses underwent a major reorganization between 3 and 1.5 million years ago. Previously, surface waters had been relatively warm across the entire tropics, but some regions then cooled as upwelling systems developed or intensified. A cooling of subsurface waters in the Indian Ocean also occurred at this time. As noted in Chapter 7, the cause of this widespread tropical shift is not yet clear, although it may have been a regional expression of a global-scale response to decreasing atmospheric CO_2 levels.

Limited amounts of long-term evidence are available from Africa and nearby oceans, including material blown off the continent. Cores from the tropical eastern Atlantic Ocean show very gradual increases in the rate of influx of continental dust (mostly quartz and clay) from North Africa after 4.5 million years ago (see Figure 16-5 right).

In East Africa, the carbon-isotopic ($\delta^{13}C$) composition of carbon left behind in soils and in the fossilized teeth of grazing animals gradually changed during the last 7–8 million years, and especially the last 3 million years (see Figure 16-5 center). This trend indicates a change from C3 vegetation—trees and shrubs—toward C4 vegetation—warm-season grasses (See Appendix B). It has been variously interpreted as caused by regional drying or by a decrease in atmospheric CO_2 concentration. Either way, tree and shrub vegetation was giving way to grass in eastern Africa. This same interval also saw the proliferation of grazing (grass-eating) animals such as antelope across a large area of Africa.

Based on these various kinds of evidence, three hypotheses involving climatic factors have been put forward to explain human evolution. The oldest, the **savanna hypothesis**, ties human evolution to long-term drying in Africa. In this view, tropical rain forests became interspersed with small patches of semiarid grasslands, which then gradually spread between groves of trees and coalesced into grassland savannas. This slow fragmentation of once-continuous forest caused our ancestors to move on the ground for ever-longer distances, forcing more rapid movement to cover longer distances and also greater resourcefulness to survive in a more exposed environment. The long-term $\delta^{13}C$ trend is consistent with this hypothesis, and the largest shift toward C4 (grassland) vegetation occurred from 3 to 1.8 million years ago (see Figure 16-5), a time of prominent diversification of the hominin line (see Figure 16-2). A cooling of the tropical Indian Ocean during this interval is consistent with a reduced flow of ocean moisture into East Africa. In addition, every hominin site investigated so far shows substantial amounts of open (grass) vegetation.

Human evolution has also been linked to the onset and intensification of northern ice sheets, perhaps transmitted by the intermediary of the North Atlantic Ocean. One of the major changes in hominin species is dated to 2.9–2.6 million years ago, the

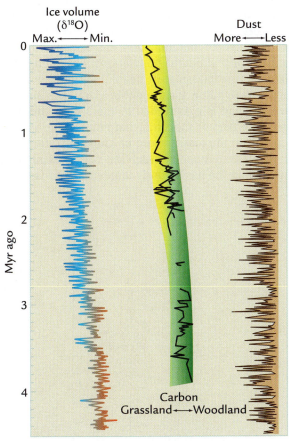

FIGURE 16-5

Long-term changes in African dust and vegetation

Over the last 4.5 Myr, a time of global cooling shown by $\delta^{18}O$ trends (left), vegetation cover in many regions gradually shifted away from trees and shrubs toward warm-season grasses (middle), and increasing amounts of dust were blown from North Africa to the tropical Atlantic (right). (ADAPTED FROM P. B. DEMENOCAL, "PLIO-PLEISTOCENE AFRICAN CLIMATE," *SCIENCE* 270 [1995]: 53–59.)

time of the first northern ice sheets, but a second interval dates to 1.9–1.4 million years ago, which falls in between the two times of major ice sheet intensification. In any case, model simulations have shown that the climatic effects of even the very large ice sheets of the last million years on North African climate are relatively weak at orbital time scales. The size of the northern ice sheets causes no substantial cooling or reduction in monsoonal precipitation during the rainy summer season. The most significant effect of the northern ice sheets on Africa in the simulations was an increase in the intensity of winter (late spring) winds. This northern overprint may contribute to correlations between ice sheet changes and dust fluxes to the ocean, which are obvious in the Indian Ocean, but weaker in the Atlantic Ocean (see Figure 16-5). In any case, the long-term link between northern ice sheets and the climate of Africa is now viewed as unconvincing.

More recently, the anthropologist Richard Potts proposed the **variability selection hypothesis**. Its basic premise is that evolution occurred at times when rapid changes in climate (temperature and/or precipitation) put new demands on our ancestors, favoring those individuals or groups who evolved traits that proved useful for adapting to an increasingly challenging environment. This hypothesis seems broadly consistent with the increased amplitude of climatic variability evident in many regions during the last 3 million years.

> **In Summary**, several hypotheses explaining early human evolution in Africa as a result of climate change are viable, yet none has been convincingly demonstrated as *the* answer. One reason for this ongoing uncertainty is the sparse nature of some environmental records. A more critical reason is the extremely sparse nature of the entire hominin record.

16-3 Testing Climatic Hypotheses with Fragmentary Records

Again and again we bump up against the problem of the fragmentary nature of the record of hominin remains. As noted earlier, preservation of fossils in Africa, Arabia, and southern Asia is sparse in part because of prevailing aridity. For hominin bones made of easily dissolved calcium phosphate (Ca_3PO_4), preservation is even worse in the acid-rich soils of the rain forests. Because of these problems, the total record of human evolution over 5 million years is based on just a few dozen fragments of skeletons, enough to reveal some of the broad outline of human evolution but few of the details.

We have already seen in Chapter 8 that sampling records of orbital-scale climate change even at intervals as close as a few thousand years can lead to gross misrepresentations of the shape of the actual climate signals. This aliasing problem becomes all the more formidable in records with just one sample every 100,000 years or so.

Aliasing can produce erroneous indications of the time of first acquisition of new physical or technical skills (such as walking or the use of tools and fire) or of the first or last appearance of a new hominin species. With only a few samples, the actual first or last appearances are likely to be missed, and instead we will see a much-reduced range (Figure 16-6). As a result, the true ranges of most hominins are probably longer than those shown in Figure 16-2. Similarly, aliasing also complicates attempts to define the relative timing between climate changes and the first use of new skills.

A second undersampling problem has to do with quantitative measurements of the evolution of physical traits. In this case, the basic problem is that a broad range of natural variation occurs within all human (or prehuman) populations. For example, the sizes of your classmates' heads vary widely around the average value for the entire class. The size of a human head is closely tied to the size of the braincase (the part of the skull that cradles the brain), a key trait in the fossil record of human evolution.

The actual range of braincase sizes present at any one time in the past is fairly large in comparison with

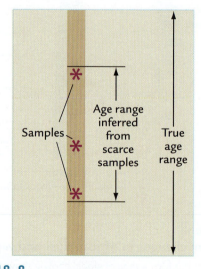

FIGURE 16-6
Undersampling of a fossil record
If few samples of the fossil record of an organism are available, the true first and last appearances of that organism will be poorly estimated.

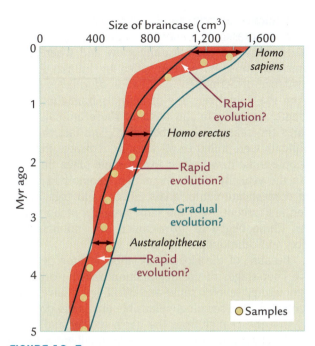

FIGURE 16-7

Undersampling of a measurable characteristic

If a slowly evolving characteristic, such as the size of the human brain, varies widely at all points in time (black double-arrows), scattered sampling (yellow circles) will permit an interpretation of either gradual or rapid evolution.

the evolution of the mean value (Figure 16-7). If the fossil record provides only one or two well-preserved specimens every hundred thousand years or so, a good chance exists that some of these specimens will not be representative of the brain size of the entire population living at that time, but will fall above or below the mean. Depending on the specific samples collected and analyzed, an inaccurate picture of the long-term trend could emerge.

Even if the available fossil record is sparse, it should be possible to obtain a general sense of the direction in which the trend is moving, especially if the net amount of evolution far exceeds the natural range of variation at any one time. But the limitations of sparse data make it impossible to define the true rates of change. Figure 16-7 shows that sparse data on brain size could be interpreted equally well as either a slow, gradual trend or as rapid bursts of change.

In Summary, major parts of the outline of human evolution have gradually fallen into place, but the sparse fossil record makes it difficult to test hypotheses about the cause or causes. As a result, the degree to which climate has affected human evolution is unknown.

The Impact of Climate on Early Farming

For almost 2.5 million years, hominins moved only slightly beyond the most primitive level of Stone Age life, adding control of fire and gradually more sophisticated stone tools to a rather meager repertoire of skills. But once our species appeared, near 200,000 to 150,000 years ago, the pace of change began to quicken. By the time of the most recent glaciation, our ancestors painted amazingly lifelike portrayals of animals on the walls of caves and rock shelters (Figure 16-8). They also made small statues of human and animal figures and created jewelry by stringing together shells, and they buried their dead with food and possessions for use in a future life. In these changes, we recognize the early origins of human "culture."

More sophisticated stone tools designed for specific functions also appeared, and, for the first time, people began to use bone, a much more "workable" substance than stone, yet hard enough for many uses: needles, awls (hole-punchers), and engraving tools. Needles made possible sewn clothing that fit closely, rather than loosely draped animal hides. With greater protection from the elements, people pushed north into higher and colder latitudes of Asia. There they built dome-shaped houses with large mammoth bones supporting the superstructure and animal hides

FIGURE 16-8

A cave painting of the glacial era

Despite the harsh glacial climate, our ancestors left beautiful, almost modern-looking paintings on the walls of caves in southern Europe. (RIEGER BERTRAND/GETTY IMAGES.)

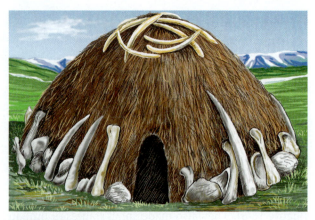

FIGURE 16-9
Early human buildings
Humans living in northern Asia during the last glaciation constructed domed dwellings of hides draped over mammoth bones. Other bones served as anchors.

draping the roof for protection from rain and snow (Figure 16-9).

People also learned how to make rope from naturally available fibers and used the rope to make snares and lines and nets to catch small animals, birds, and fish. The hunter-gatherer life became a hunter-gatherer-fisher life. People shaped bone or wood into spears that held stone spear points, and used rope to help bind the spear points to the long shafts. This new technology produced a lethal and revolutionary new way of hunting that combined a weapon that could kill efficiently with a hand that could grasp it and an arm with enough natural range of motion to throw it. Hunters could now bring down larger game from a safer distance, even mammoths, as shown in cave paintings. By the start of the present interglaciation, an explosive alteration of basic human existence was underway.

16-4 Did Deglacial Warming Lead to Early Agriculture?

The first evidence of agriculture dates to just over 12,000 calendar years ago in a region of the Middle East called the **Fertile Crescent**, encompassing present Syria, Iraq, Jordan, and Turkey. The people living in this region (called Natufians) abandoned the hunter-gatherer way of life and began to cultivate wheat, rye, barley, peas, and lentils rather than simply harvesting the same foods in the wild. Because agriculture eliminated the need for seasonal migrations to search for food, these people took up residence in permanent dwellings. Within 1,000 years, the dwellings began to cluster into permanent village settlements.

Evidence of cultivation in the Fertile Crescent is based on preserved remains of grains found in regions where they did not naturally grow and where their presence must have been aided by human efforts. Evidence of permanent occupation of villages comes from the dental remains of animals from the settlements. Layering in the teeth of these animals indicates the season when they died. The fact that the animals were killed in all seasons indicates that the people must have stayed in the same place throughout the year. By 10,000 years ago, people had begun to domesticate cattle, goats, sheep, and pigs in the Fertile Crescent. Near the same time, people in northern China began to grow millet, dry (nonirrigated) rice, and other crops.

Because of the close association in time between the later stages of the deglaciation and the origin of agriculture, several cause-and-effect links have been proposed. One seemingly plausible link is the possibility that the change from the harsh (colder and drier) glacial climate with large short-term oscillations to the more stable and accommodating (warmer and wetter) climate provided conditions more favorable for humans to begin the grand experiment of growing crops.

On the other hand, a totally different climatic hypothesis centers on the Younger Dryas climatic reversal between 13,000 and 11,700 years ago (see Chapter 14). In this view, the Younger Dryas episode intensified the already dry conditions across the eastern Mediterranean region and forced people to retreat to dependable water sources. In these more closely clustered conditions, people who harvested and ate wild grains may have accidentally scattered some grains near their threshing sites, with the discarded grains sprouting in succeeding years as a form of primitive farming. Some evidence places the time of the earliest domestication of crops during the Younger Dryas.

Neither of these directly opposing hypotheses is easy to test. One problem is that agriculture in some regions may have begun earlier than the still-incomplete records indicate. Another problem is that the beginnings of agriculture in each region were one-of-a-kind events. Many such events from many regions, each related to a similar kind of climate change, would be required to show a conclusive cause-and-effect relationship. In any case, the independent emergence of agriculture in so many places within a period of about 5,000 years seems to indicate that humans had reached a threshold of cultural sophistication that made such advances all but inevitable for many cultures.

16-5 Impacts of Climate on Early Civilizations

Climate change has been hypothesized as the cause of, or at least a major factor in, the deterioration or collapse of several early civilizations. One hypothesis focuses on the role of an early flood in the Black Sea (Box 16-1), but drought is the climatic factor most commonly invoked as influencing humans. In low-latitude regions where water was scarce, civilizations were susceptible to drought.

The first advanced civilizations of the early Egyptian dynasties developed between 6,000 and 5,000 years ago, when the North African summer monsoon was still stronger than today. Then and now, Egyptian life centered on the Nile River, fed by monsoonal rains in the Ethiopian highlands and flowing northward through hyperarid desert (see Chapter 9). When the Nile ran strong, large seasonal floods provided fertile soils and moisture for farming along the floodplain.

Climate in sub-Saharan North Africa turned much drier after 5,000 years ago as the summer monsoon weakened. This drying trend affected the civilizations that had come into existence and grown in size during the wetter monsoon climates in the preceding millennium. The weakening of the summer monsoon after 5,000 years ago greatly reduced the extent of summer flooding of the Nile. This change must have put greater stress on populations that had expanded in response to the stable food supply from large crop yields in a wetter monsoonal climate.

The Akkadian empire, centered in what is now Syria, was the dominant civilization in Mesopotamia until 4,200 years ago. Evidence from archeological investigations and from marine sediments in the Persian Gulf shows that an abrupt abandonment of major northern cities across a broad region coincided with a period of intense aridity and increased movement of windblown dust that lasted for a few hundred years. Climate change is thought to have been responsible.

Another example is the sudden collapse of the Mayan civilization on the Yucatán Peninsula in Mexico near the year 860. Evidence from lake sediments indicates that the worst part of this dislocation occurred during an interval of drought (Figure 16-10). A civilization that had built cities and massive stone temples, carved monumental statues out of stone, and developed a system of writing abandoned many of its inland cities and scattered throughout the Yucatán over an interval of 100 years or more. Some of the people moved northward to coastal regions, perhaps because shallow groundwater was still accessible in such regions. Even these coastal populations dwindled.

Near 1300, the Anasazi people, who had occupied beautiful cave dwellings cut into the sides of cliffs in the southern Colorado Plateau, abruptly abandoned the entire region. Evidence from tree ring studies indicates that their sudden departure occurred during an interval of drought.

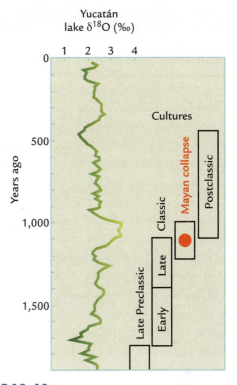

FIGURE 16-10
Did drought destroy Mayan civilization?
Lake sediments indicate periods of prolonged drought during the time that Mayan civilization disappeared. (ADAPTED FROM J.H. CURTIS, D. A. HODELL, AND M. BRENNER, "CLIMATIC VARIABILITY ON THE YUCATAN PENINSULA (MEXICO) DURING THE PAST 3,500 YEARS, AND IMPLICATIONS FOR MAYA CULTURAL EVOLUTION," *QUATERNARY RESEARCH* 46 [1996]: 37–47.)

Depletion of resources is also a plausible competing explanation for some of these cultural changes. For example, while the Anasazi did abandon their cliff dwellings in the American Southwest during a drought, previous dry intervals of comparable or nearly comparable intensity had not driven them from the region. Abandonment also coincided with the near disappearance of tree pollen from climate records, indicating that the Anasazi had cut down the juniper and piñon pine trees previously used as fuel for cooking and for winter heat on the high, cold plateau. If so, depletion of this crucial resource may also have been a major factor in the abandonment.

Similarly, the Maya of Central America may have altered their own local environment. They may have contributed to regional drought by cutting trees and reducing the positive moisture feedback from evapotranspiration. In addition, their farming methods may have exhausted the limited supply of nutrient-rich soils in a region of nutrient-poor limestone bedrock, making agriculture difficult or impossible. Wars with neighboring cultures may also have been a factor,

Looking Deeper into Climate Science

Sea Level Rise and Flood Legends

Many early cultures have a flood legend, a story about a great flood that swept away earlier people. Many of these legends share several features: punishment of the sins of the people by a higher authority, a warning that a flood was coming, and pre-flood advice to gather every kind of animal on a large vessel in order to preserve all forms of life in a post-flood world. The story of Noah in the Hebrew Old Testament is the most widely known flood legend, but much the same story is found in the older Babylonian tale of Gilgamesh and in legends of other early cultures in the Old World. Storytellers passed down these legends over many generations.

In the eighteenth and nineteenth centuries, scientists attempted to reconcile the biblical story of the flood with their emerging geological knowledge by advocating the **diluvial hypothesis**. This hypothesis called on a great worldwide flood to explain the widespread deposits of unsorted debris (everything from clay to boulders) strewn across the northern continents. Today, we recognize these deposits as moraines left by retreating ice sheets. Scientists who now interpret the biblical flood legend less literally have continued to search for evidence of a major regional-scale flooding event, and many of them focus on the Near East because of its early civilizations.

In 1998, the geophysicists Bill Ryan and Walter Pitman pulled together evidence from the Black Sea region into a dramatic **Black Sea flood hypothesis**. During the last glacial maximum, the present connection between the Black Sea and the Aegean Sea through the Sea of Marmara in Turkey did not exist. Global sea level stood 110–125 meters lower than it does today, and the level of the Aegean, linked to the global ocean as part of the Mediterranean Sea, lay well below the threshold needed to connect the two bodies of water. A freshwater lake rimmed by reedy swamps covered a much smaller part of the basin than the modern Black Sea.

Later, as the very last portions of the great northern hemisphere ice sheets melted, the meltwater they returned to the ocean pushed the rising Aegean Sea over a sill into the Sea of Marmara and then into the freshwater lake in the area of the modern Black Sea. Radiocarbon dates in sediment cores from the Black Sea suggested that an abrupt transition occurred near 7,600 calendar years ago. Based on the record of fossil shells, freshwater molluscs that lived in a lake were replaced by molluscs and plankton that lived in salty ocean water. Within a short interval, it seemed, incoming seawater transformed this freshwater lake into a salty inland sea.

Ryan and Pitman also found a huge gorge lying buried underneath a thin layer of sediments at the bottom of the strait known as the Bosporus, between the Sea of Marmara and the Black Sea. Cut into bedrock, this gorge is clear evidence that an enormous flow of water poured into the Black Sea in the past. The coarse sediments along the floor of the gorge (sand, pebbles, cobbles, and even boulders) are arrayed in great dunelike shapes tilted toward the north, consistent with a flow from the Aegean Sea into the Black Sea. At the place where the Bosporus meets the Black Sea, a deep pool is cut into the underlying rock, apparently carved by an immense waterfall created by a torrent of incoming water.

---

along with disease. Isolating climate change as the sole cause of changes in early civilizations is often difficult.

▶ Early Impacts of Humans on Climate

Eventually, the relationship between humans and their environment began to change. Instead of being passive players, humans began to actively influence their environment and Earth's climate.

16-6 Did Humans Cause Megafaunal Extinctions?

Populations of large mammals called **megafauna** decreased drastically during the most recent glaciation. Prior to 50,000 years ago, more than 150 genera of mammals larger than 45 kg (~100 pounds) existed. By 10,000 years ago, 50 or fewer genera were left. This interval of mammal extinctions was unprecedented in millions of years of prior geologic history.

Near 45,000 years ago, many of the larger marsupials in Australia became extinct within a few

Box 16-1

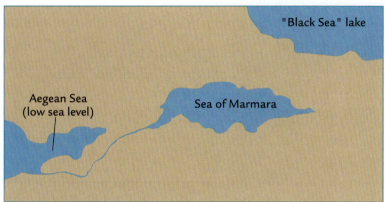

A Last glaciation

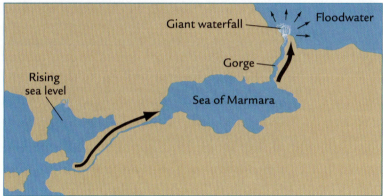

B Deglacial flood (7,600 years ago)

The Black Sea flood Slowly melting ice sheets caused the Aegean Sea to rise until it flooded into a freshwater lake that was located in the region of the modern Black Sea. (ADAPTED FROM W. RYAN AND W. PITMAN, *NOAH'S FLOOD* [NEW YORK: SIMON & SCHUSTER, 1998].)

Ryan and Pitman concluded that the rise of floodwaters caused by the torrent of seawater entering the Black Sea 7,600 years ago would have inundated settlements on the shores of the freshwater lake within a single year, displacing thousands of people. Subsequent investigations of this region have suggested that some of the flooding of the modern Black Sea region did occur 7,600 years ago, but some of it also took place earlier, near 9,400 years ago, perhaps because of runoff from the meltwater lakes to the north. The ancient flood legends may have their origin in the sea level rise at the end of the last deglaciation.

millennia, including various kinds of kangaroos and wombats and a lion, as well as nonmarsupials such as giant tortoises and flightless birds. Just before this time, humans had first entered Australia from southeast Asia, traveling by boat along islands and landmasses exposed by the lowered glacial sea level. These people, who used "fire sticks" to burn grasslands and drive game, have been proposed as the cause of the extinctions in Australia.

In the Americas, the megafaunal population was highly diverse until the late stages of the most recent deglaciation, with a particularly rich array in North America (Figure 16-11). Then, within an interval of a few thousand years centered on 12,500 years ago, over half of the large mammal species living in both North and South America became extinct. The list in North America includes giant mammoths and mastodons (larger than modern elephants), horses the size of modern Clydesdales, camels, giant ground sloths, saber-toothed tigers, and beavers as large as modern bears.

One long-standing explanation for this rapid pulse of extinction is that major climate changes at the end of the glacial maximum created new environmental combinations to which many mammals were unable

FIGURE 16-11

Mammals of the glacial maximum

A rich array of large mammals lived on the North American plains prior to the most recent deglaciation, including several forms that became extinct: wooly mammoths, saber-toothed tigers, and giant ground sloths. (© 2013 SMITHSONIAN INSTITUTION.)

to adapt. These conditions included strong summer warming and drying of the land south of the ice sheets by high summer insolation, reduction of habitat in the cooler north by the slowly retreating ice, and unusual mixtures of vegetation that developed as forests and grasslands shifted from their glacial positions to their modern locations.

Critics of this climatic hypothesis note that no comparable pulse of extinction occurred in any of the fifty or so preceding deglaciations. In fact, the number of species that went extinct near 12,500 years ago exceeds the total during all of the previous 2.75 million years. The same basic combination of changing climatic conditions—increasing summer insolation, rising CO_2 levels, and rapidly melting ice sheets—had occurred during all of the earlier deglaciations without causing pulses of extinction. So critics of the climatic explanation ask why the extinctions only occurred during this one deglaciation.

Another vulnerability of the climatic hypothesis is the fact that so many kinds of mammals throughout the Americas suffered the same fate, even though they lived in environments ranging from semiarid grasslands to rain forests. Because these different environments followed different climatic paths during deglaciation, climate change is difficult to use as an explanation for all the extinctions.

A second explanation, called the **overkill hypothesis**, put forward by the paleoecologist Paul Martin, is that human hunting caused this extinction pulse. The immediate cause of the extinctions could have been either the first arrival of humans in the Americas, the buildup of their populations to the necessary levels, or the first appearance of a new hunting technology or strategy among people already present.

Both the origin and time of arrival of the first humans in the Americas were once thought to have been resolved. They supposedly came by land from Asia near 12,500 years ago, during the late stages of the last deglaciation. They crossed into Alaska over a land bridge in the Bering Strait exposed by the lower glacial sea level and moved down the interior of North America east of the Rocky Mountains. They passed through the ice-free corridor opened by early melting and separation of the Laurentide ice sheet to the east and the smaller Cordilleran ice sheet over the Canadian Rocky Mountains (see Chapter 14, Figure 14-2).

This view is now in dispute. Evidence from several sites documents the arrival of humans by 15,000 years ago, when an ice-free corridor had not yet opened. In addition, one of the earliest sites is in southern Chile, far down the South American coast. Evidence from this site indicates that these people ate a diet based mainly on marine food resources. As a result of evidence such as this, it now seems more likely that people from Asia first arrived by water, traveling rapidly down the food-rich Pacific coasts of the Americas, and later penetrating eastward across the continents.

In any case, a new hunting technology appeared at the same time that the extinctions occurred (12,500 years ago). Many archeological sites that date to near 12,500 calendar years ago contain spears fitted with a new and elegant kind of stone spear point fashioned by humans (Figure 16-12). This new technological development could have helped people hunt large mammals more effectively.

One criticism of the overkill hypothesis is that the early Americans were too few in number to have caused so many extinctions, but studies from population models have refuted this criticism. Because reproduction (gestation) times for large mammals are long, hunters only need to cull an extra 1–2% of a total species population per year beyond natural attrition rates to cause their extinction within a millennium. In addition, these people worked in groups to drive animals to their deaths over steep cliffs at the end of narrow bluffs or in traps that permitted mass killing. Only a fraction of the number of animals

FIGURE 16-12
Pulse of mammal extinctions

Woolly mammoths and other large mammals abruptly became extinct in North America near 12,500 years ago. Distinctive grooved spear points ("Folsom points," named for the site in New Mexico where they were first found, shown here with bones) suddenly appeared during this interval of widespread extinction. (IMAGE ARCHIVES/DENVER MUSEUM OF NATURE & SCIENCE.)

killed in these places were used for food and clothing. Still another factor that may have contributed to the extinctions may have been the extensive use of fire to drive animals for hunting purposes.

Another criticism of the overkill hypothesis is that some creatures that do not seem likely to have been hunted also went extinct, including large meat-eating mammals that may have preyed on humans rather than becoming their prey. A plausible response to this criticism is that carnivores that depended on the carcasses of large mammals for food may have gone extinct because much of their natural prey had gone extinct at the hands of humans. At this point, however, the overkill hypothesis does not explain why some large mammals that would seem to have been likely targets for hunters (moose, musk ox, and one species of bison) survived.

In Summary, both explanations of the pulse of extinctions have their critics, but the lack of any extinction pulse during all of the previous ice-age cycles is a powerful argument against the hypothesis that climate was responsible. The cause appears to have been humans.

Because the rest of this book will take an increasingly historical approach, the axes of the time plots from this point on have been rotated to the typical historical perspective, with younger to the right (and warmer upward).

16-7 Did Early Farmers Alter Climate?

In the **early anthropogenic hypothesis** published in 2003, I claimed that early agriculture had a substantial impact on greenhouse gases and on global climate thousands of years ago, much earlier than previously thought. This claim was based on the fact that the concentrations of CO_2 and CH_4 in the atmosphere increased during the last several thousand years of the current interglaciation, even though they had fallen during comparable intervals in previous interglaciations (Figure 16-13). If natural processes in the

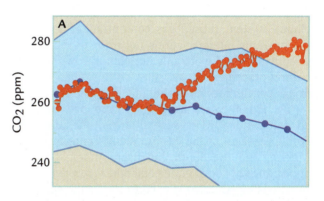

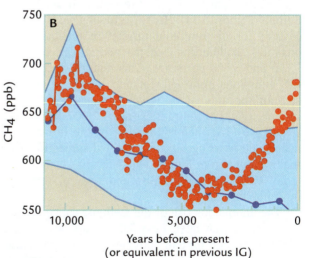

Years before present
(or equivalent in previous IG)

FIGURE 16-13
"Wrong-way" CO_2 and CH_4 trends

Greenhouse-gas concentrations in the atmosphere rose during the later part of the current interglaciation (red) but fell during previous interglaciations (mean shown in dark blue; standard deviation in light blue). (GAS TRENDS FROM EPICA COMMUNITY MEMBERS, "EIGHT GLACIAL CYCLES FROM AN ANTARCTIC ICE CORE," *NATURE* 429 [2004]: 623–628.)

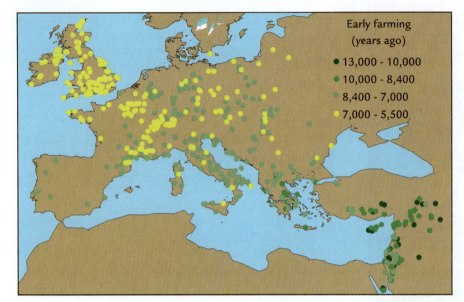

Early farming
(years ago)

● 13,000 - 10,000
● 10,000 - 8,400
● 8,400 - 7,000
● 7,000 - 5,500

FIGURE 16-14

The spread of Fertile Crescent agriculture

Several kinds of agriculture originated in the Fertile Crescent region north and east of the Mediterranean and gradually spread into Europe, North Africa, and other parts of Asia.
(ADAPTED FROM D. ZOHARY AND M. HOPF, *DOMESTICATION OF PLANTS IN THE OLD WORLD: THE ORIGIN AND SPREAD OF CULTIVATED PLANTS IN WEST ASIA, EUROPE, AND THE NILE VALLEY* [OXFORD: OXFORD UNIVERSITY PRESS, 2000].)

climate system caused gas concentrations to *fall* in the previous interglaciations (all prior to the advent of agriculture), the obvious question is how the same natural processes can be called on to explain the *rise* in gas concentrations during this interglaciation.

By 10,000 years ago, agriculture had originated independently in several areas, including the Fertile Crescent of the eastern Mediterranean, China, Mexico, and the South American Andes. The first appearance of cereal grains and other crop remains in hundreds of ^{14}C-dated lake cores shows the spread of agriculture from the Fertile Crescent into and across Europe (Figure 16-14). Because growing crops in tree-covered Europe required deforestation, this map demonstrates that cutting became widespread between 7,500 and 5,500 years ago. This regional increase in CO_2 emissions from deforestation coincides with the start of the anomalous upward trajectory in atmospheric CO_2 concentrations near 7,000 years ago (Figure 16-13A). Before the start of the Bronze Age near 5,000 years ago, agriculture had spread into virtually every part of Europe where farming is practiced today. Deforestation was also underway by this time in China and India.

The anomalous rise of the methane concentration began somewhat later, near 5,000 years ago (Figure 16-13B). Several human activities generate methane, including irrigating fields, burning biomass (weeds and crop residues), tending livestock, and production of organic waste, including human waste. In irrigated fields, both rice and weedy plants grow in the standing water and then die and rot. Once the dead plants consume the available oxygen, methane is emitted to the atmosphere.

The archeobotanist Dorian Fuller and colleagues have traced the spread of rice irrigation across Asia

based on archeological records. After originating in China between 6,000 and 5,000 years ago, rice irrigation expanded rapidly into other countries (Figure 16-15) during the interval of rising atmospheric methane concentrations. By 1,000 years ago, rice irrigation was practiced across all areas of Southeast Asia where it is in use today. Using these mapped trends of first rice arrival in each region and estimates of the subsequent increases in density of irrigation farming, Fuller estimated the total area of rice irrigated between 5,000 and 1,000 years ago (Figure 16-16A). The rising trends in area irrigated and in methane emitted resembled the CH_4 increase measured in ice-core records, with 70% of the observed methane rise accounted for by irrigation (Figure 16-16B).

Critics of the early anthropogenic hypothesis have questioned whether humans can explain the large anomalies in CO_2 and methane that developed prior to the last two centuries (see Figure 16-13). This criticism is based on the reaction that too few humans were present to have taken control of greenhouse-gas trends millennia ago. Some model simulations of forest clearance based on a direct link to population appear to confirm this view: by 1800, most of the world's forests are shown as largely intact, even in China, India, and Europe (Figure 16-17A).

Historical evidence from both Europe and China has shown, however, that the average farmer 2,000 years ago used four to five times as much land as farmers in the 1800s. This decrease during historical times resulted from the fact that population growth slowly reduced the size of land holdings, forcing farmers to obtain more produce per acre by adapting new innovations such as spreading animal manure for fertilizer and rotating crops. As a result, early farmers had cleared far more land than estimates tied directly

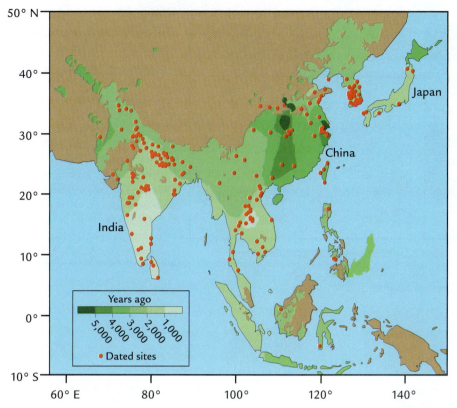

FIGURE 16-15
The spread of rice irrigation

Rice irrigation spread from China across southern Asia between 5,000 and 1,000 years ago. (ADAPTED FROM D. Q. FULLER ET AL., "THE CONTRIBUTION OF RICE AGRICULTURE AND LIVESTOCK TO PREHISTORIC METHANE LEVELS: AN ARCHAEOLOGICAL ASSESSMENT," *THE HOLOCENE* 25 [2011]: 743–759, DOI:10.1177/0959683611398052.)

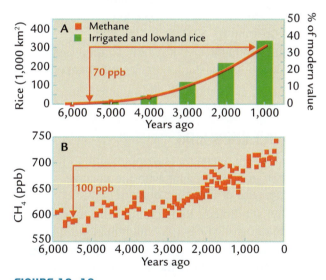

FIGURE 16-16
Rice irrigation and methane

The increasing area of irrigated rice fields emitted greater amounts of methane that contributed to the observed CH_4 increase in ice cores. (ADAPTED FROM D. Q. FULLER ET AL., "THE CONTRIBUTION OF RICE AGRICULTURE AND LIVESTOCK TO PREHISTORIC METHANE LEVELS: AN ARCHAEOLOGICAL ASSESSMENT," *THE HOLOCENE* 25 [2011]: 743–759, DOI:10.1177/0959683611398052.)

to population would suggest. For example, a survey of England in 1086 (the Domesday Book) found that 85% of the arable land was already in pasture or crops, and only 15% still in forest (mostly kept as hunting preserves for royalty and nobility). A model simulation based on historical evidence shows prevalent clearance in Europe, as well as other heavily populated areas by 1800 (Figure 16-17B), consistent with the early anthropogenic hypothesis.

One corollary claim of the early anthropogenic hypothesis is that the anomalous rises in CO_2 and CH_4 kept the atmosphere warmer than it would have been if nature had remained in control. Throughout the last several millennia, the decrease in summer insolation across the Northern Hemisphere was driving a natural cooling at high northern latitudes (see Chapter 14, Figure 14-19). If the greenhouse-gas concentrations had fallen as they had done during the previous interglaciations (see Figure 16-13), climate would have cooled even more, perhaps to the point of allowing new ice sheets to begin to form.

Sensitivity tests with general circulation models support this idea. When greenhouse-gas concentrations are reduced to the "natural" level predicted by the hypothesis (~245 ppm for CO_2 and ~450 ppb for CH_4), some regions at high northern latitudes retain snow throughout the year (Figure 16-18). Because snow that persists through summer forms a base that will be added to during the following autumn and winter, simulated year-round snow cover in these models is equivalent to a state of incipient glaciation. Year-round snow is simulated in the Arctic margins

Clearance at 1800 (HYDE)   Clearance at 1800 (Kaplan)

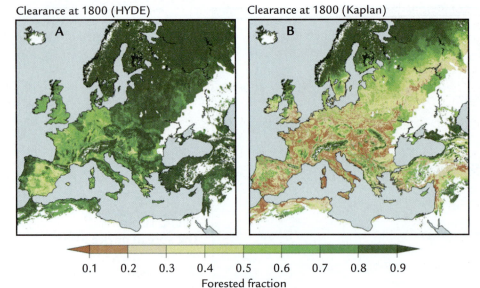

0.1 0.2 0.3 0.4 0.5 0.6 0.7 0.8 0.9
Forested fraction

FIGURE 16-17

Differing interpretations of early forest clearance

Simulations of forest clearance by the year 1800 have been based on two differing assumptions: estimates tied directly to population simulate little clearance by 1800 (A), while estimates based on historical evidence indicate much greater clearance (B). (ADAPTED FROM J. O. KAPLAN ET AL., "HOLOCENE CARBON EMISSIONS AS A RESULT OF ANTHROPOGENIC LAND COVER CHANGE," *THE HOLOCENE*, SAGE PUBLICATIONS, AUGUST 1, 2011.)

Months of snow cover (CCSM4)

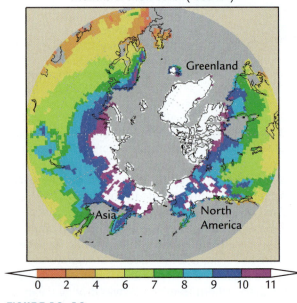

0   2   4   6   7   8   9   10   11

FIGURE 16-18

Is a glaciation overdue?

Climate model simulations with present insolation values, but with CO_2 and CH_4 concentrations lowered to the natural levels they are hypothesized to have reached during previous interglaciations, produce year-round snow cover (incipient glaciation) in northern Eurasia and North America. (ADAPTED FROM S. VAVRUS ET AL., "THE ROLE OF GCM RESOLUTION IN SIMULATING GLACIAL INCEPTION," *THE HOLOCENE* 25 [2011], DOI:10.1177/0959683610394882.)

of Canada and Russia, Baffin Island northeast of the Canadian mainland, and the northern Rocky Mountains. Each of these new ice caps would be relatively small in size, but their aggregate area is estimated to be larger than the modern Greenland ice sheet.

Key Terms

savanna hypothesis (p. 321)

variability selection hypothesis (p. 322)

Fertile Crescent (p. 324)

diluvial hypothesis (p. 326)

Black Sea flood hypothesis (p. 326)

megafauna (p. 326)

overkill hypothesis (p. 328)

early anthropogenic hypothesis (p. 329)

Review Questions

1. What are the problems with the savanna hypothesis of human evolution?

2. Why is it difficult to determine whether or not climate affected human evolution?

3. Why are climatic explanations of the megafaunal extinctions 12,500 years ago suspect?

4. Why are natural explanations for the CO_2 and CH_4 increases in recent millennia suspect?

Additional Resources

Basic Reading

Diamond, J. 1997. *Guns, Germs and Steel*. New York: W. W. Norton.

Roberts, N. 1998. *The Holocene*. Oxford: Blackwell.

Ruddiman, W. F. 2005. *Plows, Plagues, and Petroleum*. Princeton: Princeton University Press.

Ryan, W., and W. Pitman. 1998. *Noah's Flood.* New York: Simon & Schuster.

Tudge, C. 1996. *The Time Before History*. New York, Simon & Schuster.

Advanced Reading

Curtis, J. H., D. A. Hodell, and M. Brenner. 1996. "Climatic Variability on the Yucatan Peninsula (Mexico) During the Past 3,500 Years, and Implications for Maya Cultural Evolution." *Quaternary Research* 46: 37–47.

Fuller, D. Q., J. van Etten, K. Manning, C. Castillo, E. Kingwell-Banham, A. Weisskopf, L. Qin, Y.-I. Sato, R. J. Hijmans. 2011. "The Contribution of Rice Agriculture and Livestock to Prehistoric Methane Levels: An Archaeological Assessment." *The Holocene* 25: 743–759. doi:10.1177/0959683611398052.

Kaplan, J. E., K.M. Krumhardt, E. Ellis, W. F. Ruddiman, C. Lemmen, and K. K. Goldewijk. 2010. "Holocene Carbon Emissions as a Result of Anthropogenic Land Cover Change." *The Holocene* 25. doi:10.1177/0959683610386983.

Martin, P. S., and R. G. Klein. 1984. *Quaternary Extinctions*. Tucson: University of Arizona Press.

Potts, R. 1997. *Humanity's Descent: The Consequences of Ecological Instability*. New York: Avon.

Ruddiman, W. F. 2003. "The Anthropogenic Greenhouse Era Began Thousands of Years Ago." *Climatic Change* 61: 261–93.

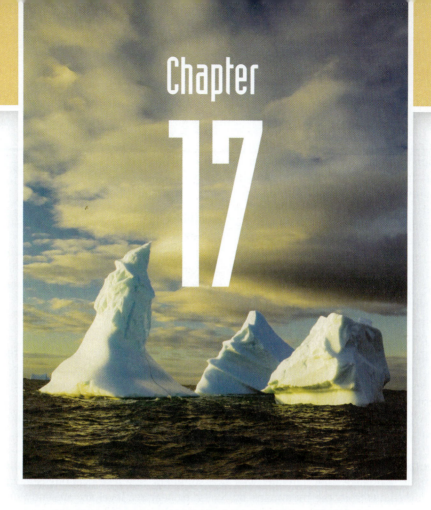

Climate Changes During the Last 1,000 Years

Detailed records of climate change during the last 1,000 years come from proxy indicators stored in various climatic archives, including mountain glaciers, tree rings, and corals. Historical observations of miscellaneous climate-related phenomena collected by humans over several centuries are also available from a few regions, but instrumental measurements began only in recent centuries.

Climate changes over the last thousand years were small and variable in pattern from region to region until the large warming of the last century. In high northern latitudes around the North Atlantic Ocean, a small cooling occurred between roughly 1,000 years ago and the time known as the **Little Ice Age** (from 1400 to 1900). The scope of this cooling elsewhere on Earth remains uncertain, as does the cause.

335

The Little Ice Age

As the last millennium began, scattered evidence from Europe and high latitudes surrounding the North Atlantic suggests a time of relatively warm climate known as the **medieval warm period**. During this interval (from 1000 to 1300), Nordic people settled southwestern Greenland along the fringes of the ice sheet and managed to grow wheat. Sea ice, common during the past century around the Greenland coasts, is rarely mentioned in chronicles from this era.

The subsequent cooling into the Little Ice Age (1400–1900) affected people in northern Europe. With colder winters and a shorter growing season, grain crops and grape harvests repeatedly failed in far northern regions where they had been successfully grown during the medieval warm period, causing local famines. Lakes, rivers, and ports froze throughout northern Europe during severe winters. Near the onset of the Little Ice Age, the settlements in Greenland were abandoned, in part because the marginal climate had become inhospitable but also perhaps because of conflicts with native Arctic peoples.

The most dramatic evidence that climate was cooler than it is today comes from the mountains of Europe. People who lived in the Alps of south-central Europe and the mountains of Norway in the fourteenth and fifteenth centuries experienced large-scale advances of glaciers down mountain valleys. They wrote about the effects of these episodes on their lives, and they sketched drawings of the advancing ice. Many of the ice advances spread over alpine meadows where livestock had once grazed, and the advancing ice destroyed farmhouses and even small villages where people had been living. These historical expansions also prompted the glacial geologist Louis Agassiz to theorize that much larger ice ages had occurred tens of thousands of years earlier.

Other evidence confirms that temperatures were often cooler in Europe and the nearby North Atlantic during the Little Ice Age. One documentary record shows the frequency of sea ice along the north and west coasts of Iceland (Figure 17-1). Because fishing was vital to the food supply on this isolated island, records were kept of times when coastal sea ice prevented ships from leaving port. This record shows the number of weeks per year in which sea ice reached and blocked the northern coast of Iceland.

Sea ice appears to have been very rare until 1200 and infrequent until 1600, although the earlier records may be less reliable than the later ones. Sea ice increased in frequency and reached a maximum during the 1800s, but then all but disappeared from the coasts during the twentieth century, except in

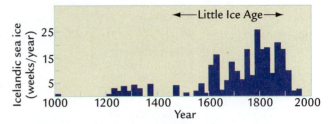

FIGURE 17-1
Sea ice on the coast of Iceland
The frequency of sea ice along the coast of Iceland increased from 1200 into the 1800s and then declined rapidly during the 1900s. (ADAPTED FROM H. H. LAMB, *CLIMATE: PRESENT, PAST AND FUTURE*, VOL. 2 [LONDON: METHUEN, 1977].)

occasional extreme years. This record documents a major contrast near Iceland between the cold conditions of the Little Ice Age and the relative warmth of the twentieth century.

Because major ice sheets did not actually form, the Little Ice Age was not a true ice age, but in some regions it was at least a small step in that direction. Much of the Canadian Arctic has long, cold winters and short, chilly summers. In a few locations, small ice caps that persist at or near sea level are presently melting rapidly in the warmth of the modern climate. Rock outcrops in these regions are covered by lichen, a primitive moss-like form of vegetation that can live on bare rock surfaces even under inhospitable conditions (Figure 17-2A). These organisms secrete acids that attack bedrock and break it down into mineral grains that provide nutrients. Small lichen growing in the Canadian Arctic today can be dated by their size (Figure 17-2B). They begin life as small specks and then slowly expand into round blobs at predictable rates.

Scattered across portions of Baffin Island, northeast of the Canadian mainland, are broad halos of dead lichen bordering small modern ice caps; other halos are found in high areas that now lack ice caps (Figure 17-2C). These halos of dead lichen are a product of the Little Ice Age. The size of these lichens indicates that they must have developed over warm intervals of several hundred years in the relatively recent past and then been subsequently killed.

Because lichens are tolerant of extreme cold, it is unlikely that frigid temperatures killed them. A more likely explanation is that they were buried beneath snowfields that completely blocked sunlight through the summer growing season. If summer melting failed to remove the previous winter's snow for many years in succession, the lichens would have died for lack of photosynthesis. The lichen halos are an indication that growing snowfields covered larger areas of high terrain on Baffin Island in the fairly recent past.

A

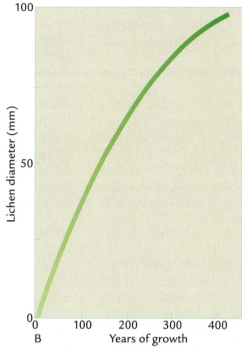

B

FIGURE 17-2
Lichen halos: a Little Ice Age snowfield?

Lichens grow on rock surfaces in the Arctic (A) at known rates (B). Halo-like areas of dead lichen in the Canadian Arctic (C) record an interval when snowfields expanded during the Little Ice Age.

(A: GRAMBO/FIRST LIGHT/CORBIS IMAGES; B: ADAPTED FROM G. H. DENTON AND W. KARLEN, "HOLOCENE CLIMATIC VARIATIONS: THEIR PATTERN AND POSSIBLE CAUSE," *QUATERNARY RESEARCH* 3 [1973]: 155–205; C: ADAPTED FROM J. T. ANDREWS ET AL., "THE LAURENTIDE ICE SHEET: PROBLEMS OF THE MODE AND INCEPTION," WMO/IMAP *SYMPOSIUM PROCEEDINGS* 421 [1975].)

☐ Present day ice
☐ Dead lichen, expanded snowfields

C

When did the lichens die off? The small young lichen now growing on top of the older dead ones have been present only during the last century or so—since the end of the Little Ice Age. Expanded snowfields during the Little Ice Age kept them from growing back until late in the nineteenth century. Scattered radiocarbon dating confirms that the original lichen formed just before the start of the Little Ice Age.

It makes sense that permanent snowfields would have begun to expand across this particular region. Baffin Island is one of three locations in which the last remnants of the great glacial ice sheets melted near 6,000 years ago (see Chapter 14, Figure 14-2). Climate model simulations also suggest that it is one of the regions in which new ice sheets probably began to readvance during the last 3 million years (see Chapter 16, Figure 16-18). In this context, the growth of snowfields on Baffin Island during the Little Ice Age was a very real, if small, step toward an ice age.

The Little Ice Age is often used as the "type example" of a cooler Earth from the recent past, with the implicit assumption that the cooling that caused it was hemispheric or even global in extent. Yet evidence in Chapter 15 showed that millennial-scale climatic changes during the last 8,000 years have been highly variable from region to region, and that no single pattern appears to have persisted on larger spatial scales. From this point of view, the cooling in Iceland and nearby areas of the North Atlantic Ocean might simply have been a local phenomenon restricted to this one region.

The next section of this chapter examines several methods that provide climate records spanning the last 1,000 years. The purpose of this exploration is to see what these methods indicate about the regional extent of the Little Ice Age cooling. By examining climate records covering major portions of the last millennium, we can also begin to address the issue of whether or not the last 100 years of climate change fit into natural longer-term trends.

▶ Proxy Records of Historical Climate

Before the nineteenth century, people rarely kept records of temperature or precipitation. As a result, climate scientists have to rely mainly on proxy records in archives like those used for earlier intervals of Earth's history, such as lake sediments and ice cores. In addition, sources of proxy records such as mountain glaciers, tree rings, and corals can also be used to reconstruct climate changes across Earth's surface during the last few centuries.

17-1 Ice Cores from Mountain Glaciers

Like the huge ice sheets on Antarctica and Greenland, glaciers in mountain valleys and small ice caps covering mountain summits are excellent climate archives (Figure 17-3). Some mountain ice dates back many thousands of years into the last glaciation, while other glaciers span only a few hundred years of climate history. Layers deposited at the surface usually contain annual signals (see Figure 17-3A), but ice flow may degrade the resolution deeper in the ice.

Retrieving ice cores from mountains is a formidable task. Heavy equipment (including solar-powered ice drills) must be hauled up to subfreezing mountain summits (see Figure 17-3B). Mountain ice caps are between 100 and 200 meters thick, and most

expeditions drill and sample the entire thickness of ice at several locations on each ice cap (see Figure 17-3C and D). Lack of oxygen at these altitudes quickly causes exhaustion and other problems. The ice cores that are extracted must be lugged down to lower elevations and then kept from melting in the warmer air.

Only a few mountain glaciers have been cored. By far the most extensive efforts have been those by the intrepid glacial geologist Lonnie Thompson. His expeditions have retrieved ice cores at elevations of 5,670 meters (about 18,500 feet) above sea level from the Quelccaya ice cap in the Peruvian Andes. Based on counting annual layers and matching volcanic ash layers with historically documented eruptions, he found that these records extend back to 1,500 years ago.

Cores taken in the 1980s show annual-scale changes in $\delta^{18}O$ values and dust concentrations that can be averaged over decadal intervals (Figure 17-4A). The variations in $\delta^{18}O$ values reflect the same processes as those affecting continent-sized ice sheets: changes in source area, transport paths, amount of water vapor carried to the glacier, and particularly temperature in the atmosphere above the ice where the snow condenses. Higher dust concentrations indicate some combination of drier source areas and stronger winds.

Early cores from the Quelccaya ice cap registered a shift toward more positive $\delta^{18}O$ values and less dust near the year 1900, implying that a change toward some combination of warmer temperatures, weaker winds, and different source areas occurred at that time (see Figure 17-4A). This $\delta^{18}O$ record also resembles the Little Ice Age pattern shown in Figure 17-1, with more positive (probably warmer) values from 1000–1400 and then more negative (probably cooler) values after 1500. In contrast, the dust record does not show the expected match before 1600; if anything, the lowest dust concentrations occurred within the early part of the Little Ice Age (1400–1600), and higher concentrations during the medieval warm period (1000–1300).

In a return expedition to Quelccaya in the early 1990s, Thompson encountered something totally unexpected. During the 1970s and 1980s, previous coring expeditions at the top of the ice cap (see Figure 17-4B) had found annual layering extending from the most recent ice all the way down to the deepest layers deposited 1,500 years ago. In the new cores, however, meltwater percolating down from the surface had begun to destroy the uppermost annual layers (see Figure 17-4C). This dramatic finding means that a tropical ice cap that had been continuously recording intact annual layering for 1,500 years had suddenly begun to melt. In this location, the warming of the late twentieth century was obviously unprecedented for the last millennium and a half.

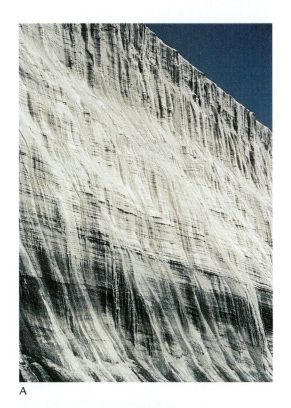
A

B

C

D

FIGURE 17-3
Coring mountain glaciers

Drilling ice cores from annually layered mountain ice (A) requires hauling equipment and supplies up and down the sides of mountains and ice caps (B), and then drilling at elevations approaching 20,000 feet (C), sometimes with towering summer storm clouds rising from lower elevations and looming in the background (D). (COURTESY OF L. G. THOMPSON, BYRD POLAR RESEARCH INSTITUTE, COLUMBUS, OH.)

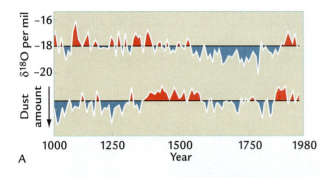

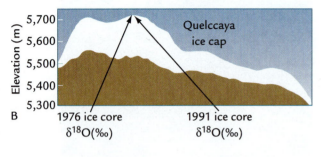

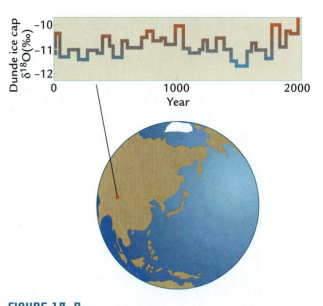

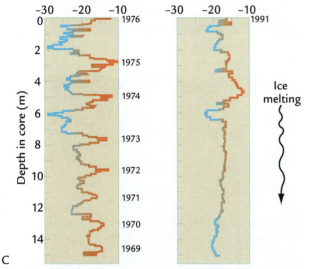

FIGURE 17-4
Quelccaya ice cap in Peru

An ice core taken in 1980 from the Quelccaya ice cap in the Peruvian Andes shows more negative $\delta^{18}O$ values and higher dust concentrations from 1600 to 1900 than in the twentieth century (A). A return expedition that cored the same location on the ice cap summit in 1993 found that the annual signal at the surface was being destroyed by melting, the first such melting event in the last 1,500 years (B, C). (A: ADAPTED FROM L. G. THOMPSON ET AL., "THE LITTLE ICE AGE AS RECORDED IN THE STRATIGRAPHY OF THE TROPICAL QUELCCAYA ICE CAP," SCIENCE 234 [1986]: 361–64; C: ADAPTED FROM L. G. THOMPSON ET AL., "RECENT WARMING: ICE CORE EVIDENCE FROM TROPICAL ICE CORES, WITH EMPHASIS ON CENTRAL ASIA," GLOBAL AND PLANETARY CHANGE 7 [1993]: 145–56.)

Thompson has also studied ice caps on sub-tropical mountains in Asia, including a much longer $\delta^{18}O$ record from Dunde ice cap in northern Tibet (Figure 17-5). Measurements averaged across

FIGURE 17-5
Dunde ice cap in Tibet

An ice core from Dunde ice cap in eastern Tibet shows lighter $\delta^{18}O$ values during the last 50 years than in any such interval of the previous 2,000 years. (ADAPTED FROM L. G. THOMPSON ET AL., "HOLOCENE-LATE PLEISTOCENE CLIMATIC ICE CORE RECORDS FROM QINGHAI-TIBETAN PLATEAU," SCIENCE 246 [1989]: 361–64.)

50-year intervals show some similarity to those in Figure 17-1, with more positive $\delta^{18}O$ values before 1500 and especially around 1000, and more negative values during much of the Little Ice Age interval. In this record, however, the transition to more positive (probably warmer) $\delta^{18}O$ values began near 1700, well before the Little Ice Age ended.

The average $\delta^{18}O$ concentration at the Dunde site during the most recent 50-year interval (1937–1987) is more positive than in any other interval of the entire 12,000 years of record. Either the temperature of snow precipitation was uniquely warm during the mid-twentieth century, or a major change has occurred in the sources or transport paths of the incoming water vapor. Other records from Tibet, however, do not show a dramatic difference during the last few decades.

In Summary, results from the Quelccaya and Dunde ice caps suggest that climate on low-latitude mountains may have been colder during the Little Ice Age and warmer (perhaps even uniquely so) during the twentieth century. In contrast, some ice cores from the Antarctic and Greenland ice sheets do not show such a pattern (Figure 17-6). These disagreements indicate that variations in climate recorded by $\delta^{18}O$ changes in ice cores during the last 1,000 years have in some cases been regional rather than global.

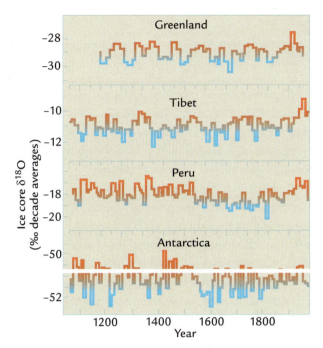

FIGURE 17-6

δ¹⁸O signals from ice cores in four regions

Cores taken from ice sheets and mountain glaciers in widely separated regions yield widely varying $\delta^{18}O$ signals, and some do not show uniquely positive (warm?) $\delta^{18}O$ values during the twentieth century. (ADAPTED FROM L. G. THOMPSON ET AL., "RECENT WARMING: ICE CORE EVIDENCE FROM TROPICAL ICE CORES, WITH EMPHASIS ON CENTRAL ASIA," *GLOBAL AND PLANETARY CHANGE* 7 [1993]: 145–56.)

17-2 Tree Rings

The use of tree rings to reconstruct climate change over the last several hundred years or more is called **dendroclimatology**. In regions of large seasonal changes, trees produce annual rings of varying distinctness depending on species. The rings shift from lighter, low-density "early wood" of spring and early summer to darker, denser bands of "late wood" at the end of the growing season (Box 17-1).

Warmth and abundant rainfall during the growing season are favorable to tree growth, while cold and drought inhibit growth. The basic strategy for tree ring studies is to search out regions where trees are most sensitive to climatic stress, usually at the limit of their natural temperature or precipitation ranges. In such regions, trees often grow in isolated clusters. Years of unfavorable climate (low temperature or precipitation) are stressful to the trees and slow their growth, producing unusually narrow rings. Changes between favorable and unfavorable growth years produce distinctive variations in the widths and other properties of tree rings.

Tree ring studies have been carried out in many areas, such as the Arctic, where the primary trees are spruce, larch, and Scots pine (Figure 17-7A).

A

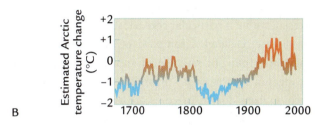

B

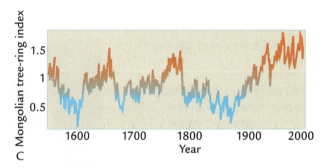

C

D

FIGURE 17-7

Arctic and Asian tree ring signals

Signals from trees on northern continents, such as Siberian larch (A), are combined to create average circumarctic temperature changes over the last several centuries (B). Similar tree ring signals for Central Asia (C) come from studies of larch and pine in the mountains of Mongolia (D). Curve B shows departures from the 1951–1980 average. Curve C shows an index of changing tree ring width in Central Asia. (A AND D: COURTESY OF G. C. JACOBY, LAMONT-DOHERTY EARTH OBSERVATORY OF COLUMBIA UNIVERSITY; B: ADAPTED FROM G. C. JACOBY AND R. D. D'ARRIGO, "RECONSTRUCTED NORTHERN HEMISPHERE ANNUAL TEMPERATURE SINCE 1671 BASED ON HIGH-LATITUDE TREE-RING DATA FROM NORTH AMERICA," *CLIMATE CHANGE* 14 [1989]: 39–49; C: ADAPTED FROM G. C. JACOBY ET AL., "MONGOLIAN TREE RINGS AND 20TH-CENTURY WARMING," *SCIENCE* 273 [1996]: 771–73.)

Tools of Climate Science

Analyzing Tree Rings

At chosen sites, a dozen or more trees are sampled by taking small radial cores about 0.5 cm in diameter. The investigators date each tree by comparing the sequences of tree rings back in time, beginning from the year the cores are taken. Taking multiple cores lets scientists detect annual rings that may be missing or falsely present within a particular core because of local damage or disease. Deliberate sampling of old trees produces sequences spanning several centuries. Width and density changes in the tree rings are measured in each core, and records from dozens of cores are averaged to create a single representative signal for each site.

The first step in tree ring analysis is to remove the gradual effects of aging. Trees grow wider rings when they are young than at maturity, and investigators have to remove this longer-term growth effect in order to focus on year-to-year changes in climate.

The next step is to relate the sequence of tree ring measurements to nearby instrumental records of climate. In the geographically remote areas often chosen for tree ring studies, instrumental records may only cover the last 50 to at most 100 years, and thus they overlap only with the lattermost part of the tree ring sequence. By examining the correlations between the width or density of the tree rings

Coring for tree ring studies To study tree rings, scientists drill into trees at sites where trees are under moderate stress because of cold or dryness (A). The cores extracted (B) are small in diameter compared to the trees (C). (COURTESY OF G. C. JACOBY, LAMONT-DOHERTY EARTH OBSERVATORY OF COLUMBIA UNIVERSITY.)

A

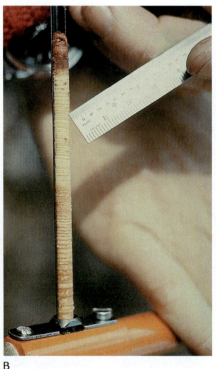

B

C

An integrated signal developed from trees distributed over the entire circumarctic region (Figure 17-7B) shows cool conditions in the late seventeenth and early eighteenth centuries, some warming in the middle and late eighteenth century, a deeper cooling in the early to mid-nineteenth century, a slow but substantial warming after 1850, a brief cooling between 1950 and 1970, and a small warming since 1980.

The 320-year trend in Figure 17-7B covers the middle and end of the Little Ice Age and the instrumentally recorded warming of the twentieth century. Estimated regional temperature variations larger than 1°C occurred around the Arctic within the Little Ice Age, which was clearly not a time of unrelieved cold. The warming of the Arctic in the mid-twentieth century reached values unique for

and the instrumental record of climate change, scientists determine what aspects of climate control tree growth.

The final step is to use correlations defined from the calibration period of tree ring and climate data to model the tree ring/climate relationship and to estimate past climate using the tree ring data as predictors. Such projections into the past assume that the climatic controls on tree ring properties found over the **calibration interval** were also in operation during the earlier interval.

In cold regions, temperature early in the growing season is often the strongest influence on tree growth, but a response to temperatures in other seasons is also present. In relatively arid regions, precipitation is more important. In some cases, precipitation in one year can affect growth in the next by providing moisture to the soil, which favors growth the following spring. On the other hand, deep winter snows can slow growth in the spring in colder areas.

Many records from the circumarctic margins show an unexpected decrease in tree ring growth during the last several decades, despite the fact that temperatures have been warming and CO_2 (fertilization) levels rising. This seemingly anomalous trend could indicate that an unusual combination of environmental changes (perhaps increased summer aridity or the effects of disease caused by insects) is now occurring in the Arctic, but this trend is not well understood.

The period of instrumental observations has also been a time of steadily rising atmospheric CO_2 levels. Controlled experiments with vegetation in greenhouses show that tree growth is enhanced by higher levels of CO_2 and photosynthesis. Some scientists speculate that correlations between tree ring properties and climate (temperature and precipitation) during the calibration interval have ignored this "fertilization" effect of rising

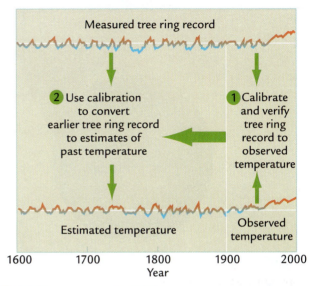

Calibrating tree ring signals Tree ring analysis is based on correlating the width (or density) of individual tree rings with monthly changes in temperature and precipitation recorded in the last half of the instrument record of climate (the last few decades). After these relationships are tested against the first half of the instrumental record, they are used to convert older tree ring characteristics into estimates of past temperature and precipitation using the entire instrumental record for calibration.

CO_2 on plants. CO_2 fertilization could masquerade as a climate signal, falsely indicating gradual warming in cold regions or a trend toward wetter climates in dry regions.

In arid regions, some evidence suggests that rising CO_2 levels and faster fertilization of leaf pores have reduced the exposure of vegetation to the dry air that normally constrains growth in these regions. If this evidence is valid, then CO_2 fertilization may have been a factor in faster tree growth in dry regions, regardless of other climate changes.

the 320 years of record, but temperatures in the late twentieth century decreased to values similar to those in the 1700s.

Another important region for tree ring studies is Central Asia. Because Asia is the largest continent, its climate is less subject to the moderating effects of ocean water than other regions. Climate signals derived from tree rings in Mongolia (see

Figure 17-7C) show a trend similar to that of the circumarctic region. Intervals of warmth occurred during the Little Ice Age, both in the mid-eighteenth century and earlier, while colder temperatures occurred in the late sixteenth, late seventeenth, and middle to late nineteenth centuries. The warming of the middle and late twentieth century appears unprecedented within the 450 years of record. These

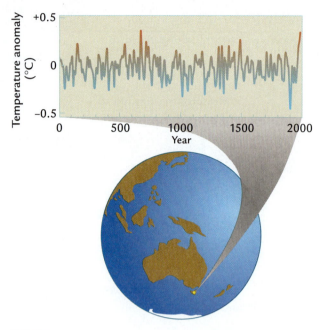

FIGURE 17-8

Tasmanian tree rings

Tree ring records from the island of Tasmania, south of Australia, show that temperatures in the twentieth century are nearly unprecedented during the last 2,000 years. (ADAPTED FROM E. COOK ET AL., "INTERDECADAL CLIMATIC OSCILLATIONS IN THE TASMANIAN SECTOR OF THE SOUTHERN HEMISPHERE: EVIDENCE FROM TREE RINGS OVER THE PAST THREE MILLENNIA," IN *CLIMATIC VARIATIONS AND FORCING MECHANISMS OF THE LAST 2000 YEARS*, ED. P. D. JONES AND R. S. BRADLEY [BERLIN: SPRINGER-VERLAG, 1996].)

signals come from larch and pine trees growing on the high flanks of mountains in Central Asia (see Figure 17-7D).

Far fewer data are available from trees at middle to high latitudes of the Southern Hemisphere, which is mainly ocean. Records from pines on the island of Tasmania, south of Australia, extend back more than 2,000 years, and the warmth of the last several decades matches any level reached during that interval (Figure 17-8). The interval of the Little Ice Age after 1500 is cooler than the late twentieth century but does not stand out distinctively. Other long tree signals from the Southern Hemisphere vary considerably in character, but tree ring records from New Zealand, Chile, and Argentina extending back a few hundred years show unique warmth in the twentieth century.

In Summary, tree rings, like ice cores, tell us that climate has varied from region to region over the last several hundred years, so that no one record

fully describes the trends in all areas. Viewed in their entirety, tree ring signals tell us that climate varied significantly within the Little Ice Age, in some regions even warming at times to levels comparable to those observed during part of the twentieth century. Many records show substantial, and in some cases unprecedented, warmth in the 1900s.

17-3 Corals and Tropical Ocean Temperatures

Observations of climate changes at annual or decadal resolution are generally not available from the deep ocean because of slow deposition and mixing of sediments by burrowing organisms. But climate scientists have exploited corals as climate archives, using annual bands in their $MgCO_3$ or $CaCO_3$ structures. Because most corals grow in warm tropical or subtropical oceans, the information they provide complements ice core and tree ring studies from higher latitudes and altitudes. Most coral studies come from the tropical Pacific, which is dotted with volcanic islands surrounded by corals.

The most widely used climatic index in corals is $\delta^{18}O$ measurements at seasonal or better resolution. The two major controls on $\delta^{18}O$ variations are temperature (warmer waters produce more negative $\delta^{18}O$ values) and salinity (heavier rainfall produces more negative $\delta^{18}O$ values). The temperature effect on coral $\delta^{18}O$ values is analogous to the changes recorded in the shells of planktic foraminifera (see Appendix A).

Longer-term $\delta^{18}O$ trends in the Pacific Ocean are overprinted by large year-to-year fluctuations in the El Niño and ENSO systems (Box 17-2). These annual $\delta^{18}O$ variations largely reflect temperature changes in the eastern tropical Pacific Ocean and combined temperature-salinity changes in the central and western Pacific Ocean.

Modern corals deposited on the coast of the Galápagos Islands in the eastern Pacific record seasonal temperature changes similar to those measured directly by thermometers in surface waters (Figure 17-9). The match between the $\delta^{18}O$ record and ocean temperature is not perfect because salinity changes also affect the $\delta^{18}O$ values. Prominent El Niño years appear as $\delta^{18}O$ minima that indicate warm temperatures.

Corals spanning several hundred years have recently become more widely available from the tropical oceans. A regional average signal created by combining records from the Pacific and Indian Oceans

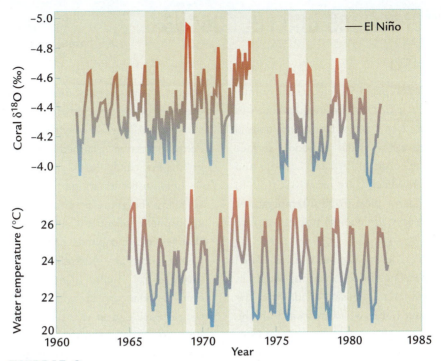

FIGURE 17-9
Coral δ¹⁸O records: the eastern tropical Pacific

Corals from the Galápagos Islands in the eastern Pacific tend to record low (negative) $\delta^{18}O$ values during warm El Niño years. (ADAPTED FROM R. B. DUNBAR ET AL., "EASTERN PACIFIC SEA SURFACE TEMPERATURES SINCE 1600 A.D.: THE $\delta^{18}O$ RECORD OF CLIMATE VARIABILITY," *PALEOCEANOGRAPHY* 9 [1994]: 291–315.)

shows that the late twentieth century is the warmest period for at least the last 400 years (Figure 17-10). An interval of heavier (colder or drier) $\delta^{18}O$ values during the 1800s was preceded by one of greater warmth during the 1700s, although the earliest part of the record is based on just a handful of sites.

17-4 Other Historical Observations

As human civilizations developed, people in some regions began to keep records of climatic phenomena for reasons unique to the cultures in which they lived. Climate scientists who attempt to use these

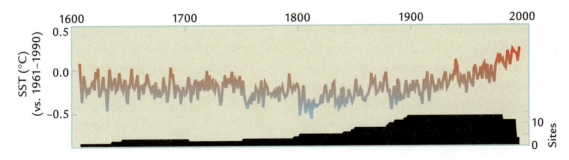

FIGURE 17-10
Stacked δ¹⁸O records from corals

A record produced by stacking individual $\delta^{18}O$ records from corals in the topical Pacific and Indian Oceans shows lighter (warmer and/or wetter) values in the 1700s, heavier values in the 1800s and very light values in the 1900s. (ADAPTED FROM R. WILSON ET AL., "TWO-HUNDRED-FIFTY YEARS OF RECONSTRUCTED AND MODELED TROPICAL TEMPERATURES," *JOURNAL OF GEOPHYSICAL RESEARCH* 111 [2006]: C10007, DOI:10.1029/2005JC003188.)

Climate Interactions and Feedbacks

El Niño and ENSO

The **El Niño** circulation pattern interrupts the normal circulation of the Pacific Ocean at irregular intervals generally ranging from 2 to 7 years. During the normal years that occur between El Niño episodes, surface temperatures along the coasts of Peru and Ecuador and in the eastern equatorial Pacific are near 18°C (50°F) in winter—far cooler than typical tropical temperatures (25°C, 77°F) and in fact the coolest tropical surface water on Earth.

Lower temperatures during normal (non-El Niño) years result from upwelling driven by strong winds in southern hemisphere winter (August). Winds from the south drive warm surface waters westward away from the coast of South America, and cooler water wells up from below. The winds then turn to the west near the equator as trade winds and drive warm surface water toward the southwest, causing cool water to well up near the equator. Upwelling water brings nutrients to the surface, supplying food to an ecosystem ranging from plant plankton to fish (anchovy and tuna), sea birds, and marine mammals (seals and sea lions). Non-El Niño years are also dry along the coast of South America because cool upwelling waters are a poor source of water vapor for the atmosphere. As a result, the coastal deserts of Peru and Chile are normally among the driest regions on Earth.

During El Niño years, strong southerly winds fail to blow in winter along the eastern and tropical Pacific, upwelling does not occur, and the surface waters along the South American coast warm by 2° to 5°C. Without upwelling, the plankton populations crash, and most fish die or migrate away. Without fish, sea birds on tropical islands cannot feed their young, and they abandon their nests to fly elsewhere in search of food. In severe El Niño years, a significant fraction of the year's populations of young sea birds and mammals dies. Ocean warming near the coastal South American deserts provides a large source of moisture, and rain falls in cloudbursts that produce flash floods in regions with little or no natural vegetation cover to absorb the water. The warm rains also favor the spread of tropical diseases such as malaria and cholera among humans.

Even though the El Niño circulation pattern reaches its height during the southern hemisphere winter in August, the first hint of unusual warming of the surface ocean is often detected during the previous southern summer, in December near Christmas. For this reason, Peruvian fishermen named this phenomenon "El Niño," or "the boy child."

El Niño events are part of a larger-scale circulation spanning the entire tropical Pacific. In the 1920s, the atmospheric scientist Gilbert Walker found matching changes in atmospheric pressure between the western Pacific (northern Australia and Indonesia) and the south-central Pacific island of Tahiti. High pressures over Australia correlate with low pressures in the south-central Pacific, and vice versa. Low atmospheric pressures are associated with rising air motion and rainfall, while high surface pressure is associated with sinking motion and dry conditions (see Chapter 2). The opposing pressure trends through time across the tropical Pacific are part of an enormous circulation cell called the **southern oscillation**. Sinking and rising motions occur at opposite times over northern Australia and Indonesia in the west and across the south-central Pacific in the east.

El Niño and the southern oscillation are linked. El Niño years, with warm ocean temperatures and heavy rains in Peru, are times of high pressure and drought over northern Australia, and of low pressures and high rainfall in the south-central Pacific. Non-El Niño years, with cool ocean temperatures near South America, are times of low pressure and increased rainfall in northern Australia, and of higher pressures and reduced rainfall in the south-central Pacific. This linked circulation is known as **ENSO** (El Niño–Southern Oscillation).

The physical link between these two systems occurs in the lower atmosphere and the upper ocean. Strong trade winds that cause upwelling in the eastern Pacific during non-El Niño years also drive warm surface water westward across the tropical Pacific. Warm water piles up in the western Pacific at a height several tens of centimeters above the level of the eastern Pacific and forms a natural source of moisture for evaporation and precipitation in northern Australia and Indonesia. Some of the rising air flows eastward at high elevations and sinks in the east-central Pacific, contributing to the normally cooler and drier conditions near South America.

Box 17-2

During El Niño years, without strong trade winds pushing water westward, some of the pool of warm water in the western Pacific flows back eastward and becomes a source of latent heat and moisture for local rains. When the flow reaches the Americas, it deflects northward and southward along the coast, bringing warmer and wetter conditions north to California and south to Peru, along with reduced upwelling. In the western Pacific, slightly cooler conditions during El Niño years result in drying in northern Australia and Indonesia. Eventually, El Niño conditions subside and the tropical ocean reverts to its normal state. During some years, it overshoots this normal state and produces abnormally cool sea-surface temperatures in the eastern Pacific. This overshoot is called **La Niña,** or "the girl child."

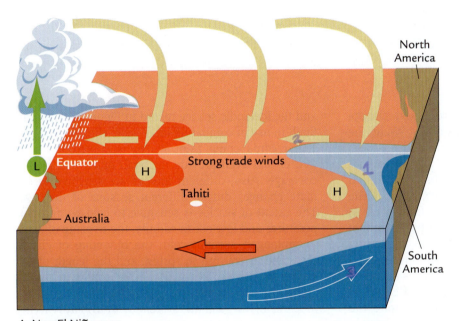

A Non El Niño year

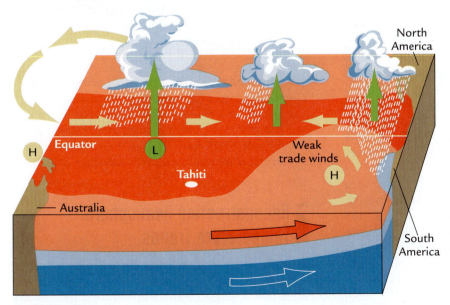

B El Niño year

Circulation in El Niño and non-El Niño years El Niño events change atmospheric and ocean circulation across the entire Pacific Ocean, from Australia and Indonesia to the west coast of South America.

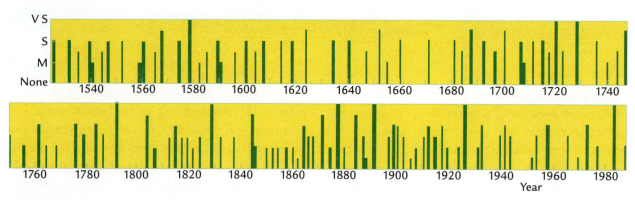

FIGURE 17-11
Historical records of El Niño

Historical chronicles of unusual phenomena along the South American coast reveal El Niño events since the early sixteenth century. (VS = very strong, S = strong, M = moderate.)

(ADAPTED FROM W. H. QUINN AND V. T. NEAL, "THE HISTORICAL RECORD OF EL NIÑO EVENTS," IN *CLIMATE SINCE A.D. 1500*, ED. R. S. BRADLEY AND P. D. JONES [LONDON: ROUTLEDGE, 1992].)

early historical records to reconstruct past climate have to weigh their reliability carefully. Were the people recording the phenomena continuously present through the years, or were they relying on second-hand information? Was the task of observing passed to dependable observers through the generations? The phenomena recorded also vary widely from place to place. They include: the frequency and timing of first and last frosts and of droughts and floods, the timing of autumn lake or river freeze-up and spring ice break-up, the dates of the first flowering of shrubs and trees (such as cherry blossoms), and the dates of harvests.

Immediately after conquering the Inca empire, the Spanish began to make environmental observations along the coast of Peru. Ships' logs are the major source of information from this region, supplemented by records kept by missionaries and others. These observations include responses now understood to result from El Niño events, including sea-surface temperatures warmer than normal, reduced catches of anchovy and other fish, departure of sea birds from coasts and islands, unusually heavy rains and floods, and outbreaks of cholera and malaria. The records start in 1525 and continue through most of the twentieth century (Figure 17-11).

El Niño events during this interval have been ranked by historians on a qualitative scale ranging from none to very severe. With 115 events in 465 years, the time between successive El Niño events averages 4 years, but the timing varies widely around this number. El Niño events can cluster within certain intervals (the late nineteenth century) but be rare in others (the mid-seventeenth century).

For a number of reasons, histories of climatic phenomena such as the El Niño record in Figure 17-11

and the Icelandic sea ice record in Figure 17-1 are difficult to use in comparing large-scale climate changes. The records come from widely scattered locations that do not even provide regional, much less global, coverage. Also, different indices are sensitive to climate changes during different seasons of the year. The extent of sea ice is sensitive to cold winter temperatures and low ocean salinity, the freezing of lakes to prolonged autumn cold, the blooming of cherry blossoms to warmth in spring, and the length of the growing season to unusual overnight frosts in spring and autumn. As a result, these indices record changes in parts of the climate system with widely varying response times. For this reason, historical observations recorded before the era of weather instruments give us only anecdotal information about climate changes during preceding centuries.

In Summary, exploration of a variety of proxy climate indicators (ice cores, tree rings, and tropical corals) and of a few historical records has improved our knowledge of climate changes that occurred during the past 1,000 years. Despite these efforts, coverage of global climatic trends falls well short of global.

▶ Reconstructing Hemispheric Temperature Trends

Several attempts have been made to synthesize high-resolution records from ice cores, tree rings, and corals into a single estimate of northern hemisphere temperature changes during the last millennium. These efforts face daunting problems. Proxy records are all linked in some way to temperature, but they are also affected in complex ways by other climatic, biological, and ecological factors. Because the links between the proxy variations and climatic variables are complicated, scientists are forced to rely on statistical approximations that simplify the true underlying relationships. Another problem is that the proxy indicators may be sensitive primarily to temperature changes in summer or winter or spring, yet the reconstructions combine them into a single "common" temperature trend. Still another problem is the scarcity of proxy data. More than 100 records are available for the last century or two, but only a few dozen for the year 1500, and little more than half a dozen for the year 1000.

Despite these complications, the reconstructed signals (Figure 17-12) all show a gradual but erratic temperature decline for almost 900 years, followed by a dramatic warming that began just before 1900. The amplitude of the changes that occurred prior to the instrumental record of the last 125 years varies by a factor of two or more among the reconstructions. Reconstructions based on records that are more heavily weighted toward higher latitudes show the largest range of variations, while those with a greater representation of lower latitudes tend to have a smaller range. This difference is consistent with the fact that high latitudes are more climatically reactive than low latitudes because of the amplifying effects of albedo feedback from reflective surfaces of bright snow and sea ice (see Chapters 2 and 3).

Because of the sparse coverage of sites, all of the reconstructions inevitably have large uncertainties. The shaded region in Figure 17-12 shows one attempt to estimate the uncertainties in the temperature reconstruction through time around the trend shown in red. The estimated uncertainty is moderately large for recent centuries, but almost ten times

the size of the estimated temperature variations prior to 1500.

The better-constrained portions of all reconstructions show a cool interval equivalent to the peak of the Little Ice Age in the 1600s, 1700s, and 1800s. In comparison, the warming since the 1880s stands out as highly unusual both in the abruptness of the rate of change and in the level of warmth attained. This unprecedented warming will be examined more closely in Chapters 18 and 19.

Evidence for the existence of a supposed medieval warm period and a subsequent cooling into the Little Ice Age is more ambiguous, in part because of the large uncertainties in the records from those intervals. Records from different regions show widely varying trends. In the reconstructions that attempt to balance the latitudinal distribution of sites, the net century-scale cooling between 1000–1300 and 1400–1900 averages about 0.2–0.3°C, with larger fluctuations during particular decades. In addition, the different reconstructions often place the beginning of this cooling at different times.

When a reasonably complete "global" reconstruction becomes available, this cooling trend may not persist, or it may be smaller. Attempts to reconstruct meaningful temperature trends for the Southern Hemisphere are not yet possible because very few records are available. One very preliminary attempt at such a reconstruction shows no obvious cooling trend between 1000–1300 and 1400–1900. When this estimate was combined with one from the Northern Hemisphere to calculate a global average, the century-scale global cooling from the interval 1000–1300 to the interval 1400–1900 was reduced to about 0.1°C. If later reconstructions confirm this result, the small northern hemisphere cooling between the intervals 1000–1300 and 1400–1900 could turn out to be mainly a northern hemisphere oscillation in the North Atlantic Ocean region.

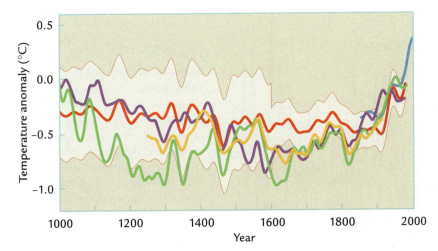

FIGURE 17-12
Northern hemisphere temperatures during the last millennium

High-resolution climatic reconstructions spanning all or part of the last millennium show a small gradual cooling for the first 900 years, followed by a large and abrupt warming in the twentieth century. Light shading indicates uncertainty in estimated temperature of reconstruction shown in red. (ADAPTED FROM P. D. JONES AND M. E. MANN, "CLIMATE OVER PAST MILLENNIA," *REVIEWS OF GEOPHYSICS* 42 [2004]: 2003RG000143.)

In Summary, reconstructions of northern hemisphere temperature during the last 1,000 years show that temperatures between 1400 and 1900 were cooler than during the last century (1900–2000). Evidence for an earlier interval of moderate warmth (a medieval warm period) between 1000 and 1300 is suspect because fewer records exist and responses vary widely in timing and amplitude from region to region. The notion of a medieval warm period, followed by a cooler Little Ice Age, is a valid description of trends that occurred from eastern Canada, across Greenland and Iceland, and east into northern Europe, but it may not characterize the changes across the rest of Earth's surface.

▶ Proposed Causes of Climate Change from 1000 to 1850

The northern hemisphere temperature reconstructions shown in Figure 17-12 show a small cooling that began between 1000 and 1300 and developed slowly until 1600 to 1800. This cooling has several possible explanations: orbital-scale forcing, millennial variability linked to a bipolar seesaw, solar variability, volcanic eruptions, and anthropogenic factors.

17-5 Orbital Forcing

Orbital forcing contributed to the gradual cooling that occurred in higher northern latitudes between 1000 and 1850. Over the last 6,000 years, portions of the circumarctic have cooled by 1–2°C because of decreasing summer insolation at both the tilt and precession cycles (see Chapter 14, Figures 14-18 and 14-19). If the rate of cooling were uniform over the entire 6,000 years, the fraction of the cooling that would have occurred between 1000 and 1850 would have amounted to between 0.15 and 0.3°C.

Allowing for the fact that temperature changes are amplified near the poles, the mean hemisphere-scale cooling would probably have been a factor of 2 or 3 smaller, or about 0.1°C, equivalent to about half the total amount observed in the reconstruction for the Northern Hemisphere (see Figure 17-12). The slow background cooling trend between 1000 and 1850 is consistent with gradual forcing from orbital changes.

17-6 The Millennial Bipolar Seesaw

Millennial oscillations have been proposed as another explanation for the northern hemisphere cooling during the last millennium, but climatic trends during the last 8,000 years have been highly irregular and local in extent, with no evidence of widespread oscillations (see Chapter 14). If the north polar cooling into the Little Ice Age was part of a small millennial oscillation, it should have been accompanied by a very small warming in the Antarctic region, based on the pattern typical of the large glacial-age oscillations (see Chapter 15). At this point, proxy coverage of the last 1,000 years in the Southern Hemisphere does not suggest such a pattern, but current coverage is insufficient to test this idea fully.

17-7 Solar Variability

The ^{14}C and ^{10}Be evidence examined in Chapter 15 suggested that solar variability does not cause changes in climate over millennial time scales, but the possibility was left open that changes in solar output could still play a significant role in changes at decadal and century scales.

One possible solar-terrestrial link for the last several centuries is tied to the 11-year **sunspot cycle**. Continuous and accurate observations of the number of dark sunspots visible on the Sun began with the invention of the telescope more than 400 years ago. The most obvious feature of the long-term record is the 11-year cycle (Figure 17-13).

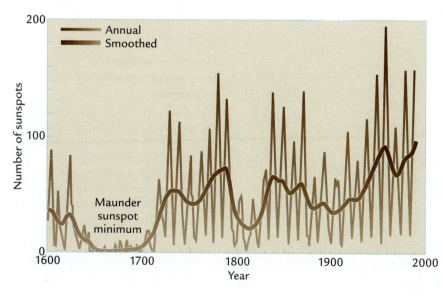

FIGURE 17-13

Sunspot history from telescopes

Measurements made with telescopes over the last several hundred years show that the number of sunspots increase and decrease at an 11-year cycle. During the interval known as the Maunder minimum, sunspots were absent for several decades.

(ADAPTED FROM C.-D. SCHONWIESE ET AL., "SOLAR SIGNALS IN GLOBAL CLIMATIC CHANGE," *CLIMATIC CHANGE* 27 [1994]: 259–81.)

Intuitively, it would seem likely that the presence of cooler dark spots on a hot bright surface would reduce the total amount of solar radiation emitted, but in fact the actual relationship is just the reverse. Most of the radiation emitted by the Sun streams out from its polar regions and from bright rings surrounding the sunspots, called **faculae** (Figure 17-14). During years when sunspots are abundant, the amount of radiation emitted in solar flares is at a maximum because mechanisms operating within the Sun simultaneously regulate both sunspots and net solar emissions. As a result, the amount of solar radiation arriving at Earth during sunspot maxima is at a maximum rather than a minimum. Based on satellite measurements in the past two decades, the range of variation in solar irradiance (other than very brief fluctuations) is about 0.11%.

A longer-term trend is also apparent in the sunspot record assembled from telescope observations, which show larger sunspot numbers in the last three centuries compared to the middle to late 1600s. An interval between 1645 and 1715 that had almost no sunspots is called the **Maunder sunspot minimum**, after the astronomer who discovered it (see Figure 17-13). A similar interval between 1460 and 1550 is called the **Sporer sunspot minimum**. Despite the relatively crude observational techniques available at those times, the existence of these two intervals is certain.

Some scientists initially proposed that variations in solar output during the last several centuries were considerably larger than the small range measured during the satellite era. They suggested that solar output might have been as much as 0.25–0.4% weaker than today during the Maunder and Sporer minima, with a gradual rise in irradiance during the last few centuries. Because these intervals of low sunspot activity persisted for decades, the slow-responding parts of the climate system would have had time to respond to them more fully compared to the brief 11-year cycle. This link seemed particularly appealing because the long sunspot minima generally occurred during times when northern hemisphere temperatures were considerably cooler than today. As a result, some scientists suggested that changes in solar irradiance accounted for 25% or more of the increase in temperature during the 1900s.

Later astronomical observations have failed to support this claim. Archival images of Sunlike stars from several observatories failed to show variations comparable to those proposed for the Sun. No evidence for variations greater than those at the 11-year cycle has been detected. Recent estimates place the contribution from solar irradiance changes since 1880 at less than 0.07°C, or at most 10% of the warming of the last century or so.

The search for possible solar effects on climate during this interval has not ended. Although most of the Sun's emissions arrive as visible or near-visible (ultraviolet and infrared) radiation, the Sun also sends out a plasma or ionized gas called "the solar wind," which interacts with Earth's stratosphere as it is deflected by Earth's magnetic field. One possibility under consideration is that the solar wind affects the formation of ozone, which in turn alters the formation of clouds in the troposphere and thereby affects climate at Earth's surface. Other possible links via the stratosphere continue to be explored.

The similarity of ^{14}C and ^{10}Be trends during the last 5,000 years seems to point to a common origin from solar changes (see Chapter 15). Scientists who have looked for a correlation between these isotopic trends and climatic proxies during recent millennia have met with mixed success. Temperature-sensitive changes in $\delta^{18}O$ within the last millennium show a substantial correlation with ^{10}Be variations, implying a solar role. On the other hand, the fact that no such relationship is evident in previous millennia of the Holocene weakens this case. The possibility also remains that the similar changes in ^{14}C and ^{10}Be are

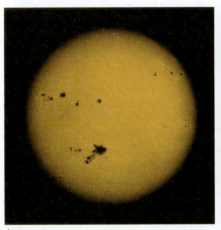
A

B

FIGURE 17-14

Sunspots and solar emissions

Intervals when sunspots were abundant (A) were also times when strong solar emissions from the bright margins of the sunspots sent increased levels of radiation to Earth (B). (NSO/AURA/NSF. REPRODUCED WITH PERMISSION FROM NATIONAL OPTICAL ASTRONOMY OBSERVATORY.)

both responses to changes occurring within the climate system (changes in ocean circulation, rate of ice accumulation, etc.).

> **In Summary**, changes in solar irradiance featured prominently in several initial attempts to explain climatic trends during the last millennium, but recent evidence suggests that irradiance changes were small and had little impact on climate during this interval.

17-8 Volcanic Explosions

Explosive eruptions of volcanoes cool climate over intervals of a few years. Volcanoes emit sulfur dioxide (SO_2) gas, which mixes with water vapor in the air and forms droplets and particles of sulfuric acid called **sulfate aerosols**. Highly explosive eruptions can reach 20 to 30 kilometers into the stratosphere, where the aerosols block some incoming solar radiation and keep it from reaching the ground. With solar radiation reduced, Earth's surface cools.

The latitude of eruption determines the geographic extent of their climatic impact. Volcanoes that erupt poleward of about 25° produce particles that mostly stay within the hemisphere in which the eruption occurs, limiting the cooling impact to that hemisphere. Explosions that occur in the tropics are redistributed by Earth's atmosphere to both hemispheres and can have a global impact on climate.

Ocean-island volcanoes with iron- and magnesium-rich compositions tend not to cause explosive eruptions but instead emit lava that flows across the land. Volcanic particles sent into the air by these eruptions stay within the troposphere and rarely reach the stratosphere. Within a few days, these particles are brought back down to Earth by precipitation. With so brief a time in the atmosphere, these particles cannot be widely enough distributed around the planet to produce large-scale effects on climate.

In contrast, volcanoes located along converging plate margins are fed by magmas richer in silica and other elements found in continental crust (see Chapter 5). Their eruptions are more explosive because the natural resistance of this kind of molten magma to internal flow causes internal pressures to build up to the point where highly explosive eruptions send volcanic particles up into the stratosphere, well above the level where precipitation can wash them out. Slow settling of these fine particles because of the pull of gravity takes several years, long enough for them to be distributed within a hemisphere or across the entire planet.

Sulfate aerosol concentrations in the stratosphere caused by explosive eruptions reach their maximum distribution within months and then begin to decrease as gravity gradually removes the particles.

This decrease follows an exponential trend: each year about half of the remaining particles settle out, and within two to three years aerosol concentrations are much reduced (Figure 17-15 top). The effect of these aerosols on temperature follows the same trend, with a maximum initial cooling that soon fades away (Figure 17-15 bottom). If several explosions follow within an interval of a few years, their impact on climate may be sustained for a decade or more.

Reconstructing the effects of older volcanic explosions on climate is difficult. Some idea about the magnitude of ancient eruptions can be obtained from the sizes of the craters left by explosions or from the volume of volcanic ash deposited nearby. The geographic area over which the ash is distributed may also provide clues about the power of the eruption and whether the volcanic particles might have reached the stratosphere.

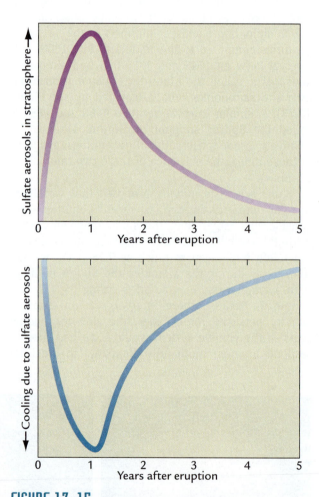

FIGURE 17-15
Volcanic explosions and cooling

Large volcanic eruptions launch sulfate aerosols into the stratosphere (top) and cool climate for a few years (bottom). (ADAPTED FROM R. S. BRADLEY, "THE EXPLOSIVE VOLCANIC ERUPTION SIGNAL IN NORTHERN HEMISPHERE CONTINENTAL TEMPERATURE RECORDS," *CLIMATIC CHANGE* 12 [1988]: 221–43.)

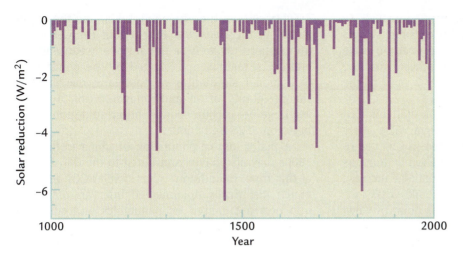

FIGURE 17-16
Volcanic explosions and solar radiation

One estimate of the effect of volcanic explosions during the last millennium in reducing the intensity of incoming solar radiation over annual to decadal intervals. (ADAPTED FROM T. CROWLEY ET AL., "MODELING OCEAN HEAT CONTENT CHANGES DURING THE LAST MILLENNIUM," *GEOPHYSICAL RESEARCH LETTERS* 30 [2003]: GL017801.)

But sulfur forms only a small and variable fraction of the total volume of erupted material, and the volume of ash cannot directly be used to estimate the amount of sulfur erupted. As a result, it is difficult to estimate the potential climatic effects of ancient eruptions reliably based on this kind of evidence.

Sulfate layers preserved in well-dated ice cores offer another way to assess past explosions (Figure 17-16). Comparison of these volcanic signals with the northern hemisphere temperature trends in Figure 17-12 indicates that sequences of large eruptions played a role in decadal-scale cooling intervals, for example during the 1200s and near 1450. Precise assessments of the temperature contributions from volcanic eruptions is difficult because of complications such as varying distances from the eruptive source to the ice sheets.

> **In Summary**, the more frequent clusters of eruptions after 1200 appear to have contributed to the small cooling into the Little Ice Age. Whether or not these episodes can account for the more gradual background cooling trend is not clear.

17-9 Greenhouse-Gas Effects on Climate

Changes in carbon dioxide concentrations in the atmosphere are another potential factor in the cooling of the last millennium prior to 1850 (Figure 17-17). CO_2 concentrations were relatively high (as much as 284 ppm) near 1,100 to 1,200, but fell as low as 274 ppm by 1600. Although this declining CO_2 trend broadly matches the reconstructed northern hemisphere cooling, the cooling was more gradual, while much of the CO_2 decrease was concentrated between 1525 and 1600. This short drop in CO_2 has generally been attributed to natural factors, but humans may also have played a role.

The natural explanation relies mainly on the fact that CO_2 solubility in seawater depends mainly on temperature, with lower temperatures increasing the solubility and allowing the ocean to hold larger amounts of CO_2. But the northern hemisphere cooling estimated for the 1525–1600 interval in most

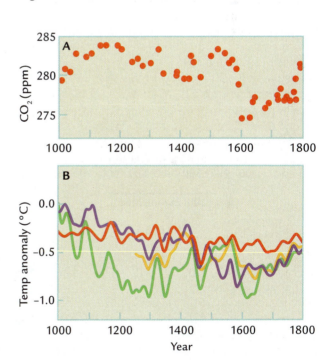

FIGURE 17-17
CO_2 and northern hemisphere temperature

Atmospheric CO_2 and northern hemisphere temperature signals show downward trends between 1000 and the 1800s, but the CO_2 drop is concentrated mainly in the interval 1525–1600. (CO_2 CONCENTRATIONS ADAPTED FROM EPICA COMMUNITY MEMBERS, "EIGHT GLACIAL CYCLES FROM AN ANTARCTIC ICE CORE," *NATURE* 429 [2004]: 623–28; RECONSTRUCTED TEMPERATURE TRENDS ADAPTED FROM P. D. JONES AND M. E. MANN, "CLIMATE OVER PAST MILLENNIA," *REVIEWS OF GEOPHYSICS* 42 [2004]: 2003RG000143.)

reconstructions in Figure 17-12 was too small to explain the CO_2 drop recorded in the ice cores. Models that simulate interactions between the carbon cycle and the rest of the climate system indicate that the 1525–1600 cooling explains less than 2 ppm of the observed 10-ppm CO_2 decrease.

The other proposed explanation is that this CO_2 decrease was anthropogenic in origin. The early anthropogenic hypothesis summarized in Chapter 16 is based on the premise that deforestation by humans was the primary cause of the anomalous increase in atmospheric CO_2 during the last 7,000 years (see Chapter 16, Figure 16-13A). From this point of view, any subsequent drop in the rising CO_2 trend could have had an origin connected to humans.

The drop in CO_2 values between 1525 and 1600 correlates with a terrible pandemic in the Americas, when a host of diseases carried by arriving Europeans killed tens of millions of Native Americans between 1492 and 1700, or 85–90% of the previous population (Figure 17-18). When the outbreaks of diseases occurred, most Native Americans had been farming the higher levees of river valleys or clearings in the forests. After the pandemics decimated the populations, the forests

grew back over the untended fields, and the carbon stored in the trees was taken from the CO_2 in the atmosphere (as well as the surface ocean and vegetation in regions not affected by the pandemics). One model simulation of past land use supports the possibility of an anthropogenic link by showing large net carbon sequestration on a global basis between 1525 and 1600.

A major loss of population in China in the early 1600s may also have contributed to the CO_2 decrease at that time. In addition, other smaller CO_2 decreases match historically documented intervals of population decrease caused by pandemics and civil strife (see Figure 17-18).

Humans may also have played a role in the cooling of the last millennium by changing the surface cover of the land they cleared. At high latitudes, replacement of darker woodlands (especially evergreen trees) by lighter pastures and croplands increases the amount of incoming solar radiation reflected back to space, especially when snows lay on the cleared croplands and pastures. The first major clearance of evergreen forests anywhere on Earth occurred in the area of the eastern Baltic and Russia during the cooling between 1000 and 1850.

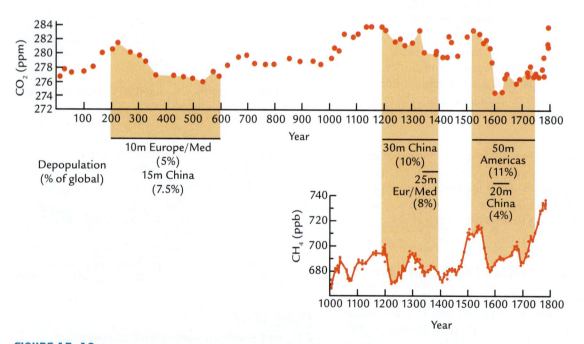

FIGURE 17-18

Atmospheric CO_2 decreases and catastrophic population losses

Times of massive human mortality caused by pandemics and civil strife correlate with intervals of decreasing CO_2 in the atmosphere. (CO₂ RECORD ADAPTED FROM C. MCFARLING MEURE ET AL., "LAW DOME CO₂, CH₄ AND N₂O ICE CORE RECORDS EXTENDED TO 2000 YEARS BP," *GEOPHYSICAL RESEARCH LETTERS* 33 [2006]: L14810, DOI:10.1029/2006GL026152; POPULATION LOSSES FROM C. MCEVEDY AND R. JONES, *ATLAS OF WORLD POPULATION HISTORY* [NEW YORK: PENGUIN BOOKS, 1978]; AND FROM W. M. DENEVAN, "THE PRISTINE MYTH: THE LANDSCAPE OF THE AMERICAS IN 1492," *ANNALS OF THE ASSOCIATION OF AMERICAN GEOGRAPHERS* 82 [1992]: 369–85.)

In Summary, the estimated cooling from 1,000 years ago through the 1800s is small—a few tenths of a degree Celsius. The slow steady background cooling at higher northern latitudes was likely due to decreasing summer insolation in the Arctic region, while volcanic eruptions account for shorter-term cooling episodes lasting a few years to a decade or more. Decreases in CO_2 caused by catastrophic pandemics and mass human mortality also overprinted the slow cooling trend during some intervals.

Before reliable cause-and-effect conclusions can be drawn, greater geographic coverage of proxy sites is needed to define the global climatic response, especially in the Southern Hemisphere. In contrast, no such ambiguity exists about the large, rapid, and global warming since 1850.

Key Terms

Little Ice Age (p. 335)

medieval warm period (p. 336)

lichen (p. 336)

dendroclimatology (p. 341)

calibration interval (p. 343)

El Niño (p. 346)

southern oscillation (p. 346)

ENSO (p. 346)

La Niña (p. 347)

sunspot cycle (p. 350)

faculae (p. 351)

Maunder sunspot minimum (p. 351)

Sporer sunspot minimum (p. 351)

sulfate aerosols (p. 352)

Review Questions

1. What evidence indicates a cooler climate in Europe and nearby regions during the Little Ice Age?

2. What evidence from ice cores suggests that the warming during the twentieth century reached levels unprecedented over the last 1,000 years?

3. Why are the rings of environmentally stressed trees ideal for detecting climate signals?

4. How could rising CO_2 levels complicate interpretations of changes recorded in tree rings?

5. What factors influence $\delta^{18}O$ values recorded in corals, and how?

6. What are the major characteristics of El Niño years in comparison with years of normal circulation?

7. How does the latitudinal distribution of proxy sites affect hemispheric temperature reconstructions?

8. What is the connection between sunspots and solar radiation sent to Earth?

9. Over what length of time can large volcanic explosions alter climate?

10. How could disease have affected climate during the last millennium?

Additional Resources

Basic Reading

Alverson, K. D., R. S. Bradley, and T. F. Pedersen. 2003. *Paleoclimate, Global Change, and the Future*. Berlin: Springer.

Bradley, R. S., and P. D. Jones. 1992. *Climate Since A.D. 1500*. London: Routledge.

Grove, J. M. 1988. *The Little Ice Age*. London: Methuen.

International Geosphere-Biosphere Program (IGBP) Web site. http://www.igbp.net. Last accessed March 13, 2013.

Past Global Changes Project (PAGES) Web site. http://www.pages-igbp.org. Last accessed March 13, 2013.

Advanced Reading

D'Arrigo, R. D., and G. C. Jacoby. 1993. "Secular Trends in High Northern Latitude Temperature Reconstructions Based on Tree Rings." *Climatic Change* 15: 163–77.

Mann, M. E., R. S. Bradley, and M. K. Hughes. 1999. "Northern Hemisphere Temperatures During the Past Millennium." *Geophysical Research Letters* 26: 759–62.

North, G. A., et al. 2006. *Surface Temperature Reconstructions for the Last 2,000 Years*. Washington, DC: The National Academies Press.

Quinn, W. H., V. T. Neal, and S. E. Antunez de Mayolo. 1987. "El Niño Occurrences over the Past Four and a Half Centuries." *Journal of Geophysical Research* 92: 14449–61.

Thompson, L. G., E. Mosley-Thompson, M. E. Davis, P. N. Lin, T. Yao, M. Sdyurgerov, and M. Dai. 1993. "Recent Warming: Ice Core Evidence from Tropical Ice Cores, with Emphasis on Central Asia." *Global and Planetary Change* 7: 145–55.

Wilson, R., A. Tudhope, P. Brohan, K. Briffa, T. Osborn, and S. Tett. 2006. "Two-Hundred-Fifty Years of Reconstructed and Modeled Tropical Temperatures." *Journal of Geophysical Research* 111: C10007, doi:10.1029/2005JC003188.

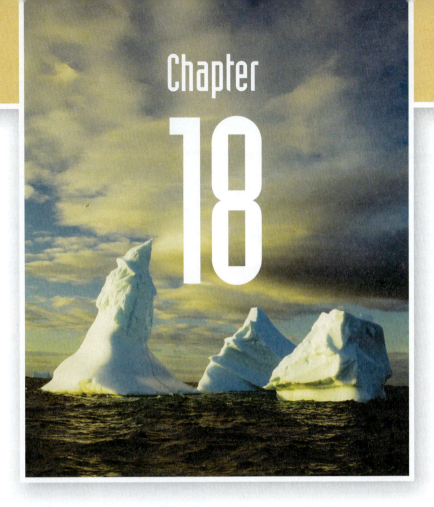

Chapter 18

Climatic Changes Since 1850

The climatic proxies summarized in Chapter 17 show that the rate and amount of warming since 1900 has been unprecedented compared to changes during the previous 1,000 years. Prior to the late 1900s, climatic changes were mostly measured with instruments at Earth's surface, including thermometers on land and at sea, and field studies of glacier flow and length using stakes driven into the ice. More recently, changes in global sea level have been reconstructed from tide gauges after correction for the lingering effects of the last deglaciation. Over the last few decades, satellite measurements of the extent of sea ice and snow cover, of growing-season length, and of ice sheet mass balance have also become available, along with other information from remotely deployed instruments. All of these measurements show that climate warmed significantly during the last century and the early part of the current one. The measurements also reveal shorter-term (multiannual to decadal) oscillations in climate across smaller regions of the globe.

357

Reconstructing Changes in Sea Level

One key source of information on global climatic trends over the last 150 years is the average level of the ocean, but reconstructing changes in global sea level is complicated by lingering effects from ice sheets that melted thousands of years ago. In addition, recent sea level changes are the result of several factors, including melting of land ice and changes in ocean temperature.

Beginning as early as the late eighteenth century, seaport towns and cities installed **tide gauges** to measure changes in the level of the ocean caused by tides and large storms. The immediate goal of these efforts was to understand local sea level changes well enough to build structures to protect communities from flooding. Tide gauges were most common in seaports in Europe and along the East Coast of the United States, areas that had begun to industrialize long before most of the rest of the world (Figure 18-1). Some 100 to 200 years later, these records show not only the short-term changes caused by tides and storms but also longer-term trends of change during the decades and centuries since their installation.

Deriving sea level trends from tide gauge records is difficult. Tide gauge records in some regions indicate rapid sea level falls, while others show a slow sea level rise, and still others indicate faster rises. At first, it doesn't seem to make sense that such different trends could occur. The world ocean is one interconnected body of water, and it would seem that long-term global sea level should rise or fall by the same amount everywhere, rather than rising in one place and falling in another.

Some of these regional differences result from modern processes that vary regionally. Areas of active tectonic uplift caused by mountain building or of active subsidence caused by recently added sediment loads (such as river deltas) have to be avoided in determining the actual change in the level of the ocean. In addition, adjustments must also be made for human effects such as subsidence caused by pumping of groundwater and impoundment of rainfall runoff in reservoirs behind dams.

18-1 Fading Memories of Melted Ice Sheets

By far the greatest problem in reconstructing past sea levels, however, is the fact that the land and ocean bedrock still retains a memory of the ice sheets from the most recent glaciation. Even though the last remnants of the glacial maximum ice sheets finished melting near 10,000 years ago in Europe and 6,000

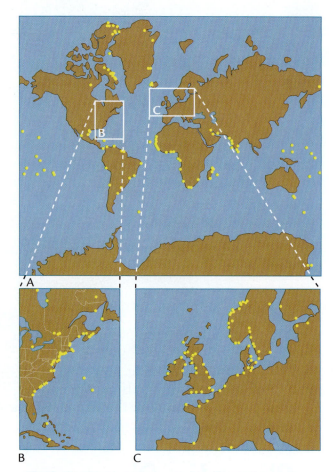

FIGURE 18-1
Tide gauge stations

Tide gauge records spanning several decades to as much as two centuries are available from hundreds of coastal locations on Earth's surface (A). The largest concentrations are in eastern North America (B) and northwestern Europe (C). (ADAPTED FROM A. M. TUSHINGHAM AND W. R. PELTIER, "ICE-3G: A NEW GLOBAL MODEL OF LATE PLEISTOCENE DEGLACIATION BASED UPON GEOPHYSICAL PREDICTIONS OF POST-GLACIAL RELATIVE SEA-LEVEL CHANGES," JOURNAL OF GEOPHYSICAL RESEARCH 96 [1991]: 4497–523.)

years ago in northern Canada, the rock in Earth's upper mantle is still in the process of adjusting to the previous load of ice. This bedrock "memory" causes different behaviors in the present movements of Earth's land and seafloor surfaces. The types of long-term change in relative sea level defined by tide gauges occur in three geographic clusters (Figure 18-2).

One group of tide gauges shows rapid drops in relative sea level in recent centuries (Figure 18-3). These gauges are located in regions that were once directly beneath the ice sheets, such as the Hudson Bay region of Canada and the Baltic Sea region of Scandinavia. Relative sea level is now falling rapidly in these regions primarily because the bedrock is still rebounding from the removal of the ice sheet load

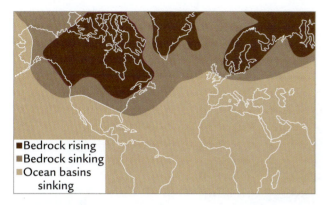

FIGURE 18-2
Patterns of sea level change

Relative sea level today is changing in several ways regionally because of bedrock movement. Bedrock is rapidly rising in areas formerly covered by thick ice and sinking in regions surrounding the former ice sheets. Farther from the ice sheets, ocean basins are sinking under the added weight of meltwater. (ADAPTED FROM A. M. TUSHINGHAM AND W. R. PELTIER, "ICE-3G: A NEW GLOBAL MODEL OF LATE PLEISTOCENE DEGLACIATION BASED UPON GEOPHYSICAL PREDICTIONS OF POST-GLACIAL RELATIVE SEA-LEVEL CHANGES," *JOURNAL OF GEOPHYSICAL RESEARCH* 96 [1991]: 4497–523.)

The enormous load of glacial ice thousands of years ago depressed bedrock surfaces beneath the central parts of the ice sheets by as much as 1 km, causing deep rock to flow slowly outward at great depths. Later, when the ice melted, rock slowly flowed back into this region, gradually raising the depressed bedrock toward its former elevation. Even now, thousands of years after this ice melted, rates of relative sea level fall caused by ongoing bedrock rebound are locally as high as 10 mm a year.

As a result of this slow bedrock rebound, ancient beach ridges surround the lower-lying parts of Hudson Bay (Figure 18-4A). The modern beach lies at sea level, with older beach ridges at successively higher elevations away from the coast of the bay (Figure 18-4B). The highest-elevation ridge dates to 7,000 ^{14}C years ago, just after the last ice melted from this area and ocean water flooded back into the depression in modern Hudson Bay that was left by the ice. The stair-step series of beach ridges shows that the land beneath the former ice sheet has been rising for thousands of years, with the rates gradually slowing as the bedrock memory of the ice sheet load has weakened.

The second major group of tide gauge responses shows a relatively fast rate of sea level rise. This group is clustered in a halo pattern surrounding the former ice sheets but extending well beyond the ice margins (see Figure 18-2). In North America, these gauges are found along the east coast from southern New England south to Florida, and in Europe in a

thousands of years ago. A local rise in bedrock means a local fall in *relative* (but not global) sea level.

Bedrock at depths of 100 to 200 km has a slow viscous component of response, and it takes many thousands of years to recover fully from loads imposed upon it or removed from it (see Chapter 10).

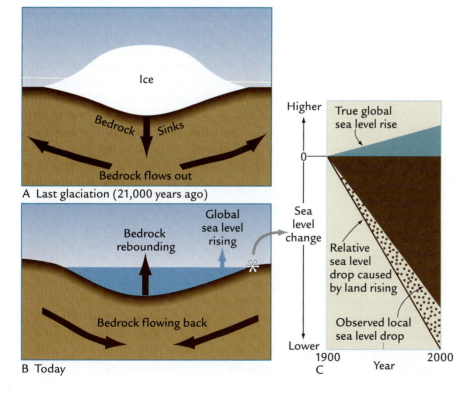

A Last glaciation (21,000 years ago)

B Today

C

FIGURE 18-3
Bedrock rebound and sea level fall

In the regions where ice sheets once were present, relative sea level is rapidly falling today. Bedrock in these areas is still rebounding in response to the earlier melting of ice, and the rebound of the land overwhelms the true global rise of sea level.

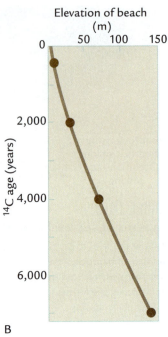

A   B

FIGURE 18-4
Old beach ridges

A series of old beach ridges surrounds Hudson Bay (A). The beaches increase in age with
elevation because the land has been slowly rising for 7,000 years (B). (A: COURTESY OF CLAUDE HILLAIRE-
MARCELL, UNIVERSITY OF QUEBEC, MONTREAL; B: ADAPTED FROM W. R. PELTIER AND J. T. ANDREWS, "GLACIAL-ISOSTATIC
ADJUSTMENT. I. THE FORWARD PROBLEM," *GEOPHYSICAL JOURNAL OF THE ROYAL ASTRONOMICAL SOCIETY* 46 [1976]: 605–46.)

narrower band across England, France, and northern
Germany.

The rapid present rise of sea level in these regions
is also caused by a memory of the glacial maximum
ice sheets, even though these areas were not located
directly beneath the ice loads. During glacial times,
the deep rock displaced from beneath the center of
the ice sheet load had to go somewhere. Flowing outward from the margins of the ice sheets, it caused an
increase in the elevation of the land, called a **peripheral forebulge** (Figure 18-5). As a rough analogy,
the weight of a person sitting on a partly inflated air
mattress in water tends to submerge the center of the
air mattress, but the excess air pushed to the edges
of the mattress causes the edges to bulge up out of
the water.

After the ice sheets melted, the rock displaced
beyond the ice margins gradually flowed back into
the region where the ice sheet had formerly been.
This return flow caused the peripheral forebulge to
collapse and the land surface to sink (see Figure
18-5). Because the land in the region of the collaps-
ing forebulges is still sinking, this local bedrock effect
roughly doubles the rate of relative sea level rise
in these regions. Dating of older (now submerged)
beaches in these regions indicates that this pattern of

unusually rapid rise in relative sea level has persisted
for thousands of years.

The third group of tide gauge responses comes
from coastlines located far from the northern hemisphere ice sheets (Figure 18-6). Relative sea level in
these regions is rising at rates slightly less than the
global rate of sea level rise. It might seem that regions
located so far from the glacial ice sheets should be
free of memory effects from the ice, but they are not
because the return of glacial meltwater to the oceans
has added an extra load on the bedrock beneath the
ocean floor in these regions.

At maximum size, the ice sheets extracted from
the world ocean a layer of water some 110–125 m
thick. With this water load removed, the average level
of the crust in the ocean basins rose by over 30 m
compared to the level of the nearby continents that
were not directly affected by the weight of ice sheets
or ocean water. When this layer of water returned
to the oceans during ice melting, it loaded down the
ocean crust and caused it to sink (see Figure 18-6).
This slow sinking of the ocean crust is still going on
today (because of the viscous memory effect), and it
counteracts a small part of the true rise of global sea
level, producing a slightly reduced rise in relative sea
level in regions far from the ice sheets.

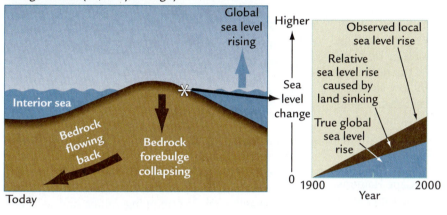

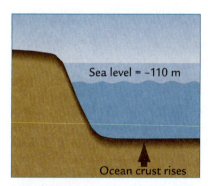

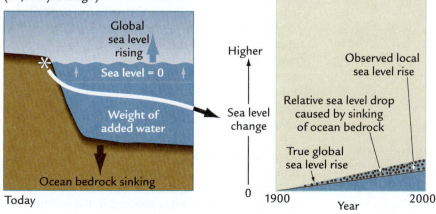

FIGURE 18-5
Bedrock sinking and sea level rise

In regions surrounding glacial ice sheets, relative sea level is rising rapidly today. The continued flow of bedrock away from these areas and into the former ice sheet centers causes the land to sink and adds to the true global rise of sea level.

FIGURE 18-6
Ocean bedrock sinking and sea level rise

In coastal regions far from glacial ice sheets, relative sea level is not rising as fast as the true global average. The continued sinking of ocean bedrock under the weight of meltwater counteracts a small fraction of the true global rise in sea level.

In Summary, while the great glacial ice sheets are long gone, they are not forgotten, at least not by the bedrock and the shorelines. Bedrock memories of these shifting loads of ice and water are the major obstacle to determining the true rate of global sea level rise during the past century or more. Unfortunately, many of the longest and most reliable tide gauge records were located in just those regions of Europe and North America where the lingering overprints from the ice sheets are largest.

Attempts to remove all these complications from the melted ice sheets indicate that sea level rose by about 17 cm during the 1900s (Figure 18-7). Humans have been adjusting to this slow rise of the ocean for years. In coastal plain regions with very low slopes, a 17-cm rise in sea level within a century can translate into an advance of 1,700 cm (17 m, or almost 60 ft) across the land. Many lighthouses built along coastal land a century or more ago now sit marooned in the ocean, protected by constructed boulder walls, at least for the near future. Some, such as Cape Hatteras Light in North Carolina, have been moved inland to better-protected positions.

Because this rise in sea level has been tied to several aspects of the warming that has occurred during the last century or more, the next two sections in this chapter explore the instrumental and satellite evidence that document this recent warming and the more recent rate of sea level rise.

Other Instrumental Records

The "modern" era of instrumental measurements of climate can be divided into two parts: (1) an earlier period in which measurements of climate were made using land-based instruments invented during the era of scientific exploration that began centuries ago; and (2) an era extending back about four decades during which satellites in space have sensed climatic changes remotely.

18-2 Thermometers: Surface Temperatures

Thermometers have been used to measure air temperature at a few locations in Eurasia and North America for over 200 years (Figure 18-8). During this time, the surface temperature of the ocean has been measured along heavily traveled shipping routes and at ocean islands, but with large gaps in coverage at middle and high latitudes of the Southern Hemisphere because of frequent storms and extensive sea ice. Only since the late 1800s have enough stations been recording temperature to permit reasonably reliable estimates of the surface temperature of the entire planet.

Measurements both on land and at sea have been difficult. Ocean temperatures were once measured by scooping up seawater in a canvas bucket and inserting a thermometer. If a few minutes elapsed between collecting the water and measuring its temperature, evaporation cooled the water by several tenths of a degree Celsius. More reliable measurements came later from thermometers embedded in the outer parts of the intake valves that drew in seawater to cool the ship's engines.

On land, the largest complication has been the growth in populations around many land stations. Near large towns and cities, the spread of asphalt surfaces and the loss of vegetation has led to increased absorption of solar radiation during the day and greater back radiation of heat at night. These changes caused an extra warming effect at urban stations. Although this warming reflects real temperature changes at these stations, it is not characteristic of changes across much more extensive rural areas. Care must be taken not to project this bias, called the **urban heat island** effect, into rural regions. Fortunately, stations in regions of little or no population growth provide a check on the heat island effect. Adjusting for this bias has reduced initial estimates of warming trends during the last century.

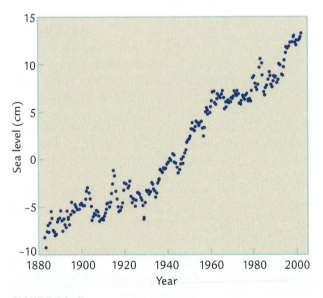

FIGURE 18-7
Global rise in sea level during the twentieth century

Sea level rose by ~17 cm during the 1900s because land ice has melted and seawater has warmed and expanded. (ADAPTED FROM S. JEVREJEVA ET AL., "NONLINEAR TRENDS AND MULTIYEAR CYCLES IN SEA LEVEL RECORDS," *JOURNAL OF GEOPHYSICAL RESEARCH* 111 [2006]: DOI:10.1029/2005JC003229.)

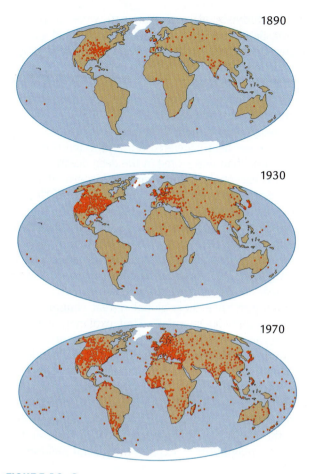

FIGURE 18-8
Temperature stations
Coverage of land-based stations that measure surface temperature increased significantly during the twentieth century. (NATIONAL CLIMATE DATA CENTER, NOAA, ASHEVILLE, NC.)

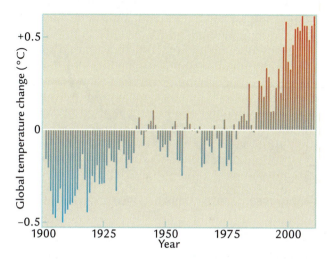

FIGURE 18-9
Change in surface temperature since 1900
Reconstructions of global surface temperature based on surface station thermometer measurements show a warming trend of 0.8°C since 1900, interrupted by a small cooling from the late 1940s to the mid-1970s. (NATIONAL CLIMATE DATA CENTER, NOAA, ASHEVILLE, NC.)

Reconstructions of global temperature during the last century by different groups are very similar, mainly disagreeing in the early 1900s when station coverage was still sparse. Temperatures have warmed by about 0.8°C over the last century (Figure 18-9). Temperatures were considerably cooler before the early 1900s, rose quickly during the 1920s to early 1940s, stabilized or fell slightly from the late 1940s through the late 1970s, and have again risen abruptly since 1980. The decade between 2000 and 2010 was warmer than the 1990s, which were warmer than the 1980s, which were warmer than the 1970s.

18-3 Subsurface Ocean Temperatures

The ocean has the capacity to store enormous amounts of heat for long intervals of time (see Chapter 2). Changes in surface climate do not easily penetrate below the wind-mixed upper 100 meters, but important information on deeper ocean trends during the last half-century can be extracted by painstakingly examining millions of subsurface temperature profiles.

These observations have provided a detailed picture of the slow penetration of heat from the atmosphere into the subsurface layers. Slow downward molecule-by-molecule diffusion has transferred some of the surface heat to depths of a few hundreds of meters. In addition, near-horizontal movement of water sinking at higher latitudes has transferred even more heat into the somewhat deeper subsurface ocean. Still slower penetration to depths of 1,000 m or more has also occurred in areas of deep overturning like the subpolar North Atlantic Ocean.

Large increases in stored heat are evident in records spanning the last half-century (Figure 18-10). This trend, expressed in joules (the amount of work required to produce one watt of power for one second) is equivalent to a net ocean temperature increase of ~0.07°C. Although the ocean warming since the middle 1900s is much smaller than the rise in surface air temperature, the total amount of heat stored in the ocean exceeds that in the atmosphere by more than a factor of ten. This large storage of ocean heat is evidence of a marked change in the heat balance of the entire climate system compared to earlier decades.

18-4 Mountain Glaciers

Today, mountain glaciers cover 680 km² of Earth's land surface and represent about 4% of the total surface area of land ice on Earth today (Figure 18-11). Mountain glaciers at middle and high latitudes

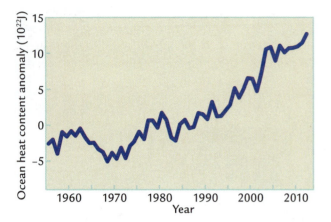

FIGURE 18-10
Subsurface ocean warming
The mean ocean heat content of the upper 700 m of the world ocean has increased since 1955. (ADAPTED FROM S. LEVITUS, NODC.)

respond to local climate, primarily changes in summer temperature, and also to variations in winter snowfall. At lower latitudes, solar radiation and precipitation are also important. Because of these differences in sensitivity, mountain glaciers in different regions show varying behavior. Individual glaciers may respond to climate over intervals that range from about a decade to as much as several hundred years. The response times of most mountain glaciers fall in the range of 10 to 40 years.

Despite this wide range of possible responses, historical observations of the lower limits of glaciers on mountain sides between 1860 and 1900 show that thirty-five of the thirty-six glaciers examined were already in retreat. Between 1900 and 1980, 142 of the 144 glaciers analyzed were retreating. The average retreat of all glaciers between 1850 and 2000 was ~1,750 meters, or just over a mile (Figure 18-12). In some cases, studies of glacier lengths have been supplemented by analyses of ice thickness that permit calculations of full glacier volume. The energy used to melt these ice sheets absorbed part of the excess heat generated during the industrial era, but far less than was stored in the deep ocean.

Rare exceptions to this general pattern of retreat exist. Some glaciers in the mountains of Norway advanced during the 1960s and 1970s during an interval of cooling in the Norwegian Sea, but the prevailing trend during the twentieth century has been one of melting. In recent decades, the rate of melting has accelerated for many glaciers. All tropical mountain glaciers studied are in retreat, and some have disappeared entirely.

This pervasive, near-global retreat cannot be the result of reduced snowfall. That explanation would require an average drop in precipitation of 25% in many sites across the globe. Instrumental evidence indicates that precipitation changes in most regions are much smaller than 25%, with increases in some glacier areas and reductions in others. In contrast, the temperature increases observed during the late 1800s and the 1900s have been both global in scale and of the right magnitude to explain the glacial melting.

18-5 Ground Temperature

Heat probes inserted into soils or bedrock boreholes reaching hundreds of meters below the surface

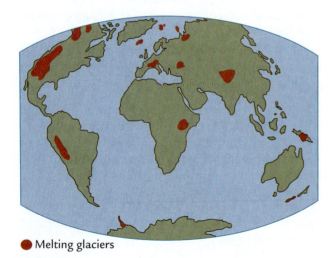

● Melting glaciers

FIGURE 18-11
Locations of retreating mountain glaciers
Mountain glaciers in many regions retreated during the 1900s.
(ADAPTED FROM M. F. MEIER, "CONTRIBUTION OF SMALL GLACIERS TO GLOBAL SEA LEVEL," *SCIENCE* 226 [1984]: 1418–21.)

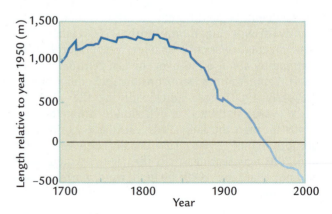

FIGURE 18-12
Retreat of mountain glaciers since the 1800s
Mountain glaciers around the world have retreated by an average of more than 1.5 km since the 1800s. (ADAPTED FROM J. OERLEMANS, "EXTRACTING A CLIMATE SIGNAL FROM 169 GLACIER RECORDS," *SCIENCE* 308 [2005]: 675–7.)

measure past changes in temperature that have slowly penetrated from the atmosphere into subsurface layers. These profiles are sensitive to longer-term (century-scale) temperature changes at the surface, but smooth out the shorter decadal-scale variations. Subsurface temperature records have been taken at hundreds of stations in both hemispheres (Figure 18-13A), and most profiles show warmer

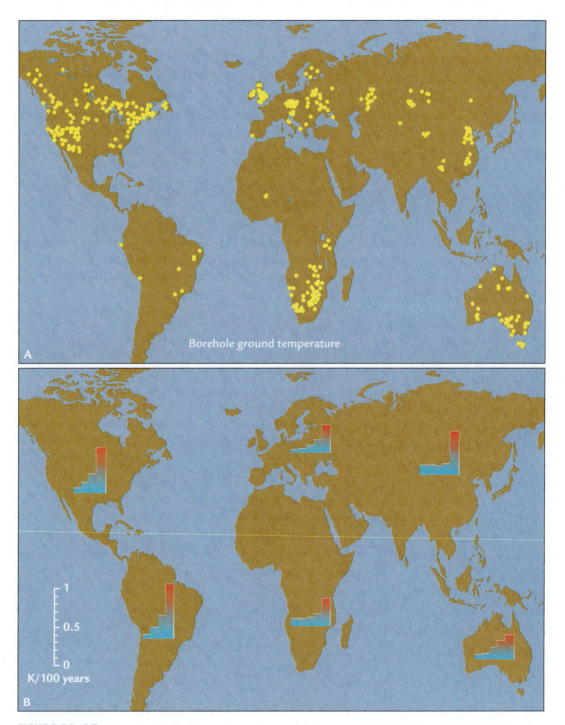

FIGURE 18-13
Ground temperature

Ground temperature profiles have been measured at hundreds of stations at middle and high latitudes (A). Model simulations based on these profiles show a warming (measured in degrees Kelvin or Celsius) on all continents from the 1500s (far-left bar) to the 1900s (far-right bar) (B).

(ADAPTED FROM S. HUANG ET AL., "TEMPERATURE TRENDS OVER THE PAST FIVE CENTURIES RECONSTRUCTED FROM BOREHOLE TEMPERATURES," *NATURE* 403 [2000]: 756–8.)

temperatures in the near-surface layers than a few tens of meters below. The measurements indicate that a warming has occurred at the surface in the last century or two and that it is in the process of penetrating to deeper layers. This warming of the subsurface across the continents has also absorbed a small fraction of the excess heat generated during the industrial era.

Models that simulate the penetration of the temperature anomalies beneath the surface indicate that the warming during the last two centuries lies at or slightly above the upper end of the range of surface-temperature reconstructions based on climatic proxies (see Figure 18-13B). This mismatch in part reflects the fact that both the ground temperature profiles and the proxy reconstructions showing larger variations tend to be based on high-latitude and/or high-altitude sites, where temperature responses are larger than the global average. Other complications include the depth of snow, which may shield the ground from extreme temperature changes in winter, and clearance of forests, which can cool local temperatures because of the higher reflectivity of open land surfaces.

In Summary, instrumental measurements of temperature at the surface, in the subsurface ocean, and in the ground, along with measurements of the mass balance of mountain glaciers, provide consistent evidence that a large warming has occurred during the last 125 years.

▶ Satellite Observations

The range of measurements of Earth's climate increased markedly with the advent of satellite sensors in the late 1900s. Different satellite sensors have come on line at different times, including a few within the last decade.

18-6 Disagreement Between Satellite and Ground Stations Resolved

As recently as the early 2000s, measurements of temperature from satellites did not appear to agree with the warming trend measured in stations at Earth's surface. The satellite data suggested no warming had occurred since 1980 and perhaps even a small cooling, in contradiction to the warming shown by surface stations. Scientists skeptical about global warming pointed to these measurements as evidence that the temperature trend assembled from the surface stations must be in error. Their case seemed to

be strengthened by the fact that some temperature measurements from **radiosondes** (metal-enclosed devices suspended from balloons sent into the lower atmosphere) agreed better with the satellite data than with the surface station data, mainly in the tropics.

Satellite estimates of the temperature of the lower atmosphere are based on measurements of the *brightness* (energy emission level) of vibrating molecules of oxygen (O_2), the second largest constituent of Earth's atmosphere. This brightness parameter correlates with the temperature of the oxygen molecules in the air, and of the air itself. The satellite sensors integrate energy emissions from oxygen molecules across the entire lower atmosphere (0–10 km, the troposphere) as well as the lower part of the stratosphere. As Earth's troposphere warms, the stratosphere cools. As a result, changes in the temperature of the lower troposphere from satellite measurements cannot be estimated without first removing the cooling trend in the lower stratosphere.

These apparent mismatches have been resolved in recent years in favor of the ground station records. Subsequent reexamination found that incorrect adjustments had been used to remove the changes in stratospheric temperatures in order to isolate the temperature changes in the troposphere. In effect, the cooling effect in the stratosphere had been overestimated, which resulted in too large a correction (subtraction) to derive the tropospheric signal. Additional problems were caused by the effect of friction, which caused a gradual drift of the satellite orbits toward lower altitudes and a slightly later arrival over particular locations on Earth's surface. With all these complications taken into account, the satellite data came into basic alignment with the surface station temperature trend shown in Figure 18-9.

The problem with the radiosonde data turned out to be similar: an overcorrection for solar heating of the metal shields. Data from radiosondes sent aloft at night agreed with the evidence from ground stations, but data from daytime releases did not show the warming recorded at the surface. Because radiosondes are made of metal, their measurements of air temperature have to be corrected for heating of their metal shields by the Sun. The overcorrection for this solar heating effect imposed a false cooling signal on the measurements. A more accurate correction for the Sun's heating showed temperatures that agreed well with ground stations.

With these problems resolved, those who argued that global warming was not real lost one of their last supporting arguments. Satellite and radiosonde data now confirm that a major warming has occurred in the last century or more.

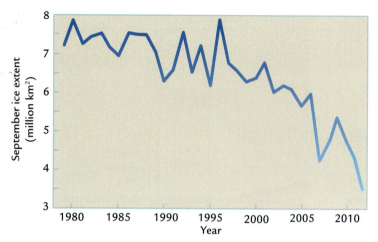

FIGURE 18-14

Decrease in Arctic sea ice cover

Satellite measurements show almost a 40% decrease in the extent of September sea ice cover in the Arctic Ocean since the early 1970s. (ADAPTED FROM INTERGOVERNMENTAL PANEL ON CLIMATE CHANGE (IPCC), *CLIMATE CHANGE 2007: THE PHYSICAL SCIENCE BASIS* [CAMBRIDGE, UK AND NEW YORK: CAMBRIDGE UNIVERSITY PRESS, 2007].)

18-7 Circumarctic Warming

The most dramatic climatic responses during recent decades are those in the Arctic Ocean and over nearby continents. This warming trend has been accompanied by large changes in a number of climatically sensitive indices.

Satellites in use since 1979 map the extent of Arctic sea ice by distinguishing between brighter sea ice and darker open ocean water. Each year, the minimum annual extent of sea ice occurs in early September at the end of the summer melt season. In the decades since the 1970s, the extent of September sea ice has decreased by almost 40%, with a very rapid loss since 2006 (Figure 18-14). In recent years, the northwest ocean passage through the northern Canadian archipelago has become ice-free for a few weeks in September. The retreat of the ice limits in September mostly reflects melting of 1-m thick ice that had formed during the previous winter (first-year ice).

Ice in the central Arctic survives into subsequent years (multi-year ice) and can attain thicknesses of 3–4 m before being transported out of the Arctic Ocean into the Atlantic. Measurements made by upward-looking sonar soundings from submarines during the middle 1900s and more recently from downward-looking radar surveys from satellites show that the average thickness of all ice in the Arctic (first-year and multi-year combined) has shrunk by more than half since the middle to late 1900s. The combination of rapid retreat of marginal first-year ice, along with thinning of multi-year ice in the central Arctic, has produced an even more dramatic loss of total ice volume during all seasons of the year, ranging from about 40% in March to more than 70% in September (Figure 18-15).

Satellite measurements from the higher latitudes of the Northern Hemisphere show other trends that are consistent with major regional warming. The extent of northern hemisphere snow cover is decreasing (Figure 18-16), with earlier melting of snow in spring and later initiation of snow cover in autumn.

An additional indication of warming at middle and high northern latitudes comes from satellite and surface station measurements of the length of the growing season. Surface measurements in central Alaska indicate an irregular increase in the growing season by two weeks over the past 50 years (Figure 18-17). Satellite sensing of the chlorophyll produced by vegetation north of 45°N has also shown that by the mid-1990s the growing season was beginning a week earlier in spring than it had in the early 1980s and that it was ending half a week later in autumn.

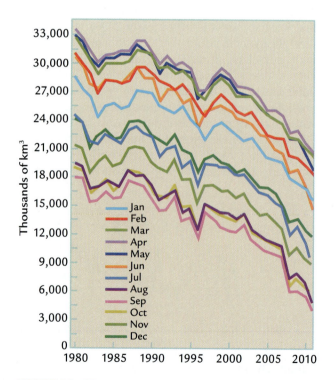

FIGURE 18-15

Decrease in volume of Arctic sea ice

Satellite and submarine measurements reveal very large decreases in Arctic sea-ice volume since the 1970s. (ADAPTED FROM UNIVERSITY OF WASHINGTON POLAR SCIENCE CENTER.)

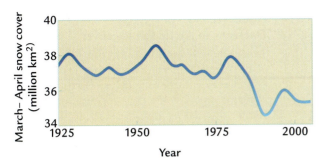

FIGURE 18-16
Decrease in snow over the Northern Hemisphere

In the last several decades, satellite measurements show a gradual decrease of snow cover in the Northern Hemisphere. (ADAPTED FROM IPCC, *CLIMATE CHANGE 2007: THE PHYSICAL SCIENCE BASIS* [CAMBRIDGE, UK AND NEW YORK: CAMBRIDGE UNIVERSITY PRESS, 2007].)

In contrast to the Arctic, the much larger amount of sea ice surrounding Antarctica shows no major trend in recent decades. The unusually complex interactions between the atmosphere, sea ice, and underlying ocean water in the Southern Ocean may be masking climatic trends.

18-8 Ice Sheets

Accurate measurements of the volume of ice sheets first became possible in the early 2000s when satellites began to measure the elevation of ice surfaces with sufficient accuracy (Figure 18-18). Multiple passes across the ice sheets, combined with computer analysis of the radar images received, made it possible to obtain accurate measurements across the entire surfaces of the ice sheets.

Because the elevation of bedrock under the ice can also change through time, ice thickness and ice volume cannot be determined solely from measurements

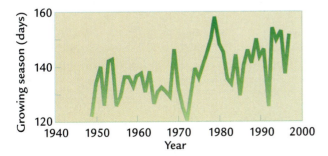

FIGURE 18-17
Longer growing season in Alaska

Surface temperature measurements indicate the length of the growing season increased in Alaska during the last half of the twentieth century. (ADAPTED FROM S. W. RUNNING ET AL., "RADAR REMOTE SENSING PROPOSED FOR MONITORING FREEZE-THAW TRANSITIONS IN BOREAL REGIONS," *EOS* 80 [1999]: 213–21.)

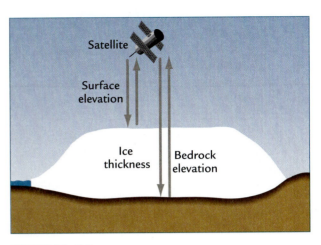

FIGURE 18-18
Changes in ice volume

Changes in the thickness and volume of ice sheets are monitored from satellites using radar and gravity measurements.

of surface ice elevation. Until recently, the elevation of bedrock beneath the ice was measured by labor-intensive efforts in which recording stations were moved across the ice sheets, sending out radar waves that bounced off the bedrock and returned to the station. Because these radar pulses traveled at known velocities through the ice, the elapsed time until the return pulses from underlying bedrock provided a measure of ice thickness. But this time-consuming technique could only be used to measure profiles across limited parts of the ice sheets. More recently, radar measurements of bedrock elevation have been made quickly and inexpensively from satellites, providing full coverage of these continent-sized masses of ice.

In addition, surface stations installed in critical locations with global positioning receivers linked to satellites now measure both the elevation of the ice and the gravity field (the strength of the pull of Earth's gravity on the satellite receiver). Because rock is almost three times denser than ice, the gravity field at each station primarily measures changes in the elevation of the underlying bedrock in response to ice melting or growth. Thinning of the ice allows the bedrock to rebound quickly because of the near-instantaneous elastic part of the response of Earth's mantle to unloading (see Chapter 10). This gravity-data information complements satellite and radar measurements of ice elevation and thickness.

Based on a decade of measurements of the Greenland ice sheet, the highest and coldest part of the ice at elevations above 2 km have been slowly gaining ice as new snow falls, as typically occurs for ice sheets that are at equilibrium. Ice sheets that are stable accumulate snow at high altitudes and transport it as ice flow to lower altitudes.

But very recent trends have disproven the idea that the Greenland ice sheet is at a stable overall mass balance. The upper limit of the line dividing net ice gain at higher elevations from ice loss at lower elevations has moved higher on the ice sheet in recent years, leaving a smaller area of net ice accumulation. And many coastal margins in Greenland have begun losing ice rapidly. Many thinner ice margins along coastal outlet glaciers in the warmer southern half of Greenland are retreating by hundreds of meters each year. Overall, the annual loss of ice on Greenland has accelerated slightly since 2002 (Figure 18-19). Years of additional observations will be needed to place these changes in a longer-term context.

Changes in Antarctic ice have been less dramatic and more variable from region to region. The narrow Antarctic Peninsula that protrudes north toward the southern tip of South America has been rapidly shedding ice during the last decade because that area has warmed very rapidly, much faster than the global average. In contrast, the huge mass of ice in the East Antarctic ice sheet shows both thickening and thinning in different areas, and the year-to-year variability in temperatures and ice budgets is large. As a whole, the ice sheets in Antarctica appear to have lost a small amount of ice during the last two decades, but many more years of observation are needed to resolve the long-term trend. Part of this trend may depend on narrow ice streams that drain ice from the interior of the East Antarctic ice sheet at rates as much as 100 times faster than the slow flow across the rest of the ice sheet.

In Summary, satellite measurements show that north polar regions have warmed dramatically and lost large amounts of sea ice, snow, and land ice. The north polar trends add to the instrumental evidence that the planet has warmed over the last 150 years, particularly during the last 30–40 years. South polar regions have so far shown smaller warming trends.

Sources of the Recent Rise in Sea Level

We now have the information needed to return to the problem of the origin of the ~17-cm rise in the average level of the ocean estimated for the last century (see figure 18-7). This increase has primarily resulted from three factors (Table 18-1).

Because water expands slightly when heated above 4°C, warming of the ocean causes sea level to rise. The subsurface ocean warming trend shown in Figure 18-10, along with less complete data from earlier decades, indicate that thermal expansion can explain about 4 cm of sea level rise since 1960. In addition, even though mountain glaciers and ice caps only account for about 1% of the total amount of ice on land, 99% of them have been melting and retreating and adding water to the ocean since the middle 1800s (see Figure 18-12). These glaciers account for at least another 5 cm of the rise in sea level during the last century. Current estimates are that about 2 cm of sea level rise during the 1900s came from melting of the Greenland and Antarctic ice sheets, although these estimates are not well constrained for earlier decades.

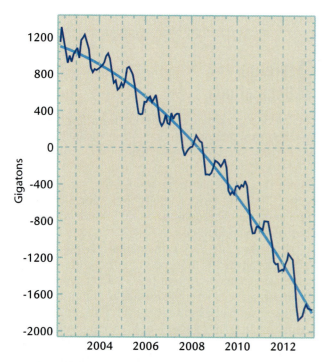

FIGURE 18-19
Recent loss of Greenland ice mass
A trend of accelerated melting of Greenland ice appear in records collected by GRACE satellites beginning in 2002.
(©2013 COOPERATIVE INSTITUTE FOR RESEARCH IN ENVIRONMENTAL SCIENCES, UNIVERSITY OF COLORADO AT BOULDER.)

Table 18-1	Factors in the Rise of Sea Level in the Twentieth Century (in cm)
Ocean thermal expansion	+4
Mountain glaciers	+5
Greenland and Antarctic ice	+2
All factors	+11
Observed sea level rise	+17

In Summary, the combination of thermal expansion, melting of mountain glaciers, and melting of ice on Greenland and Antarctica can account for ~11 cm of the estimated sea level rise of ~17 cm during the twentieth century. This calculation leaves a substantial gap unexplained, but uncertainties in all of the contributing factors are large prior to recent decades. Some scientists infer that most of the remaining imbalance came from greater-than-expected melting of the Greenland ice sheet. Other scientists regard this discrepancy as an important "enigma" and have concluded that the 17-cm estimate may be too high because of errors in correcting tide gauge records for delayed responses to past ice sheets (see Section 18-1).

Measurements from a range of satellites indicate that the rate of annual sea level rise has increased from the ~1.7 mm average during the 1900s to 2.5–3.0 mm per year since the middle 1990s. Even for this more highly instrumented interval, large uncertainties remain for several contributing factors, although some estimates suggest that no "gap" is present. At this point, mountain glaciers and thermal expansion remain the largest sources of meltwater, but in the future most mountain glaciers will disappear and accelerated melting of the Greenland ice sheet will become more important.

In future years, estimates of sea level change will be much better constrained because of new instrumentation deployed in recent years (Figure 18-20). In addition to satellites that measure the height of the sea surface directly, another technological innovation targets the portion of sea level change caused by warmer ocean temperatures and thermal expansion. The velocity of sound waves moving through the ocean is dependent on water temperature: the velocity averages about 1,500 meters per second but increases by 4.6 meters per second for each 1°C of warming of the water. The SOFAR (SOund Fixing And Ranging) channel at a water depth near 1 km is particularly favorable for transmission of sound waves. Sound waves moving through the overlying and underlying ocean layers are gradually bent into this channel from above and below because its temperature and density make the waves move slightly faster than in the surrounding layers. Scientists can use these far-traveled sound waves to measure the average sound velocity across the paths the waves follow, and therefore the average temperature, across large stretches of the subsurface ocean. In addition, thousands of ARGO floats that have been deployed in a multinational effort covering most the world's oceans will deliver greatly improved resolution of changes in ocean temperature and salinity that cause sea level changes.

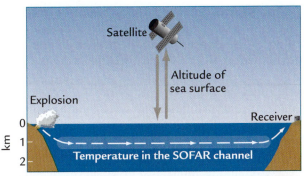

FIGURE 18-20
Changes in subsurface ocean temperature
Warming of the ocean is monitored by measuring increases in the height of the ocean caused by thermal expansion of seawater and increases in the velocity of sound traveling through subsurface layers.

Internal Oscillations

Oscillations in climate that persist for a few years to as much as a few decades can affect distinct regions on Earth's surface. These fluctuations arise from natural variability within the climate system, but in a form that is distinct from short-term (year-to-year) noise. Efforts to assess longer-term climatic trends need to take into account the contributions from these shorter-term oscillations.

The large-scale ENSO fluctuations in the tropical Pacific Ocean are the most prominent short-term oscillation in the climate system (see Chapter 17). During El Niño years, sea level pressure falls and ocean temperatures warm across the east-central tropical Pacific Ocean, while sea level pressure rises and precipitation decreases in the far western tropical Pacific Ocean and over New Guinea and northern Australia (Figure 18-21). El Niño events recur irregularly within a broad band of 2 to 7 years, and each fluctuation lasts for about a year. In addition to the ENSO changes, other oscillations affect smaller regions. These oscillations appear as changes in surface pressure, temperature, and winds that may persist for many years.

A

FIGURE 18-21
El Niño and the Southern Oscillation

Temperatures are warmer off western South America and across the eastern tropical Pacific during El Niño years (A). Warm El Niño years occur on average every 2 to 7 years (B). (ADAPTED FROM E. M. RASMUSSEN, "EL NIÑO AND VARIATIONS IN CLIMATE," *AMERICAN SCIENTIST* 73 [1985]: 108–77.)

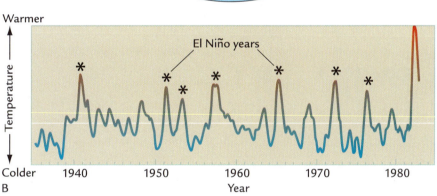

B

The **Pacific Decadal Oscillation** (PDO) has a spatial distribution similar to ENSO. During "positive" PDO phases, sea-surface temperatures are warmer in the tropics and along the Pacific coast of western North America and cooler in the west-central North Pacific (Figure 18-22A). "Negative" PDO phases have the opposite pattern. Based on a little more than a century of high-resolution instrumental observations, PDO patterns can persist for intervals ranging from several years to decades (Figure 18-22B). A prominent change from a negative to a positive PDO pattern occurred in 1976, near the time that the global temperature trend shifted from a few decades of stable or cooling climate to the rapid warming that continues today. But the positive PDO pattern has been less persistent during the last 20–25 years, and yet global warming has continued. The century-length record is too short to determine whether the PDO has a cyclical signature of multi-decadal or century length.

The **North Atlantic Oscillation** (NAO) is a fluctuation in atmospheric pressure between a subpolar low-pressure center near Iceland and a high-pressure center in the Azores-Gibraltar region. The NAO is best developed in winter. The "positive" NAO mode features lower pressure over Iceland, higher pressure over the Azores, and a strengthening of westerly winds across the intervening latitudes of the Atlantic Ocean between the two pressure centers (Figure 18-23A). As a result, the subtropical Atlantic Ocean is warmer in a large region extending from the mid-Atlantic and southeast coast of the United States eastward to the Azores Islands. The warm, moisture-bearing winds arriving from this part of the Atlantic make Europe warmer and wetter than during years of negative NAO. Cooler temperatures occur off the west coast of Africa, where strong trade winds also send extra amounts of dust out across the ocean toward the Caribbean Sea.

The NAO has varied in strength over multi-year time scales, with more frequent positive years in the early 1900s, many negative years between 1940 and 1975, and a return to more positive years during the

A

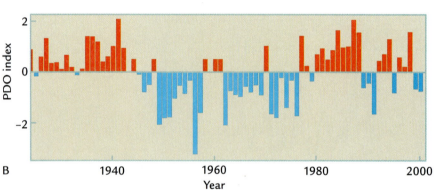

B

FIGURE 18-22
Pacific Decadal Oscillation (PDO)
Like El Niño years, positive PDO years have warm temperatures off the west coast of the United States, but cool temperatures in the northwest Pacific ocean (A). PDO phases can persist for many years to decades (B). (ADAPTED FROM N. J. MANTUA ET AL., "A PACIFIC INTERDECADAL CLIMATIC OSCILLATION WITH IMPACTS ON SALMON PRODUCTION," *BULLETIN OF THE AMERICAN METEOROLOGICAL SOCIETY* 78 [1997]: 1069–79, AND FROM K. E. TRENBERTH AND J. W. HURRELL, "DECADAL ATMOSPHERE-OCEAN VARIATIONS IN THE PACIFIC," *CLIMATE DYNAMICS* 9 [1994]: 303–19.)

1980s and early 1990s (Figure 18-23B). Some scientists have claimed that the rapid retreat of sea ice during the 1990s was affected by the NAO. However, the NAO weakened during the late 1990s, yet the retreat of sea ice has not only continued but even intensified.

Still another possibility is that recent climate may have been influenced by solar forcing at the relatively weak and short-term cycles of 440 years or less that were identified by differences between tree ages derived by counting rings and by ^{14}C dating (see Chapters 15 and 17). Some high-resolution records show small tendencies toward cyclic behavior at these periods, but many of them also show similar tendencies at other periods unconnected to solar changes. In addition, many other records show no evidence of short-term solar forcing. At this point, the possibility of short-term cycles in response to solar forcing has not been settled.

In Summary, it is premature to dismiss the possibility that short-term oscillations have played some role in the global trend toward greater warmth since the late 1800s, but the numerous climatic sensors now in place have shown a persistent warming that looks more and more like a long-term trend, rather than a response to short-term oscillations.

Key Terms

tide gauges (p. 358)
peripheral forebulge (p. 360)
urban heat island (p. 362)

radiosondes (p. 366)
Pacific Decadal Oscillation (p. 371)
North Atlantic Oscillation (p. 371)

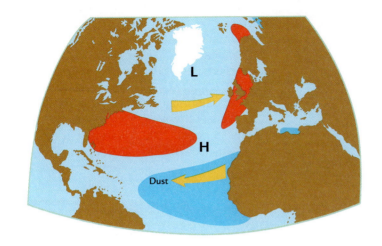

A

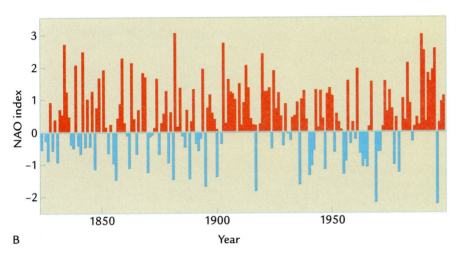

B

FIGURE 18-23
North Atlantic Oscillation (NAO)

During positive NAO years, the western subtropical North Atlantic Ocean is warm, and strong winds blow this warmth and moisture into north-central Europe (A). NAO conditions (positive or negative) can persist for many years (B). (ADAPTED FROM J. W. HURRELL ET AL., "AN OVERVIEW OF THE NORTH ATLANTIC OSCILLATION," IN THE NORTH ATLANTIC OSCILLATION: CLIMATE SIGNIFICANCE AND ENVIRONMENTAL IMPACT, GEOPHYSICAL MONOGRAPH SERIES 134 [2003]: 1–35.)

Review Questions

1. How do ice sheets that melted many thousands of years ago complicate efforts to determine the global sea level rise during the past century?

2. What is the urban heat island effect? How does it complicate attempts to synthesize trends of regional, hemispheric, or global temperature change?

3. Name four kinds of satellite evidence that support a gradual warming of high northern latitudes in the last two decades.

4. If Arctic sea ice has retreated by 25% and thinned by 40% in the last 50 years, what has been the percentage loss in its volume?

5. Why is it difficult to determine whether or not ice sheets are growing or shrinking?

6. How does warming of the ocean affect sea level?

7. How do the North Atlantic Oscillation and Pacific Decadal Oscillation (NAO and PDO) complicate efforts to detect global warming?

Additional Resources

Basic Reading

International Geosphere-Biosphere Programme (IGBP) Web site. http://www.igbp.net. Last accessed March 13, 2013.

World Climate Research Program (WCRP) Web site. http://www.wcrp-climate.org/. Last accessed March 17, 2013.

National Climatic Data Center Web site. "Global Warming FAQs." http://www.ncdc.noaa.gov/oa/climate/globalwarming.html. Last accessed March 17, 2013.

Intergovernmental Panel on Climate Change (IPCC) Web site. http://www.ipcc.ch. Last accessed March 17, 2013.

PAGES—Past Global Changes Project Web site. http://www.pages-igbp.org. Last accessed March 17, 2013.

Advanced Reading

Huang, S. H., N. Pollack, and P.-Y. Shen. 2000. "Temperature Trends over the Past Five Centuries Reconstructed from Borehole Temperatures." *Nature* 403: 756–8.

Hurrell, J. W., 1995. "Decadal Trends in the North Atlantic Oscillation: Regional Temperatures and Precipitation." *Science* 269: 676–9.

Intergovernmental Panel on Climate Change (IPCC). 2007. *Climate Change 2007: The Physical Science Basis* Cambridge, UK and New York: Cambridge University Press.

Levitus, S., et al. 2000. "Warming of the World Ocean," *Science* 287: 285–93.

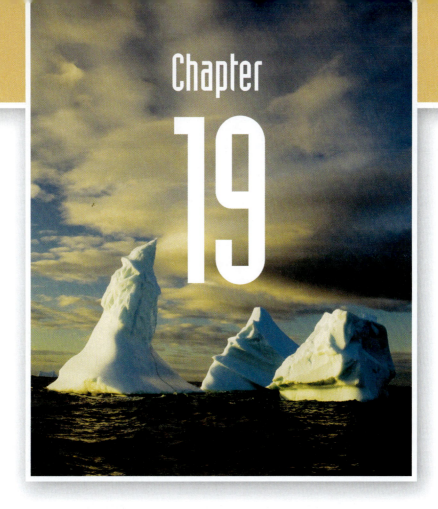

Chapter 19

Causes of Warming over the Last 125 Years

nstrumental measurements show that Earth's average surface temperature has warmed by 0.8°C in the last century (see Chapter 18). During the past two decades, climate scientists have explored how much of this warming has been caused by natural factors and how much by the buildup of CO_2, CH_4, and other greenhouse gases in Earth's atmosphere as a result of human activities. Most climate scientists now agree that natural changes at tectonic, orbital, and millennial time scales do not explain this warming. Although solar forcing may have played a small role, the rapid warming of the last century or more is largely the result of humans.

▶ Natural Causes of Recent Warming

A key question in the 0.8°C global warming since the late 1800s is the role of natural changes in climate. Here, we examine possible sources of natural variations in climate during the last 125 years, proceeding from the longer-term to the shorter-term factors.

19-1 Tectonic, Orbital, and Millennial Factors

As shown in Chapters 5 to 7, changes in climate over *tectonic* time scales are clearly irrelevant to the changes of the last 125 years. During the transition from greenhouse (ice-free) conditions to the current icehouse state, Earth's climate cooled by at most 5°C to 10°C over 50 to 100 million years. The average rate of cooling (~0.00001°C per century) falls far short of the rates needed to alter climate measurably in 125 years.

As shown in Chapter 8, changes in Earth's tilt and precession over *orbital* time scales have altered the amount of insolation received at different latitudes and in different seasons, but orbital forcing is also not a viable explanation of the recent warming. Global average temperature during the last 6,000 years has cooled by less than 1°C, at an average rate of 0.00016°C per century or less, much too slow to account for the 0.8°C warming during the last century, as well as opposite in direction.

Millennial-scale oscillations were large when northern ice sheets existed (see Chapter 15), but they weakened as the ice melted. During the last 8,000 years of the current interglaciation, climatic oscillations have been small and highly irregular in character from region to region. In contrast, the warming of the last 125 years has been global (or very nearly so) in extent, with ever-fewer regions trending counter to the warming pattern. This global response does not match the pattern of the glacial millennial oscillations, with Antarctic responses opposite in sense to those in the North Atlantic region. Millennial-scale oscillations do not appear to be a factor in the recent warming.

19-2 Century- and Decadal-Scale Factors: Solar Forcing

Satellite measurements of the amount of radiation arriving from the Sun began in 1978 and now extend over three 11-year cycles (Figure 19-1). During those cycles, solar radiation has varied by a little over 1 W/m^2, equivalent to about 0.1% of the global average value. Prior to 2011, satellites had indicated a global average value of 1,365–1,366 W/m^2, but improvements in technology have now reduced that value to 1,361–1,362 W/m^2. The amplitude of the 11-year cycles remains very nearly the same.

Climate models indicate that a change of ~0.1% in the strength of radiation arriving from the Sun could alter global mean temperature by as much as 0.2°C if it persisted for many decades. Half of an 11-year cycle, however, does not give the climate system enough time to register its equilibrium response. As a result, Earth's mean temperature in the models warms and cools by less than 0.1°C in response to the 11-year solar variations. Temperature changes this small are difficult to distinguish from the natural variability in Earth's climate system, and very few observational records show convincing evidence of an 11-year temperature signal.

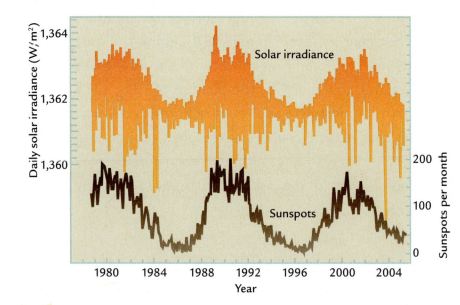

FIGURE 19-1
Solar radiation and sunpots

Since 1978, satellites have measured changes in the solar radiation arriving at the top of the atmosphere (top), and these changes correlate with the observed numbers of sunspots (bottom). (SOLAR IRRADIANCE VALUES ADAPTED FROM HTTP://WWW.PMODWRC.CH/, AND SUNSPOT DATA FROM HTTP://SIDC.OMA.BE/INDEX.PHP3.)

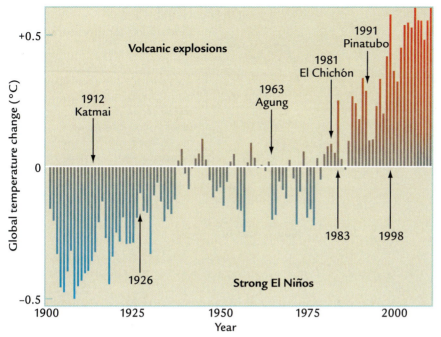

FIGURE 19-2

Brief volcanic coolings and El Niño warmings

Large volcanic explosions and major El Niño events cause short-term changes that can be large enough to be detected in the instrumental temperature record but leave the long-term baseline trend unaffected. (ADAPTED FROM NATIONAL CLIMATE DATA CENTER, NOAA, ASHEVILLE, NC.)

As noted in Chapter 17, until a few years ago scientists estimated that the Sun may have been 0.25–0.4% weaker than it is now during several long sunspot minima that occurred prior to the industrial era, such as the Maunder minimum (1645–1715). Such a change would be two to three times the range measured by satellites since 1979. If changes this large persisted for decades, the climate system would have had time to come much closer to its full equilibrium response. Model simulations at that time indicate that temperatures could have risen by as much as 0.3°C to 0.4°C above those during the 1800s. If so, almost half of the 0.8°C warming during the 1900s might have been accounted for by solar forcing.

This hypothesized Sun–climate link has not held up well under closer scrutiny because observations of Sun-like stars have not detected variations over these longer multidecadal intervals. Current estimates suggest that changes in solar irradiance have accounted for less than 0.1°C of the 0.8°C warming since the late 1800s. During the 35 years of direct satellite observations, when temperatures have climbed especially rapidly, the solar contribution has been negligible. A proposed Sun–climate link operating through the stratosphere remains a possibility, but it remains largely speculative.

19-3 Annual-Scale Forcing: El Niños and Volcanic Eruptions

Two factors have had measurable effects on global climate over brief intervals lasting about a year: major El Niño events and large volcanic explosions. During the last century, both factors have altered global temperature by less than 1°C, and their effects have disappeared into the background noise of the climate system within a year or two.

Large El Niño episodes can warm the eastern tropical Pacific sea surface by 2°C to 5°C and raise average global temperature by 0.1°C to 0.2°C. The major El Niños that occurred in 1983 and 1998 produced one-year warm spikes in temperature (Figure 19-2), but their effects disappeared by the following year.

Several large volcanic eruptions occurred during the era of instrumental temperature records, including Katmai (1912), Agung (1963), and El Chichón (1981). Although these eruptions probably cooled global climate by ~0.1°C for a year or two (see Figure 19-2), the amount of cooling is difficult to determine because of uncertainties about the sulfur content of each eruption and the height in the atmosphere to which the SO_2 was injected.

The large Mount Pinatubo eruption in the Philippines in 1991 was the first volcanic explosion measured in sufficient detail to assess its effect on global climate. Global climate cooled by 0.6°C during the summer after the eruption and by an average of ~0.3°C for the first full year after the eruption. By the second year, the cooling effect of Pinatubo disappeared into the background noise of natural year-to-year temperature variability. If the brief effects of the warm El Niño years and the cool volcanic years are removed from the temperature records, the steady background warming trend of the last decades becomes even more obvious (Figure 19-3).

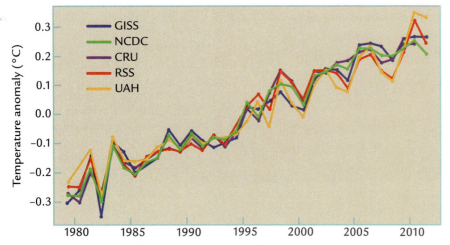

FIGURE 19-3

Underlying temperature trend

With the brief (one-year) effects of volcanic explosions and El Niño years removed, the warming trend in global temperature during the last several decades is even more obvious. (ADAPTED FROM G. FOSTER AND S. RAHMSTORF, "GLOBAL TEMPERATURE EVOLUTION 1979-2010," *ENVIRONMENTAL RESEARCH LETTERS* 6 [2011]: 044022, DOI:10.1088/1748-9326/6/4/044022.)

In Summary, the gradual 0.8°C increase in global temperature over the last 125 years cannot be explained by natural forcing operating at tectonic, orbital, or millennial time scales, nor is it the result of short-term forcing from volcanic explosions or El Niño events. Up to 10% of the warming (0.07°C) could result from changes in solar irradiance.

▶ Anthropogenic Causes of the Recent Warming

With natural causes all but ruled out, the bulk of the recent warming must be anthropogenic in origin. Here we explore several kinds of emissions from human activities that have altered climate: the greenhouse gases carbon dioxide and methane, chlorine-bearing chemicals of various kinds, and sulfate and carbon aerosols.

19-4 Carbon Dioxide (CO₂)

Bubbles of ancient air trapped in ice and direct measurements of air by the geochemist Charles Keeling begun in 1958 show an accelerating rise in the CO_2 concentration during the last two centuries. By 2010 the CO_2 concentrations had risen to 390 ppm, well above the 180–300 ppm range of natural (glacial-interglacial) variations (Figure 19-4).

The additional carbon emitted from human activities has come mainly from two sources (Figure 19-5). Throughout the late 1700s and most of the 1800s, the main source of carbon was clearing of forests to meet human needs: farmland for agriculture, wood for home heating, and charcoal to fuel the furnaces of the early part of the Industrial Revolution. Cutting and burning of forests in eastern North America was

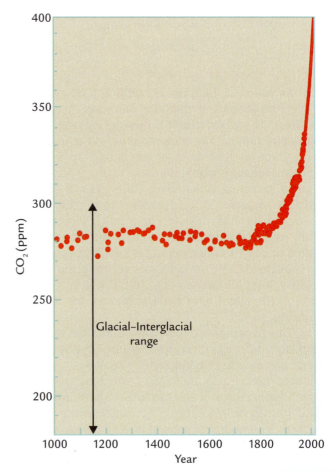

FIGURE 19-4

Preindustrial and anthropogenic CO₂

The combined atmospheric CO_2 record from bubbles in ice cores and from instrument measurements since 1958 shows an accelerating increase in the last 200 years from a baseline of ~280 ppm. (ADAPTED FROM K. D. ALVERSON ET AL., EDS., *PALEOCLIMATE, GLOBAL CHANGE AND THE FUTURE* [BERLIN: SPRINGER-VERLAG, 2003].)

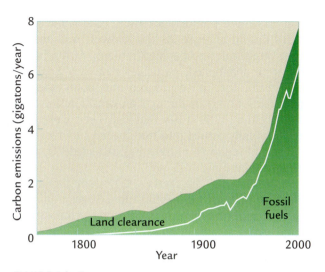

FIGURE 19-5

Human production of CO₂

Two factors account for the increase in atmospheric CO_2 caused by human activities in the last 250 years: (1) burning of carbon in trees to clear land for agriculture; and (2) burning of carbon in fossil fuels—coal, oil, and gas. (ADAPTED FROM H. S. KHESHGI ET AL., "ACCOUNTING FOR THE MISSING CARBON SINK WITH THE CO₂-FERTILIZATION EFFECT," *CLIMATE CHANGE* 33 [1996]: 31-62, FROM DATA IN T. A. BODEN ET AL., *TRENDS '91: A COMPILATION OF DATA ON GLOBAL CHANGE*, ORNL/CDIAC-46 [OAK RIDGE, TN: OAK RIDGE NATIONAL LABORATORY, 1991].)

a particularly important source of CO_2 emissions during that interval.

After 1900, most of the extra carbon added to the atmosphere came from extraction of fossil fuels buried beneath Earth's surface. At first, coal was the main fuel, but later on oil and natural gas became major sources of energy. Gradually, the carbon released by fossil fuels came to exceed the amount produced

by land clearance. Today, industrial carbon emissions (mostly in the Northern Hemisphere) account for most of the fossil-fuel total, while cutting and burning of tropical rain forests are the largest land-clearance contribution. Although volcanoes are the major source of CO_2 over long (multi-million-year) time scales (see Chapter 5), average annual emissions from human activities are larger by a factor of more than 100.

In recent decades, 55% of the excess carbon produced each year has ended up in the atmosphere and another 25–30% has been added to the ocean (Figure 19-6). Because the atmosphere is relatively well mixed as a result of its rapid circulation, a single measurement or a small number of them can easily characterize its CO_2 concentration. But because the slower-circulating ocean is not as well mixed, many measurements in many areas are needed to characterize the average change in its CO_2 content.

The excess CO_2 from human activities has already been well mixed into the upper tens of meters of the ocean, which quickly exchange molecules of gas with the atmosphere (see Chapter 11). In contrast, the shallow subsurface ocean below 100 m is more out of touch with the atmosphere, and most of the deep ocean below 1 km is even more isolated from the surface. As a result, smaller amounts of the excess CO_2 produced in the last two centuries have penetrated below 100 m, and very little has entered the deep ocean except in regions of active turnover like the subpolar North Atlantic Ocean.

In the surface layer of the ocean, the CO_2 exchanges with the atmosphere vary by region and by season. On an annual average, cold high-latitude ocean water acts as a net CO_2 sink and takes CO_2 from the atmosphere, while warm low-latitude ocean

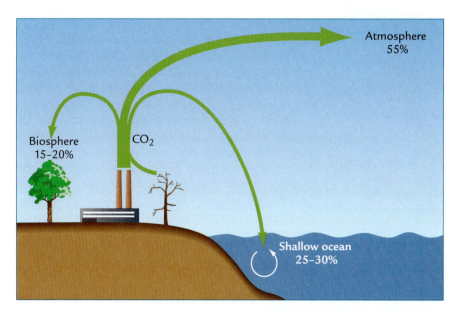

FIGURE 19-6

Where does the CO₂ produced by humans go?

Of the carbon added to the climate system by humans, 55% ends up in the atmosphere, 25–30% enters the surface ocean, and the rest is stored in the biosphere (vegetation on land, litter, and organic carbon in estuaries).

surfaces act as a CO_2 source and give some of it back (Figure 19-7). One reason for this pattern is that CO_2 gas is more easily dissolved in cold water than in warm water. Local air-sea exchanges are also governed by the relative concentration of CO_2 in the surface ocean compared to that in the overlying atmosphere, and by other physical and biochemical processes that control carbon exchanges with subsurface waters.

For all of these reasons, huge numbers of measurements made over vast regions of the ocean during all seasons are required to quantify the slow penetration of CO_2 into and beneath the ocean surface. Despite these problems, growing numbers of measurements confirm that the ocean takes up about 25–30% of the total human carbon input.

If 55% of the excess carbon ends up in the atmosphere and 25–30% in the oceans, where does the other 15–20% go? The only major reservoir left is the biosphere, both live vegetation (trees and grasses) and dead organic litter in soils and coastal estuaries (see Figure 19-6). Although burning of forests to clear land has been a major source of extra carbon for the atmosphere and the ocean for millennia, in some areas carbon has been returning to the vegetation. One way this can happen is by regrowth of forests in previously cleared regions.

Across nearly all of eastern North America, forests had been almost completely cut by the late 1800s for farming and for fuel. Little over a century later, these regions now look completely different. Forests grow in areas where photographs from 1900 show landscapes devoid of trees. As the Midwest was opened up to large-scale mechanized farming, farms in the East were abandoned, particularly in New England. Rock walls that once marked the boundaries of open fields now run through growing forests. Near eastern cities, rural areas that had been cleared of trees gradually turned into tree-shaded suburbs. This widespread regrowth of trees has extracted CO_2 from the atmosphere.

A second way to remove CO_2 from the atmosphere is through CO_2 fertilization. Vegetation uses CO_2 during photosynthesis to create the cellulose that forms leaves, blades of grass, tree trunks, and roots. Greenhouse experiments show that most plants obtain carbon more easily from a CO_2-rich atmosphere and grow faster as a result. Scattered evidence suggests that the 35% rise in atmospheric CO_2 in the last 200 years has increased this fertilization effect and taken more carbon from the atmosphere. The vegetation grows faster; it becomes more varied in composition and grows more densely; the amount of woody material in tree branches, trunks, and roots increases; and trees and shrubs shed more fresh carbon litter into soils and coastal estuaries.

In Summary, ice core and instrumental measurements show that atmospheric CO_2 levels have risen by almost 40% in the last 150 years. Because greenhouse gases trap outgoing radiation from Earth's surface, the rising CO_2 level has warmed the planet.

19-5 Methane (CH_4)

The concentration of the greenhouse gas methane in the atmosphere has also increased as a result of human activity. The influence of humans is again evident from trends measured both in ice core air bubbles and (since 1983) from instrumental observations (Figure 19-8). Since the 1800s, the methane concentration has risen to over 1,750 ppb, well above the natural range of 350–700 ppb during the previous 800,000 years based on measurements in ice cores.

Methane comes from sources that are rich in organic carbon but lacking in oxygen: from swampy bogs where plants decay; from the guts of cattle and other grazing and browsing animals that digest vegetation; from animal and human waste; and from burning of grassy vegetation in smoldering piles where all of the oxygen is consumed but burning continues. In the absence of oxygen, bacteria break down the vegetation and extract the carbon, which combines with hydrogen to form methane gas.

Increased emissions of CH_4 during the last 200 years have resulted from the explosion in human

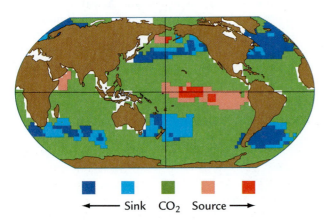

FIGURE 19-7
Ocean sources and sinks of CO_2

Annually averaged CO_2 concentrations in ocean surface waters are close to those in the overlying atmosphere, but the cool higher-latitude oceans act as net sinks that absorb carbon from the atmosphere, while the warm tropical oceans are net sources that give some of it back. (ADAPTED FROM T. TAKAHASHI ET AL., "GLOBAL AIR-SEA FLUX OF CO_2: AN ESTIMATE BASED ON MEASUREMENTS OF SEA-AIR PCO2 DIFFERENCES," IN *CARBON DIOXIDE AND CLIMATE CHANGE* [WASHINGTON, D.C.: NATIONAL ACADEMY OF SCIENCES, 1997].)

← Sink CO_2 Source →

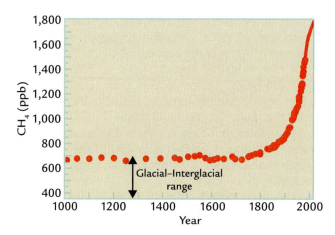

FIGURE 19-8
Preindustrial and anthropogenic CH₄

The combined atmospheric CH₄ record from bubbles in ice cores and from instrument measurements since 1958 shows an accelerating increase in the last 200 years from a baseline of ~675 ppb. (ADAPTED FROM K. D. ALVERSON ET AL., EDS., *PALEOCLIMATE, GLOBAL CHANGE AND THE FUTURE* [BERLIN: SPRINGER-VERLAG, 2003].)

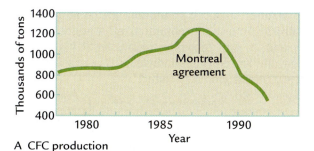

A CFC production

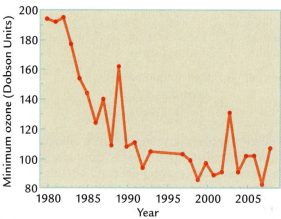

B Ozone concentration

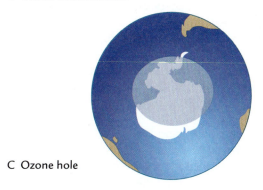

C Ozone hole

FIGURE 19-9
Anthropogenic CFC increases

CFC concentrations in the atmosphere prior to the 1987 Montreal agreement rose because of increased human production (A). Between 1979 and the late 1980s, the spring minimum value of ozone in a column of air over Antarctica decreased by almost half (B), creating an ozone hole (C).
(A: ADAPTED FROM NASA, 2009; HTTP://OZONEWATCH.GSFC.NASA.GOV/; B: AFTER A. J. KRUEGER, GODDARD SPACE FLIGHT CENTER, NASA; C: ADAPTED FROM T. E. GRAEDEL AND P. J. CRUTZEN, *ATMOSPHERE, CLIMATE, AND CHANGE*, SCIENTIFIC AMERICAN LIBRARY, ©1997 BY LUCENT TECHNOLOGIES.)

population. Ever-increasing areas of tropical land have been put into rice paddy cultivation in Southeast Asia, and these artificial wetlands produce methane. The growing numbers of cattle and other livestock have increased the amount of methane gas sent to the atmosphere. Today, the amount of CH₄ produced by these and other human activities is more than twice as large as that from natural sources.

Methane also acts as a greenhouse gas by trapping outgoing radiation from Earth's surface. Although its concentration is much lower than that of CO₂, it is far more effective on a molecule-by-molecule basis in trapping radiation than CO₂. The enormous increase in methane concentration during the last two centuries has also caused the planet to warm.

19-6 Increases in Chlorofluorocarbons

Another group of chemical compounds has increased in abundance in the atmosphere: the **CFCs**, or **chlorofluorocarbons**. These compounds contain elements chlorine (Cl), fluorine (Fl), and bromine (Br) that have important impacts on upper-atmospheric chemistry.

CFCs have for decades been produced for use as refrigerator and air conditioner coolants, chemical solvents, fire retardants, and foam insulation in buildings. Released at ground level, they are slowly mixed upward through Earth's atmosphere. Because CFCs stay in the atmosphere for an average of 100 years, they eventually reached the stratosphere, where their concentrations increased during the late 1900s (Figure 19-9A).

Several decades ago, the biologist James Lovelock, originator of the Gaia hypothesis, inferred that the rising production of the CFCs might be causing dangerous increases in their concentration in the stratosphere. In the 1970s, the atmospheric chemists Sherwood Roland and Mario Molina confirmed that the CFCs were indeed having a dangerous impact because of their interaction with stratospheric ozone in the Antarctic region.

Ozone (O_3) occurs naturally in the stratosphere, with the largest concentrations at altitudes between 15 and 30 km. Incoming ultraviolet (UV) radiation from the Sun liberates individual O atoms from oxygen (O_2) and produces ozone:

$$Radiation + O_2 \rightarrow O + O$$
$$(UV)$$
$$O + O_2 \rightarrow O_3$$

Ozone is naturally converted back to oxygen (O_2) in the atmosphere by a similar process, but in this case the radiation source can be light in either ultraviolet or visible wavelengths:

$$O_3 + Radiation \rightarrow O_2 + O$$
$$(UV \text{ or visible})$$
$$O + O_3 \rightarrow 2O_2$$

With visible radiation far more abundant than ultraviolet radiation, ozone is normally destroyed much faster than it is produced. As a result, ozone is a short-lived gas. In addition, the rate of conversion of ozone back to O_2 increases when certain chemicals are present to speed up the reaction. Chlorine reacts with ozone and destroys it, forming chlorine monoxide (ClO):

$$Cl + O_3 \rightarrow ClO + O_2$$

Chlorine then reacts with free oxygen molecules and is liberated from ClO:

$$ClO + O \rightarrow Cl + O_2$$

These liberated chlorine atoms then begin a new cycle of ozone destruction. This cycle is important to humans because ozone in the stratosphere forms a natural protective barrier that shields life-forms from levels of ultraviolet radiation that would otherwise produce cell mutations including skin cancers.

In the last century, humans have greatly accelerated the natural destruction of ozone by adding extra chlorine to the stratosphere. Measurements from 1979 to 1990 showed that the amount of ozone in a column of air over Antarctica had decreased considerably in the region where stratospheric chlorine is unusually abundant (Figure 19-9B).

The largest decreases occurred high in the Antarctic stratosphere during the spring season. Isolation of Antarctic polar air from the rest of Earth's atmosphere through the winter allows CFCs to accumulate to high levels that rapidly destroy ozone when solar radiation increases early in the following spring. The region over Antarctica in which stratospheric ozone is much less abundant than elsewhere is called the **ozone hole** (Figure 19-9C).

This clear connection between CFC emissions and ozone depletion caused so much alarm that the world's nations signed a treaty in Montreal in 1987 to reduce and ultimately eliminate the use of CFCs. Production of CFCs immediately began to decline (Figure 19-9A), and the concentrations of the type of CFCs that industries found easiest to replace stabilized and began a slow decline. Other CFCs that are still in widespread use have continued to increase, but at slower rates. Stratospheric ozone levels have now stopped falling, but have not yet begun a major recovery toward natural levels (Figure 19-9B). Rebounds in ozone concentration are expected to begin within a decade or two.

Ozone also occurs naturally in much greater abundance in the lower troposphere. It originates from both natural and anthropogenic processes, including biomass burning and oil production in refineries. At these lower levels in the atmosphere, ozone generally plays a positive environmental role by cleansing carbon monoxide (CO) and sulfur dioxide (SO_2) from the air.

At high concentrations, however, ozone is toxic to plants and an irritant to human eyes and lungs. In the lowermost atmosphere, ozone trends have moved in the opposite direction during the industrial era. Human activities have caused large ozone *increases* that have produced periodic smog alerts in many large cities. Slow-moving air masses settle over urban areas in summer and allow concentrations of ozone and other pollutants to build to levels that are dangerous to humans with respiratory problems. Tropospheric ozone at lower levels has a warming effect like the greenhouse gases CO_2 and CH_4.

In Summary, the increase in CFC concentrations during the late twentieth century has added to the greenhouse-gas warming, but the destruction of stratospheric ozone has cooled the planet slightly. Increases in tropospheric ozone have contributed to the warming.

19-7 Sulfate Aerosols

Industrial era smokestacks emit the gas sulfur dioxide (SO_2) as a by-product of smelting operations in furnaces and from burning coal. SO_2 reacts with water vapor and is transformed into sulfate particles, called sulfate aerosols. Because these aerosols stay within the lower several kilometers of the atmosphere, their primary impact on climate is regional in scale.

Until the 1950s, smokestacks in Europe and North America were small and most SO_2 emissions stayed close to ground level, producing thick industrial hazes and sulfur-rich acidic air that etched building facades and cemetery monuments made of limestone and marble. In the 1970s, taller smokestacks were built

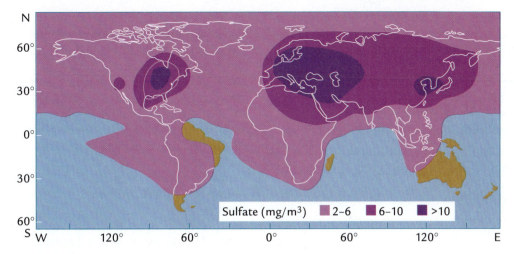

FIGURE 19-10
Sulfate aerosol plumes

Sulfur dioxide emissions from industrial smokestacks in the North American Midwest, Eastern Europe, and China produce sulfate aerosols that prevailing winds carry eastward in large plumes. (ADAPTED FROM R. J. CHARLSON ET AL., "PERTURBATION OF THE NORTHERN HEMISPHERE RADIATIVE BALANCE BY BACKSCATTERING FROM AEROSOLS," *TELLUS* 43 [1991]: 152–63.)

to disperse SO_2 emissions higher in the atmosphere (up to 3 km). This effort dramatically improved air quality in many cities, but created a different problem in more distant areas.

The sulfate particles that are now being sent higher in the atmosphere are carried by fast-moving winds across broad areas. Although sulfates stay in the atmosphere for only a few days before rain removes them, they can be carried 500 or more kilometers downwind from source regions. Today large plumes of sulfate aerosols are carried far from sources in Eastern Europe, east-central North America, and China (Figure 19-10).

Greenland ice contains a record of industrial SO_2 emissions from regions upwind (Figure 19-11). Throughout the 1800s, industrial SO_2 emissions had been smaller than those from natural sources, but by the middle 1900s industrial emissions had become the dominant SO_2 source. Concentrations then began to drop sharply after 1980, when the United States acted to limit its SO_2 emissions.

Climate scientists have inferred that the large plumes of sulfate aerosols cause regional cooling downwind from smokestack sources. Like the sulfate aerosols created by volcanic explosions, industrial sulfate particles reflect and scatter some of the incoming solar radiation back to space and keep it from reaching Earth's surface. The reduction in radiation cools climate regionally, and these effects show up in tabulations of global mean temperature change.

A second potential climatic effect of sulfate aerosols is less well understood. Tiny aerosol particles form centers (nuclei) around which water vapor condenses, forming droplets and then clouds. Clouds can

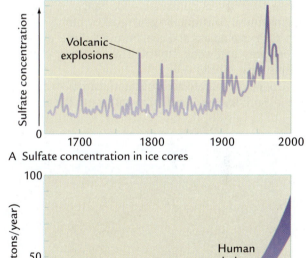

A Sulfate concentration in ice cores

B Estimated global SO_2 emissions

FIGURE 19-11
Preindustrial and anthropogenic sulfates

Measured sulfate concentrations in Greenland ice reveal a significant regional increase since the middle 1800s (A). Estimated global SO_2 emissions from smokestacks now exceed natural emissions (B). (A: ADAPTED FROM P. A. MAYEWSKI ET AL., "AN ICE-CORE RECORD OF ATMOSPHERIC RESPONSES TO ANTHROPOGENIC SULPHATE AND NITRATE," *ATMOSPHERIC ENVIRONMENT* 27 [1990]: 2915–19; B: ADAPTED FROM R. J. CHARLSON ET AL., "CLIMATE FORCING BY ANTHROPOGENIC AEROSOLS," *SCIENCE* 255 [1992]: 423–30.)

have two opposing effects on climate: the surfaces of thick clouds can reflect more incoming solar radiation and cool climate, but thinner higher clouds can absorb more outgoing radiation from Earth's surface and increase the greenhouse effect. Because sulfate aerosols mainly affect low-level clouds, their indirect effects appears to cool climate, but the amount of cooling is highly uncertain.

In Summary, the net global effect of increased sulfate aerosols from industrial activities is a cooling, but estimates of its size vary widely.

19-8 Brown Clouds, Black Carbon, and Global Dimming/Brightening

In addition to sulfate aerosols, human activities send a wide range of other pollution into the atmosphere. Anyone traveling on airplanes can see low-lying hazes hanging over many regions, including large cities and rural regions where burning is frequent. These haze layers contain a wide range of physical and chemical constituents generated by human activities such as biomass and coal burning, diesel exhaust, forest fires and cook stoves. The resulting climatic effects vary widely.

Unlike the sulfate aerosols, dark carbon-rich aerosols absorb the Sun's radiation and warm the air layer in which they lie. Scientists initially suspected that these hazes heated the lower atmosphere and added to the net amplitude of global warming on a regional basis. But investigations during the late 1990s and early 2000s led to a surprisingly different interpretation of hazes in Southeast Asia called **brown clouds**. Most of the carbon-rich aerosols in these hazes come from small cook stoves in which people burn organic matter for fuel, including cow dung. The aerosols in these hazes do absorb radiation, but their effects are concentrated in a layer of air two to three kilometers above the surface. Although this layer warms, the haze blocks a portion of the incoming solar radiation and prevents it from reaching Earth's surface, which cools. In regions of severe brown-cloud hazes, the reduction in solar radiation during peak seasons is almost an order of magnitude larger than the global average increase from the greenhouse-gas effect. The effects of the brown clouds have also been found to extend thousands of kilometers downwind from source regions (Figure 19-12).

In contrast, particles of **black carbon** produced by various kinds of burning can have a warming effect on climate in far-northern regions. When dark particles ("soot") fall out of the atmosphere onto bright surfaces of snow and sea ice, they absorb solar radiation that would otherwise have been reflected. This absorbed heat melts snow and sea ice, in some cases exposing the darker land and ocean surfaces below, which further warm regional climate by albedo

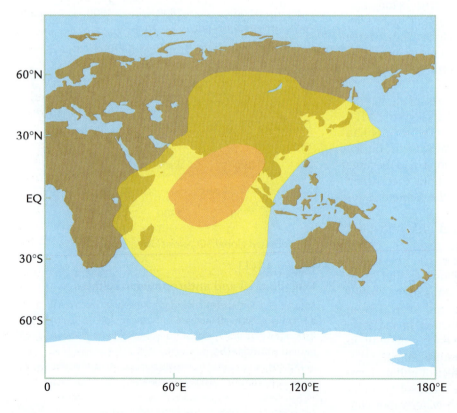

FIGURE 19-12
Brown clouds

Brown-cloud hazes in the lower atmosphere shroud regions downwind of large populations in India and China. (ADAPTED FROM C. E. CHUNG AND V. RAMANATHAN, "SOUTH ASIAN HAZE FORCING: REMOTE IMPACTS WITH IMPLICATIONS TO ENSO AND AO," *JOURNAL OF CLIMATE* 16 [2003]: 1791–1806.)

feedback. Much of the soot that falls on Arctic snow and sea ice appears to originate in biomass burning of crops and grasslands, not forests. Some estimates place the global warming role of black carbon at levels near or above that from methane, but it remains difficult to evaluate this warming role against the cooling effect of brown clouds (in which black carbon also plays a part). The role of black carbon in "seeding" aerosols that form clouds is also unclear.

During the 1950s to 1980s, measurements taken at Earth's surface in many mid-latitude regions of the Northern Hemisphere detected a phenomenon called **global dimming**, in which the amount of solar radiation reaching the ground slowly decreased. Toward the end of that period, satellite measurements of the brightness of Earth's surface from space confirmed a decrease of several percent. This decrease was thought to be the result of many factors, including sulfate aerosols, brown clouds, contrails emitted by jets, and other emissions. The cooling caused by reduced solar radiation is thought to have been large enough during the 1940s to 1970s to cancel the greenhouse-gas warming trend during that interval.

Since the 1980s, aerosol emissions in several industrial regions (excluding India and China) have been reduced, reversing the long-term dimming trend and allow the surface to brighten. This trend toward increased solar radiation is likely responsible for part of the rapid warming during that interval. But the dimming trend continues over Southeast Asia and other regions where emissions from urbanization and other sources like biomass burning are still growing.

19-9 Land Clearance

Human activities, primarily the cutting of forests for agriculture, have altered much of the land surface of the planet. At high and middle latitudes, forest clearance causes a net increase in albedo, as darker forests are replaced by brighter pastures and croplands. With more solar radiation reflected back into space, the surface cools. At tropical and subtropical latitudes, forest clearance also reduces the amount of evapotranspiration. With reduced moisture availability, land surfaces dry out and bake in the intense summer Sun, warming regional climate. Trends of land clearance during the last 150 years have been complex. Most active deforestation in recent decades has occurred in tropical South America and tropical islands of Southeast Asia, but most northern hemisphere regions that were deforested until the 1800s or early 1900s have been reforesting for many decades. On a global average basis, the net effect of land clearance has been a small cooling of the planet.

▶ Earth's Sensitivity to Greenhouse Gases

Our understanding of past, present, and future climate hinges in part on Earth's sensitivity to greenhouse gases. Two independent sources of evidence put constraints on this sensitivity: (1) numerical models of the climate system (mainly general circulation models); and (2) climate reconstructions of intervals from Earth's history when greenhouse-gas concentrations differed from those today.

19-10 Sensitivity in Climate Models

For convenience, Earth's sensitivity to changes in greenhouse gases is often quantified as the global average change in surface temperature caused by a doubling of CO_2 concentrations beyond the modern (preindustrial) level of 280 ppm. With a doubled CO_2 concentration used as an initial boundary condition, the models are run until the simulated temperature comes into equilibrium (or near-equilibrium) with this higher CO_2 level (usually over many decades). The global average increase in simulated temperature is the **2 × CO_2 sensitivity** for that model.

The CO_2 concentration has increased by ~40% since the start of the industrial era (from 280 ppm to nearly 400 ppm). Climate scientists also include the combined heat-trapping effects of the other greenhouse gases by converting changes in these gases to equivalent changes in CO_2. For example, the 150% increase in methane concentrations over the last 150 years has had an additional greenhouse effect equivalent to a 12% increase in CO_2. This is counted as a 12% increase in **equivalent CO_2**. The 40% increase in CO_2, combined, with the 12% equivalent increase in methane and the 13% equivalent increase in other greenhouse gases, has produced a combined increase equivalent to a CO_2 rise of ~65%.

Calculating the potential effect of the greenhouse-gas increases on Earth's climate involves two steps. Step 1 is to determine the **radiative forcing** provided by the gases, using the same W/m^2 units with which incoming solar radiation and Earth's back radiation are measured (Box 19-1). This radiative forcing excludes the complicating effect of the many natural feedbacks operating in the climate system. It is sometimes called "clear-sky forcing," meaning the amount of warming caused by gases in the absence of any changes in clouds or other factors produced as an indirect effect. A range of observational and modeling evidence indicates that doubling CO_2 levels (and/or the equivalent amounts of other greenhouse gases) would increase global temperature by ~1.2°C through this radiative effect alone.

Climate Interactions and Feedbacks

Radiative Forcing of Recent Warming

The movement of heat through Earth's climate system can be traced from the incoming solar radiation, through the subsequent reflection and absorption of that radiation, the trapping and redistribution of Earth's back radiation by greenhouse gases, and the outgoing radiation back into space (see Chapter 2). Each of these processes redistributes radiative energy and each can be measured in units of W/m².

Of the incoming solar radiation, an average of 240 W/m² (about 70%) penetrates into the climate system. Prior to the industrial era, naturally occurring greenhouse gases such as water vapor, CO_2, and CH_4 trapped ~150 W/m² of back radiation from Earth's surface in a natural greenhouse effect. This trapping of energy, and the internal feedbacks that resulted, helped to make Earth 33°C warmer than it would have been without greenhouse gases (and kept it from freezing).

Greenhouse gases produced by humans since 1850 have added to this natural greenhouse effect. The radiative forcing effect of these added gases is their impact on climate in the absence of feedback effects from clouds, water vapor, snow and ice albedo, and other factors. Estimates place the radiative effect of the industrial era buildup of greenhouse gases at 2.7 W/m², just over 1% of the received solar radiation of 240 W/m². CO_2 has

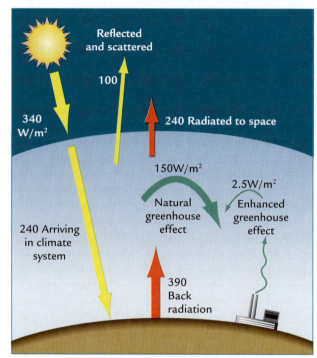

Effects of increases in greenhouse gases on radiation Human activities since the start of the industrial era have increased greenhouse-gas concentrations enough to enhance the natural greenhouse effect by 1.8%. (ADAPTED FROM IPCC SCIENTIFIC ASSESSMENT WORKING GROUP, *RADIATIVE FORCING OF CLIMATE CHANGE*, ED. J. T. HOUGHTON ET AL. [CAMBRIDGE: CAMBRIDGE UNIVERSITY PRESS, 1994].)

Step 2 is to assess the effects of natural feedbacks within the climate system that can either amplify or oppose the radiative warming. Positive feedbacks add to the warming produced by the baseline radiative forcing, while negative feedbacks reduce it.

Together, these two factors (radiative forcing and feedbacks) determine Earth's average temperature sensitivity to changes in greenhouse gases. General circulation model experiments over many decades have produced a wide range of estimates of the temperature sensitivity to the 2 × CO_2 (or equivalent CO_2) level (Figure 19-13). These estimates have generally varied between 2°C and 5°C.

Uncertainties about the size of feedbacks in the climate system are the reason for this wide range of estimates. The most prominent feedbacks come from changes in water vapor, in the albedo of snow and ice, and in clouds and cloud aerosols.

Water vapor, the major greenhouse gas in Earth's atmosphere today, provides positive feedback to the warming initiated by increases in greenhouse gases. In a clear, cloudless sky, the amount of water vapor that air can hold increases exponentially with higher temperatures (see Chapter 2). Today, this feedback helps to make the tropics warmer than the poles and summers warmer than winters. The positive feedback effect from water vapor also occurs in response to temperature changes initiated by increases in CO_2 and other gases. Current estimates are that the initial 1.2°C radiative global warming caused by doubling CO_2 levels should cause an additional warming of between 1.1°C and 1.5°C through water vapor feedback (Figure 19-14).

Another positive feedback results from reflection of incoming solar radiation by bright surfaces covered by sea ice and snow. If a warming initiated

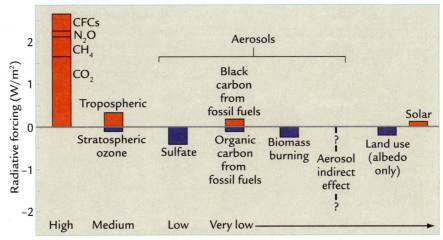

Radiative effects of greenhouse gases Several greenhouse gases have contributed a total of 2.7 W/m2 to the greenhouse effect since 1850. The contributions of aerosols and other factors are less certain. (ADAPTED FROM IPCC, *CLIMATE CHANGE 2007: THE PHYSICAL SCIENCE BASIS* [CAMBRIDGE, UK AND NEW YORK: CAMBRIDGE UNIVERSITY PRESS, 2007].)

contributed 60% of the total increase in radiative forcing, with CH_4, CFCs, and N_2O contributing the remaining 40%. Black carbon and solar forcing also contribute to the positive radiative forcing, while industrial aerosols and land-use changes act in the opposite direction.

Because 2.7 W/m^2 represents a 1.8% addition to the natural greenhouse effect of 150 W/m^2, it is referred to as an **enhanced greenhouse effect.** In response to this anthropogenic enhancement, Earth's surface and lower atmosphere have warmed, and the atmosphere now radiates additional heat to space to compensate for the warming. During times of rapidly rising greenhouse-gas levels, however, Earth's full temperature response lags decades behind the gas levels.

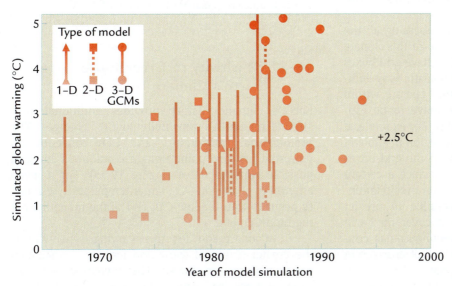

FIGURE 19-13

Model simulations of 2 × CO$_2$ sensitivity

Simulations with several kinds of climate models in recent decades have yielded estimated sensitivities of global mean temperature to doubled concentrations of atmospheric greenhouse gases in the range of +0.5°C to +5°C. (ADAPTED IN PART FROM J. ADEM AND R. GARDUNO, "FEEDBACK EFFECTS OF ATMOSPHERIC CO$_2$–INDUCED WARMING," *GEOFÍSICA INTERNACIONAL* 37 [1998]: 55-70.)

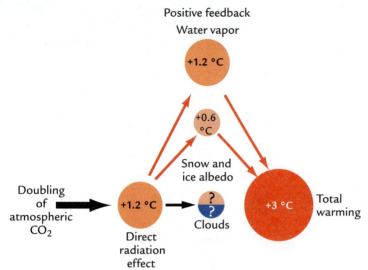

Positive feedback
Water vapor

+1.2 °C

+0.6 °C

Snow and
ice albedo

Doubling
of
atmospheric
CO_2

+1.2 °C

?
?
Clouds

+3 °C

Total
warming

Direct
radiation
effect

FIGURE 19-14
Components of 2 × CO_2 warming
Higher concentrations of greenhouse gases in the atmosphere alter global temperature both by increasing the amount of heat trapped in a clear (cloud-free) atmosphere and by activating feedbacks (water vapor, snow and ice, and clouds) that amplify or reduce the amount of temperature change.

by greenhouse gases causes a retreat of snow and ice toward the poles (see Chapter 18), the reduced extent of these high-albedo surfaces will increase the absorption of solar radiation at high latitudes and warm them further. The resulting positive feedback should increase the initial 2 × CO_2 global warming by about 0.6°C.

Feedback from clouds is the major uncertainty in climate models. Different types of clouds vary in the amount of solar radiation they reflect (which cools climate) compared to the amount of back radiation from Earth's surface they absorb and retain in the climate system (which warms climate). High, wispy clouds tend to warm Earth's climate, because they are composed of ice crystals that are better at absorbing outgoing radiation than at reflecting incoming radiation. Thicker, lower clouds cool climate because they are better at reflecting incoming solar radiation than at trapping outgoing radiation. The cooling effect of lower clouds is obvious in our daily lives; the warming effect of high clouds is not.

The problem facing scientists is assessing all the changes in the many types of clouds as Earth's climate warms. Even the best climate models have grid boxes much too large to simulate in a realistic way individual clouds and the processes that operate within them. These small-scale processes have to be estimated based on statistical probability. For example, if the air temperature across a specific region in a model simulation cools to a particular value, the grid boxes within that region are directed to produce a certain fraction of cloud cover that may deliver precipitation.

At present, it is not even possible to predict whether the total amount of cloud cover on Earth would increase or decrease as greenhouse-gas concentrations increase, and whether or not changes in one cloud type will be larger than another. The reason for this uncertainty is that a warmer atmosphere will

evaporate more water vapor from tropical oceans, thereby increasing the amount of water vapor available to form clouds, but a warmer atmosphere also gains in its capacity to hold water vapor, which reduces the likelihood that vapor will condense into clouds. Which of these two competing effects will win out in a warming world remains unclear.

As a result of all these uncertainties, the net feedback from clouds remains highly uncertain. If the 2 × CO_2 climate sensitivity is 3°C, then the net feedback from clouds should lie close to zero (Figure 19-14). But considering the full range of possible climate sensitivities, the cloud feedback could either be strongly negative (for a 2°C sensitivity) or strongly positive (for a 5°C sensitivity).

In Summary, the current best estimate places the combined effect of all feedbacks in the climate system at 1.8°C for doubled CO_2 values, which adds to the initial radiative (clear-sky) warming of 1.2°C.

19-11 Sensitivity to Greenhouse Gases: Earth's Climate History

Earth's climatic history provides additional constraints on Earth's sensitivity to CO_2 changes. The preindustrial temperature and CO_2 concentration represent just one point along a continuous curve of possible climatic states (Figure 19-15). For a given amount of change in CO_2, larger temperature responses occur toward the low-CO_2 end of the range than at the high-CO_2 end.

A major reason for this varying sensitivity is the greater extent of snow and ice cover when CO_2 concentrations are lower. The increased areas of snow and sea ice in these colder climates provide a

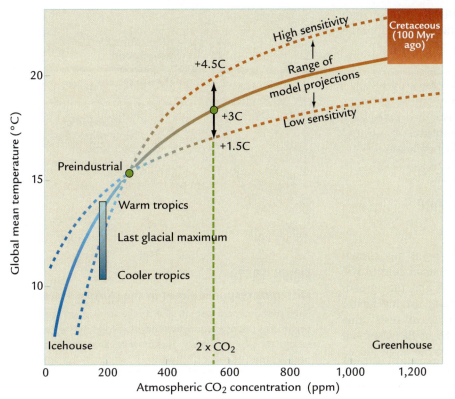

FIGURE 19-15

Estimates of 2×CO₂ sensitivity from Earth's history

Estimated changes in global temperature and measured or estimated changes in atmospheric CO_2 concentrations for the same intervals of Earth's history fall within the range of sensitivities indicated by climate models. (ADAPTED IN PART FROM R. J. OGLESBY AND B. SALTZMAN, "SENSITIVITY OF THE EQUILIBRIUM SURFACE TEMPERATURE OF A GCM TO CHANGES IN ATMOSPHERIC CARBON DIOXIDE," *GEOPHYSICAL RESEARCH LETTERS* 17 [1990]: 1089–92.)

larger albedo-temperature feedback to the initial CO_2 changes. In contrast, at the high-CO_2 end of the range, snow and ice are reduced in extent, and little albedo feedback occurs. A second reason for the varying sensitivity is that the trapping of Earth's back radiation becomes less efficient at higher CO_2 concentrations as the atmosphere gradually becomes saturated with CO_2.

The model used to generate Figure 19-15 had a 3°C increase in global temperature for a CO_2 doubling from 280 to 560 ppm. The overall range of uncertainty shown by other GCM models (dashed lines) can be tested by examining intervals from Earth's climate history when past CO_2 levels and global mean temperature are either well known or can be reasonably well constrained. The temperature changes at these times should represent the response to changes in CO_2 and other greenhouse gases.

THE LAST GLACIAL MAXIMUM The most informative interval from the past is the last glacial maximum, 21,000 years ago. At that time atmospheric CO_2 values were near 190 ppm, about 30% lower than the preindustrial level of 280 ppm, and methane values were 350 ppb, about 50% lower than their preindustrial value. The methane reduction translates into a 15% loss in CO_2, bringing the net drop in equivalent CO_2 concentration to 45%. As discussed in Chapter 13, greenhouse gases must have been the major factor affecting ocean temperatures at tropical and lower subtropical latitudes because the ice sheets were too

distant to have had large direct effects through changes in atmospheric circulation patterns, and because insolation values were close to those today.

Estimates of the amount of glacial cooling in the tropics range from as small as −1.5°C (CLIMAP) to as large as −4°C or −5°C (see Chapter 13). This range of estimates indicates that Earth's sensitivity to CO_2 largely falls within the range indicated by climate models, but toward its higher end (see Figure 19-15). Incorporating the cooling effects of heavy glacial dust concentrations in the atmosphere and higher albedo due to reduced vegetation would bring the range of estimated sensitivities for the last glacial maximum within the 2°C to 5°C range.

THE CRETACEOUS (100 MILLION YEARS AGO) The Cretaceous world of 100 million years ago can also be compared to the modern world. Estimates of greater warmth during this greenhouse interval range from +5°C to +11°C, with recent estimates favoring the middle end of this range (see Chapter 6). Unfortunately, CO_2 values for the Cretaceous are not tightly constrained. Several techniques based on analysis of carbon isotopes in the remains of fossil plants and soils suggest values considerably higher than those today, but the estimates range from four to twelve times the preindustrial CO_2 level. The upper limits of possible changes in CO_2 concentrations lie well off the limits plotted in Figure 19-15. With Cretaceous temperatures warmer by substantially more than 5°C,

Earth's sensitivity appears to fall toward the high end of the wide range of uncertainty indicated by the dashed lines in Figure 19-15.

> **In Summary**, analyses of past intervals generally support the range of CO_2 sensitivity estimated from climate models. For the best-constrained case, the most recent glacial maximum, the inferred sensitivity largely lies within the range of model estimates.

Why Has the Warming Since 1850 Been So Small?

The evidence summarized to this point has shown that natural factors have played a small role in the global warming trend since ~1880 and that greenhouse gases have been the dominant driving factor. Yet some climate skeptics continue to resist this conclusion based on the small size of the observed global warming (0.8°C) compared to the large (40%) rise in CO_2, and the even larger (65%) rise in equivalent CO_2. They argue that an increase in greenhouse gases of this magnitude should have warmed Earth's surface by ~1.6°C to 1.7°C, far more than the 0.8°C increase actually observed. Based on this analysis, they claim that Earth's sensitivity to CO_2 and other greenhouse gases is well below the lower end of the range indicated by climate models. But mainstream climate scientists have shown that this analysis is invalid, because it ignores two other factors that have also affected global temperature since 1880.

19-12 Delayed Warming: Ocean Thermal Inertia

One important factor is the effect of thermal inertia in delaying the full response of the climate system to greenhouse-gas emissions in recent decades. When external factors cause climate change, some parts of the climate system react more slowly than others because of their greater thermal inertia (Figure 19-16).

The most important source of thermal inertia for the climate system over intervals of a century or so is the ocean, which covers 70% of Earth's surface and stores enormous amounts of heat. The upper layer of the ocean (from 0 to 100 m in depth) is stirred by lower-atmospheric winds and acts as a relatively fast-responding part of the climate system within a decade or two. The ocean layers lying several hundred meters below the wind-mixed layer and less affected by changes at the surface still play a role in the ocean's response, but over longer time intervals (up to 50–75 years). As a result of this slower response, the amount of increase in global surface temperature measured at any particular time within the last 125 years

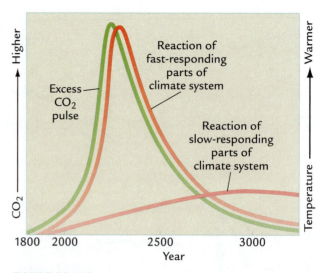

FIGURE 19-16
Different response times in the climate system
Parts of the climate system, such as the atmosphere, can respond to imposed changes in days or weeks, while ice sheets require thousands of years. The upper ocean response occurs over decades.

(including the present day) represents only a part of the warming that is eventually going to occur when those deeper layers respond, even with no further increase in gas concentrations (Figure 19-17).

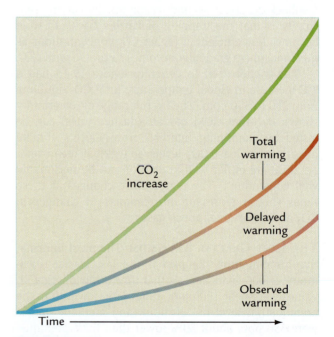

FIGURE 19-17
Delayed warming in the climate system?
During an interval of rapidly rising CO_2 concentrations, the warming measured at any point in time is smaller than the warming that will be realized when Earth's delayed response is fully registered.

The implication of this delayed warming is that Earth's $2 \times CO_2$ sensitivity must be considerably larger than the value that would be derived by comparing the CO_2 increase between the 1800s and the early 2000s against the amount of warming that took place during that interval. Because the rise in greenhouse-gas concentrations during the last half-century has been especially rapid, some of the warming they will eventually cause is still "in the pipeline" (not yet realized).

19-13 Cooling from Anthropogenic Aerosols

In the late 1970s, the climate scientist Murray Mitchell proposed that the size of the greenhouse-gas warming was being significantly reduced by an offsetting cooling effect caused by smokestack emissions of SO_2, leading to production of sulfate particles in the atmosphere (Figure 19-18). At that time, SO_2 emissions in most areas were rising with an exponential trend similar to the CO_2 emissions because both were by-products of the era of heavy industrialization. Today, three major sulfate plumes lie over and downwind of the eastern United States, Eastern Europe, and China (see Figure 19-10).

Current estimates are that the direct effect of sulfate aerosols from smokestacks and carbon-rich aerosols from biomass burning and cook stoves have reduced incoming solar radiation enough to offset about 25% of the radiative forcing caused by greenhouse-gases. Another potentially larger negative feedbacks arises from the indirect effect of aerosols in seeding cloud nuclei (see Box 19-1). By increasing

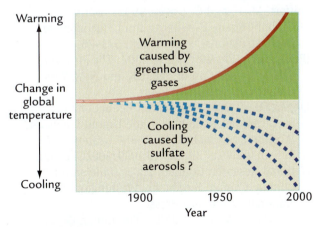

FIGURE 19-18

Aerosol cooling counteracts parts of the greenhouse-gas warming

The warming effect of greenhouse gases is partly canceled by the cooling effect of sulfates produced by SO_2 emitted in smokestacks. (ADAPTED FROM J. M. MITCHELL, "THE NATURAL BREAKDOWN OF THE PRESENT INTERGLACIAL AND ITS POSSIBLE INTERVENTION BY HUMAN ACTIVITIES," *QUATERNARY RESEARCH* 2 [1973]: 436–45.)

low-level cloud cover, this feedback could promote surface cooling, but the amplitude is highly uncertain. Ignoring these offsetting cooling effects from various anthropogenic activities makes the sensitivity of the climate system appear smaller than it actually is.

In Summary, the combined effects of oceanic thermal inertia and anthropogenic aerosols can reconcile the moderate amount of global warming measured at present relative to the large greenhouse-gas buildup during the last century.

► Global Warming: Summary

By 2012, global climate had warmed by 0.8°C above the level in the late 1800s. Observations show that atmospheric concentrations of CO_2 and other greenhouse gases produced by human activities increased markedly during that interval, and most climate scientists agree that the gas increases have been the major reason for the warming.

As recently as the late 1990s (when the first edition of this book was written), it was still possible to question some aspects of anthropogenic global warming, but subsequent findings have undercut the central arguments that once supported this position. Satellite-measured temperature trends that seemed not to show a warming since 1980 were found to be invalid because of incorrect adjustments for various complicating effects (see Chapter 18). Earlier hypotheses that the variability of Sun-like stars might account for half or more of the observed warming were undercut by observations that Sun-like stars do not show the large variations in irradiance proposed. Earlier speculation that carbon aerosols may have been falsely boosting the apparent size of the true greenhouse-gas warming on a regional basis were upended by the finding that brown-cloud hazes have actually kept large areas of Earth's surface cooler and countered part of the greenhouse-gas warming.

An enormous range of evidence now supports the view that Earth's climate has warmed in the last 125 years mainly because of humans. Meanwhile, Earth gradually continues to register the many effects of global warming. Even though natural variability in the climate system makes year-to-year trends very noisy, the decade 2000–2010 was warmer than the decade of the 1990s, which was warmer than the decade of the 1980s, which was warmer than the 1970s. Circumarctic snow and sea ice are shrinking to limits that are unprecedented during the satellite era. Almost every mountain glacier on Earth is melting, and many will disappear within 25 years at their current melting rates. The margins of the

Greenland ice sheet are melting at increasing rates. Humanity's great experiment with the climate system is well underway, and the ongoing increases in greenhouse gases promise much more of the same in the future.

Key Terms

chlorofluorocarbons, CFCs (p. 381)

ozone (p. 382)

ozone hole (p. 382)

brown clouds (p. 384)

black carbon (p. 384)

global dimming (p. 385)

$2 \times CO_2$ sensitivity (p. 385)

equivalent CO_2 (p. 385)

radiative forcing (p. 385)

enhanced greenhouse effect (p. 387)

Review Questions

1. What human activities produce CO_2 and how have they changed in the last 200 years?

2. Where does the CO_2 produced by humans go?

3. How high in the atmosphere do sulfate aerosols from smokestacks reach?

4. Why do chlorofluorocarbons (CFCs) reach much higher in the atmosphere than sulfate aerosols?

5. What are the strongest positive and negative feedbacks on changes in Earth's temperature?

6. In a net sense, do feedbacks increase or decrease the direct radiative effects of greenhouse gases on global temperature?

7. What factors complicate attempts to estimate Earth's sensitivity to CO_2 by directly comparing the observed twentieth-century warming to the measured rise in greenhouse gases?

8. Some climate skeptics point out that temperatures were warmer in north polar regions 6,000 years ago, and conclude that modern greenhouse-gas concentrations have not produced warmth that is unusual by natural standards. Evaluate the relevance of this conclusion based on what you have learned from this book.

Additional Resources

Basic Reading

Henson, R. 2006. "The Rough Guide to Climate Change." London: Rough Guides, Ltd.

Karl, T. R., and K. E. Trenberth. 1999. "The Human Impact on Climate." *Scientific American* (December), 100–105.

International Geosphere-Biosphere Program (IGBP) Web site. http://www.igbp.net. Last accessed March 13, 2013.

World Climate Research Program (WCRP) Web site. http://www.wcrp-climate.org/. Last accessed March 17, 2013.

National Climatic Data Center Web site. "Global Warming FAQs." http://www.ncdc.noaa.gov/oa/climate/globalwarming.html. Last accessed March 17, 2013.

Intergovernmental Panel on Climate Change (IPCC) Web site. http://www.ipcc.ch. Last accessed March 17, 2013.

PAGES—Past Global Changes Project Web site. http://www.pages-igbp.org. Last accessed March 17, 2013.

Advanced Reading

Archer, D. 2011. *Global Warming: Understanding the Forecast*. Wiley.

Charlson, R. J., S. E. Schwartz, J. M. Hales, R. D. Cess, J. A. Coakley, J. E. Hansen, and D. J. Hoffman. 1992. "Climate Forcing by Anthropogenic Aerosols." *Science* 225: 423–30.

Foukal, P., G. North, and T. Wigley. 2004. "A Stellar View on Solar Variations and Climate." *Science* 307: 68–9.

Hansen, J. E., A. Lacis, D. Rind, G. Russell, P. Stone, I. Fung, K. Ruedy, and J. Lerner. 1984. "Climate Sensitivity: Analysis of Feedback Mechanisms." *American Geophysical Union Geophysical Monograph Series* 29: 49–52.

Hoffert, M. I., and C. Covey. 1992. "Deriving Global Climate Sensitivity from Paleoclimate Reconstructions." *Nature* 360: 573–76.

IPCC. 2007. *Climate Change* 2007: *The Physical Science Basis*. Cambridge, UK and New York: Cambridge University Press.

Liepert, B. G. 2002. "Observed Reductions in Surface Solar Radiation in the United States and Worldwide from 1961 to 1990." *Geophysical Research Letters* 29. doi:10.1029/2002GL014910.

National Research Council. 2003. *Understanding Climate Change Feedbacks*. Washington, D.C.: The National Academies Press.

Ramanathan, V., et al. 2001. "The Indian Ocean Experiment: An Integrated Assessment of the Climate Forcing and Effects of the Great Indo-Asian Haze." *Journal of Geophysical Research* 106: 28371–99.

Wigley, T. M. L., P. J. Jaunman, B. D. Santer, and K. E. Taylor. 1998. "Relative Detectability of Greenhouse Gas and Aerosol Climate Change Signals." *Climate Dynamics* 14: 781–90.

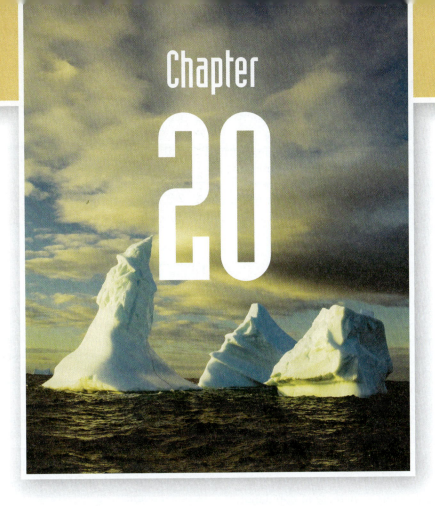

Chapter 20

Future Climatic Change

The climatic effects of rising CO_2 levels will become increasingly significant in the future. Consumption of estimated reserves of fossil fuels in the next two or three centuries will add far more CO_2 to the atmosphere than has been emitted to date unless ambitious conservation efforts or technological innovations reduce the influx. Atmospheric CO_2 concentrations will increase to levels at least twice and possibly four times the highest amounts measured in the last 800,000 years spanned by ice-core records. This warming will overwhelm natural variations in climate and cause climatic and environmental changes unprecedented in human experience. Atmospheric CO_2 levels will eventually decrease as the ocean slowly absorbs much of the excess carbon, but this process will acidify the oceans. After the main pulse of fossil-fuel emissions has passed, atmospheric CO_2 values will decrease back toward preindustrial levels, but enough CO_2 will remain in the atmosphere to prevent future glaciations for tens of thousands of years.

▶ Future Human Impacts on Greenhouse Gases

Over the next few centuries, natural factors will continue to affect climate change, but their impacts will remain small (as they have been during recent millennia). The largest driver of future climate will be emissions of greenhouse gases and aerosols from human activities.

20-1 Factors Affecting Future Carbon Emissions

In recent decades, atmospheric CO_2 concentrations have been rising at rates of ~2 ppm/year, mainly because of burning of fossil fuels, and to a lesser extent because of deforestation in some tropical regions. This rise will continue in the future, but the rates are difficult to predict. The uncertainties in this prediction center mainly on the amount of carbon we will emit but also on the way the climate system will distribute the added CO_2 among its carbon reservoirs.

For as long as fossil fuels remain reasonably abundant, future carbon emissions can be approximated by multiplying three factors:

increase in carbon emissions = increase in population × change in emissions per person × changes in efficiency of carbon use

This equation could also be expressed as: population growth × economic growth × technology. The number of humans is critical to these projections for obvious reasons: expanding numbers of humans require more fossil fuel for industry, transportation, and home heating, and in some regions increasing populations will cut more forest to clear land for farming and urban/suburban growth. As medical advances have extended life expectancy, the number of humans has increased from 1.5 billion in 1990 to over 7 billion today.

Attempts to project future population increases are complicated by the tendency of birth rates to fall as per capita income rises and by centralized efforts to slow population growth in China. Global population is projected to rise rapidly from 7 billion in the year 2010 to 9 or 10 billion by 2050, but with a leveling off of the trend after mid-century (Figure 20-1). The explosive growth that has occurred between 1900 and 2010 will slow as developing countries reach higher levels of wealth and parents in those countries choose to have fewer children. This trend has already been underway for decades in fully industrialized countries.

The change in carbon emissions per person is primarily linked to the average standard of living. Historical examples show that demand for more

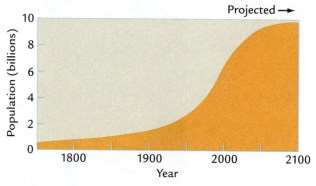

FIGURE 20-1
Future population

A United Nations projection of human population during the twenty-first century at first continues the explosive increase of the twentieth century but later levels off at ~10 billion people. (MODIFIED FROM F. PRESS AND R. SIEVER, *UNDERSTANDING EARTH*, 2ND ED. © 1998 BY W. H. FREEMAN AND COMPANY.)

carbon-based fuel for industrialization, transportation, and home heating and cooling has increased as standards of living have risen. In the near future, the largest changes of this kind will occur in the many countries that shift from semi-industrialized to industrialized status.

The third term in the equation—the average efficiency of carbon use—is by far the most difficult to constrain because of uncertainties about several factors, including the mix of carbon-based fuels. The next few decades will see peaks in annual production of oil followed by a decline in annual production. Natural gas will follow a similar path, but decades later because more gas will become accessible to new recovery techniques.

Compared to coal, oil and gas are relatively "clean" fuels that emit smaller amounts of CO_2 per unit of energy produced. When oil and gas use declines, coal and other "dirty" sources will increasingly become the primary carbon-based fuels until the end of the fossil-fuel era. Most high-grade anthracite coal was burned in the early years of industrialization, leaving mostly lower-grade bituminous coal in the ground. Because bituminous coal produces far more CO_2 per unit of usable energy, burning coal will increase global carbon emissions long into the future. Oil sands found in vast deposits under the western Canadian prairies are another potentially enormous contributor of carbon emissions. If the bulk of the carbon stored in tar sands becomes commercially worth exploiting, CO_2 emissions will rise even higher.

The best hope for reducing (or limiting increases in) future carbon emissions lies in technology, a part of the "efficiency" term in the equation. Humans are a uniquely innovative species, and our ingenuity will inevitably make future breakthroughs in improving the efficiency with which we use fossil carbon sources. Signs of this kind of innovation have begun to appear during the early 2000s as conventional fossils fuels have become more expensive. Promising areas include: solar and wind energy, biofuels (organically based fuels derived from agriculture), hydrogen-powered vehicles, and nuclear power plants that do not produce CO_2 or other greenhouse gases. Future increases in the cost of fossil fuels will help to accelerate the technology-based search for alternative energy.

20-2 Projected Carbon Emissions and CO_2 Concentrations

The Intergovernmental Panel on Climate Change (IPCC) has predicted a wide range of possible carbon emission trends for the rest of this century. These projections are based on tons of carbon emitted; expressed as tons of CO_2 emitted, the values would be 3.66 times as large. The two projections shown in Figure 20-2 encompass the range of the IPCC estimates. Because of the huge inherent uncertainties in the distant future, projections beyond the year 2100 would be nothing more than guesses.

The higher projection in Figure 20-2 is based on the pessimistic assumption that efforts to curb emissions will have little or no effect, because assessments of economic benefits will guide national and individual decisions about energy use. In this projection, emission rates reach a level of almost 30 billion tons of carbon per year (four times the current rate) near 2100, and then fall to lower values in later centuries.

The lower projection in Figure 20-2 assumes that individuals, civic groups, cities, and nations will take strong action to curb carbon emissions and that advances in technology aid this effort. As a result, carbon emissions build up more slowly, reach a smaller peak, and are spread out over a much longer interval. Based on these optimistic assumptions, the lower curve shows worldwide carbon emissions peaking at ~9 billion tons per year near 2050, only about 20% above the modern rate, and then falling below the current rate late in this century.

Converting these best-guess estimates of carbon emissions into possible trends in future CO_2 concentrations in the atmosphere is even more difficult because of several additional complications. For example, in addition to the uncertainties in the basic emissions trend, questions exist about the way the climate system will redistribute the pulse of excess CO_2 among its carbon reservoirs.

The simplest assumption is that the atmosphere will continue to receive just over half of the total carbon emissions in the future, as it has been doing

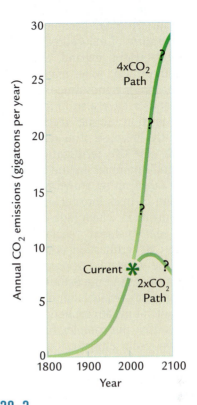

FIGURE 20-2
Projected carbon emissions

Projections of future carbon emissions are highly speculative because of uncertainties in future populations, living standards, conservation efforts, and technological innovations. An optimistic projection (the $2 \times CO_2$ scenario) shows an earlier reduction of emission rates because of strong international efforts, but the $4 \times CO_2$ scenario shows an ongoing rise until emissions decline because fossil-fuel reserves are consumed. (EMISSIONS TRENDS ADAPTED FROM IPPC, *CLIMATE CHANGE 2007: THE PHYSICAL SCIENCE BASIS* [CAMBRIDGE, UK AND NEW YORK: CAMBRIDGE UNIVERSITY PRESS, 2007].)

for decades, with the rest entering the ocean and the biosphere in the same proportions as today. This projection assumes that the climate system will continue to operate as it does today, but it may not.

The part of the CO_2 emissions that will be taken up by the ocean in the future is difficult to predict. Geochemists have gained some insight into this problem in recent decades by tracking the gradual penetration into the ocean of tracers produced by human activities. One tracer is the pulse of extra ^{14}C produced by nuclear tests in the middle 1900s. This bomb-produced ^{14}C has penetrated well into the subsurface ocean in many regions. Other measurements show that a small part of the growing pulse of fossil-fuel CO_2 in the atmosphere has also begun reaching into the deep ocean in high-latitude regions where cold water sinks. But will this absorption of atmospheric CO_2 continue in the same way in the near future? As the surface ocean becomes warmer, it will hold less CO_2 because of reduced gas solubility. And as surface waters warm, expand, and become less dense, formation of deep water may weaken, reducing CO_2 transfer into the subsurface ocean.

Climate scientists have also found evidence that continental vegetation during recent decades has absorbed extra CO_2 because of increased fertilization of plant growth—the **CO_2 fertilization effect.** If this fertilization trend continues in the future, CO_2 will continue to be taken up by growing vegetation, and by the expansion of trees into regions of Arctic tundra. But experiments in which high-CO_2 air has been blown across semi-enclosed vegetation plots have found that the CO_2 fertilization effect drops after a few years. As the vegetation approaches or reaches its full natural response to the extra CO_2, it no longer plays a significant CO_2-fertilization role.

Projections of future atmospheric CO_2 concentrations also depend on the complex interplay between ongoing anthropogenic emissions of excess CO_2 to the atmosphere and its slow removal by the ocean and the biosphere. The two projections of future atmospheric CO_2 concentrations shown in Figure 20-3 differ in the amplitude and timing of the CO_2 peaks attained. In the upper curve, which assumes rapid burning of all fossil-fuel reserves, the CO_2 concentration rises to more than 1,100 ppm just before 2200, reaching a level roughly four times the preindustrial value of 280 ppm. The concentration then begins a slow decline that lasts centuries (millennia). In the lower curve, which assumes much more gradual (and less complete) burning of fossil-fuel reserves, the CO_2 concentration rises more slowly to a peak value of twice the preindustrial level late in this century and later begins a gradual decline. Again, projections beyond 2100 are basically just guesses.

Atmospheric CO_2 levels anywhere in the range of two to four times the preindustrial value would be unprecedented in the 100,000 to 200,000 years that our species has existed and in the millions of years in which our human ancestors (hominins) evolved. The CO_2 history recorded in Antarctic ice cores shows no values above 300 ppm for at least the last 800,000 years (see Chapter 11, Figure 11-6). By some estimates, concentrations as high as 560 ppm have not existed since 5 to 20 million years ago, and concentrations as high as 1,100–1,200 ppm may not have existed for many millions of years (see Chapter 7).

The projected CO_2 trends in Figure 20-3 reach their peak values later than the projected CO_2 emissions trends in Figure 20-2, and they remain at high levels for a much longer time. The reason for this relative delay in timing is the slow rate of CO_2 removal by the ocean. Although individual CO_2 molecules move back and forth between the air and the surface ocean within just a few years, it will take several centuries for the slowly circulating deep ocean to absorb most of the excess CO_2 emitted by humans (Figure 20-4A), and 10–15% of the fossil-fuel pulse will stay in the atmosphere for tens of thousands of years or more.

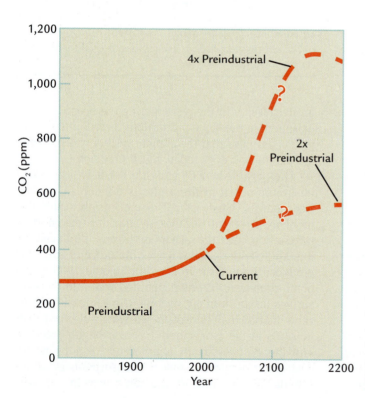

FIGURE 20-3

Projected CO_2 concentrations

Concentrations of CO_2 in the atmosphere are projected to reach levels somewhere between twice ($2 \times CO_2$) and four times ($4 \times CO_2$) the preindustrial value of 280 ppm within the next two centuries. (CO_2 TRENDS THROUGH 2100 ADAPTED FROM IPPC, *CLIMATE CHANGE 2007: THE PHYSICAL SCIENCE BASIS* [CAMBRIDGE, UK AND NEW YORK: CAMBRIDGE UNIVERSITY PRESS, 2007].)

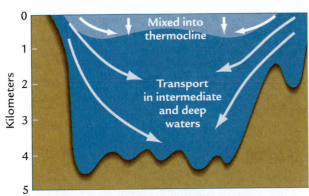

A Mixing into deep ocean (hundreds of years)

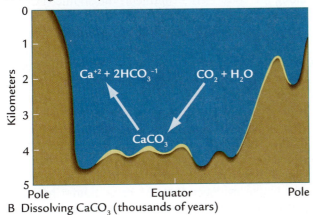

B Dissolving $CaCO_3$ (thousands of years)

FIGURE 20-4
Fate of the CO_2 pulse created by humans

Over decades to centuries, most of the excess CO_2 in the atmosphere will be mixed into the subsurface ocean (A). Over centuries to millennia, the excess CO_2 will make seawater more acidic and will dissolve seafloor $CaCO_3$ (B).

As a result, new CO_2 emitted to the atmosphere in the future will be added to older CO_2 left from emissions in prior decades and centuries. Even after CO_2 emissions generated by humans begin to drop, the remaining emissions will continue to push the cumulative atmospheric concentration to higher levels. Not until well after the peak in CO_2 emissions is reached does the rate of additional CO_2 input by humans drop below the rate of removal by the ocean. At this point, the CO_2 concentration begins to fall.

Over time, the acidity produced by the CO_2 absorbed in the deep ocean dissolves some of the $CaCO_3$ on the seafloor and neutralizes CO_2 in the ocean (see Figure 20-4B). Ultimately, the fate of our pulse of excess atmospheric CO_2 will be a slow-acting chemistry experiment carried out in the deep ocean.

Both of the projections in Figure 20-3 ignore the "equivalent CO_2" increases in other greenhouse gases that have amounted to almost half of the industrial era emissions to date. Methane production is likely to increase, but by relatively small amounts. Most of the land that can be used for growing rice is already in irrigation, and future increases in the extent of CH_4-emitting "rice wetlands" are likely to be negligible. As the number of humans on Earth levels out after mid-century (see Figure 20-1), the number of CH_4-emitting livestock that farmers tend should also stabilize. CFC emissions have already begun to decrease and should continue to do so in the future. As a result, the future trend will increasingly be dominated by CO_2.

► Effects of Future CO_2 Increases on Climate and the Environment

As atmospheric CO_2 levels rise in the future, Earth's climate will continue the warming already underway. Attempts to estimate the amount of future warming are limited mainly by uncertainties in three factors: the amount of CO_2 emitted by humans, the levels of atmospheric CO_2 reached as the excess carbon is redistributed among various carbon reservoirs, and Earth's sensitivity to higher CO_2 concentrations. Emissions of other greenhouse gases are also a source of uncertainty.

The two projections shown in Figure 20-3 span the likely range of increased CO_2 concentrations during the next few centuries. Here, we assume that the actual path will fall midway between the "optimistic" ($2 \times CO_2$) and "pessimistic" ($4 \times CO_2$) projections, reaching a $3 \times CO_2$ peak of 800–900 ppm, or three times the 280-ppm preindustrial level. As shown in Chapter 19, the best current estimate of Earth's sensitivity to a doubling of CO_2 is 3°C. For a peak of $3 \times CO_2$, the global average temperature increase should be 4.5°C to 5°C, of which less than a degree of that total has been registered since 1850 and roughly another 4°C is yet to come. Temperature increases at polar and near-polar latitudes will be twice that large.

20-3 A World in Climatic Disequilibrium

The high-CO_2 world of the future will be a place of major climatic disequilibrium because the high-CO_2 pulse will arrive too quickly for all parts of the climate system to come into full temperature equilibrium (see Chapter 1). Our future world will be characterized by a strange mix of faster-responding parts of the climate system like the atmosphere, land surfaces, vegetation, and upper ocean, and slower-responding components like the deep ocean and ice sheets. The faster-responding parts will register the CO_2-driven warming within just a few decades, the slower-responding deep ocean will react much more slowly, and the ice sheets on Greenland and Antarctica will not have fully reacted to the high-CO_2 levels even by the time the pulse begins to decline. Near the slow-responding ice sheets, which strongly influence regional climates by their high albedo and their effect on atmospheric winds, the lingering cold caused by the ice will suppress part of the fast response of the atmosphere and the nearby surface ocean to the higher CO_2 levels. Farther from the ice, the faster-responding parts across most of the climate system will react strongly to the new warmth. Because this kind of disequilibrium has not occurred in Earth's past, we have no perfect analogs for future climate.

20-4 Fast Climatic Responses in a 3 × CO_2 World

Several characteristics of a future 3 × CO_2 world are hinted at by trends that are already underway. Other aspects of future climate can be inferred from general lessons from Earth's long climatic history.

In a 3 × CO_2 world, north polar regions will warm enough to eliminate shallow permafrost, tundra, all summer sea ice and probably most winter sea ice (Figure 20-5). Both the extent and volume of Arctic sea ice have been falling during the 30 years of reliable satellite measurements, and doing so at accelerating rates in recent years. Summer ice volume has fallen by 70%, and winter ice volume by almost 40% (Figure 20-6).

If these trends continue, the Arctic Ocean will be completely ice-free in summer within a few decades, long before the peak 3 × CO_2 concentration is reached.

With little or no sea ice through the warmer part of the year when the Sun is above the horizon, solar radiation will deliver more heat to the ice-free Arctic, allowing the cold-season atmosphere to extract the stored heat and keeping the winter Arctic much warmer than it is now. The trend in winter sea ice volume also projects to zero in the near future. Small amounts of ice will probably continue to reform each winter in the central Arctic in a 3 × CO_2 world, but thick multi-year ice will no longer exist.

In the Antarctic, where most sea ice disappears during modern summers except in a narrow strip along the coast of the continent, the Southern Ocean will likely become entirely ice-free in summer. The extent of winter sea ice is difficult to predict, but it will likely be less extensive than it is now.

As a consequence of warmer Arctic winters and the absence or near-absence of winter sea ice, the north polar continents will be transformed by forests moving northward into areas that are now tundra. The climatic vulnerability of this region is evident both in the northward advance of forests 6,000 years ago caused by summer insolation values several percent higher than today (see Chapter 14, Figure 14-19) and in the evidence that extensive tundra has only existed in the Arctic for about the last 3 million years (see Chapter 7). Tundra will likely disappear in a 3 × CO_2 world, replaced by evergreen forests extending to the Arctic coast.

At present, near-surface and subsurface permafrost (permanently frozen ground) is widely distributed in

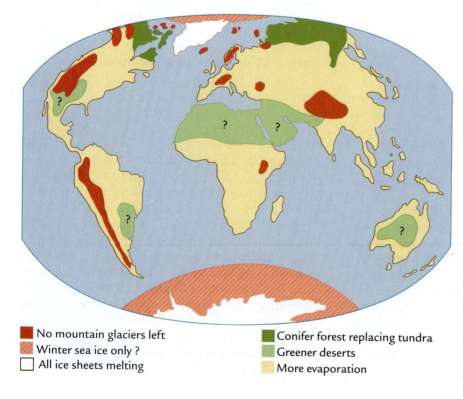

No mountain glaciers left
Winter sea ice only ?
All ice sheets melting
Conifer forest replacing tundra
Greener deserts
More evaporation

FIGURE 20-5
The 3 × CO_2 world

The 3 × CO_2 world that may come into existence in two or three centuries will be a disequilibrium world of fast- and slow-responding parts of the climate system.

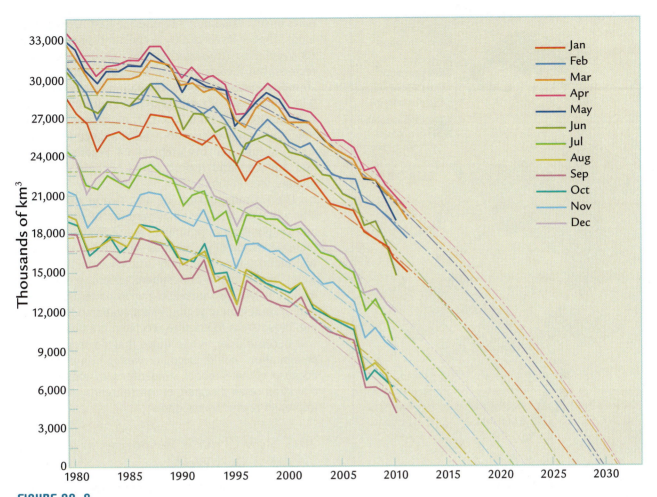

FIGURE 20-6
Projected volume of Arctic sea ice

If current trends continue, late-summer sea ice will disappear in a decade or two, and winter sea ice will reach very low volumes not long after. (ADAPTED FROM THE UNIVERSITY OF WASHINGTON POLAR SCIENCE CENTER PIOMAS DATA, WITH TREND DATA AFTER 2012 PROJECTED BY THE ARCTIC SEA ICE BLOG.)

the Arctic (Figure 20-7). In recent years, the discontinuous permafrost belt, much of which may have formed during the Little Ice Age, has begun to melt and will likely disappear in a $3 \times CO_2$ world. The much colder band of continuous permafrost belt farther north remains unthawed at present. In a $3 \times CO_2$ world, melting of its upper layers will occur, but the thaw is unlikely to reach most of the deeper layers.

No evidence of mountain glaciers older than about 7 million years has been found in North or South America, Africa, or Asia (see Chapter 7). The only place where mountain glaciers may have existed was in far northern Scandinavia and on Greenland, where the combined effect of high latitudes and high altitudes could have made temperatures cold enough for ice to persist. Almost every mountain glacier on Earth has already been retreating for a century or more, and the rates are accelerating (see Chapter 18, Figure 18-12). At the prevailing lapse rate of 6.5°C per km,

a 5°C warming in the future should cause a vertical retreat of glaciers up the sides of mountains by more than 800 m (roughly 2,500 ft), with larger changes at higher latitudes because of amplified warming. Because mountain glaciers can begin to respond to climate changes within just a few decades, most of them should have completely disappeared by the time we reach a $3 \times CO_2$ world.

Coupled models of climate and vegetation simulate a temperature-driven northward shift of mid-latitude hardwood trees like maple and beech during the next century, with warm-adapted trees like oak and hickory shifting north into the areas they abandon. By one reckoning, mid-latitude tree types already need to shift northward at an average rate of 10 m (> 30 ft) per day to stay within their areas of optimal growth conditions during the current warming. Most species can match this rate of movement by dispersal of pollen, seeds, and cones by winds and animals.

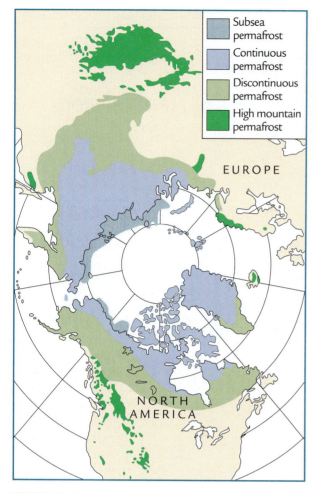

FIGURE 20-7
Melting permafrost
The large modern ring of permafrost around the Arctic Ocean will become vulnerable to gradual melting in the warmth of a $2 \times CO_2$ world. (FROM F. PRESS AND R. SIEVER, *UNDERSTANDING EARTH*, 2ND ED. © 1998 BY W. H. FREEMAN AND COMPANY.)

Legend:
- Subsea permafrost
- Continuous permafrost
- Discontinuous permafrost
- High mountain permafrost

With the larger and faster shifts expected in the next two centuries, however, some species may not be able to keep up with the northward displacement of their optimal environment.

Fauna and flora on mountainsides will also be affected by future warming. In order to remain in an optimal temperature regime, both will have to shift to higher elevations, and the transition will be easier for relatively mobile life-forms than less mobile ones. In some cases, the warming may push the preferred environment "off the top" of the mountains and cause species extinctions.

In the tropics and subtropics, scrub and tree vegetation were more prevalent 10 million years ago in several arid regions: the sub-Himalayan region of India and Pakistan, the western American High Plains, the South American pampas region of Argentina, and parts of sub-Saharan Africa and the East African highlands. Higher levels of CO_2 in the atmosphere allowed C3 vegetation (trees and shrubs) to survive in arid regions, but the gradual CO_2 lowering during the last 10 million years made C3 vegetation less competitive with C4 grasses. In the next two centuries, we will pass through the same $3 \times CO_2$ threshold, but heading in the opposite direction and at a much faster rate. As C3 shrubs and trees replace C4 vegetation, some arid and semiarid regions could become somewhat greener (see Figure 20-5).

The warming during the next two or three centuries will also alter regional patterns of precipitation and evaporation in significant ways. Evaporation will increase worldwide because warmer temperatures will permit air to hold more water vapor. With more water vapor in the atmosphere, global average precipitation will also increase, but in patterns that will vary from region to region. With evaporation increasing, those areas that fail to receive more precipitation will become drier, while those that do receive more precipitation could become wetter. Unfortunately, because climate model simulations of regional precipitation often disagree, moisture trends are difficult to predict region by region.

20-5 Slow Climatic Responses in a $3 \times CO_2$ World

The deep ocean is one of the slower-responding parts of the climate system, in part because of the 1,000 years or so it takes for an average molecule of water to circulate through the world ocean. Changes in the pattern and rate of this circulation in future centuries and millennia are difficult to predict, but in any case, the ocean will definitely be affected by **ocean acidification**. The ocean has an "alkaline" (non-acidic) pH that ranges today between 7.8 and 8.5 regionally, well above the 7.0 pH boundary that separates acidic and alkaline conditions. Because the pH scale is logarithmic, a shift from 8.0 to 7.0 in pH units represents a tenfold increase in relative acidity (or decrease in alkalinity).

During the 150 years of the industrial era, burning of fossil fuels and forests has added roughly 300 billion tons of carbon to the ocean, pushing the average pH of the ocean ~0.1 unit lower (toward more acidic values), equivalent to an alkalinity decrease of ~30%. Over the previous several thousand years, gradual deforestation by farmers had already added even more carbon to the ocean (see Chapter 16).

Several thousand billion tons of carbon remain buried in Earth's carbon reservoirs. If CO_2 emissions trends continue into the distant future (see Figure 20-2), most of this carbon will end up in the ocean, intensifying the acidification.

As a result, some coral species will have trouble forming their reef structures, which are made of aragonite, a form of $CaCO_3$. This will add to the problem already faced by many coral reefs: vulnerability to dying during years of unusual warmth caused by large El Niños. Other ocean organisms that will also have trouble making $CaCO_3$ shells include several kinds of shellfish and tiny snaillike organisms called pteropods that are one part of the plankton floating in the ocean.

Acidification of the oceans will proceed as long as we keep burning carbon, and it will occur whether we burn it slowly (using conservation measures) or quickly (ignoring the environmental consequences). The only way to reduce acidification is to leave more fossil-fuel carbon in the ground. Current projections are that the pH will have shifted at least another 0.2 units lower by 2100, and that the oceans may become so close to acidic by 2300 that coccoliths, one of the most abundant types of plankton (see Chapter 3) may no longer be able to form $CaCO_3$ shells. If this happens, transformations of the entire biochemistry of the ocean would occur.

An imperfect analog of our large future releases of carbon occurred some 55 million years ago (see Chapter 6). Estimated releases of 2,500–7,000 billion tons of carbon at that time killed most of the fauna in the deepest ocean and dissolved $CaCO_3$ that had been deposited on the seafloor in previous millennia. Our future carbon inputs to the ocean could approach the size of this earlier episode if we attempt to burn the vast reserves stored in oil shale and tar sands, but our emissions will occur at a much faster rate (over a few hundred, rather than a few thousand, years).

Other threats to the ocean have only recently been identified. The amount of oxygen in the world ocean has decreased in recent years, partly because warm water holds less oxygen in solution, and partly because warming of surface waters results in lower densities and less sinking of oxygen-rich water to great depths. This general ocean-wide loss of oxygen adds to a problem now underway in many coastal regions where excess runoff of chemical fertilizers from agriculture produces algal "blooms" that strip all of the oxygen out of the near-surface layers and leave seasonal "dead zones."

By far the slowest-responding parts of the climate system are the ice sheets on Greenland and Antarctica. Prior to ~7 million years ago, no ice sheet existed on Greenland, apparently because temperatures were too warm to permit ice accumulation at near-polar latitudes. In the future, we face a different situation: we start off in a world with the Greenland ice sheet already in existence (Figure 20-8), and we need to predict how it will respond to a warmer world.

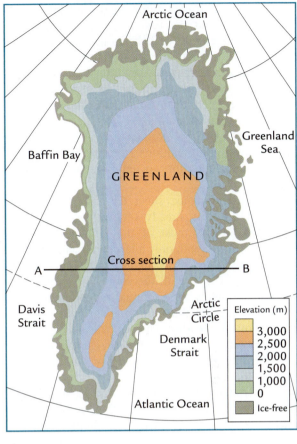

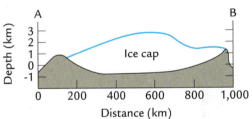

FIGURE 20-8

The Greenland ice sheet

Lower margins of the Greenland ice sheet will melt rapidly in the warmth of a 3 × CO_2 world, but the highest central surface of the ice sheet may be less affected. (FROM F. PRESS AND R. SIEVER, *UNDERSTANDING EARTH*, 2ND ED. © 1998 BY W. H. FREEMAN AND COMPANY; FROM R. F. FLINT, *GLACIAL AND QUATERNARY GEOLOGY*. © 1971 BY WILEY.)

Most climate scientists think that much of the Greenland ice sheet will survive the pulse of high CO_2 levels during the next few centuries and persist into the subsequent era of decreasing CO_2 concentrations (see Figure 20-3), but some disagree. Although the 2007 IPCC estimate suggested only a modest level of melting of Greenland ice during the rest of this century, this projection is now widely seen as an underestimate. Ablation monitored by satellites during recent summers has begun occurring at high elevations on the ice sheet, suggesting a time in the relatively near future when the entire ice surface will lie in a summer melting regime. Satellite data also show general acceleration in melting and retreat of coastal ice margins caused by rapid outflow in marginal ice streams (Figure 20-9). Present ice sheet models are ill-equipped to handle the many complexities of ice stream flow such as the role of marginal ice shelves, iceberg calving, and complications from bedrock topography, although attempts to improve the models are underway.

The enormous ice sheet on eastern Antarctica (Figure 20-10) was already very large by 13 million years ago. Today, most of this frigid ice sheet is starved for snow, with only a few centimeters per year falling across most of its high-elevation surface. In a warmer $3 \times CO_2$ world, the supply of snow should increase and allow faster annual accumulation of ice in the interior. But this change toward positive mass balance in the ice sheet interior will be opposed by faster flow in marginal ice streams. Current estimates are that the marginal ice losses will outweigh the gains in the interior.

The smaller West Antarctic ice sheet was less extensive 10 million years ago than it is now, if it existed at all. The extensive ice shelves that now fringe this ice sheet are vulnerable to destabilization because they are in contact with an ocean that will probably become warmer. Destabilization of the ice shelves can accelerate movement in ice streams flowing to the ocean and draw ice out of the interior of the continent (see Figure 20-9). This evidence suggests that the western Antarctic ice sheet will be considerably more vulnerable to melting in a $3 \times CO_2$ world, but the amount of melting is uncertain.

Projections of future sea level rise are very uncertain, mainly because of large disagreements about the amount of melting of Greenland and West Antarctic ice. The rate of sea level rise for most of the 1900s averaged 1.7 mm/year, or a total of about 17 cm for the full century (see Chapter 18, Figure 18-7). The 2007 IPCC report projected that the rate of rise by the year 2100 would range between 1.8 and 5.9 mm/year, but most scientists now regard this as a considerable underestimate. Since the late 1990s, the rate of rise has already accelerated to 2.5–3 mm/year, well above the low end of the IPCC projection.

Thermal expansion of a warming ocean is the largest factor in the 2007 IPCC estimate for the year 2100 at 1–4.1 mm/year. The IPCC also estimated a Greenland meltwater contribution of 0.1–1.2 mm/year, but several estimates that incorporate faster responses along coastal margins predict rates in the range of 1.6–5.4 mm/year. Along with a larger contribution from the West Antarctic ice sheet because of similarly revised analyses, the total rate of sea level rise could be as high as 0.78–2.08 mm/year by 2100. Over the rest of this century, sea level could rise by as much as half a meter or more, with further acceleration occurring during the 2100s. A sea level rise this large would impose major economic costs on many low-lying cities and on structures built along beach coasts. Constraining rates of future rise is a major problem, not just for climate scientists, but also for the world at large.

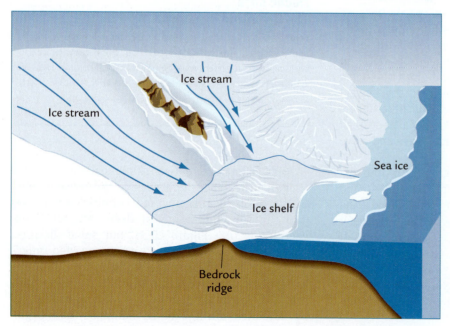

FIGURE 20-9
Vulnerable ice shelves
Ice from the interior of Greenland and Antarctica flows in ice streams to the shelves along the margin. In a warmer world, ice shelves may be vulnerable to destruction by rising ocean temperatures, which may in turn accelerate flow in the ice streams. (ADAPTED FROM R. A. BINDSCHADLER ET AL., "WHAT IS HAPPENING TO THE WEST ANTARCTIC ICE SHEET?" *EOS* 79 [1998]: 256–65.)

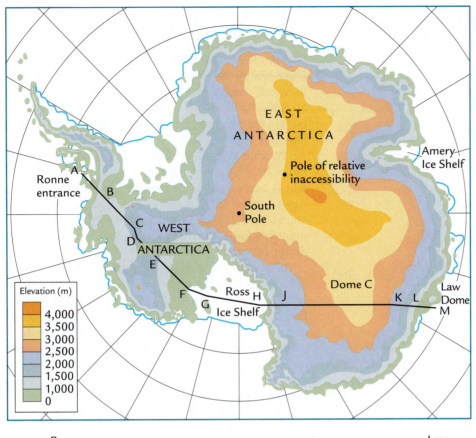

FIGURE 20-10
The Antarctic ice sheet
Fringing ice-shelf margins of the small, low-altitude West Antarctic ice sheet may begin to melt in a $3 \times CO_2$ world, but the higher, colder East Antarctic ice sheet will be less vulnerable. (FROM F. PRESS AND R. SIEVER, *UNDERSTANDING EARTH*, 2ND ED. © 1998 BY W. H. FREEMAN AND COMPANY; AFTER U. RADOK, "THE ANTARCTIC ICE," *SCIENTIFIC AMERICAN*, AUGUST [1985]: 100, BASED ON DATA FROM THE INTERNATIONAL GLACIOLOGICAL PROJECT.)

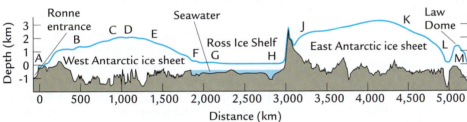

20-6 How Will the Greenhouse World Change Human Life?

The projected climatic transformations in the $3 \times CO_2$ worlds are large and in some cases dramatic, but what about the everyday perspective on how climate will change for people living at middle latitudes, including the children and grandchildren of students reading this book?

Consider the ~4.5°–5°C (9°F) global mean temperature increase in a $3 \times CO_2$ world. Poleward amplification of this temperature change by albedo feedback from snow and sea ice retreat will boost the warming at upper to middle latitudes (45° to 55°N) where many humans live to average increases of at least ~7°C (10°F) and more in winter months (Figure 20-11).

One way to evaluate these changes is in the context of modern seasonal changes. Today, on landmasses located at 45° to 55°N latitude, mean daily temperatures of ~0°C (32°F) in January rise to ~25°C (77°F) by July, and fall back again the following winter. This 25°C shift over six months amounts to an average change of 4.25°C (7.5°F) per month. In comparison, the anticipated 7°C change in temperature at these latitudes in a future $3 \times CO_2$ world will feel like a shift of more than a month and a half in both spring and autumn. Future Aprils will be like modern Mays/Junes, and future Novembers like modern Septembers/Octobers (Figure 20-12). Summer will last an extra three months, and winters will be shorter and less frigid.

These changes will come on slowly over the next two or three centuries and will be masked by normal year-to-year variability. As a result, the underlying trend will be only slightly apparent to a person whose life spans many decades. But if that same person fell into a Rip Van Winkle sleep of 50–100 years, the change in average climate upon awakening would be striking, although no doubt overwhelmed by far more bewildering societal and cultural transformations.

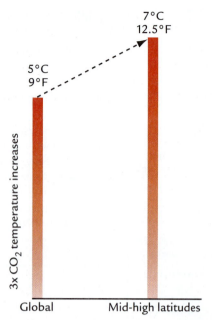

FIGURE 20-11
Larger temperature increases at high latitudes
Temperature changes caused by a tripling of atmospheric CO_2 concentrations will be larger than the global average at high mid-latitudes.

Some of the most striking changes will occur in regions where snow and ice retreat northward. For modern ski resorts situated at higher elevations in the lower mid-latitudes, a loss of three months of

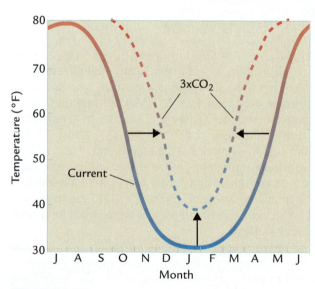

FIGURE 20-12
Changes in length of seasons
The warming caused by 3 × CO_2 levels will shorten winters and lengthen summers by about one and a half months each at high middle latitudes (45–60°N).

subfreezing winter cold will truncate the snowmaking season enough to make the ski business impossible. On the other hand, in far-northern regions of Hudson Bay and the Arctic coast of Siberia, the sea-ice retreat will open up new year-round shipping lanes. In addition, less energy will be required for winter heating at middle and higher latitudes, but more energy will be needed for summer cooling at lower and middle latitudes.

Warmer winters at high latitudes should extend the northern limit for growing many crops, especially wheat across broad regions of Canada and Russia. Also, higher CO_2 levels in the atmosphere should allow some plants to obtain the CO_2 necessary for photosynthesis more quickly without exposure to the drying effects of evaporation.

Unusually hot climate extremes will become common in a 3 × CO_2 world. Because the climate system is highly variable ("noisy") from year to year, record-breaking cold and warm temperature extremes occur in a random way whether or not the background climate is changing. But recent studies in regions such as the United States show that the balance between new record highs and record lows has already begun to shift (Figure 20-13). In the 1960s and 1970s, record highs and lows were more or less evenly balanced, but by the 2000s, record heat extremes outnumbered record cold by more than a factor of two. By this measure, the global warming signal has begun to emerge from the natural noise level in the climate system.

Another way to assess the extreme climate trend is to map the frequency of local heat waves or cold waves that lie well outside the normal range of variability for

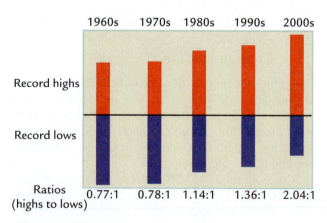

FIGURE 20-13
More hot temperature records than cold ones
Record hot temperatures in the United States have become much more common than cold ones in recent decades.
(ADAPTED FROM "RECORD HIGH TEMPERATURES FAR OUTPACE RECORD LOWS ACROSS U. S.," HTTP://WWW.UCAR.EDU/NEWS/RELEASES/2009/MAXMIN.JSP.)

each region (Figure 20-14). During the 1970s, extreme temperatures at Earth's surface were more or less evenly split between cold extremes (blue) and warm extremes (orange). But by the late 2000s, regions with blue extremes had shrunk considerably, mostly replaced by areas with warm extremes, and even some very hot extremes (dark red). Although year-to-year climate changes are still noisy and move around from region to region, the underlying global warming signal is now showing more clearly through the noise. Future climate will have ever-increasing areas of record-breaking heat across ever-larger regions.

In the tropics and subtropics, where 80% of humans live, moisture balances will also be greatly altered. The greater warmth will increase the global mean rate of evaporation, putting greater stress on water sources in already dry regions. In regions where evaporation increases but precipitation does not, agriculture will become more dependent on irrigation. People in many semiarid regions have for decades been pumping subsurface water for irrigation and have seen the water table drop. With further pumping, these regions may run out of usable groundwater, because of increasing salt content at depth.

Areas that depend on runoff from meltwater from seasonal snows and shrinking glaciers for irrigation late in the growing season will also feel the effects of decreases in both sources of water. In many arid areas, winter rains deliver water during a time when crops are not or cannot be grown. Today, large amounts of water during the first part of the growing season come from melting of winter snowpacks on nearby mountains. In the future, mountain snowpacks will be smaller because more winter precipitation will fall as rain and run off during times when crops cannot be grown. In addition, more of the run off from snows that still accumulate will arrive

in the spring before the crops need irrigating. As for mountain glaciers, runoff from gradual melting in the modern world arrives in late summer, a very arid time of year in many regions. As mountain glaciers shrink and eventually disappear, this late-summer source of irrigation water will decrease and in most places come to an end.

Because the warmer atmosphere takes up more water due to higher evaporation rates, precipitation will increase in some areas and on a global average basis. Some areas will benefit from increased rainfall, although more future precipitation will fall as localized summer cloudbursts, rather than as the longer-lasting precipitation associated with passage of frontal systems during cooler seasons.

Future sea level rise caused by melting of land ice and expansion of ocean water will have an adverse effect on coastal populations. The same factors that caused the 17-cm rise in sea level during the 1900s will continue to operate: thermal expansion of ocean water, melting of mountain glaciers, and partial melting of the Greenland and the West Antarctic ice sheets. A century-long sea level rise of 50 cm to as much as a meter doesn't sound very large, but in the flat coastal regions where many people live this estimate could translate into inland advances of the ocean margin of 500 meters to a kilometer. In heavily populated Bangladesh, a sea level rise within this range would displace tens of millions of people currently living less than one meter above current sea level. People in low-lying rural areas will have to retreat well inland as the sea invades.

In built-up coastal regions such as the eastern and Gulf of Mexico coasts in the United States, ever-greater engineering efforts (construction of sea walls and frequent pumping of sand to replenish

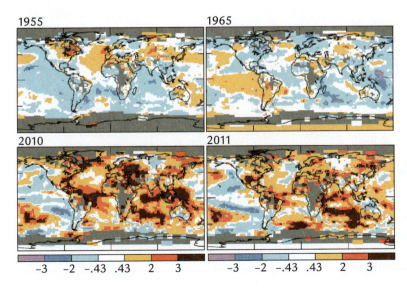

1955   1965

2010   2011

−3 −2 −.43 .43 2 3   −3 −2 −.43 .43 2 3

FIGURE 20-14

More hot temperature extremes than cold ones

Extreme temperatures far above the average have become more common globally in recent decades. (ADAPTED FROM J. HANSEN ET AL., "PUBLIC PERCEPTION OF CLIMATE CHANGE AND THE NEW CLIMATE DICE," *PROCEEDINGS NATIONAL ACADEMY OF SCIENCE*, IN PRESS.)

beaches) will become necessary and in fact have been underway for years in many regions. Insurance companies alert to the growing vulnerability of built structures have already begun to increase premiums sharply for homes and other buildings in such areas. Many of the world's great cities will have to build massive walls for protection against a large rise in sea level. Some scientists have also suggested that the intensity of the largest hurricanes will increase with global warming, although that issue is being debated.

Some of the largest changes caused by future global warming may have relatively little broad economic impact on humans and yet still be crucial from an ecosystem perspective. Large-scale melting of sea ice and permafrost and loss of tundra around the Arctic margins will hardly be a central issue in the lives of people who have never traveled to that frigid region. But the shrinking of sea-ice habitats could devastate polar bears and other marine mammals, as well as many other parts of the terrestrial polar ecosystem. These changes will in turn alter the lives and cultures of the few native people who still rely on hunting polar-adapted animals for survival. Loss of species in lower-latitude regions because of rapid northward or upward dislocation of preferred environments is also likely to be large. Future climate change will add to the ongoing destruction of ecosystems already underway because of widespread loss and fragmentation of preferred habitats.

Greenhouse Surprises?

The unprecedented mixture of slow and fast responses to the large $3 \times CO_2$ pulse of the future will create a climatic disequilibrium unprecedented in Earth's history. As a result, interactions within the climate system may produce unanticipated phenomena, or "greenhouse surprises." Several possibilities are summarized in this section.

20-7 Methane Clathrate Releases?

Methane exists as a gas in the atmosphere, but in Earth's colder regions it also occurs in a frozen form known as **methane clathrate,** a mixture of methane and slushy ice. Clathrates occur in deep-ocean sediments along continental margins, where the pressure produced by overlying water and sediments makes CH_4 stable at temperatures well above freezing (5°C or more). Clathrates also occur in the Arctic, both in shallow ocean sediments and in permafrost layers on land. The volume of CH_4 stored in these reservoirs is

enormous, far exceeding all carbon reservoirs in wetlands and livestock combined.

Without large changes in climate, most methane clathrates will remain trapped in their present slushy deposits, but with the large future warming projected for north polar regions, could large amounts of this potent greenhouse gas be liberated? Some scientists have expressed concern that if this were to happen, the emitted methane would further warm climate and set off a positive feedback loop that could drive far more global warming than our future CO_2 emissions (Figure 20-15).

Melting of the more vulnerable shallow permafrost and clathrates in less frigid regions will certainly release methane, but most scientists expect that the vast amounts of methane buried as clathrates at great depths in the most frigid regions of the high Arctic will remain largely unaffected. Because permafrost thaws from the top down, the rates of penetration of atmospheric heat from anthropogenic global warming will be slow. It would take many centuries for the highest surface temperatures caused by the future CO_2 pulse to penetrate deep into permafrost and ocean sediments, and before that happens CO_2 levels will be decreasing and the planet will have begun to cool. In this view, methane clathrates will not cause a large-scale greenhouse surprise. But we may yet have much to learn about this matter.

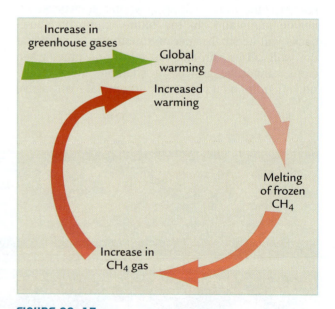

FIGURE 20-15
Methane clathrate feedback
Future warming of the coastal Arctic oceans and melting of Arctic permafrost could release enough methane to cause additional global warming in a positive feedback loop.

20-8 Chilling of the North Atlantic and Europe?

One frequently mentioned possibility is that faster melting of Greenland ice could send enough freshwater to the North Atlantic Ocean to lower its salinity and slow or stop formation of deep water. A relatively small drop in salinity in the Labrador Sea beginning in the 1970s lowered the density of the surface waters enough to prevent them from sinking during winters over the next two decades. Large-scale melting of the Greenland ice sheet or increased precipitation over the North Atlantic caused by greenhouse warming could conceivably add enough low-salinity water to slow formation of deep water.

A likely consequence of such a circulation change would be colder temperatures in northern Europe. Today, the heat extracted from North Atlantic surface waters during formation of deep water is carried eastward into Europe and helps to keep its climate warmer in winter than Canada or Siberia at the same latitude. Without this ocean heat, Europe would become colder than it is now. This scenario was the basis of the Hollywood movie *The Day After Tomorrow*, although it was carried to ridiculous extremes in the movie.

Recent reassessments suggest, however, that a major cutoff of deep-water formation in the Atlantic is unlikely. The planetary wind system will continue to drive relatively warm and salty surface water northward from the tropical Atlantic Ocean, where relatively cold winter air masses will continue to extract heat. Partial reductions (or relocations) of deep-water formation could occur with minor lowering of salinity, but any cooling effect on Europe is now considered unlikely to offset more than a fraction of the larger greenhouse-gas warming in a $3 \times CO_2$ world.

20-9 A Different Kind of Anthropogenic Climate Surprise: Nuclear Cooling?

One possible future climatic change does not depend on the greenhouse-gas emissions we will put in the atmosphere but instead depends on the dismal prospect of a future nuclear war. The primary result of a nuclear exchange would be firestorms caused by burning of fuel-rich urban and manufacturing areas that would send dark, carbon-rich smoke high into the atmosphere. The soot reaching high into the stratosphere would linger for five years or more and absorb part of the Sun's incoming radiation. As a result, solar radiation reaching the surface would be reduced and temperatures would cool.

Climate modelers have attempted to simulate nuclear exchanges on a variety of scales. Full nuclear exchanges between major nuclear powers would kill hundreds of millions of people immediately and would be catastrophic for even more of those left alive. Global temperatures would fall below those typical of the last glacial maximum 20,000 years ago (more than 5°C below present values) and stay there for 5 to 10 years. As a result, the growing season would shrink on every continent, disappearing entirely in the Ukraine and northernmost agricultural regions of North America. At the same time, global precipitation would fall by about half, further limiting agriculture. The resulting food shortages would result in mass starvation that would multiply the (already unimaginable) level of mortality in regions far from the nuclear explosions and firestorms. An immediate disaster of this kind stands in stark contrast with the large, but slowly arriving consequences of anthropogenic global warming.

Simulations indicate that even a more limited exchange between nations with smaller nuclear arsenals (such as India and Pakistan, which are highly hostile) would kill tens of millions of people immediately and result in a cooling deeper than any level reached during the last thousand years. Model results suggest that global precipitation would fall by 10–20%, and the strength of the wet summer monsoons in Africa and Asia could be cut in half, further increasing human mortality in many regions.

Early attempts during the 1970s to model nuclear exchanges produced effects so large that they were described as causing "nuclear winter." Later models scaled back the size of the changes somewhat to a level called "nuclear autumn." Recent models that incorporate a more complete range of interactive climatic processes and feedbacks have pushed the estimates back toward larger nuclear winter impacts. In any case, the climatic and other consequences of nuclear exchanges are too terrifying to be fully grasped.

▶ Climate Modification?

Until recently, the possibility of altering Earth's future climate by engineering was rarely discussed among mainstream climate scientists, some of whom feel strongly that we have an ethical obligation to minimize future CO_2 increases by vigorous conservation efforts before even considering other alternatives. But this subject is no longer entirely taboo, in part because international efforts to reduce emissions have produced so little progress.

20-10 Reducing Greenhouse-Gas Emissions to the Atmosphere

Limiting future emissions of greenhouse gases is a preemptive form of climate modification. By not emitting greenhouse gases in the first place, or not as much as some current projections suggest, we can alter future climate.

Compared to CO_2, methane has a smaller net greenhouse effect, but it is of interest in the context of climate modification because its 10-year average residence time in the atmosphere is much shorter than that of carbon dioxide. As a result, reductions in methane emissions would have nearly immediate effects in reducing warming. Possible mechanisms for reducing methane emissions include trapping emissions from landfills more efficiently, reducing the amount of methane escaping from coal mines and natural gas drilling, building more efficient diesel engines, and minimizing emissions from rice paddies by altering irrigation patterns and kinds or amounts of fertilizer.

One idea to avoid adding huge amounts of CO_2 to the atmosphere is to pipe excess industrial CO_2 into the ocean or into old oil fields. As with many technological solutions, cost is a major issue. Many CO_2-emitting sites lie far from the ocean or from oil reservoirs, and some of the CO_2 would have to be piped hundreds to thousands of miles at considerable cost. Pumping CO_2 into the ocean would obviously exacerbate the problem of acidifying the ocean.

Another suggestion is to use genetic manipulation to develop microbes that feed on CO_2 and turn it into some form of carbon other than greenhouse gases. If such a transformation turns out to be possible, we might be able to get rid of one greenhouse gas and even turn it into a source of usable energy.

Yet another suggestion focuses on CO_2 already in the atmosphere. One way that CO_2 leaves the atmosphere during glacial cycles is by planktic organisms fixing carbon in their shells and soft parts, dying, and sinking into the deep ocean (see Chapter 11). The element iron is central to this mechanism, as first proposed by marine biologist John Martin. Although not very abundant in ocean waters, iron acts as a limiting nutrient on plankton growth. In many areas of the world ocean where major nutrients like phosphorous and nitrogen are readily available, production of planktic organisms would increase if more iron were available. More than a dozen ship-based experiments that have spread iron across patches of ocean water covering 1–10 km² have shown a short-term increase in many kinds of plant plankton, and even blooms of a major group—diatoms (see Chapter 3). While these results are promising, it is not yet clear how they would apply to slower long-term additions of iron over vastly larger areas.

Using less energy through conservation measures is another way to reduce emissions. Several energy-saving measures that are inexpensive or relatively inexpensive have increasingly come into use, but these alone can't dramatically reduce atmospheric CO_2 emissions. Conservation methods that work on a

larger scale are more expensive and require support from local or national governments. The higher the costs of action rise, the more vocal the opposition from citizens and politicians becomes.

20-11 Reducing the Effects of the Sun's Heating

Future aerosol additions to the atmosphere from human activities are difficult to predict. The positive impact of current and future environmental cleanup efforts in some industrialized nations will be countered to an unknown extent by increased emissions in nations still industrializing.

Humankind faces the interesting dilemma suggested by the imaginary sequence in Figure 20-16. If all current emissions of industrial sulfates to the lower

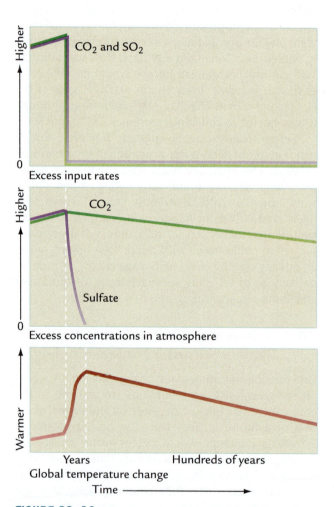

FIGURE 20-16
What if we abruptly ended CO_2 and SO_2 emissions?

If humans instantly eliminated all industrial and other emissions of greenhouse gases and SO_2, the sulfates and their cooling effect would soon disappear, but the CO_2 and its warming effect would linger for centuries.

layers of the atmosphere were suddenly eliminated, precipitation would wash out most of the aerosol load within a few weeks, eliminating their cooling effect. As a result, attempting to clean up our current sulfate aerosol emissions would cause a large and rapid additional warming.

In a sense, we have already run a small-scale version of this experiment. After the massive industrialization during World War II, sulfate aerosol loads in the atmosphere increased very rapidly (see Chapter 19, Figure 19-11), and the warming trend that had been underway for several decades slowed or was even reversed between the 1940s and 1970s (see Chapter 19, Figure 19-3). This small net cooling is generally attributed to increased aerosol levels in the atmosphere countering or even outweighing the effects of rising greenhouse-gas concentrations. Then, in the late 1970s, laws to reduce smokestack emissions began to be enacted, aerosol concentrations in the atmosphere fell, and global temperature began a rapid rise. This timing suggests that cleaning the air removed part of the previous cooling effect of the aerosols and allowed the underlying global warming trend to emerge.

In contrast, a simultaneously abrupt and total end to anthropogenic CO_2 emissions would lead to a much slower reduction in atmospheric CO_2 levels (see Figure 20-16). The ocean would take up half of the pulse of excess CO_2 that had been produced by previous human activities within 50 years or less, but 10–20% of the total pulse of emissions would remain in the atmosphere for many millennia. As a result, part of the anthropogenic warming would persist as long as this extended tail end of the excess CO_2 pulse lingered in the atmosphere. The net effect of reducing aerosol and CO_2 emissions would be additional warming.

Atmospheric chemist Paul Crutzen proposed that the warming effect of greenhouse gases could be countered by sending sulfate aerosols high into the stratosphere. At those levels, industrial aerosols would mimic the effects of sulfate aerosols from major volcanic eruptions by blocking solar radiation and cooling climate (see Chapter 19). Adding enormous amounts of aerosols might cool Earth's climate even in the absence of effective actions to reduce emissions of CO_2 and other greenhouse gases.

Because gravity pulls stratospheric aerosols down into the troposphere within a few years (where they are then soon removed by rainfall), sustaining a high enough concentration at those altitudes to offset global warming would require a constant supply of aerosols, by one estimate half a million tons per year. Proposed ways of delivering such huge amounts of aerosols include: balloons, artillery shells fired into space, or a fleet of high-flying planes much larger in size than modern commercial airliners. Aside from issues of cost, critics note that the added sulfates would return

to Earth as acid rain and increase the acidification of terrestrial environments and the ocean.

Another proposal is to send trillions of tiny mirrors into orbit to reflect a few percent of the incoming solar radiation back into space. This mechanism would of course add to all the debris ("space junk") we have already put in orbit.

Yet another idea is to reduce the amount of black carbon soot that settles on snow and sea ice at high-Arctic latitudes. These particles accelerate melting of ice and snow because of their dark, heat-absorbing properties. Because most black carbon in the Arctic comes from annual burning of crops and pasture grasses at high latitudes, the darkening effect of these aerosols could be reduced by delaying burning until after most of the late-spring snow melted. This approach is a possible short-term "fix" that could buy some time while other efforts continue to offset the larger effect of rising CO_2 concentrations.

Because all of the proposed climate-engineering mechanisms cost money, each will face hard questions about cost-effectiveness. Interactions between climate scientists, engineers, and professionals in disciplines like economics have begun to produce evaluations of the ratio between costs and benefits (and risks), as shown in Figure 20-17. The proposed

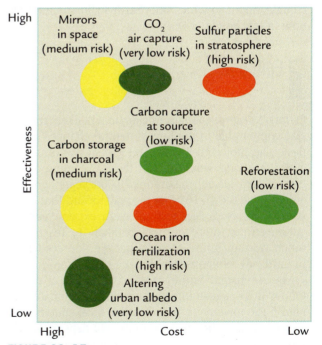

FIGURE 20-17

Cost-effectiveness of climate modifications

Present projections of the effectiveness of proposed climate modifications versus their cost range widely. (ADAPTED FROM THE ROYAL SOCIETY, "GEOENGINEERING THE CLIMATE: SCIENCE, GOVERNANCE AND UNCERTAINTY," SEPTEMBER 2009, HTTP://ROYALSOCIETY.ORG/ GEOENGINEERING-THE-CLIMATE/.)

measures plotted in this figure relate to current science-based knowledge, but projections of this kind are very unlikely to stay the same in the future. Technological developments occur so fast that today's wisdom is out of date tomorrow.

On a philosophical level, if we (meaning humanity) can manipulate climate, and if we choose to do so, what climate will we pick? Our first instinct may be that the safest path will return Earth to a "natural" climate, but what is "natural"? This is not a simple question.

We are not likely to go back to the preagricultural natural world that existed 10,000 years ago. The massive clearance of forests and plowing of prairies during the last century and a half will not be reversed because humanity needs the cropland and pastureland to grow enough food for the current global population, not to mention the 10 billion people expected by the middle of this century (see Figure 20-1).

Perhaps a consensus may emerge that the optimal "natural" world was the not-too-warm, not-too-cold climate just prior to the industrial era, even though Earth's surface had already been massively transformed by that time. Or perhaps humanity will be unable to agree on a desirable future climate. In any case, humankind in the near future will likely be weighing the merits of whether to engineer a different climate, and if so which one. Earth's climatic future is now in our hands.

Epilogue

Estimated future changes in climate will likely reach a size comparable to the largest natural changes of the past. The projected 5°C average global warming in a $3 \times CO_2$ world would match the 5°C cooling at the most recent glacial maximum 20,000 years ago. We are now ~0.8°C (~16%) of the way toward that ~5°C warmer future. Unless technology or extreme conservation efforts intervene, we are likely to reach that much warmer future at rates that are unprecedented in Earth's 4.5 billion year climatic history.

In the distant future, most of the pulse of excess CO_2 will disappear into the ocean, and climate will cool back toward its preindustrial level, but not all the way back. The rate of decline will depend mostly on how much of the carbon reservoir buried in the ground we extract between now and then.

Part of the excess CO_2 pulse (about 15%) will linger in the atmosphere for tens of thousands of years and keep climate too warm for new ice sheets to accumulate on North America or Eurasia (Figure 20-18). From the perspective of Earth's past climatic history, it is astounding to realize that we have now

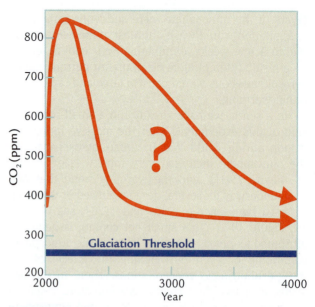

FIGURE 20-18
Long after the fossil-fuel pulse

As the pulse of industrial CO_2 is absorbed in the ocean, atmospheric concentrations will drop, but they will not return to levels that permit future northern hemisphere glaciations.

put an end to northern hemisphere glacial cycles that had been underway for almost 3 million years.

What other major transformations will we cause in the future?

Key Terms

CO_2 fertilization effect (p. 396)

ocean acidification (p. 400)

methane clathrate (p. 406)

Review Questions

1. What factors will determine how much CO_2 humans add to the atmosphere in the future?

2. Where will all of this excess carbon eventually go?

3. In what way will the future CO_2 warming be like and unlike past CO_2 warmings?

4. In a $3 \times CO_2$ world, where will ice of any kind still be found on Earth?

5. Will future temperature changes be readily apparent to the average person? Why or why not?

6. What are the disadvantages in drastically reducing our industrial emissions of sulfur and carbon?

7. Yes or no: Do you think that future global warming is a major concern?

Additional Resources

Basic Reading

Henson, R. 2006. "The Rough Guide to Climate Change." London, Rough Guides, Ltd.

World Climate Research Program (WCRP) Web site. http://www.wcrp-climate.org/. Last accessed March 17, 2013.

National Climatic Data Center Web site. "Global Warming FAQs." http://www.ncdc.noaa.gov/oa/ climate/globalwarming.html. Last accessed March 17, 2013.

Intergovernmental Panel on Climate Change (IPCC) Web site. http://www.ipcc.ch. Last accessed March 17, 2013.

Advanced Reading

Archer, D., and A. Ganapolski. 2005. "A Movable Trigger: Fossil Fuel CO_2 and the Onset of the Next Glaciation," *Geochemistry, Geophysics, Geosystems* 6, QO5003, doi:10.1029/2004GC000891.

Graedel, T. E., and P. J. Crutzen. 1997. *Atmosphere, Climate, and Change.* New York: Scientific American Library.

IPCC. 2007. *Climate Change 2007: The Physical Science Basis.* Cambridge, UK and New York: Cambridge University Press.

National Research Council. 2003. *Understanding Climate Change Feedbacks.* Washington, D. C. The National Academies Press.

Appendix 1: Isotopes of Oxygen

Oxygen is present in abundance in several key parts of Earth's climate system: as the second most abundant gas (O_2) in the atmosphere, as water vapor (H_2O_v) in the atmosphere, as a component of water molecules (H_2O) in the ocean and in lakes, and as frozen water in ice sheets. These reservoirs interact with each other and exchange oxygen.

Oxygen occurs in nature mainly as two isotopes. The lighter ^{16}O isotope accounts for almost 99.8% of the total amount, and the heavier ^{18}O isotope for most of the rest. The ratio of ^{18}O to ^{16}O is approximately 1/400, equivalent to a value of about 0.0025.

Individual measurements of the $^{18}O/^{16}O$ ratio in natural materials are reported as departures in parts per thousand (‰) from a laboratory standard:

$$\delta^{18}O \atop (\text{in ‰}) = \frac{\left(^{18}O/^{16}O\right)_{\text{sample}} - \left(^{18}O/^{16}O\right)_{\text{standard}}}{\left(^{18}O/^{16}O\right)_{\text{standard}}} \times 1,000$$

All measurements of samples are referenced to a standard in order to establish a common reference point for analyses in all laboratories. The reason that the measured ratio is multiplied by 1,000 is simply for convenience: it converts the extremely small variations in an already small ratio to a more workable numerical form (values that range between about +3‰ and −55‰).

Samples with relatively large amounts of ^{18}O (compared with ^{16}O) are said to have more positive $\delta^{18}O$ values and are referred to as ^{18}O-enriched (or ^{16}O-depleted). Samples with relatively small amounts of ^{18}O have more negative $\delta^{18}O$ values and are referred to as ^{18}O-depleted (or ^{16}O-enriched).

Modern δ¹⁸O Values in Water, Ice, and Water Vapor

Scientists are mainly interested in the ocean and the ice sheets because these reservoirs contain large amounts of oxygen and have exchanged it over time (Figure 1). The average $\delta^{18}O$ value of ocean water is set by definition near 0‰ (more specifically −0.1‰), but the ratio actually varies between about −2‰ and +3‰ across ocean regions and at different depths.

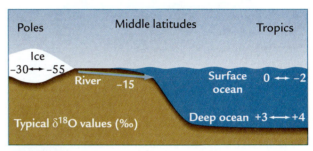

FIGURE-1

δ¹⁸O values in the modern world

In the modern ocean, $\delta^{18}O$ values vary from 0 to −2‰ in warm tropical surface waters to as much as +3 to +4‰ in cold deep-ocean waters. In present ice sheets, typical $\delta^{18}O$ values reach −30‰ in Greenland and −55‰ in Antarctica.

Most of this range results from changes in the temperature of ocean water. For each 4.2°C increase in temperature, the $\delta^{18}O$ ratio decreases by 1‰ (that is, ^{18}O becomes less abundant in relation to ^{16}O). As a result, warm waters in the surface ocean have more negative $\delta^{18}O$ values that range from 0‰ to −2‰, while colder waters deeper in the ocean have more positive values that range from +3‰ to +4‰.

The ice sheets are the other major oxygen reservoir of interest. The water vapor that supplies snow to the ice sheets comes from the ocean. The tropical atmosphere is rich in water vapor (H_2O_v) evaporated from the warm tropical ocean (see Chapter 2). The natural circulation of the atmosphere transports water vapor to higher latitudes and higher altitudes, where it condenses and falls to Earth's surface as precipitation.

The transport of water vapor occurs through repeated cycles of evaporation and precipitation (Figure 2). Because the lighter ^{16}O isotope evaporates more readily, it tends to be preferentially extracted from the low-latitude ocean and sent toward higher latitudes. This transfer leaves the tropical ocean enriched in ^{18}O. In addition, the heavier ^{18}O isotope is more easily removed from the atmosphere when condensation and precipitation occur, leaving the water vapor that remains in the atmosphere even more enriched in ^{16}O and the low-latitude ocean still more enriched in ^{18}O. These enrichment processes are called **fractionation**.

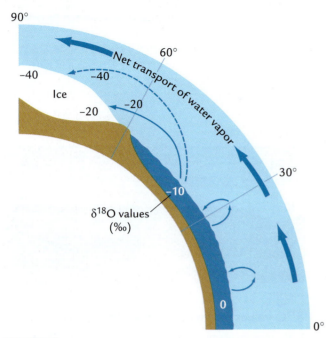

FIGURE-2

Isotope fractionation

As water vapor moves from the tropics toward the poles, it is enriched in the ^{16}O isotope during each step of evaporation and condensation. This fractionation process makes the $\delta^{18}O$ values of snow falling on (and stored in) ice sheets more negative (^{16}O-rich).

Each step in the fractionation process decreases the $\delta^{18}O$ value of the water vapor by ~10‰ in relation to that of the ocean water left behind. Because the water vapor that reaches the ice sheets is highly enriched in ^{16}O, the $\delta^{18}O$ composition of the ice sheets is very negative: −30‰ over much of Greenland and −55‰ over central Antarctica. Fractionation also occurs at high altitudes in lower and middle latitudes, because air that reaches high elevations has been through the same processes and has become similarly enriched in ^{16}O. For this reason, glacial ice high on tropical mountains has relatively negative $\delta^{18}O$ values.

The process of isotopic fractionation is also accompanied by the progressive removal of water vapor from the air, because cooler air holds much less water vapor than warmer air. As a result, air masses that are the most enriched in ^{16}O contain the smallest amount of water vapor.

Local Complications

As a result of the fractionation effect, $\delta^{18}O$ values in today's surface ocean do not follow the trend that would be expected if temperature were the only controlling factor. Modern tropical surface waters at 25°C have $\delta^{18}O$ values near 0‰. Using the temperature/$\delta^{18}O$ relationship defined earlier, high-latitude surface waters at temperatures of 0°C (just above the −1.8°C freezing point of seawater) should have $\delta^{18}O$ values of about 15‰. Instead, these waters have $\delta^{18}O$ values not much different from those of tropical surface waters.

The reason for these negative values is that high-latitude rivers carry water fed by precipitation with $\delta^{18}O$ values averaging near −15‰. The $\delta^{18}O$ value of each river depends on the degree of fractionation that has occurred in the precipitation that reaches its watershed. This annual delivery of ^{16}O-rich river water amounts to just a small fraction of the total volume of the high-latitude surface ocean, but it drives the oceanic $\delta^{18}O$ composition toward more negative values than expected from the temperature relationship. Coastal surface waters heavily affected by such rivers are more negative in $\delta^{18}O$ than the tropical ocean, and high-latitude surface waters even in regions well away from rivers have values comparable to those of the tropical ocean. This dilution effect by river water is also closely related to similar effects on ocean salinity: each 1.0‰ decrease in the $\delta^{18}O$ value of ocean water is accompanied by a 0.5‰ decrease in salinity due to delivery of fresh (nonsaline) water.

Climatic Application 1: Changes in Seawater $\delta^{18}O$

Through time, the $\delta^{18}O$ composition of ocean water is affected by changes in ocean temperature and in the amount of ^{16}O-rich water extracted and stored in the ice sheets. The effect of past temperature changes on the $\delta^{18}O$ values of ocean water is the same as that in the modern ocean: a 1‰ decrease for each 4.2°C warming (and conversely). Because the size of past temperature variations has varied from region to region, the effect on $\delta^{18}O$ has also varied on a local basis.

During colder climates such as the last glacial maximum 20,000 years ago, the large ice sheets on North America and Europe held enough ^{16}O to leave the ocean enriched in ^{18}O by an average of about 1‰. In contrast, when no ice sheets were present on Earth, the ^{16}O-rich ice that is now trapped in the Greenland and Antarctic ice sheets was instead water in the ocean. At that time, the mean ocean $\delta^{18}O$ value was about 1‰ lower than it is now.

These changes in past $\delta^{18}O$ values are recorded in the $CaCO_3$ shells of two kinds of foraminifera. Planktic foraminifera live mainly in the upper 100 meters, and their shells contain oxygen taken from bicarbonate ions (HCO_3^{-1}) in the near-surface ocean. When these floating organisms die, their shells fall to the seafloor and accumulate as a permanent record of past values of seawater $\delta^{18}O$ at the surface.

In comparison, benthic foraminifera live on the seafloor and within the uppermost layers of deep-ocean sediment, and their shells contain oxygen taken from bicarbonate ions in deep water.

Sediment samples taken from the ocean are sieved to remove the mud and silt in order to isolate the sand-sized fraction from which foraminifera are individually picked. Typical $\delta^{18}O$ analyses require a few milligrams of sample, usually less than a dozen foraminifera. The $CaCO_3$ shells are dissolved in acid to produce CO_2 gas, which is then analyzed in a **mass spectrometer**, an instrument capable of detecting the small difference in atomic mass between the ^{16}O and ^{18}O isotopes.

Climatic Application 2: Changes in Ice Sheet $\delta^{18}O$

Changes in $\delta^{18}O$ values of layers in the ice sheets are also important to studies of past climate. Samples are drilled out of ice cores and melted, and the water is vaporized to form the gas H_2O_v for analysis on mass spectrometers. The $\delta^{18}O$ values within ice sheets can vary by 5‰ or more as climate changes. In the ice, more negative values are typical of colder climates, and less negative values indicate warmer climates. In Greenland, ice from interglacial times typically varies between −30‰ and −35‰, while glacial-age ice falls between −35‰ and −40‰. In Antarctica, interglacial values are typically between −50‰ and −55‰, while glacial values fall between −55‰ and −60‰. Note that the $\delta^{18}O$ responses of the ocean and the ice sheets are opposite in direction: as marine $\delta^{18}O$ values become more positive during glacial climates, the $\delta^{18}O$ values in ice cores become more negative (and conversely).

These $\delta^{18}O$ changes in the ice sheet layers reflect several influences (Figure 3). One factor is the temperature of the snow that settles on the ice sheets. "Warm" (wet) snow can form at temperatures near freezing, but cold (dry, powdery) snow forms at much colder temperatures. The $\delta^{18}O$ value of snow falling on Greenland today trends 0.7‰ more negative for each 1°C drop in temperature of the air in which it forms. This temperature/$\delta^{18}O$ relationship holds for modern seasonal changes in the temperature of the air masses that deliver the snow, and also for the cooling of air at higher elevations on the ice sheets. But this modern relationship did not always apply in the past: $\delta^{18}O$ values of glacial-age ice in Greenland suggest temperatures about 10°C colder than those today, but direct measurements of the temperature of the ice indicate a cooling of 15°C or more.

Several other factors can affect the $\delta^{18}O$ values recorded in ice cores, such as changes in the source of the water vapor, in the path of transport to the ice sheets, and in the season when the precipitation falls (Table 1). These complications arise because the water vapor that supplies snow to the ice sheets comes from several sources, each with a different initial $\delta^{18}O$ value, and it follows different paths of transport. The longer the distance the water vapor travels, the more negative is its $\delta^{18}O$ value, because it has evaporated and condensed repeatedly along the way (see Figure 2). Changes in the relative amounts of water vapor coming from different sources can alter the mean $\delta^{18}O$ value of the snow that falls on the ice through the span of a year. Changes in the seasonal balance of water vapor delivery can also affect mean annual $\delta^{18}O$ values in the ice: more snow in the colder winter season results in lower $\delta^{18}O$ values.

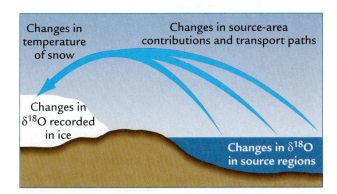

FIGURE-3

Controls on variations in ice-core $\delta^{18}O$

The $\delta^{18}O$ values recorded in ice cores vary with changes in the local temperature of the snow that falls on the ice, changes in the relative contributions among the source areas of water vapor, and $\delta^{18}O$ changes within the source regions.

Table 1	Causes of $\delta^{18}O$ Changes Recorded in Ice Cores	
Negative ⟵	Change in $\delta^{18}O$ values ⟶	Positive
Colder	Air temperature over ice	Warmer
Distant	Proximity of source region	Close
Low $\delta^{18}O$	$\delta^{18}O$ composition of source	High $\delta^{18}O$
High	Elevation of ice	Low
Winter	Primary season of precipitation	Summer

Cave Calcite $\delta^{18}O$ Deposits (Speleothems)

In recent years, high-resolution measurements of $\delta^{18}O$ in stalactites and stalagmites from caves have become an important new source of information on past climates. The oxygen in these $CaCO_3$ deposits originates from rainwater that was deposited at the surface and dripped down through the soil and bedrock into the caves. As a result, the relative abundances of the two isotopes of oxygen (^{18}O) and ^{16}O) in these $CaCO_3$ deposits are an index of processes in the overlying atmosphere.

Cave calcite deposits are most frequently used as an index of the intensity of past monsoon circulations. When monsoons are at maximum strength, they deliver enormous amounts of highly fractionated (^{16}O-rich) water vapor that has been evaporated from the ocean and carried for large distances. When the monsoons are weaker, other local sources deliver ^{18}O-rich rainwater to the caves. Temperature has little effect on cave deposits, because subsurface temperatures vary only within a narrow range.

Cave deposits can be precisely measured by the Th/U radiometric method. The high level of accuracy—to within a couple of thousand years over the last several hundred thousand years—is better than that in marine sediments and ice cores, except for annually layered sediments and ice spanning the last 10,000 years.

Both ^{13}C and ^{12}C are stable (nonradioactive) isotopes of carbon that occur naturally in Earth's vegetation, water, and air. The ^{12}C isotope accounts for more than 99% of all the carbon present on Earth, and ^{13}C accounts for most of the rest. A small amount exists as radioactive ^{14}C. Geochemists who analyze material for its carbon isotope composition measure small variations of less than 0.01 around the average $^{13}C/^{12}C$ ratio.

Similar to the convention used for oxygen isotopes, measurements of $^{13}C/^{12}C$ ratios are reported as departures in parts per thousand (‰) from a laboratory standard:

$$\delta^{13}C \atop (\text{in ‰}) = \frac{\left(^{13}C/^{12}C\right)_{sample} - \left(^{13}C/^{12}C\right)_{standard}}{\left(^{13}C/^{12}C\right)_{standard}} \times 1{,}000$$

All measurements are referenced to standards supplied by the National Bureau of Standards for use as a common reference point. Like the oxygen isotope ratios, carbon isotope ratios are multiplied by 1,000 to convert the very small measured variations in an already small ratio to a more handy numerical form. As a result, $\delta^{13}C$ values for carbon that occurs in oxygen-rich conditions fall between −25‰ for some kinds of vegetation on land to +2‰ for carbon dissolved in ocean surface waters in some regions.

For carbon that forms in the absence of oxygen (in "reducing conditions"), $\delta^{13}C$ values can be far more negative, around −50‰ to −60‰.

Carbon samples with relatively large amounts of ^{13}C compared with ^{12}C have more positive $\delta^{13}C$ values and are referred to as ^{13}C-enriched (or ^{12}C-depleted). Samples with relatively small amounts of ^{13}C compared with ^{12}C have more negative $\delta^{13}C$ values and are referred to as ^{13}C-depleted (or ^{12}C-enriched).

Fractionation during photosynthesis causes changes in $\delta^{13}C$ values (Figure 1). As the plants take inorganic carbon and turn it into organic carbon, they incorporate the ^{12}C isotope into their living tissue more easily than the ^{13}C isotope. This discrimination in favor of ^{12}C shifts the $\delta^{13}C$ of organic matter toward $\delta^{13}C$ values that are more negative (^{12}C-rich) than the initial inorganic carbon source.

For example, plant plankton in the ocean take inorganic carbon from seawater with a $\delta^{13}C$ value near 0‰ and convert it to organic carbon with a $\delta^{13}C$ value near −22‰. Some of the organic carbon is sent to the deep ocean and remains there, but most is oxidized back to inorganic form and recycled at shallower depths. The net export of a small fraction of ^{12}C-rich organic carbon to the deep ocean leaves the remaining surface water enriched in ^{13}C, with $\delta^{13}C$ values of 1‰ or slightly higher.

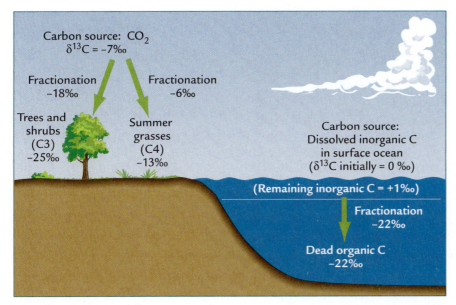

FIGURE-1

Photosynthesis and carbon isotope fractionation

Photosynthesis on land and in the surface ocean converts inorganic carbon to organic form and causes large negative shifts in $\delta^{13}C$ values of the organic carbon produced.

Overall, organic carbon forms a small fraction of the total carbon reservoir in the ocean, where inorganic carbon predominates. The combined effect of the two reservoirs (a small amount of organic carbon with a $\delta^{13}C$ value near $-22‰$ and a large amount of inorganic carbon with a $\delta^{13}C$ value near $+1‰$) yields a mean $\delta^{13}C$ value of $\sim0‰$ for all of the carbon in the ocean.

Inorganic CO_2 in the atmosphere with a $\delta^{13}C$ value of $-7‰$ is the carbon source for photosynthesis by plants. All trees, as well as most shrubs and cool-climate grasses, use a type of photosynthesis called the C3 pathway, which produces organic tissue with $\delta^{13}C$ values in the range of $-21‰$ to $-28‰$, and an average value of $-25‰$. Some shrubs and most grasses that grow in hot climates during the summer season use a different kind of photosynthesis called the C4 pathway, which produces vegetation with less negative $\delta^{13}C$ values, ranging between $-11‰$ and $-15‰$ and averaging $-13‰$. By far the largest amount of organic biomass on Earth's land surfaces resides in C3 trees, and as a result the average $\delta^{13}C$ of vegetation on land (and of litter in soils) is $-25‰$.

For purposes of reconstructing past climates, the $\delta^{13}C$ composition of a wide range of fossilized materials can be analyzed on mass spectrometers. The same $CaCO_3$ shells of marine foraminifera that yield oxygen for $\delta^{18}O$ analyses also provide carbon for $\delta^{13}C$ analyses. Acid can be used to dissolve the shells and produce CO_2 for analysis. Living vegetation or its organic residue in soils as well as organic matter in plankton and ocean sediments can also be analyzed on mass spectrometers after the organic carbon is burned to produce CO_2 gas.

Glossary

(Terms are defined as they are used in climate studies.)

ablation The loss of snow or ice from a glacier by melting, calving, or other processes.

accumulation The addition of snow to a glacier.

adiabatic Having to do with an increase in pressure that raises the temperature of a parcel of air (adiabatic warming) or a decrease in pressure that lowers that temperature (adiabatic cooling).

aerosols Extremely small particles or droplets carried in suspension in the air.

albedo The decimal fraction or percentage of incoming solar radiation reflected from a surface.

albedo-temperature feedback A positive feedback that amplifies an initial temperature change by altering the amount of snow cover or sea ice and changing the amount of solar radiation absorbed by Earth's surface.

aliasing The misleading (because unrepresentative) signals that result from sampling a record of climate change at too low a resolution.

alkenones Complex organic molecules found in fossil shells of plant plankton and used to reconstruct past temperature.

amplitude Half the height between peaks and troughs in a regular wave form.

Antarctic bottom water A dense cold-water mass that forms near the Antarctic continent by extreme chilling of surface waters, sinks, and flows along the seafloor below a depth of 4 km.

Antarctic intermediate water A water mass that forms in the Southern Ocean by chilling of seawater exposed in or near sea ice, sinks, and flows northward at depths of 1 to 2 km.

anthropogenic CO_2 increase The steadily increasing concentration of CO_2 in the atmosphere over the last 200 years due to human activities.

anthropogenic forcing All human-related factors that cause climate change.

aphelion The point in Earth's slightly eccentric orbit at which it is farthest from the Sun.

asthenosphere A partially molten layer of rock in the upper mantle that is weak enough to flow and cause movement of the overlying lithospheric plate.

axial precession The wobbling movement of Earth's axis of rotation, which causes it to point in different directions over a cycle of 26,000 years.

back radiation Electromagnetic energy at long (infrared) wavelengths emitted from any material with a temperature above absolute zero (0K).

Barents ice sheet An ice sheet that covered the present Barents Sea, north of Scandinavia, during orbital-scale glaciation cycles.

basal slip The rapid sliding of an ice sheet across its water-lubricated bed, especially in a region where ice streams lie above water-saturated sediment.

bedrock pinning point A high-standing protrusion of bedrock that lies beneath the margin of an ice sheet and slows the flow of ice into the ocean by frictional resistance.

benthic foraminifera Sand-sized organisms (protozoans) that live on and in the seafloor and form shells of $CaCO_3$.

biomass The amount of living matter in a region; also, organic matter used as a source of energy.

biome A region on Earth with a distinctive community of plants.

biome model A vegetation model that simulates the major vegetation type (for example, grassland or desert scrub) that can exist in a region under a given set of climatic conditions.

biosphere The part of the Earth system that supports life, including the oceans, land surfaces, soils, and atmosphere.

biotic proxy An index of past climate change based on measurable variations in the type or abundance of climate-sensitive organisms.

bipolar seesaw The tendency for climatic fluctuations in the Northern Hemisphere and the Southern Hemisphere to occur with opposite phasing.

black carbon Dark particles of organic origin; basically "soot."

Black Sea flood hypothesis The hypothesis that melting ice sheets caused rising ocean waters to flood into an ancient glacial lake, displacing humans and forming the Black Sea.

boundary conditions The initial configuration of Earth's properties chosen for a model simulation (such as land-sea distribution, mountain elevation, and atmospheric CO_2 concentration).

brown cloud A haze of carbon aerosols emitted by human activities, especially from cities in Asia.

burial flux The rate of deposition of a substance in a sedimentary reservoir measured in units of mass per unit of area per unit of time.

C3 pathway The means by which trees and most shrubs (about 95% of all land plants) obtain CO_2 from the air during the initial step of photosynthesis.

C4 pathway The means by which grasses that grow during the warm season (about 5% of all land plants) obtain CO_2 from the air during the initial step of photosynthesis.

calibration interval The interval of time (usually 50–100 years) over which the width or density of tree rings can be correlated with historical observations of temperature and precipitation change.

calorie The amount of energy required to raise the temperature of 1 g of water by 1°C.

calving The process by which a large block of ice breaks off from the margin of a glacier and forms an iceberg that floats in the ocean or in a lake.

carbon isotopes Isotopes of the element carbon with different atomic masses, used to trace the movement of different kinds of carbon through Earth's climate system (^{12}C and ^{13}C) or to measure elapsed time indicated by radioactive decay (^{14}C and ^{12}C).

cardinal points The two equinoxes (spring and autumn) and the two solstices (summer and winter) in Earth's annual revolution around the Sun.

Celsius scale A temperature scale on which water freezes at 0° and boils at 100°.

channeled scablands A region in Idaho and eastern Washington State in which water impounded in glacial lakes suddenly rushed out and reshaped the landscape, perhaps repeatedly.

chemical weathering Dissolving or other alteration of minerals to a different form by chemical reactions in the presence of water.

chlorofluorocarbons (CFCs) Synthetic chemical compounds generated by human activity and containing chlorine or fluorine that can destroy ozone in the stratosphere.

CLIMAP (Climatic Mapping and Prediction Project) A large cooperative research group during the 1970s and 1980s that first mapped the surface of the ice-age Earth.

climate Fluctuations in Earth's air, water, ice, vegetation, and other properties on time scales longer than 1 year.

climate data output The climatic properties (such as temperature, precipitation, and winds) that are produced by simulations with climate models.

climate point The point where the equilibrium line that separates net ice melting from net ice accumulation intercepts sea level.

climate proxy A quantifiable indicator of climate change contained in a climate archive and covering an interval that precedes direct instrument measurements of climate.

climate science The study of climate changes and their causes.

climate simulation The use of a numerical model to reproduce climate for a specified set of boundary conditions.

climate system The components of Earth (air, water, ice, vegetation, and land surfaces) that participate in climate change.

clipped responses Climatic responses that are truncated, or cut off, in one direction.

closed system A system that does not exchange matter across its boundaries.

coccoliths Tiny disklike plates of $CaCO_3$ produced by algae living in surface waters.

COHMAP (Cooperative Holocene Mapping Project) A cooperative research effort in the 1980s and 1990s that evaluated the causes of climate change during the most recent deglaciation by comparing geologic data with climate model simulations.

conifer forest Forest composed of evergreen, needle-bearing trees.

continental collision The occasional result of plate tectonic processes in which two continents are carried into each other by plate movements, creating high plateaus.

continental crust A layer of rock averaging 30 km thick, having the composition of granite, that constitutes the continents.

continental ice sheet A mass of ice kilometers thick covering a continent or a large portion of a continent and moving independently of the underlying bedrock topography.

continental shelf A shallowly submerged extension of a continent beneath the ocean.

continental slope A ramplike structural edge of a continent that slopes into the deep ocean.

control case A model simulation run to reproduce Earth's present climate from its present boundary conditions.

convection The rising motion of a fluid (air or water) produced when its bottom layer is heated, accompanied by sinking of cooler, denser fluid elsewhere.

convergent margin A boundary between two lithospheric plates that move toward each other and cause a collision of continents or subduction of one plate beneath the other.

coral bands Annual banding in the structure of corals caused by seasonal changes in sunlight, water temperature, and nutrient content.

Cordilleran ice sheet A small ice sheet that covered western Canada and the far northwestern United States during orbital-scale glaciation cycles.

Coriolis effect The apparent deflection of a fluid (air or water) from a straight-line path because of Earth's rotation. The deflection is to the right in the Northern Hemisphere and to the left in the Southern Hemisphere.

CO_2 fertilization effect The increased growth rate of plants caused by the addition of CO_2 to the atmosphere.

CO_2 saturation The point at which the concentration of CO_2 in the atmosphere reaches so high a level that additional amounts do not increase the greenhouse effect.

Cretaceous An interval of warmer climates and higher sea level between 135 and 65 Myr ago.

Dansgaard-Oeschger oscillations Oscillations in various properties (including dust and isotopes of oxygen) recorded in Greenland ice at intervals of 2,000 to 7,000 years during glacial intervals.

daughter isotope An isotope produced by radiometric decay of another isotope.

deforestation The cutting of forests by humans to clear land for agriculture and other activities.

deglacial two-step The irregular melting of ice sheets during the most recent deglaciation: fast-slow-fast.

$\delta^{13}C$ aging The gradual shift toward more negative $\delta^{13}C$ values in slow-moving deep water caused by the downward rain of ^{12}C-rich organic matter from overlying surface water.

dendroclimatology The methods used to extract climate signals from changes in the width or density of tree rings, both of which characteristics are sensitive to extremes of temperature and precipitation.

dew point The temperature at which cooling air becomes fully saturated with water vapor and permits condensation.

diatoms Silt-sized algae that live in surface waters of lakes, rivers, and oceans and form shells of opal ($SiO_2 \cdot H_2O$).

diffusion The transfer of a property such as heat by random, small-scale movements from a region of higher to a region of lower concentration.

diluvial hypothesis The hypothesis that unsorted sediments found on northern continents resulted from a great flood; these deposits are now recognized as deposits from glaciers.

dissolution A form of chemical weathering in which rocks such as limestone ($CaCO_3$) or rock salt ($NaCl$) are dissolved by water and produce ions that are removed by rivers.

divergent margin A boundary between two lithospheric plates that are moving apart, usually at the crest of an ocean ridge.

early anthropogenic hypothesis The hypothesis that the early spread of agriculture emitted greenhouse gases (CO_2 and CH_4) that warmed preindustrial climate.

Earth system The complex system of Earth's atmosphere, hydrosphere, biosphere, and lithosphere, through which energy and matter circulate.

eccentricity The extent to which Earth's orbit of the Sun departs from a perfect circle.

elastic Capable of deforming rapidly under pressure (as is bedrock under the pressure of a glacier) and rebounding when the pressure is removed.

electromagnetic radiation Self-propagating electricmagnetic waves, which include visible light as well as infrared and ultraviolet waves.

electromagnetic spectrum The complete range of electromagnetic radiation at differing wavelengths.

El Niño A climatic pattern that recurs at intervals of 2 to 7 years and is marked by warm sea-surface temperatures in the eastern tropical Pacific, off the west coast of South America.

enhanced greenhouse effect The trapping of Earth's back radiation by greenhouse gases produced by humans in addition to the warming caused by natural greenhouse gases.

ENSO (El Niño Southern Oscillation) The combined oscillations of El Niño (temperature changes in the eastern Pacific) and the southern oscillation (atmospheric pressure changes in the western and south-central Pacific).

eolian sediments Fine sediments deposited by the action of wind.

equilibrium A state of climatic stability toward which the climate system is moving and at which it will eventually remain unless disturbed.

equilibrium line The level in the atmosphere separating the zones of net addition and net loss of ice.

equinoxes The two times during each year (spring and autumn) when the lengths of days and of nights are equal.

equivalent CO₂ Changes in all greenhouse gases expressed in terms of an equivalent change in atmospheric CO_2 concentrations.

eustatic Characterized by changes in sea level that are global in scale rather than resulting from local factors such as tectonic uplift or subsidence of the land.

evaporites Minerals or rocks formed by precipitation of crystals from water evaporating in restricted basins in arid climates.

evolution The process by which particular forms of life give rise to other similar forms by gradual genetic changes.

faculae Bright rings that surround sunspots and emit large amounts of solar radiation.

Fahrenheit scale A temperature scale on which water freezes at 32° and boils at 212°.

faint young Sun paradox The paradox that astronomical models indicate a much weaker Sun through Earth's early history but geologic evidence shows that Earth never froze.

feedback A process internal to Earth's climate system that acts either to amplify changes in climate (positive feedback) or to moderate them (negative feedback).

Fertile Crescent A region in the Middle East where plants and animals were domesticated and agriculture began.

filtering The technique of extracting and isolating the shape of cycles at specific wavelengths or periods from complex signals.

fluvial sediments Sediments deposited by the action of water.

forcing Any process or disturbance that drives changes in climate.

fractionation A process favoring the transfer of one isotope of an element more than another.

frequency The number of full wave forms (each with one peak and one trough) that occur within a defined interval of time (usually 1 year). Also, the inverse of the period.

Gaia hypothesis A hypothesis that life regulates climate on Earth.

general circulation model (GCM) A three-dimensional computer model of the global atmosphere (or ocean) that simulates temperature, precipitation, winds, and atmospheric pressure.

geochemical model A model that quantifies the movement of geochemical tracers (minerals, elements, or isotopes) among reservoirs in the climate system.

geochemical tracer A chemical element or isotope whose movement between reservoirs in the climate system can be quantitatively tracked.

geological-geochemical proxy An index of past climate change based on measurable variations in physical or chemical properties of sediments, ice, or other archives.

global dimming A trend, observed during the 1950s to 1980s in many regions, in which less solar radiation reached Earth's surface.

Gondwana The large continent that existed in the Southern Hemisphere before the creation of the giant continent Pangaea.

greenhouse debate The controversy over the extent to which rising levels of atmospheric greenhouse gases have warmed Earth's climate during the last 200 years.

greenhouse effect The warming of Earth's surface and lower atmosphere that occurs when Earth's emitted infrared heat is trapped and reradiated downward by greenhouse gases.

greenhouse era An interval of warm climate on Earth, such as the Cretaceous, with ice sheets absent even in polar regions.

greenhouse gases Gases such as water vapor (H_2O_v), carbon dioxide (CO_2), and methane (CH_4), which trap outgoing infrared radiation emitted by Earth's surface and warm the atmosphere.

greenhouse surprise A climate change in the greenhouse world of the future that cannot be predicted.

grid boxes Geometric units within climate models that have uniform climatic characteristics and exchange heat, energy, and other properties with adjoining grid boxes.

Gulf Stream A narrow current of warm water that emerges from the Gulf of Mexico through the Florida Straits and flows northward along the southeastern coast of the United States.

gyre A spinning cell of water in an ocean basin, particularly at a subtropical latitude.

Hadley cell An atmospheric circulation cell in which air rises in the tropics, flows to the subtropics, sinks near 30° latitude, and flows back toward the tropics as surface trade winds.

half-life The time required for half the number of atoms of a radioactive isotope to decay.

harmonics Secondary cycles related to a wavelike climatic response with a period N, occurring at periods of $N/2$, $N/3$, $N/4$, and so on.

hardwood forest Forest composed of leaf-bearing (deciduous) trees.

heat capacity The amount of heat energy required to raise the temperature of 1 g of a substance by 1°C.

Heinrich event An interval of rapid flow of icebergs from the margins of ice sheets into the North Atlantic Ocean, causing deposition of sediment layers rich in debris eroded from the land.

historical archives Sources of information on climate based on human observations of natural phenomena made before the era of instrument measurements.

hot spot A point on the surface of a lithospheric plate where magma rising from below causes frequent volcanic activity.

hydrologic cycle The movement of water and water vapor among the atmosphere, land, and ocean through evaporation, precipitation, runoff, and subsurface groundwater flow.

hydrolysis A form of chemical weathering in which water reacts with silicate minerals (those rich in silicon and oxygen) to produce dissolved ions removed in rivers and clays left on the landscape.

hydrothermal Characterized by the circulation of hot fluids through rocks in Earth's outer crust, as at the Mid-Ocean Ridge system.

hypothesis An explanation of observations based on physical principles.

hypsometric curve A graph that summarizes the proportions of Earth's surface that lie at various altitudes above and depths below sea level.

ice-albedo-temperature feedback. Changes in surface vegetation that amplify initial changes in climate by altering the amount of incoming solar radiation reflected at the surface.

ice dome A high, gently sloping central region of an ice sheet in which snow accumulates and away from which ice flows slowly.

ice-driven response A climate change produced by fluctuations in the size of an ice sheet.

ice-elevation feedback The positive feedback that results when an ice sheet grows to a higher elevation at which accumulation exceeds ablation.

ice flow model A model that simulates ice sheet processes of snow accumulation, internal ice flow, and ablation.

icehouse era An interval of cold climate on Earth, such as the present, with ice sheets present in polar regions.

ice lobe A rounded or arc-shaped outward protrusion of the margin of an ice sheet.

ice-rafted debris Sediments of widely ranging sizes eroded from the land by ice, carried to the ocean, and deposited on the seafloor.

ice saddle A ridge that connects multiple domes of an ice sheet at a slightly lower elevation.

ice shelf A wide body of ice usually hundreds of meters thick that is fed by ice flowing off a continent, partially floats on seawater, and produces icebergs as blocks of ice break off.

ice stream A region of an ice sheet in which the motion of ice is unusually rapid, generally because of water-saturated sediments at the base of the ice.

igneous rock Rock formed by the cooling and solidification of molten magma.

insolation The amount of solar radiation arriving at the top of Earth's atmosphere by latitude and by season.

instrumental records Records of climate change measured by devices made by humans, from early thermometers through modern satellite-mounted instruments.

Intergovernmental Panel on Climate Change (IPCC) A large international group of scientists who reflect the current scientific consensus on the impact of greenhouse gases.

intertropical convergence zone (ITCZ) A narrow region within the tropics where warm moist air rises, cools, and loses its water vapor in heavy tropical rainfall.

iron fertilization hypothesis The hypothesis that iron-rich dust blown from the continents during glaciations enhances the productivity of the surface ocean, sends CO_2 into the deep ocean, and reduces the concentration of CO_2 in the atmosphere.

jet stream A narrow meandering stream of air moving rapidly (generally from west to east) at a high latitude and at an altitude averaging 10 km.

Kelvin scale A scale on which temperature is measured in Celsius degree intervals and on which water freezes at 273K and boils at 373K, and all motion ceases at 0K.

La Niña A pattern opposite from that of El Niño, in which sea-surface temperatures in the tropical eastern Pacific, off the west coast of South America, become unusually cold.

lapse rate The rate at which temperature falls with elevation in the atmosphere, averaging 6.5°C of cooling for each kilometer of altitude, but more for dry air and less for moist air.

latent heat The quantity of heat gained or lost as a substance changes state (liquid, solid, or gas) at a given temperature and pressure.

latent heat of melting The amount of heat gained as ice melts or released as water freezes.

latent heat of vaporization The amount of heat gained when water turns to water vapor or released when water vapor condenses back to water.

Laurentide ice sheet The largest of the northern hemisphere ice sheets that grow and shrink at orbital cycles, covering east-central Canada and the northern United States east of the Rockies.

lichen Primitive mosslike vegetation that lives on bare rock surfaces, uses sunlight to secrete acids and weather the rock, and slowly grows in a nearly circular shape.

lithosphere The outer rigid shell of Earth (including the upper mantle and oceanic and continental crust), characterized by strong, rocklike properties and divided into plates that move as rigid units during plate tectonic processes.

Little Ice Age An interval between approximately A.D. 1400 and 1900 when temperatures in the Northern Hemisphere were generally colder than today's, especially in Europe.

loess Silt-sized windblown glacial sediment.

longwave radiation Energy emitted in the infrared part of the electromagnetic spectrum by materials having a temperature above absolute zero (0K).

macrofossils Larger fragments of vegetation (such as needles, twigs, or cones) that prove the local presence of vegetation on past landscapes, used as a supplement to pollen analysis.

magnetic field The lines of force that are generated by motion in Earth's outer core and cause iron-bearing materials to align in specific directions in relation to the magnetic north pole.

magnetic lineations Long, stripelike regions of ocean crust marked by stronger or weaker magnetism acquired when the crust formed by cooling from molten magma at a ridge crest.

mantle The middle (and thickest) layer of the solid Earth, lying beneath the crust and above the core and composed of silicate minerals.

marine ice sheet An ice sheet whose base lies below sea level, as in western Antarctica.

mass balance The method of tracking the movement of materials within Earth's climate system by applying the law of conservation of mass.

mass balance model A model that tracks the movement of materials within Earth's climate system using conservation of mass to balance inputs and outputs among different reservoirs.

mass spectrometer An instrument that measures different isotopes of the same element (such as ^{16}O vs. ^{18}O and ^{12}C vs. ^{13}C) by separating them by mass.

mass wasting A downhill movement of rock or soil under the force of gravity.

Maunder sunspot minimum An interval between A.D. 1645 and 1715 when astronomers observed very few sunspots on the Sun's surface.

medieval warm period An interval between A.D. 1100 and 1300 in which some northern hemisphere regions were warmer than in the Little Ice Age that followed.

Mediterranean overflow water A water mass that forms in the northern Mediterranean Sea by winter chilling of salty surface waters and flows into the North Atlantic at a depth of 1 km.

megafauna Large mammals and marsupials, generally weighing more than 60 kilograms (100 pounds).

methane clathrate A partly frozen slushy mix of methane gas (CH_4) and ice.

Mg/Ca ratios The ratio of the elements Mg and Ca in shells of foraminfera, an index of past temperature changes.

Milankovitch theory The theory that orbitally controlled fluctuations in high-latitude solar radiation (insolation) during summer control the size of ice sheets through their effect on melting.

millennial oscillations Fluctuations in climate lasting thousands of years and generally larger during glacial than interglacial intervals.

modulation The tendency for peaks and troughs in a wave form to vary in size in a regular way, such that clusters of large peaks and troughs alternate with clusters of smaller ones.

moraine A pile of unsorted rubble (till) deposited by a glacier at its margin.

monsoon Winds that reverse direction seasonally, blowing onshore in summer and offshore in winter because of different rates of heating and cooling of land and water.

Monterey hypothesis The hypothesis that increased rates of burial of organic carbon on the margins of the Pacific Ocean 17 Myr ago reduced CO_2 levels in the surface ocean and atmosphere.

mountain glacier A body of ice tens to hundred of meters thick and kilometers in length confined to a valley at a high elevation. A mountain ice cap is a similar-sized body of ice lying on the rounded summit of a mountain.

no-analog vegetation A combination of types of vegetation in the past for which no similar (analogous) combination exists today.

nonlinear response A climatic response that occurs on other than a simple one-for-one basis in relation to the forcing.

North Atlantic deep water A water mass that forms in the high-latitude North Atlantic Ocean by winter chilling of salty surface water, sinks, and flows southward at depths of 2 to 4 km.

North Atlantic drift A warm, multipart current flowing northeastward into the high latitudes of the North Atlantic as a continuation of the Gulf Stream.

North Atlantic Oscillation A fluctuation in subtropical and North Atlantic temperatures that persists for more than a year.

ocean acidification The trend in the pH of the ocean to grow more acidic as a result of absorption of CO_2 emitted by humans.

ocean carbon pump hypothesis The hypothesis that changes in the amount of organic carbon taken up by ocean plankton during photosynthesis and exported to the deep ocean after they die control CO_2 levels in the surface ocean and atmosphere.

ocean crust A layer of rock averaging 7 km thick, having the average composition of basalt, composing the ocean floor.

ocean heat transport hypothesis The hypothesis that changes in the amount of heat transported toward polar regions by the ocean cause changes in polar climate.

oceanic gateway A narrow passage between continents that opens or closes and thereby alters ocean circulation.

orbital monsoon hypothesis The hypothesis that orbitally controlled changes in summer insolation at low latitudes drive the strength of the tropical summer monsoon.

orbital tuning The process of constructing a time scale by using the link between astronomically dated changes in solar radiation and the rhythmic climatic responses they cause on Earth.

orographic precipitation Precipitation on the upwind side of a mountain or plateau caused by the forced ascent of warm air to cooler elevations, where the entrained water vapor condenses.

outwash Layered sediments deposited by meltwater streams emerging from a glacier.

overkill hypothesis The hypothesis that the sudden extinction of many mammals 12,500 years ago resulted from human hunting rather than climatic stress.

oxidation A chemical reaction in which electrons are lost from an atom and its charge becomes more positive; also, the addition of oxygen to an element.

ozone A triple molecule of oxygen (O_3) formed by the collision of cosmic particles with normal (O_2) oxygen. Ozone in the stratosphere blocks harmful ultraviolet radiation from the Sun.

ozone hole A region centered over the Antarctic continent in which ozone (O_3) levels in the upper atmosphere (stratosphere) drop to very low values in the spring.

Pacific Decadal Oscillation A fluctuation in tropical and North Pacific temperatures that persists for more than a year.

paleomagnetism The study of patterns of ancient magnetism recorded in rocks or sediment.

pandemics Diseases that kill many millions of people on several continents.

Pangaea The giant supercontinent that existed between 300 and 175 Myr ago and consisted of all landmasses present on Earth.

parent isotope A radioactive isotope that naturally decays to a daughter isotope.

peat A deposit of decayed carbon-rich plant remains in a wetland environment with little oxygen.

peatlands hypothesis The hypothesis that the increase in methane levels between 5,000 and 250 years ago was caused by expanded areas of methane-producing bogs in north polar regions.

perihelion The point in Earth's slightly eccentric orbit at which it is closest to the Sun.

period The time interval between successive peaks or troughs in a series of regular wave forms.

peripheral forebulge A region in which the weight of glacial ice sheets caused bedrock to flow out to the ice margins at great depths and produced a broad upward bulge of the land.

permafrost A permanently frozen mixture of rocks and soil occurring in very cold regions.

phase lag The amount by which one cyclic signal lags behind another signal of the same wavelength (or period).

photosynthesis The process by which plants use nutrients and solar energy to convert water and CO_2 to plant tissue (carbohydrates) and thereby produce oxygen.

physical climate model A numerical model that simulates Earth's climate on the basis of physical principles of fluid motion and transfers of radiative heat energy and momentum.

physical weathering Any mechanical process by which rocks are broken into smaller fragments of the same material.

phytoplankton Small floating organisms (usually algae) that use energy from the Sun and nutrients from the water for photosynthesis.

plane of the ecliptic The plane within which Earth revolves around the Sun.

planktic foraminifera Sand-sized organisms (protozoans) that live in ocean surface waters and form shells made of $CaCO_3$.

plankton Organisms that float in the upper layers of oceans or lakes.

plate tectonics Tectonic interactions resulting from the movement of lithospheric plates.

polar alkalinity hypothesis The hypothesis that the concentration of CO_2 in the atmosphere is affected by changes in deep-water circulation

polar front A sharp boundary zone in a polar ocean between cold, low-salinity waters and warmer, saltier waters; similarly, in the atmosphere, a sharp temperature boundary.

polar position hypothesis The hypothesis that ice sheets exist during intervals in Earth's history when landmasses are moved into polar regions by plate tectonic processes.

power spectrum A graphic display of the distribution of power (the square of wave amplitude) against the period (or frequency) of each cycle present in a signal.

precessional index The mathematical product ($\varepsilon\sin\omega$) of Earth's sine wave motion ($\sin\omega$) around the Sun and the eccentricity of its orbit (ε)

precession of the ellipse The slow turning of Earth's elliptical orbit in space.

precession of the equinoxes The movement of the solstices and equinoxes around Earth's elliptical orbit over cycles of 23,000 and 19,000 years.

preindustrial CO_2 level The concentration of CO_2 in the atmosphere (280 ppm) that existed for several thousand years before the Industrial Revolution.

productivity The amount of organic matter synthesized by organisms from inorganic substances per unit of area per unit of time.

proglacial lake A short-lived lake that develops after the retreat of an ice sheet in the bedrock depression left by the weight of the ice.

radiation Electromagnetic energy that drives Earth's climate system. Ultraviolet and visible radiation emitted by the Sun affects climate on Earth, which emits infrared back radiation to space.

radiative forcing The effect of greenhouse gases in trapping (or blocking) solar radiation, expressed in the same units as those of solar energy: watts per square meter (W/m^2).

radiocarbon dating Dating of relatively young carbon-bearing geologic materials by means of ^{14}C, a radioactive isotope that decays with a half-life of 5,700 years.

radiolaria Sand-sized organisms that live in surface waters and form shells of opal ($SiO_2 \cdot H_2O$).

radiometric dating Determining the ages of rocks or sediments by measuring the amount of naturally occurring radioactive parent isotopes and their nonradioactive daughter products.

reconstruction A simulation run with a climate model by altering several boundary conditions in an effort to reproduce a climate that existed at some time in the past.

red beds Sediments or rocks with a red color caused by the oxidation of iron (similar to rust).

red noise Erratic fluctuations in a system that tend to be concentrated at low frequencies (longer wavelengths).

relaxation oscillator A system that slowly moves away from its equilibrium state before rapidly springing back to it.

regressions Relative motions of the ocean down and off the margins of the land.

reservoir A place of residence for an element or isotope that moves in a cycle.

residence time The average amount of time a tracer of any substance spends in a reservoir.

resolution The degree of detail detected in a climate signal by sampling at a particular interval.

resonant response A strong cyclic response of the climate system to perturbations occurring at the same or other cycles.

response Any change in the climate system caused by a change in climate forcing.

response time The time required for a climatic response to move a defined fraction of the way from its existing value to the value it would hold at its full equilibrium response.

salinity A measure of the salt content of seawater in parts per thousand (‰).

salt rejection The salt left in seawater when sea ice forms on the ocean.

sapropels Black organic-rich muds deposited on the Mediterranean seafloor as a result of strong inflow from the Nile River, which stifles delivery of oxygen to the deep parts of the basin.

saturation vapor density The maximum amount of water vapor that air can hold at a given temperature.

savanna A semiarid region of grasses and scattered trees.

savanna hypothesis The hypothesis that long-term drying in Africa drove evolution by forcing humans from forested regions to tree/grass savanna.

Scandinavian ice sheet An ice sheet that covered most of Norway and Sweden as well as the northern part of Germany and France during orbital-scale glaciation cycles.

seafloor spreading The mechanism by which new seafloor is created at an ocean ridge as the adjacent plates move away from the ridge crest at a rate of centimeters per year.

sediment drift A lens-shaped pile of fine sediment (clay and silt) picked up in a region where bottom currents move swiftly and then deposited in a region of the seafloor where currents slow.

sensible heat Heat energy carried by water or air in a form that can be easily felt or sensed, rather than hidden in latent form.

sensitivity test A simulation run with a climate model in which one boundary condition is altered from the (modern) control case to test its effect on climate.

shortwave radiation Electromagnetic energy emitted by the Sun in visible and ultraviolet wavelengths that deliver heat to Earth's climate system.

silicate minerals Minerals rich in silicon and oxygen, which account for most of the rocks in Earth's crust and mantle.

sine wave A perfectly regular wave form in which successive peaks and troughs are evenly spaced, each peak reaching a value of $+1$ and each trough a value of -1.

sintering The process by which bubbles of air become sealed off and preserved as snow turns to ice at depths of 50 to 100 m within an ice sheet.

snowball Earth hypothesis The hypothesis that Earth was frozen even in the tropics sometime in the interval between 850 and 550 Myr ago.

solstices The times during Earth's yearly revolution around the Sun when the days are longest (summer solstice) and shortest (winter solstice).

southern oscillation Naturally occurring fluctuations in which changes in lower atmospheric surface pressure in the far western Pacific, near northern Australia, are opposite in sense to those in the south-central Pacific, near Tahiti.

specific heat The amount of heat required to raise the temperature of 1 g of a substance by 1°C.

spectral analysis A numerical technique for detecting and quantifying the distribution of regular (periodic) behavior in a complex signal.

speleothems Cave deposits made of $CaCO_3$ containing climatic signals.

Sporer sunspot minimum An interval between A.D. 1460 and 1550 when very few sunspots were observed on the Sun's surface.

spreading rate hypothesis The hypothesis that tectonic-scale climate changes are driven by variations in the global average rate of seafloor spreading, which alter the amount of CO_2 introduced into the atmosphere.

stochastic resonance. Variations in a system that combine aspects of random noise and cyclic behavior.

stratosphere The stable layer of the atmosphere lying between 10 and 50 km above Earth's surface and containing most of Earth's ozone.

subduction The slow downward sinking of an ocean plate beneath a continent or an island arc as a result of plate tectonic processes.

sulfate aerosols Fine particles produced in the atmosphere from SO_2 gas emitted by volcanoes or by industrial smokestacks. These particles can block incoming solar radiation.

sunspots Dark areas of temperatures lower than average on the surface of the Sun.

sunspot cycle A natural 11-year cycle in the number of dark spots visible on the face of the Sun, reliably recorded by astronomers for more than four centuries.

surge A sudden and rapid forward movement of the margin of a glacier.

tabular iceberg A large flat-topped slab of ice produced by calving from the seaward margin of an ice shelf.

tectonic plates Divisions of the upper solid Earth (the lithosphere) that are 100 km thick and thousands of kilometers in lateral extent and move as rigid units in plate tectonic processes.

termination A 10,000-year interval of rapid melting of ice sheets that brings to an end a longer (90,000-year) interval of slower ice growth.

theory A hypothesis that has survived repeated testing.

thermal expansion coefficient The volumetric expansion of seawater when it warms, and contraction when it cools, by 1 part in 7,000 per degree C (for temperatures above 4°C).

thermal inertia The resistance of a component of the climate system to temperature change.

thermocline A layer of water in which temperature changes rapidly in a vertical direction.

thermohaline flow The vertical and lateral movement of subsurface waters in the ocean as a result of contrasts in density caused by differences in temperature and salinity.

thermostat A mechanism that senses changes in temperature and acts to moderate them.

threshold A level at which a sudden change in the basic nature of a climatic response occurs.

tilt The angle between Earth's equatorial plane and the plane of its orbit of the Sun, also equivalent to the angle between Earth's axis of rotation and a line perpendicular to its axis of rotation around the Sun. Also referred to as *obliquity*.

time-dependent model A geochemical model that tracks changes in the rate of movement of tracers among reservoirs within the climate system over time.

time-series analysis A group of techniques for extracting periodic signals (cycles) from complex signals and quantifying their strength, relative timing, and correlation.

transform fault margin A boundary between lithospheric plates at which the plates slide past each other.

transgression Relative movement of the ocean up and across the margins of the land.

transpiration The release of water vapor by plants into the atmosphere.

tree rings Annual bands formed by trees in regions of seasonal climate, with lighter layers formed during rapid growth in the spring and darker layers at the end of growth in the autumn.

troposphere The layer of the atmosphere (10 km or more in thickness) just above Earth's surface in which weather occurs.

tundra A high-latitude or high-altitude environment in which the ground freezes deeply in winter but thaws at the surface in summer, permitting low-growing plants to flourish.

$2 \times CO_2$ sensitivity The amount of warming produced by an increase in atmospheric CO_2 levels from the preindustrial level (280 ppm) to a level of 560 ppm.

uplift weathering hypothesis The hypothesis that tectonic-scale climate changes are caused when uplift of plateaus and mountains alters the amount of CO_2 removed from the atmosphere by chemical weathering of fragmented rock.

upwelling The rise of cool, nutrient-rich subsurface water to the ocean surface to replace warm, nutrient-poor surface water.

urban heat island An urban area where asphalt and other heat-absorbing surfaces absorb solar radiation during the day and radiate it back at night, keeping the area unusually warm.

variability selection hypothesis. The hypothesis that increasing long-term climatic variability in Africa drove human evolution by favoring those who best adapted to change.

varves Alternating layers of dark and light (or coarse and fine) sediment that accumulate in annual couplets in lakes or in ocean margin basins with no water turbulence near the bottom.

vegetation-albedo feedback A positive feedback that amplifies an initial temperature change through vegetation changes that alter Earth's surface albedo and the absorption of solar radiation.

vegetation-precipitation feedback A positive feedback that amplifies an initial precipitation change through vegetation changes that alter the transpiration of water vapor and the availability of moisture.

viscous Characterized by a consistency such that the material in question deforms slowly under pressure (as does bedrock under the pressure of a glacier) and only slowly regains its initial shape when the pressure is removed.

warm saline deep water Deep water proposed to have formed in warm salty tropical seas in the past, in contrast to modern sources of deep water in cold polar regions.

water vapor feedback A positive feedback that amplifies an initial change in temperature by altering the amount of water vapor held by the atmosphere and changing the amount of Earth's back radiation trapped by the atmosphere.

wavelength The distance between successive peaks or troughs in a series of regular wave forms.

weather Fluctuations in temperature, precipitation, and winds on time scales of less than a year.

Younger Dryas An interval during the middle of the last deglaciation (near 12,000 years ago) marked by slower melting of ice sheets and a major cooling in the North Atlantic region.

Index

Note: Page numbers in *italic* indicate figures; those followed by b indicate boxed material; those followed by t indicate tables.